# ARITHM

## (SUBJECTIVE AND OBJECTIVE)

**For Competitive Examinations**

**[Banks, L.I.C., G.I.C., Excise, Income Tax, U.P.S.C., Railways, Forest Services, P.C.S., Defence Services & Other Services.]**

**R.S. AGGARWAL**
M.Sc. Ph.D.
*Deptt. of Mathematics*
*N.A.S. College, Meerut (U.P.)*



S. CHAND
AN ISO 9001: 2000 COMPANY

**S. CHAND & COMPANY LTD.**

**RAM NAGAR, NEW DELHI-110 055**

**S. CHAND & COMPANY LTD.**
**(An ISO 9001 : 2000 Company)**
*Head Office* : 7361, RAM NAGAR, NEW DELHI - 110 055
Phones : 23672080-81-82; Fax : 91-11-23677446
Shop at: **schandgroup.com;** E-mail: **schand@vsnl.com**

*Branches* :
- 1st Floor, Heritage, Near Gujarat Vidhyapeeth, Ashram Road, **Ahmedabad**-380 014. Ph. 27541965, 27542369.
- No. 6, Ahuja Chambers, 1st Cross, Kumara Krupa Road, **Bangalore**-560 001. Ph : 22268048, 22354008
- 152, Anna Salai, **Chennai**-600 002. Ph : 28460026
- S.C.O. 6, 7 & 8, Sector 9D, **Chandigarh**-160017, Ph-2749376, 2749377
- 1st Floor, Bhartia Tower, Badambadi, **Cuttack**-753 009, Ph-2332580; 2332581
- 1st Floor, 52-A, Rajpur Road, **Dehradun**-248 011. Ph : 2740889, 2740861
- Pan Bazar, **Guwahati**-781 001. Ph : 2522155
- Sultan Bazar, **Hyderabad**-500 195. Ph : 24651135, 24744815
- Mai Hiran Gate, **Jalandhar** - 144008 . Ph. 2401630
- 613-7, M.G. Road, Ernakulam, **Kochi**-682 035. Ph : 2381740
- 285/J, Bipin Bihari Ganguli Street, **Kolkata**-700 012. Ph : 22367459, 22373914
- Mahabeer Market, 25 Gwynne Road, Aminabad, **Lucknow**-226 018. Ph : 2626801, 2284815
- Blackie House, 103/5, Walchand Hirachand Marg , Opp. G.P.O., **Mumbai**-400 001. Ph : 22690881, 22610885
- 3, Gandhi Sagar East, **Nagpur**-440 002. Ph : 2723901
- 104, Citicentre Ashok, Govind Mitra Road, **Patna**-800 004. Ph : 2300489, 2302100

***Marketing Offices :***
- 238-A M.P. Nagar, Zone 1, **Bhopal** - 462 011. Ph : 5274723
- A-14 Janta Store Shopping Complex, University Marg, Bapu Nagar, **Jaipur** - 302 015, Phone : 2709153



*Subsequent Editions and Reprints 1989, 90, 92, 93, 94, 95, 96, 97, 98 99, 2000, 2001, 2002, 2003, 2004 (Twice), 2005*
*Reprint 2006*

ISBN : 81-219-0742-X

PRINTED IN INDIA

*By Rajendra Ravindra Printers (Pvt.) Ltd., Ram Nagar, New Delhi-110 055 and published by S. Chand & Company Ltd., 7361, Ram Nagar, New Delhi-110 055*

# Preface

The tremendous response to the fifth edition of the book has encouraged me to bring out this thoroughly revised and enlarged edition.

This book is really an asset to those who plan to appear in a competitive examination conducted by Banks, L.I.C., G.I.C., Excise & Income Tax, Railways, U.P.S.C. and other departments for clerical Grade, Assistant Grade, A.A.O., Inspectors, Probationary Officers, N.D.A., C.D.S., M.B.A., C.A.T. and other executive posts.

This book carries both subjective and objective type of questions.

Most of the books in the market carry objective type questions with their answers. But, how to get these answers is not given therein. More-over, how to solve a question is not significant in such examinations. The most important aspect is to solve a question in a fraction of a minute using short cut methods.

This book contains a huge accumulation of objective type questions with their solutions by short cut methods. The referennce of questions asked in various examinations, collected from various examinees on their memory basis, have been given.

It is very much hoped that the subject matter will create a confidence among the candidates and the book will help them like an ideal teacher.

I convey my gratitude to Shri R.K. Gupta, Director and Shri T.N. Goel, Manager, S. Chand and Co., Delhi for taking all pains and interest in the publication of the revised manuscript.

For good type setting I am thankful to Mr. Mukesh Maheshwari of Brillient Computers, Meerut. For good printing I am thankful to Mr. Ravi Gupta.

**R. S. Aggarwal**
*Veenalaya*
**F-80, Shastri Nagar, Meerut (U.P.)**

# Contents

# Contents

# 1

# NUMBERS

## (FOUR FUNDAMENTAL RULES)

{+, –, ×, ÷}

## NUMBERS

It is known that in **Hindu-Arabic system,** we use ten symbols 0, 1, 2, 3, 4, 5, 6, 7, 8 and 9, called *digits* to represent any number. Expressing a number in words is called *numeration* and representing a number in figures is called *notation*.

For a given numeral, we start from the extreme right and call the digits to be at *unit's place, ten's place, hundred's place* etc. We use place system with base ten, as shown in the following table representing 547063819.

| Ten crores $10^8$ | Crores $10^7$ | Ten Lacs *(millions)* $10^6$ | Lacs $10^5$ | Ten Thousands $10^4$ | Thousands $10^3$ | Hundreds $10^2$ | Tens $10^1$ | Ones $10^0$ |
|---|---|---|---|---|---|---|---|---|
| 5 | 4 | 7 | 0 | 6 | 3 | 8 | 1 | 9 |

Thus **'547063819'** is a numeral for the number, *'Fifty four crores, seventy lacs, sixty three thousand, eight hundred and nineteen'* We may write,

$$547063819 = (9 \times 10^0) + (1 \times 10^1) + (8 \times 10^2) + (3 \times 10^3) + (6 \times 10^4)$$
$$+ (0 \times 10^5) + (7 \times 10^6) + (4 \times 10^7) + (5 \times 10^8).$$

clearly, the place value of 9 is 9 units = 9,

the place value of 1 is 1 ten = $1 \times 10^1 = 10$,

the place value of 8 is 8 hundreds = $8 \times 10^2 = 800$,

the place value of 3 is 3 thousands = $3 \times 10^3 = 3000$, and so on

**Remark.** The face **value** of a digit is the value of that digit, may be at any place. For example, the face value of 4 in the numeral 547063819 is 4, while its place value is 4 crores.

**ROMAN SYSTEM :**

In Roman sytem, we use symbols I, V, X, L, C, D and M to express the numerals 1, 5, 10, 50, 100, 500 and 1000 respectively.

**Remarks. (i)** *A smaller number placed to the right of larger one indicates addition, whereas, if it is placed to the left, it indicates subtraction.*

(*ii*) *A bar on any symbol represents thousand times the number.*

Thus XI = 10 + 1 = 11, IV = 5 – 1 = 4, XCI = 100 – 10 + 1 = 91,

LXXIV = 50 + 10 + 10 + (5 – 1) = 74, $\overline{\text{V}}$ = 5000,

CCLI = 100 + 100 + 50 + 1 = 251,

$\overline{\text{L}}$CDLIX = 50000 + (500 – 100) + 50 + (10 – 1) = 50459.

## SOLVED PROBLEMS

**Ex. 1.** ***Write in figures :***

***(i) Four lacs and thirty five (ii) Four crores, four lacs, four thousand and four (iv) Ten crores, five lacs, six thousand and one.***

**Sol.** (*i*) 400035 (*ii*) 40404004 (*iii*) 100506001.

**Ex. 2.** ***Write in words :***

***(i) 20500056 (ii) 10003007 (iii) 190405006.***

**Sol.** (*i*) Two crores, five lacs and fifty six.

(*ii*) One crores, three thousand and seven.

(*iii*) Nineteen crores, four lacs, five thousand and six.

**Ex. 3.** ***Write down the face value and the place vlaue of 7 in the numeral 2370546.***

**Sol.** The face value of 7 in given numeral is 7, but the place value of 7 in given numeral is $7 \times 10^4 = 70{,}000$.

**Ex. 4.** ***Express the numeral, 43782503 in the expanded form, showing each digit with its place value.***

**Sol.** $43782503 = (3 \times 10^0) + (0 \times 10^1) + (5 \times 10^2) + (2 \times 10^3) + (8 \times 10^4) + (7 \times 10^5) + (3 \times 10^6) + (4 \times 10^7).$

**Ex. 5.** ***Express the following numbers in Roman notations :—***

***(i) 77 (ii) 96 (iii) 353 (iv) 2246 (v) 5081 (vi) 48301***

**Sol.** (*i*) LXXVII. (*ii*) XCVI. (*iii*) CCCLIII.

(*iv*) MMCCXLVI. (*v*) $\overline{\text{V}}$LXXXI. (*vi*) $\overline{\text{XLVIII}}$ CCCI.

**Ex. 6.** ***Write the numerals for :***

***(i) LXVI, (ii) CCXVII. (iii) CDVI.***

***(iv) CMIX. (v) MMCCXLI (vi) $\overline{\text{L}}$CDLIX.***

**Sol.** (*i*) LXVI = 50 + 10 + 5 + 1 = 66.

(*ii*) CCXVII = 100 + 100 + 10 + 5 + 1 + 1 = 217.

(*iii*) CDVI = (500 – 100) + 5 +1 = 406.

(*iv*) CMIX = (1000 – 100) + (10 – 1) = 909.

(*v*) MMCCXLI = 1000 + 1000 + 100 + 100 + (50 – 10) + 1 = 2241.

(*vi*) $\overline{\text{L}}$CDLIX = (50,000) + (500 – 100) + 50 + (10 – 1) = 50459.

## PROBLEMS ON ADDITION AND SUBTRACTION (Short Cut Methods)

**Ex. 1.** ***In a single operation, subtract the sum of 329, 2958, 41367 and 10651 from 98213.***

**Sol.** Actually, we are to find a number, which when added to the bracketed numbers, gives 98213.

$$\begin{array}{r} 98213 \\ \left.\begin{array}{r} 329 \\ 2958 \\ 41367 \\ 10651 \end{array}\right\} \\ \hline 42908 \end{array}$$

We find it in the following way :—

*1st column* : 9 + 8 + 7 + 1 − 25.

Now, to have 3 in the units place, the least number to be added is 8. We write 8 at the units place in the answer. Now 25 + 8 = 33. We carry over 3.

*Second column* : 2 + 5 + 6 + 5 + 3 = 21. So, to have 1 at the ten's place, we must add 0 to it. So, we write 0 at the ten's place in the answer Now, 21 + 0 = 21. We carry over 2

*Third column* : 3 + 9 + 3 + 6 + 2 = 23. So, to have 2 at the hundred's place, the least number to be added is 9. We write 9 at the hundred's place in the answer. Now, 23 + 9 = 32. We carry over 3.

*Fourth column* : 2 + 1 + 0 + 3 = 6. So, to have 8 at the thousand's place, the least number to be added is 2. We write 2 at the thousand's place in the answer. Now, 6 + 2 = 8.

*Fifth column* : 4 + 1 = 5. So, to have 9 at the ten thousand's place, the least number to be added to 5 is 4. So, we write 4 at the the thousands place in the answer. Hence, the answer is 42908.

**Ex. 2.** ***Fill in the blanks :***

***7691 − (58 + 374 + 1693 + 2085) = ..... ? .......***

**Sol.** Clearly, a sum of 58, 374, 1693 and 2085 is to be subtracted from 7691. This is equivalent to finding a number which when added to the sum of given numbers, gives 7691.

$$\begin{array}{r} 7691 \\ \left.\begin{array}{r} 58 \\ 374 \\ 1693 \\ 2085 \end{array}\right\} \\ \hline 3481 \end{array}$$

∴ The required number is 3481.

**Ex. 3.** ***Find the missing number :***

$$(4300521) — (\ldots\ldots ? \ldots\ldots) = 1867138.$$

**Sol.** Let the required number be $x$. Then,

$$(4300521) - x = 1867138$$

or $\quad x = 4300521 - 1867138 = 2433383.$

## PROBLEMS ON MULTIPLICATION (SHORT CUT METHODS).

When a number $x$ is multiplied by a number $y$, then $x$ is called the *multiplicand* and $y$ is known as a *multiplier*.

**To multiply a given number by any of the numbers 9,99,999,9999 etc.**

**Rule.** *Place as many zeros to the right of the multiplicand as is the* *number of nines and from this number, subtract the multiplicand to get the answer.*

**Ex. 1.** ***Multiply 342197 by 9999.***

**Sol.** $342197 \times 9999 = 342197 \times (10000 - 1)$
$= 3421970000 - 342197$
$= 3421627803.$

*Short cut Method :* Annex four zero to right of multiplicand, as there as four nines in the multiplier. Now, from this number, subtract the multiplicand.

*i.e.* $342197 \times 9999 = 3421970000 - 342197 = 3421627803.$

**Ex. 2.** ***Multiply 5947 by 999999.***

**Sol.** $5947 \times 999999 = 5947000000 - 5947 = 5946994053.$

**To multiply a given number by any power of 5 :**

**Rule.** Put as many zeros to the right of the multiplicand as is the number of powers of 5 in the multiplier. Divide the number so obtained by 2 raised to the same power as is the number of power of 5.

**Ex. 3.** ***Multiply 634397 by 625.***

**Sol.** $634397 \times 625 = 634397 \times 5^4$

$$= \frac{6343970000}{2^4} = \frac{6343970000}{16}$$

$= 396498125.$

**Ex. 4.** ***Multiply 8546 by 15625.***

**Sol.** $8546 \times 15625 = 8546 \times 5^6$

$$= \frac{8546000000}{2^6} = \frac{8546000000}{64} = 133531250.$$

**Ex. 5.** ***Multiply 39472 by 639 in two lines.***

**Sol.** $639 = 630 + 9$ and $630 = 9 \times 70.$

So, multiply the given number by 9.

Multiply this product by 70 to obtain the product of original number by 630. Now, add the two products to get the answer.

| | |
|---|---|
| | 39472 |
| | × 639 |
| Multiplication by 9 → | 355248 |
| Multiplication by 630 → | 24867360 |
| Multiplication by 639 → | 25222608 |

$\therefore$ $39472 \times 639 = 25222608.$

**Ex. 6.** ***Multiply in three lines : 65734 by 369324.***

**Sol.** $369324 = (360000 + 9000 + 324)$

Also, $360000 = 9000 \times 40$ and $324 = 36 \times 9$

*Method.* First multiply the given number by 9000. In second line, multiply the first product, by 40 to have a product of the given number by 360000. However, in this product, if we neglect the last four zeros, then we get the product of given number with 36. Now, multiply this product by 4 to obtain the product of given number and $36 \times 9$ *i.e.* 324 and put this product in third line. Add these three products to get the required product.

```
       65734
    × 369324
 ------------
   591606000  ← Multiplication by 9000
 23664240000  ← Multiplication by 9000 × 40 i.e. 360000
    21297816  ← Multiplication by 36 × 9 i.e. 324
 ------------
 24277143816  ← Multiplication by 369324.
```

**Ex. 7. *Supply the mising digits in the multiplication process given below :—***

```
    9873
   × ***
  ------
   *****
  *****
 *****
 -------
 ****248
```

**Sol.** Since the first digit in the product is 8 and $3 \times 6 = 18$, so the first digit of the multiplier is 6. Taking the unit digit of multiplier as 6, we find the product as shown herewith.

Now, the second digit in the product is 4 and 3 + 1 = 4. So,, there must be 1 as the second digit in second row.

But, $7 \times 3 = 21$.

$\therefore$ Second digit of multiplier is 7. Filling it up and take the product, as shown herewith.

```
    9873
   × **6
  ------
   59238
  *****
 *****
 -------
 ****248
```

Now, we observe that the third digit in the product is 2 and 2 + 1 = 3. So, least number to be added to it is 9, *i.e.,* the third digit in the third row is 9. But $3 \times 3 = 9$. So, the third digit of multiplier is 3. Filling it, we have

```
    9873
   × *76
  ------
   59238
  69111
 *****
 -------
 ****248
```

```
    9873
   × 376
  ------
   59238
  69111
 29619
 -------
 3712248
```

## DIVISION

*It is the process of finding how many times the given number called the* ***'divisor'*** *is contained in another given number, called the* ***'dividend'****. This number* ***'so many times'****, is called the* ***'quotient'****. The excess of dividend over the divisor taken the greatest number of times that the dividend contains exactly, is called* ***'the remainder'****.*

**Ex.** If we divide 47 by 6, then

*dividend* = 47, divisor = 6,

*quotient* = 7 and remainder = 5.

$$\begin{array}{r|l|l} 6 & 47 & 7 \\ & 42 & \\ \hline & 5 & \end{array}$$

Clearly, $47 = (6 \times 7) + 5$.

In general :—

**Dividend = (Divisor × Quotient) + Remainder.**

## SOLVED PROBLEMS

**Ex. 1.** ***On dividing 69371 by a certain number, the quotient is 26 and the remainder is 35. Find the divisor.***

**Sol.** $\text{Divisor} = \dfrac{\text{Dividend} - \text{Remainder}}{\text{Quotient}}$

$= \dfrac{69371 - 35}{216} = 321.$

**Ex. 2.** ***What least number must be subtracted from 5731625, to get a number exatly divisible by 3546 ?***

**Sol.** On dividing 5731625 by 3546, the remainder is 1289.

Hence, the number to be subtracted is 1289.

**Ex. 3.** ***What least number must be added to 954131, to get a number exactly divisible by 548 ?***

**Sol.** On dividing 954131 by 548, the remainder is 63.

∴ The least number to be added = 548 – 63 = 485.

**Ex. 4.** ***Find the greatest number of 5 digits, which is exactly divisible by 547.***

**Sol.** The greatest number of 5 digits is 99999. On dividing 99999 by 547, we get 545 as the remainder.

∴ The required number = (99999 – 545) = 99454.

**Ex. 5.** ***Find the least number of 5 digits which is exactly divisible by 456.***

**Sol.** The least number of 5 digits is 10000. On dividing 10000 by 456 we get 424 as remainder. So, to get a number exactly divisible by 456, we mus add (456 – 424) or 32 to 10000.

Hence, the required number = (10000 + 32) = 10032.

**Ex. 6.** ***Find the number which is nearest to 68624 and exactly divisible by 587.***

**Sol.** On dividing 68624 by 587, the remainder obtained is *532, which is more than half the divisor* and so the nearest requisite number can be obtained by adding (587 – 532) or 55 to the dividened.

∴ *The required number = (68624 + 55) = 68679.*

**Ex. 7.** ***Find the number nearest to 144759 and exactly divisible by 927.***

**Sol.** On dividing 144759 by 927, the remainder obtained is 147, which is less than half the divisor and therefore, the reuisite type of number can be obtained by subtracting 147 from the dividend.

∴ The required number = (144759 – 147) = 144612.

**Ex. 8.** ***A number, when divided by 779 gives a remainder 47. What remainder would be obtained by dividing the same number by 19 ?***

**Sol.** Let the number, when divided by 779 gives the quotient $k$ and remainder 47. Then,

the number $= (779\,k + 47)$

$= (19 \times 41\,k) + (19 \times 2) + 9$

$= [\,19 \times (41\,k + 2)\,] + 9$

$= [\,19 \times (\text{new quotient})\,] + 9.$

So, this number, when divided by 19 gives 9, as the remainder.

**Ex. 9.** ***When a certain number is multiplied by 13, the product consists of entirely of fives; find th smallest such number.***

**Sol.** By trial, we find that the smallest number consisting entirely of fives and exactly divisible by 13 is 555555. On dividing 555555 the quotient is 42735.

Hence, the required number = 42735.

**Ex. 10.** ***Divide 22197 by 147, using factors.***

**Sol** $147 = 3 \times 7 \times 7$

Now

| | |
|---|---|
| 3 | 22197 |
| 7 | 7399 |
| 7 | 1057 |
| | 151 |

∴ (22197 + 147) = 151.

**Ex. 11.** ***Complete the division given below :—***

| | |
|---|---|
| **7** | **.....................** |
| **9** | **..............– 5 (Remainder)** |
| | **156 – 6 (Remainder)** |

**Sol.** The quotient obtained on dividing the original number by 7 is $(9 \times 156) + 6$ or 1410.

$\therefore$ original dividend = $(7 \times 1410) + 5 = 9875$.

Thus, *the given problem is*

| | |
|---|---|
| 7 | 9875 |
| 9 | 1410 – 5 (remainder) |
| | 156 – 6 (remainder) |

## VARIOUS TESTS FOR DIVISIBILITY

(*i*) **Divisibility by 2.** *A given number is divisible by 2, if the unit digit in the number is any one of 0, 2, 4, 6, 8.*

**For example,** each one of the numbers 729560, 417352, 368194, 62436, 35278 is divisible by 2, since they end in 0, 2, 4, 6 and 8 respectively.

**Remark.** A number divisible by 2 is known as an **Even number** and that not divisible by 2 is called an **odd number.**

(*ii*) **Divisibility by 5.** *A given number is divisible by 5, if the unit digit in the number is 0 or 5.*

**For example,** each one of the numbers 39425 and 924720 is divisible by 5, since they end is 5 and 0 respectively.

(*iii*) **Divisibility by 10.** *A given number is divisible by 10, if the unit digit of the number is* 0. *e.g.* 3697850.

(*iv*) **Divisibility by 4.** *A given number is divisible by 4, if the number formed by last two digits*, (namely the ten's and the unit's digits) **is divisible by 4.**

**For example,** if we cosider 45971364, then the number formed by last 2 digits is 64, which is divisible by 4. Hence 45971364 is divisible 4.

On the other hand, if we cosnider 39876562, then the number formed by last 2 digits is 62, which is not divisible by 4. Hence 398765462 is not divisible by 4.

(*v*) **Divisibility by 8.** *A given number is divisibe by 8, if the number formed by the last 3 digits of the given number, is divisible by 8.*

**For example,** if we consider 61974512, then the number formed by last 3 digits is 512, which is divisible by 8. Hence, the given number is divisible by 8.

However, 61974532 is not divisible by 8, since 532 is not divisible by 8

(*vi*) **Divisibility by 3.** *A given number is divisible by 3, if the sum of the digits of the given number is divisible by 3.*

**For example,** in the number 631425, the sum of the digits is $(6 + 3 + 1 + 4 + 2 + 5)$ or 21, which is divisible by 3. Hence 631425 is divisible by 3.

Again, if we consider 35216, then the sum of the digits is 17, which is not divisible by 3. Hence 35216 is not divisible by 3.

(*vii*) **Divisibility by 9.** *A given number is divisible by 9, if the sum of the digits of the number is divisible by 9.*

**For example,** if we consider the number 932625, then the sum of the digits of the given number is 27, which is divisible by 9. Hence, 932625 is divisible by 9.

However, the sum of the digits in 35265 is 21, which is not divisible by 9. So, 35265 is not divisible by 9.

(*viii*) **Divisibility by 11.** *A given number is divisible by 11; if the difference of the sum of the digits in odd places and the sum of its digits in even plces, is either 0 or a number divisible by 11.*

**For example,** Consider the number 4832718.

Sum of its digits in odd places = 8 + 7 + 3 + 4 = 22,

Sum of its digits in even places = 1 + 2 + 8 = 11.

∴ Difference of two sums = (22 – 11) = 11, which is divisible by 11.

Hence, 4832718 is divisible by 11.

## PRIME AND COMPOSITE NUMBERS

*A number other than 1 is said to be prime if its only factors are 1 and the number itself.*

*A number other than 1, which is not prime, is knwon as a* ***composite number.***

**Remarks.** (*i*) 1 is neither prime nor composite.

(ii) The only even prime number is 2.

(iii) The prime numbers upto 100 are listed below :—

2, 3, 5, 7, 11, 13, 17, 19, 23, 29, 31, 37, 41, 43, 47, 53, 59, 61, 67, 71, 73, 83, 89, 97.

## SOLVED PROBLEMS

**Ex. 1.** ***Resolve 39270 into prime factors.***

| | |
|---|---|
| 2 | 39270 |
| 5 | 19635 |
| 3 | 3927 |
| 11 | 1309 |
| 17 | 119 |
| 7 | 7 |

∴ *39270 = 2 × 5 × 3 × 11 × 17 × 7.*

**Ex. 2.** ***Without actual division prove that :—***

***(i) 50376 is divisible by 6.***

***(ii) 1952256 is divisible by 24.***

***(iii) 214668 is divisible by 36.***

***(iv) 47784 is divisible by 88.***

*(v) 3825360 is divisible by 792.*

**Sol.** *(i)* Since $6 = 2 \times 3$, so we must test the divisibility of the gien number by 2 and 3.

Since, the unit digit in the given number is 6, which is divisible by 2, so the given number is divisible by 2.

Also, sum of digits in given number is $5 + 0 + 3 + 7 + 6 = 21$, which is divisible by 3. So, the number is divisible by 3.

Hence, 50376 is divisible by $2 \times 3$ or 6.

*(ii)* Since $24 = 8 \times 3$, so we must test the divisibility of given number by 8 and 3.

The number formed by last 3 digits of given number is 256, which is divisible by 8. So, given number is divisible by 8.

Again, the sum of digits in given number is $1+9+5+2+2+5+6=30$, which is divisible by 3. So, the given number is divisible by 3.

Hence, 1952256 is divisible by $8 \times 3$ or 24.

*(iii)* Since $36 = 4 \times 9$, so we must test the divisibility of given number by 4 and 9.

Now, the number formed by the last 2 digits of the given number is 68, which is divisible by 4. So, the given number is divisible by 4.

Sum of digits in the given number is $2 + 1 + 4 + 6 + 6 + 8 = 27$, which is divisible by 9. So, the given number is divisible by 9.

Hence, 214668 is divisible by $9 \times 4$ or 36.

*(iv)* Since $88 = 8 \times 11$, so we mus test the divisibility of the given number by 8 and 11.

Now, the number formed by last three digits of given number is 784, which is divisible by 8. So, the given number is divisible by 8.

Again, sum of digits at odd places = $4 + 7 + 4 = 15$

Sum of digits at even places = $8 + 7 = 15$

$\therefore$ Difference of the two sums is 0. So, the given number is divisible by 11.

Hence, 47784 is divisible by $8 \times 11$ i.e., 88.

*(v)* Since $792 = 8 \times 9 \times 11$, so we must test the divisibility of the given number by 8, 9 and 11.

Number formed by last 3 digits of the given number is 360, which is divisible by 8. So, the given number is divisible by 8.

Again, the sum of digits in given number = $3+8+2+5+3+6+0=27$, which is divisible by 9. So, the given number is divisible by 9.

Also, sum of digits at odd places is $0 + 3 + 2 + 3 = 8$, even

and, sum of digits at even places is $6 + 5 + 8 = 19$.

Difference of two sums = $19 - 8 = 11$, which is divisible by 11.

So, the given number is divisible by 11.

Consequently, 3825360 is divisible by $8 \times 9 \times 11$ *i.e.*, 792.

**Ex. 3.** ***Show that the following numbers are prime*** **:—**

*(i) 83* *(ii) 911* *(iii) 3517*

**Sol.** (*i*) By trial we find that $10 \times 10 = 100 > 83$.

Now test the divisibility of 83 by each prime number less that 10 viz. 2, 3, 5 & 7.

None of these numbers divides 83.

Hence 83 is a prime number.

(*ii*) By trial we find that, $31 \times 31 > 911$.

Now test the divisibility of 911 by each prime number less than 31 viz, 2, 3, 5, 7, 11, 13, 17, 19, 23, 29. It is found that none of these numbers divides 911. Hence 911 is a prime number.

(*iii*) Since $60 \times 60 = 3600 > 3517$.

Now, test the divisibility of 3517 by each prime number less than 60 namely, 2, 3, 5, 7, 11, 13, 17, 19, 23, 29, 31, 37, 41, 43, 47, 53 and 59.

Since, none of these numbers divides 3517, so it is prime.

## VARIOUS TYPES OF NUMBERS

(*i*) **Natural numbers and whole numbers.** *Counting numbers are known as Natural numbers.* Thus, **N** = {*1, 2, 3, 4, 5, .....*} denotes the set of all natural numbers. The set, **W** = {*0, 1, 2, 3, 4, 5, ......*} *is called the set of whole numbers.*

(*ii*) **Integers.** *All counting numbers, together with their negatives and 0 constitute the set* **I** = {*......., – 3, – 2, – 1, 0, 1, 2, ......*} *of integers.*

As an example, 1857 is a positive integer, – 1857 is a negative integer and 0 is an integer, which is neither positive nor negative.

(*iii*) **Fractions.** *The numbers of the form,* $\frac{p}{q}$, *where p and q are natural numbers, are called fractions, e.g.* $\frac{3}{5}, \frac{12}{7}$ *etc.* In particular, each natural number is a fraction, since we may express 5 as (5/1).

A fraction in which the numerator and denominator have no common factors, is said to be in the simplest form.

A fraction in which the numerator is less than the denominator, is known as a *proper fraction*, otherwise it is *improper*.

**Remark.** Negative numbers are not fractions.

(*iv*) **Rational numbers.** *The numbers of the form (p / q), where p and q are integers and* $q \neq 0$, *are called rational numbers.*

(*v*) **Irrational Numbers.** A number, which when expressed in decimal form, is expressible neither in terminating decimal nor in repeating decimal, is known as an irrational number, e.g., $\sqrt{2}, \sqrt{3}, \sqrt{5}, \pi, e$ etc.

(*vi*) **Real numbers.** Totality of rationals and irrationals forms the set R of all real numbers.

## EXERCISE 1 A (Subjective)

**1.** In a division sum find the unknown :

(*i*) divisor = 156, quotient = 141, remainder = 34, dividend = ?

(*ii*) dividend = 36120, quotient = 137, remainder = 89, divisor = ?

(*iii*) dividend = 55390, divisor = 299, remainder = 75, quotient = ?

**2.** Find the least number, which when subtracted from 45378 gives a number exactly divisible by 379.

**3.** Find the least number, which must be added to 36520 to give a number exactly divisible by 187.

**4.** Find the number nearest to 13601 which is exactly divisible by 87.

**5.** Find the number nearest to 17289 which is exactly divisible by 193.

**6.** Find the least number of 6 digits, which is exactly divisible by 79.

**7.** Find the greatest number of 5 digits, which is exactly divisible by 136.

**8.** When a certain number is multiuplied by 13, the product entirely consisits of nines, find the smallest such number.

**Hint.** *By trial, we find that the smallest number consisting of all nines and divisible by 13 is 999999.*

**9.** Find in the missing figures

```
(i)      4975              (ii)      336
       × * * *                     * * 7
       -------                    ------
        * * 825                    2352
      * * * * *                   * * *
     * * * * *                   * * *
     ---------                   -------
      * * * 575                  7 * * 12
```

**10.** Multiply in two lines :

(*i*) 1378 by 568 (*ii*) 2503 by 963 (*iii*) 3058 by 109.

**Hints.** (*i*) *568* = (*560* + *8*) = & *560* = *8* × *70*.

(*ii*) *963* = (*900* + *63*) & *63* = *9* × *7*.

(*iii*) *109* = (*100* + *9*) & *9* = *9* × *1*.

**11.** Multiply in 3 lines :

(*i*) 75321 by 24696 (*ii*) 91457 by 3311132.

**Hint.** (*i*) *Take the sum of the products of 75321 with 600, 24000 and 96, keeping in view that 24 = 6 × 4 and 96 = 24 × 4.*

(ii) *Take the sum of the products of 91457 with 11000, 3300000 and 132, keeping in view that 33 = 11 × 3 and 132 = 33 × 4.*

12. Without actual division, prove that :
 (i) 175904 is divisible by 32.
 (ii) 643962 is divisible by 33.
 (iii) 637142580 is divisible by 45.
 (iv) 467016 is divisible by 132. **[Hint.** *132 = 3 × 4 × 11*].
 (v) 8532 is divisible by 54.

 **Hint.** *8532 is clearly divisible by 3 and 2. Dividing 8532 by 6, we get 1422 as quotient. This quotient is divisible by 9. So, 8532 is divisible by (3 × 2 × 9), i.e. 54.*

13. Prove that the following numbers are prime :
 (*i*) 313 (*ii*) 257 (*iii*) 2407 (*iv*) 1297.

14. A boy was asked to divide 15980 by 119. He made some mistake in copying the divisor and obtained quotient = 123 and remainder = 113. What mistake did he make ?

 **Hint.** *As obtained by the boy, divisor* $= \frac{15980 - 113}{123} = 129$.

15. Write the following numbers in Romans :
 (*i*) 2247 (*ii*) 274 (*iii*) 5960 (*iv*) 21540

16. Write numerals for the following Roman numbers :
 (*i*) MCDXL VI (*ii*) LXXIV CCIX (*iii*) CCXLI
 (*iv*) MCMXCII (*v*) LCDLXXXV (*vi*) IXCMXI.

17. Fill in the blanks :

| *Sr. No.* | *Dividend* | *Divisor* | *Quotient* | *Remainder* |
|---|---|---|---|---|
| (*i*) | .... | 36 | 47 | 8 |
| (*ii*) | 1950 | .... | 43 | 15 |
| (*iii*) | 1196 | 35 | .... | 6 |

## ANSWERS (EXERCISE 1A)

**1.** (*i*) 22030 (*ii*) 263 (*iii*) 185 **2.** 277
**3.** 132 **4.** 13572 **5.** 17370 **6.** 100014
**7.** 99960 **8.** 76923
**9.** (*i*) multiplier = 657, product = 3268575
 (*ii*) multiplier = 217, product = 72912
**10.** (*i*) 782704 (*ii*) 241089 (*iii*) 33322
**11.** (*i*) 1860127416 (*ii*) 302826199324 **14.** 129
**15.** (*i*) MMCCXLVII (*ii*) CCLXXIV (*iii*) VCMLX (*iv*) XXIDXL
**16.** (*i*) 1446 (*ii*) 74209 (*iii*) 241 (*iv*) 1992
 (*v*) 50485 (*vi*) 9911
**17.** (*i*) 1700 (*ii*) 45 (*iii*) 34

## EXERCISE 1B (Objective Type Questions)

1. The face value of 6 in the numeral 4682 is :
(*a*) 6 (*b*) 600
(*c*) 46 (*d*) none of these

2. The place value of 7 in the numeral 47921 is :
(*a*) 7 (*b*) 47
(*c*) 70000 (*d*) 7000

3. What least value must be given to * so that the number 84705*2 is divisible by 9 :
(*a*) 0 (*b*) 1
(*c*) 3 (*d*) 2

4. What least value must be given to * so that the number 3791*2 is divisible by 3 :
(*a*) 0 (*b*) 1
(*c*) 2 (*d*) none of these

5. What least value must be given to * so that the number 3*63504 is divisible by 11 :
(*a*) 0 (*b*) 2
(*c*) 3 (*d*) 4

6. $587 \times 999 = ?$
(*a*) 586413 (*b*) 615173
(*c*) 587523 (*d*) 614823

7. The least prime number is :
(*a*) 0 (*b*) 1
(*c*) 2 (*d*) 3

8. Which of the following numbers is prime ?
(*a*) 119 (*b*) 187
(*c*) 247 (*d*) none of these

9. Which of the following numbers is exactly divisible by 99 ?
(*a*) 114345 (*b*) 3572404
(*c*) 135792 (*d*) 913464

10. On dividing 55390 by 299, the remainder is 75. The quotient is :
(*a*) 187 (*b*) 185
(*c*) 245 (*d*) 374

11. On dividing a number by 999, the quotient is 377 and the remainder is 105. The required number is :
(*a*) 359738 (*b*) 376538
(*c*) 376728 (*d*) 476727

12. $173 \times 240 = 48 \times ?$

(*a*) 495 (*b*) 545
(*c*) 685 (*d*) 865

**13.** A number when divided by 357 leaves a remainder 39. By dividing the same number by 17, the remainder obtained is :
(*a*) 0 (*b*) 1
(*c*) 4 (*d*) 5

**14.** What is the sum of all prime numbers between 100 and 120 ?
(*a*) 424 (*b*) 533
(*c*) 648 (*d*) 650

**15.** The ratio between two numbers is 3 : 4 and their sum is 420. The greater of the two numbers is :
(*a*) 175 (*b*) 200
(*c*) 240 (*d*) 315 **(RRB Exam. 1991)**

**16.** The difference between the squares of two consecutive numbers is 35. The numbers are :
(*a*) 14, 15 (*b*) 15, 16
(*c*) 17, 18 (*d*) 18, 19 **(RRB Exam. 1991)**

**17.** The difference of two numbers is 11 and (1/5)th of their sum is 9. The numbers are :
(*a*) 31, 20 (*b*) 30, 19
(*c*) 29, 18 (*d*) 28, 17 **(RRB Exam. 1991)**

**18.** The sum of two numbers is 100 and their difference is 37. The difference of their squares is :
(*a*) 37 (*b*) 100
(*c*) 63 (*d*) 3700

**(Clerks' Grade Exam. 1991)**

**19.** Three fourth of one fifth of a number is 60. The number is :
(*a*) 300 (*b*) 400
(*c*) 450 (*d*) 1200 **(P.O. Exam. 1990)**

**20.** A fraction becomes 4 when 1 is added to both the numerator and denominator; and it becomes 7 when 1 is subtracted from both the numerator and denominator. The numerator of the given fraction is :
(*a*) 2 (*b*) 3
(*c*) 7 (*d*) 15 **(N.D.A. Exam. 1990)**

**21.** If 1 is added to the denominator of a fraction, the fraction becomes (1/2). If 1 is added to the numerator, the fraction becomes 1. The fraction is :
(*a*) $\frac{4}{7}$ (*b*) $\frac{5}{9}$

(*c*) $\frac{2}{3}$ (*d*) $\frac{10}{11}$ **(C.D.S. Exam. 1991)**

**22.** The sum of two numbers is 15 and sum of their squares is 113. The numbers are :

(*a*) 4, 11 (*b*) 5, 10

(*c*) 6, 9 (*d*) 7, 8 **(C.D.S. Exam. 1991)**

**23.** The sum of two numbers is twice their difference. If one of the numbers is 10, the other number is :

(*a*) $3\frac{1}{3}$ (*b*) 30

(*c*) 30 or $-3\frac{1}{3}$ (*d*) 30 or $3\frac{1}{3}$ **(RRB Exam. 1991)**

**24.** $\frac{4}{5}$ of a certain number is 64. Half of that number is :

(*a*) 32 (*b*) 40

(*c*) 80 (*d*) 16 **(BSRB Exam. 1991)**

**25.** $\frac{1}{4}$ of a number subtracted from $\frac{1}{3}$ of the number gives 12. The number is :

(*a*) 144 (*b*) 120

(*c*) 72 (*d*) 63 **(Hotel Management, 1991)**

**26.** If one fifth of a number decreased by 5 is 5, then the number is :

(*a*) 25 (*b*) 50

(*c*) 60 (*d*) 75 **(Clerks' Grade Exam. 1991)**

**27.** The sum of squares of two numbers is 80 and the square of their difference is 36. The product of the two numbers is :

(*a*) 22 (*b*) 44

(*c*) 58 (*d*) 116 **(Clerk's Grade Exam. 1991)**

**28.** The product of two numbers is 120. The sum of their squares is 289. The sum of the two numbers is :

(*a*) 20 (*b*) 23

(*c*) 169 (*d*) none of these

**(Clerks' Grade Exam. 1991)**

**29.** A number is as much greater than 21 as is less than 71. The number is :

(*a*) 39 (*b*) 41

(*c*) 46 (*d*) 49

**30.** If one fourth of one third of one half of a number is 15, the number is :

(*a*) 72 (*b*) 120

(*c*) 180 (*d*) 360

**31.** 11 times a number gives 132. The number is :

(*a*) 11 (*b*) 12
(*c*) 13.2 (*d*) none of these

**(Clerks' Grade Exam. 1991)**

**32. A number when divided by 6 is diminished by 40. The number is :**

(*a*) 72 (*b*) 84
(*c*) 60 (*d*) 48

**33. Four fifth of a number is more than three fourth of the number by 4. The number is :**

(*a*) 64 (*b*) 72
(*c*) 80 (*d*) 84

**34. Four fifth of a number is 10 more than two third of the number. The number is :**

(*a*) 45 (*b*) 55
(*c*) 60 (*d*) 75

**35. A number whose fifth part increased by 5 is equal to its fourth part diminished by 5, is :**

(*a*) 160 (*b*) 180
(*c*) 200 (*d*) 220

**36. A number is 25 more than its $\frac{2}{5}$th. The number is :**

(*a*) $\frac{125}{3}$ (*b*) $\frac{125}{7}$
(*c*) 60 (*d*) 80

**(Clerks' Grade Exam. 1991)**

**37. $\frac{2}{3}$ of a number is 20 less than the original number. The number is :**

(*a*) 40 (*b*) 60
(*c*) 80 (*d*) 90

**38. The difference between two numbers is 5 and the difference between their squares is 65. The numbers are :**

(*a*) 15, 10 (*b*) 14, 9
(*c*) 12, 7 (*d*) 9, 4

**39. 24 is divided into two parts such that 7 times the first part added to 5 times the second part makes 146. The first part is :**

(*a*) 11 (*b*) 13
(*c*) 16 (*d*) 17 **(RRB Exam. 1991)**

**40. The sum of three numbers is 132. If the first number be twice the second and third number be one third of the first, then the second number is :**

(*a*) 32 (*b*) 36
(*c*) 48 (*d*) 60

**41.** $\frac{4}{5}$ of a number exceeds its $\frac{2}{3}$ by 8. The number is :

(*a*) 30 (*b*) 60

(*c*) 90 (*d*) none of these

**(RRB Exam. 1989)**

**42.** The ratio between two numbers is 2 : 3. If the consequent is 24, the antecedent is :

(*a*) 36 (*b*) 16

(*c*) $\frac{48}{5}$ (*d*) $\frac{72}{5}$

**43.** The difference between squares of two numbers is 256000 and sum of the numbers is 1000. The numbers are :

(*a*) 628, 372 (*b*) 600, 400

(*c*) 640, 630 (*d*) none of these

**(G.I.C.A.A.O. Exam. 1988)**

**44.** Of the two numbers, 4 times the first is equal to 5 times the other and the sum of 3 times the first and 5 times the second is 105. The first number is :

(*a*) 10 (*b*) 12

(*c*) 13 (*d*) 15

**45.** Three numbers are in the ratio 3 : 4 : 5. The sum of the largest and the smallest equals the sum of the third and 52. The smallest number is :

(*a*) 20 (*b*) 27

(*c*) 39 (*d*) 52 **(Accountants' Exam. 1986)**

**46.** Of the two numbers, 4 times the smaller one is less than 3 times the larger one by 5. But the sum of the numbers is larger than 6 times their difference by 6. The larger number is :

(*a*) 43 (*b*) 53

(*c*) 59 (*d*) 63

**47.** A positive number when decreased by 4, is equal to 21 times the reciprocal of the number. The number is :

(*a*) 3 (*b*) 5

(*c*) 7 (*d*) 9 **(N.D.A. Exam. 1987)**

**48.** The sum of three consecutive odd numbers is 21. The middle one is :

(*a*) 11 (*b*) 9

(*c*) 7 (*d*) 5

**49.** If 10 be added to four times a certain number, the result is 5 less than 5 times the number. The number is :

(*a*) 35 (*b*) 25

(*c*) 20 (*d*) 15

**50.** **Of the three numbers, the sum of first two is 45; the sum of the second and the third is 55 and the sum of the third and thrice the first is 90. The third number is :**

(*a*) 20 (*b*) 25
(*c*) 30 (*d*) 35

**51.** **There are two numbers such that the sum of twice the first and thrice the second is 18, while the sum of thrice the first and twice the second is 17. The larger of the two is :**

(*a*) 4 (*b*) 6
(*c*) 8 (*d*) 12

**52.** **If a number is subtracted from the square of its one half, the result is 48. The square root of the number is :**

(*a*) 4 (*b*) 5
(*c*) 6 (*d*) 8

**53.** **If 3 is added to the denominator of a fraction, it becomes $\frac{1}{3}$ and if 4 be added to its numerator, it becomes $\frac{3}{4}$. The fraction is :**

(*a*) $\frac{4}{9}$ (*b*) $\frac{3}{20}$
(*c*) $\frac{7}{24}$ (*d*) $\frac{5}{12}$

**54.** **If from twice the greater of the two numbers, 20 is subtracted, the result is the other number. If from twice the smaller number 5 is subtracted, the result is the first number. The smaller number is :**

(*a*) 6 (*b*) 8
(*c*) 10 (*d*) 12

**55.** **One number is greater by 2 than thrice the other number and 6 times the smaller number exceeds the greater by 1. The smaller number is :**

(*a*) 6 (*b*) 5
(*c*) 3 (*d*) 1

**56.** **A certain number of 2 digits is three times the sum of its digits and if 45 be added to it, the digits will be reversed. The number is :**

(*a*) 32 (*b*) 72
(*c*) 27 (*d*) 23

**57.** **The sum of 3 numbers is 68. If the ratio between first and second be 2 : 3 and that between second and third be 5 : 3, then the second number is :**

(*a*) 30 (*b*) 20
(*c*) 58 (*d*) 48 **(S.S.C. Exam. 1986)**

**58. Two numbers are such that the ratio between them is 3 : 5; but if each is increased by 10, the ratio between them becomes 5 : 7. The numbers are :**

(*a*) 3, 5 (*b*) 7, 9

(*c*) 13, 22 (*d*) 15, 25 **(RRB Exam. 1989)**

**59. Divide 50 into two parts so that the sum of their reciprocals is (1/12) :**

(*a*) 20, 30 (*b*) 24, 26

(*c*) 28, 22 (*d*) 36, 14 **(RRB Exam. 1988)**

**60. The sum of seven numbers is 235. The average of the first three is 23 and that of the last three is 42. The fourth number is :**

(*a*) 40 (*b*) 126

(*c*) 69 (*d*) 195

**(Clerks' Grade Exam. 1991)**

**61.** (...?...) – (1936248) = (1635773)

(*a*) 3572021 (*b*) 3561231

(*c*) 3562121 (*d*) 3536021

**62.** 35999 – 17102 – 8799 = (...?...)

(*a*) 27696 (*b*) 10098

(*c*) 20318 (*d*) none of these

**63.** 12846 × 593 + 12846 × 407 = (...?...)

(*a*) 24064000 (*b*) 12846000

(*c*) 24038606 (*d*) 24203706

**64.** 469157 × 9999 = (...?...)

(*a*) 4586970843 (*b*) 4686970743

(*c*) 4691100843 (*d*) 4586870843

**65.** 935421 × 625 = (...?...)

(*a*) 575648125 (*b*) 584638125

(*c*) 585628125 (*d*) 584649125

**66.** (387 × 387 + 114 × 114 + 2 × 387 × 114) = (...?...)

(*a*) 250001 (*b*) 251001

(*c*) 260101 (*d*) 261001

**67.** 1014 × 986 = (...?...)

(*a*) 998904 (*b*) 999804

(*c*) 998814 (*d*) 998804

**68.** 1299 × 1299 = (...?...)

(*a*) 1585301 (*b*) 1684701

(*c*) 1685401 (*d*) 1687401

**69.** 106 × 106 + 94 × 94 = (...?...)

(*a*) 21032 (*b*) 20032

(*c*) 23032 (*d*) 20072

**70.** $(475 + 425)^2 - 4 \times 475 \times 425$ is equal to :

(*a*) 3600 (*b*) 3500

(*c*) 2500 (*d*) 3160

**71.** 5358 × 51 = (...?...)

(*a*) 273358 (*b*) 273258

(*c*) 273348 (*d*) 273268

**72.** 1307 × 1307 = (...?...)

(*a*) 1601249 (*b*) 1607249

(*c*) 1701249 (*d*) 1708249

**73.** The value of

$$\left(\frac{343 \times 343 \times 343 + 257 \times 257 \times 257}{343 \times 343 - 343 \times 257 + 257 \times 257}\right) \text{ is}$$

(*a*) 8600 (*b*) 800

(*c*) 600 (*d*) 2600

**74.** The value of

$$\left(\frac{117 \times 117 \times 117 - 98 \times 98 \times 98}{117 \times 117 + 117 \times 98 + 98 \times 98}\right) \text{ is}$$

(*a*) 215 (*b*) 311

(*c*) 19 (*d*) 29

**75.** $\frac{137 \times 137 + 137 \times 133 + 133 \times 133}{137 \times 137 \times 137 - 133 \times 133 \times 133} = ?$

(*a*) 4 (*b*) $\frac{1}{4}$

(*c*) 270 (*d*) $\frac{1}{270}$

**76.** On dividing 55390 by 299, the remainder is 75. The quotient obtained is :

(*a*) 187 (*b*) 185

(*c*) 245 (*d*) 374

**77.** A number, when divided by 783 gives a remainder 48. What remainder would be obtained by dividing the same number by 29 :

(*a*) 17 (*b*) 15

(*c*) 21 (*d*) 19

**78.** What least number must be subtracted from 13601 to get a number exactly divisible by 87 :

(*a*) 49 (*b*) 23

(*c*) 29 (*d*) 31

**79.** What least number must be added to 1056 to get a number exactly divisible by 23 :

(*a*) 21 (*b*) 25
(*c*) 3 (*d*) 2

**80.** The least number of 5 digits which is exactly divisible by 456 is :
(*a*) 10142 (*b*) 10232
(*c*) 10032 (*d*) 10012

## HINTS & SOLUTIONS (EXERCISE 1B)

1. The face value of 6 in the given number is 6.
2. The place value of 7 in the given number is 7000.
3. $(8 + 4 + 7 + 0 + 5 + 2) = 26$. The nearest number divisible by 9 is 27. So, the required digit $= (27 - 26) = 1$.
4. $(3 + 7 + 9 + 1 + 2) = 22$. The nearest number divisible by 3 is 24. So, the required digit $= (24 - 22) = 2$.
5. $(4 + 5 + 6 + 3) - (0 + 3 + x) = 15 - x$. Now, $15 - x$ must be divisible by 11. So, $x = 4$.
6. $587 \times 999 = 587 \times (1000 - 1) = (587000 - 587) = 586413$.
7. The least prime number is 2.
8. None.
9. The number divisible by 11 as well as by 9 is 913464.
10. Quotient $= \left(\dfrac{55390 - 75}{299}\right) = 185$.
11. The required number $= (999 \times 377 + 105) = 376728$.
12. Let $173 \times 240 = 48 \times x$. Then, $x = \dfrac{173 \times 240}{48} = 865$.
13. Let number $= 357k + 39$
$= (17 \times 21k + 34) + 5 = 17 \times (21k + 2) + 5$.
$\therefore$ Required remainder = 5.
14. The prime numbers between 100 and 120 are :
101, 103, 107, 109, 113
Their sum is 533.
15. $3x + 4x = 420 \Rightarrow x = 60$.
$\therefore$ Greater number $= 4 \times 60 = 240$.
16. Let the consecutive numbers be $x$ and $(x + 1)$.
Then, $(x + 1)^2 - x^2 = 35 \Rightarrow x = 17$.
17. Let the numbers be $x$ and $(x - 11)$.
$\dfrac{1}{5} \times (x + x - 11) = 9 \Rightarrow x = 28$.
So, the numbers are 28, 17.
18. Let the numbers be $x$ and $y$.

Then, $x + y = 100$ and $x - y = 37$.

Now, $x^2 - y^2 = (x + y) \times (x - y) = 100 \times 37 = 3700$.

**19.** $\frac{3}{4} \times \frac{1}{5} \times x = 60 \Rightarrow x = \frac{60 \times 4 \times 5}{3} = 400$.

**20.** Let the required fraction be $\frac{x}{y}$.

Then, $\frac{x+1}{y+1} = 4 \Rightarrow x - 4y = 3$ ...(i)

And, $\frac{x-1}{y-1} = 7 \Rightarrow x - 7y = -6$ ...(ii)

Solving (i) and (ii), we get $x = 15, y = 3$.

**21.** Let the required fraction $= \frac{x}{y}$.

$\therefore \frac{x}{y+1} = \frac{1}{2} \Rightarrow 2x - y = 1$ ...(i)

And, $\frac{x+1}{y} = 1 \Rightarrow x - y = -1$ ...(ii)

Solving (i) and (ii), we get $x = 2, y = 3$.

$\therefore$ The fraction is $\frac{2}{3}$.

**22.** Let the numbers be $x$ and $(15 - x)$.

Then, $x^2 + (15 - x)^2 = 113$ or $x^2 - 15x + 56 = 0$

$\therefore x = 8$ or $7$.

**23.** Let the other number $= x$.

$10 + x = 2(x - 10) \Rightarrow x = 30$.

**24.** $\frac{4}{5} \times x = 64 \Rightarrow x = \frac{64 \times 5}{4} = 80$.

$\therefore \frac{1}{2} \times x = \frac{1}{2} \times 80 = 40$.

**25.** $\left(\frac{1}{3} \times x - \frac{1}{4} \times x\right) = 12 \Rightarrow \frac{1}{12}x = 12 \Rightarrow x = 144$.

**26.** $\left(\frac{1}{5} \text{ of } x\right) - 5 = 5 \Rightarrow \frac{x}{5} = 10 \Rightarrow x = 50$.

**27.** Let the numbers be $x$ and $y$. Then,

$x^2 + y^2 = 80$ and $(x - y)^2 = 36$.

$\therefore x^2 + y^2 - 2xy = 36 \Rightarrow 80 - 2xy = 36 \Rightarrow xy = 22$.

**28.** Let the numbers be $x$ and $y$. Then,

$xy = 120$ and $x^2 + y^2 = 289$.

Now, $(x+y)^2 = (x^2 + y^2 + 2xy) = 289 + 240 = 529.$

$\therefore x + y = 23.$

**29.** Let the number be $x$.

$(x-21) = (71-x) \Rightarrow x = 46.$

**30.** $\frac{1}{4} \times \frac{1}{3} \times \frac{1}{2} \times x = 15 \Rightarrow x = (15 \times 4 \times 3 \times 2) = 360.$

**31.** $11x = 132 \Rightarrow x = 12.$

**32.** $\frac{x}{6} = (x-40) \Rightarrow x = 48.$

**33.** $\frac{4}{5}x - \frac{3}{4}x = 4 \Rightarrow x = 80.$

**34.** $\frac{4}{5}x - \frac{2}{3}x = 10 \Rightarrow \frac{2}{15}x = 10 \Rightarrow x = 75.$

**35.** $\frac{1}{5}x + 5 = \frac{1}{4}x - 5 \Rightarrow \left(\frac{1}{4} - \frac{1}{5}\right)x = 10 \Rightarrow x = 200.$

**36.** $x - \frac{2}{5}x = 25 \Rightarrow \frac{3x}{5} = 25 \Rightarrow x = \frac{125}{3}.$

**37.** $x - \frac{2}{3}x = 20 \Rightarrow \frac{x}{3} = 20 \Rightarrow x = 60.$

**38.** Let the numbers be $x$ and $y$. Then,

$x - y = 5$ and $x^2 - y^2 = 65.$

$\therefore x + y = \frac{x^2 - y^2}{x - y} = \frac{65}{5} = 13.$

Solving $x - y = 5, x + y = 13$, we get $x = 9, y = 4$.

**39.** Let these parts be $x$ and $(24 - x)$. Then,

$7x + 5(24 - x) = 146 \Rightarrow x = 13.$

So, the first part is 13.

**40.** Let the numbers be $x, y, z$. Then,

$x = 2y$ and $z = \frac{1}{3}x$.

So, the numbers are $x, \frac{1}{2}x$ and $\frac{1}{3}x$.

$\therefore x + \frac{1}{2}x + \frac{1}{3}x = 132 \Rightarrow x = 72.$

So, second number $= \frac{1}{2}x = 36.$

**41.** Let the number be $x$. Then,

$\frac{4}{5}x - \frac{2}{3}x = 8 \Rightarrow \frac{2}{15}x = 8 \Rightarrow x = 60.$

**42.** Let the numbers be $2x$ and $3x$.

$3x = 24 \Rightarrow x = 8$.

$\therefore$ antecedent $= 2x = 16$.

**43.** Let the numbers be $x$ and $y$. Then,

$x^2 - y^2 = 256000$ and $x + y = 1000$.

$$\therefore x - y = \frac{x^2 - y^2}{x + y} = \frac{256000}{1000} = 256.$$

Solving $x + y = 1000$, $x - y = 256$, we get $x = 628$, $y = 372$.

**44.** Let the numbers be $x$ and $y$. Then,

$4x = 5y$ and $3x + 5y = 105$.

$\therefore 3x + 4x = 105 \Rightarrow x = 15$. So, $y = \frac{4 \times 15}{5} = 12$.

**45.** Let the numbers be $3x$, $4x$ and $5x$.

$5x + 3x = 4x + 52 \Rightarrow x = 13$.

$\therefore$ Smallest number $= 3x = 39$.

**46.** Let the numbers be $x$ and $y$. Then,

$3y - 4x = 5$, $(x + y) - 6(y - x) = 6$.

$\therefore 3y - 4x = 5$, $7x - 5y = 6$.

Solving these equations, we get $y = 59$.

**47.** Let the number be $x$. Then,

$x - 4 = \frac{21}{x} \Rightarrow x^2 - 4x - 21 = 0 \Rightarrow x = 7$.

$\therefore$ Required number $= 7$.

**48.** Let the numbers be $x$, $x + 2$, $x + 4$.

Then, $x + x + 2 + x + 4 = 21 \Rightarrow x = 5$.

$\therefore$ Middle number $= 7$.

**49.** Let the number be $x$. Then,

$4x + 10 = 5x - 5 \Rightarrow x = 15$.

$\therefore$ The number $= 15$.

**50.** Let the numbers be $x$, $y$, $z$. Then,

$x + y = 45$, $y + z = 55$, $z + 3x = 90$.

Now, $y = (45 - x)$ and $z = 55 - y = 55 - (45 - x) = 10 + x$.

$\therefore 10 + x + 3x = 90 \Rightarrow x = 20$.

So, third number $= 10 + x = 30$.

**51.** Let the numbers be $x$ and $y$. Then,

$2x + 3y = 18$, $3x + 2y = 17$.

Solving, we get $x = 3$, $y = 4$.

$\therefore$ Larger number $= 4$.

**52.** Let the number be $x$. Then,

$$\left(\frac{x}{2}\right)^2 - x = 48 \Rightarrow \frac{x^2}{4} - x = 48 \Rightarrow x^2 - 4x - 192 = 0 \Rightarrow x = 16.$$

The square root of the number is 4.

**53.** Let the fraction be $\frac{x}{y}$. Then,

$$\frac{x}{y+3} = \frac{1}{3} \Rightarrow 3x - y = 3\ .$$

And, $\frac{x+4}{y} = \frac{3}{4} \Rightarrow 4x - 3y = -16.$

Solving these equations, we get $x = 5, y = 12$.

$\therefore$ Required fraction $= \frac{5}{12}$.

**54.** Let the numbers be $x$ and $y$. Then,

$2x - 20 = y \Rightarrow 2x - y = 20$ .

And, $2y - 5 = x \Rightarrow x - 2y = -5$ .

Solving these equations, we get $x = 15, y = 10$.

$\therefore$ Smaller number = 10.

**55.** Let the numbers be $x$ and $y$. Then,

$x - 3y = 2, 6y - x = 1.$

Solving these equations, we get $x = 5, y = 1$.

The smaller number = 1.

**56.** Let unit digit = $x$ & ten's digit = $y$.

$3(x + y) = 10y + x, 10y + x + 15 = 10x + y$

$2x - 7y = 0, 9x - 9y = 45$ or $x - y = 5$.

Solving these equations, we get $x = 7, y = 2$.

$\therefore$ Required number = 27.

**57.** Let the numbers be $x, y, z$. Then,

$$\frac{x}{y} = \frac{2}{3}, \frac{y}{z} = \frac{5}{3} \Rightarrow 3x = 2y \text{ and } 5z = 3y.$$

$$\therefore y = \frac{3}{2}x, z = \frac{3}{5}y = \frac{3}{5} \times \frac{3}{2}x = \frac{9}{10}x.$$

$$\therefore x + \frac{3}{2}x + \frac{9}{10}x = 98 \Rightarrow 34x = 680 \Rightarrow x = 20.$$

So, second number $= \frac{3}{2}x = \left(\frac{3}{2} \times 20\right) = 30.$

**58.** Let the numbers be $3x$ and $5x$.

$$\therefore \frac{3x+10}{5x+10} = \frac{5}{7} \Rightarrow x = 5.$$

Hence, the numbers are 15, 25.

**59.** Let the numbers be $x$ and $(50-x)$. Then,

$$\frac{1}{x}+\frac{1}{50-x}=\frac{1}{12}\Rightarrow\frac{50-x+x}{x(50-x)}=\frac{1}{12}\Rightarrow x^2-50x+600=0\Rightarrow x=30 \text{ or } 20.$$

**60.** $(23\times 3+x+42\times 3)=235\Rightarrow x=40.$

$\therefore$ Fourth number $=40$.

**61.**

```
  1635773
+ 1936248
---------
  3572021
---------
```

**62.**

```
 35999
.17102]
  8799}
------
 10098
------
```

**63.** $12846\times 593+12846\times 407$

$=12846\times(593+407)$

$=12846\times 1000=12846000.$

**64.** $469157\times 9999$

$=4691570000-469157=4691100843$ (see rule)

**65.** Since $625=5^4$. Put 4 zeros to the right of 935421 and divide 9354210000 by $2^4$, *i.e.* 16.

$\therefore$ Required result $=9354210000\div 16=584638125.$

**66.** Use the formula, $(a^2+b^2+2ab)=(a+b)^2$.

$387\times 387+114\times 114+2\times 387\times 114$

$=(387)^2+(114)^2+2\times 387\times 114$

$=(387+114)^2=(501)^2=(500+1)^2$

$=(500)^2+(1)^2+2\times 500\times 1$

$=250000+1+1000=251001.$

**67.** $1014\times 986=(1000+14)\times(1000-14)$

$=(1000)^2-(14)^2$ [Use, $(a+b)(a-b)=(a^2-b^2)$]

$=(1000000-196)=999804.$

**68.** $1299\times 1299=(1299)^2=(1300-1)^2$

$=(1300)^2+(1)^2-\times 2\times 1300\times 1$

$=1690000+1-2600=1687401.$

**69.** $2(a^2+b^2)=(a+b)^2+(a-b)^2$

$\therefore 2[(106)^2+(94)^2]=[(106+94)^2+(106-94)^2]$

$=[(200)^2+(12)^2]=40000+144=40144.$

So, $[(106)^2+(94)^2]=20072.$

**70.** Given expression $=(a+b)^2-4ab=(a-b)^2$

$=(475-425)^2=(50)^2=2500.$

**71.** $5358\times 51=5358\times(50+1)$

$=(5358\times 50)+(5358\times 1)$

$=267900+5358=273258.$

**72.** $1307\times 1307=(1307)^2=(1300+7)^2$

$=(1300)^2+(7)^2+2\times 1300\times 7$

$=1690000+49+18200=1708249$

**73.** Given expression $=\dfrac{(a^3+b^3)}{(a^2-ab+b^2)}=(a+b)$

$=(343+257)=600.$

**74.** Given expression $=\dfrac{(a^3-b^3)}{(a^2+ab+b^2)}=(a-b)$

$=(117-98)=19.$

**75.** Given expression $=\dfrac{(a^2+ab+b^2)}{(a^3-b^3)}=\dfrac{(a^2+ab+b^2)}{(a-b)(a^2+ab+b^2)}$

$=\dfrac{1}{(a-b)}=\dfrac{1}{(137-133)}=\dfrac{1}{4}.$

**76.** Quotient $=\left(\dfrac{55390-75}{299}\right)=185.$

**77.** Let $k$ be the quotient, when the given number is divided by 789.

Then, the number $=(783k+48)=(29\times 27k)+(29\times 1)+19$

$=29\times(27k+1)+19$

$=[29\times(\text{new quotient})]+19.$

Hence, the remainder is 19.

**78.** On dividing 13601 by 87, the remainder is 29.

So, the number to be subtracted = 29.

**79.** On dividing 1056 by 23, the remainder is 21.

$\therefore$ The number to be added $=(23-21)=2.$

**80.** The least number of 5 digits is 10000.

On dividing 10000 by 456, the remainder is 424.

So, we must add $(456-424)$ i.e. 32 to 10000.

∴ Required number = 10032.

## ANSWERS (EXERCISE 1B)

| | | | | | |
|---|---|---|---|---|---|
| **1.** (*a*) | **2.** (*d*) | **3.** (*b*) | **4.** (*c*) | **5.** (*d*) | **6.** (*a*) |
| **7.** (*c*) | **8.** (*d*) | **9.** (*d*) | **10.** (*b*) | **11.** (*c*) | **12.** (*d*) |
| **13.** (*d*) | **14.** (*b*) | **15.** (*c*) | **16.** (*c*) | **17.** (*d*) | **18.** (*d*) |
| **19.** (*b*) | **20.** (*d*) | **21.** (*c*) | **22.** (*d*) | **23.** (*b*) | **24.** (*b*) |
| **25.** (*a*) | **26.** (*b*) | **27.** (*a*) | **28.** (*b*) | **29.** (*c*) | **30.** (*d*) |
| **31.** (*b*) | **32.** (*d*) | **33.** (*c*) | **34.** (*d*) | **35.** (*c*) | **36.** (*a*) |
| **37.** (*b*) | **38.** (*d*) | **39.** (*b*) | **40.** (*b*) | **41.** (*b*) | **42.** (*b*) |
| **43.** (*a*) | **44.** (*d*) | **45.** (*c*) | **46.** (*c*) | **47.** (*c*) | **48.** (*c*) |
| **49.** (*d*) | **50.** (*c*) | **51.** (*a*) | **52.** (*a*) | **53.** (*d*) | **54.** (*c*) |
| **55.** (*d*) | **56.** (*c*) | **57.** (*a*) | **58.** (*d*) | **59.** (*a*) | **60.** (*a*) |
| **61.** (*a*) | **62.** (*b*) | **63.** (*b*) | **64.** (*c*) | **65.** (*b*) | **66.** (*b*) |
| **67.** (*b*) | **68.** (*d*) | **69.** (*d*) | **70.** (*c*) | **71.** (*b*) | **72.** (*b*) |
| **73.** (*c*) | **74.** (*c*) | **75.** (*b*) | **76.** (*b*) | **77.** (*d*) | **78.** (*c*) |
| **79.** (*d*) | **80.** (*c*) | | | | |

# 2

# H.C.F. & L.C.M. OF NUMBERS

**Factors and Multiples.** *If a number m divides another number n exactly, then we say that m is a* ***factor*** *of n and that n is a* ***multiple*** *of m.*

For example, 2 is a factor of 6 and therefore 6 is a multiple of 2.

**Highest Common Factor.** *The Highest Common Factor* **(*H.C.F.*)** *or Greatest Common Divisior* **(*G.C.D.*)** *or Greatest Common Measure* **(*G.C.M.*)** *of two or more than two numbers is the greatest number that divides each one of them exactly.*

**Ex.** Consider the numbers 24 and 56.

All factors of 24 are 1, 2, 3, 4, 6, 8, 12, 24

and all factors of 56 are 1, 2, 4, 7, 8, 14, 28, 56.

Common factors of 24 and 56 are 1, 2, 4, 8.

∴ H.C.F. of 24 and 56 is 8.

**Rule For Finding Out H.C.F. of Given Numbers.** Suppose two numbers are given. Divide the larger number by smaller one. Now, divide the divisor by the remainder. Go on dividing the preceding divisor by the remainder last obtained, till a remainder 0 is obtained. The last divisor so obtained is the required H.C.F. of two given numbers.

**In case more than two numbers are given,** then choose any two numbers and find their H.C.F. The H.C.F. of these two numbers and the third number, gives the H.C.F. of these three numbers and so on.

**Co-primes.** *Two numbers are said to be co-prime, if their H.C.F. is 1.*

**Remark.** Co-primes are not necessarily primes. For example, 8 and 9 are co-primes, but none of them is a prime number.

**Ex. 1.** ***Find the H.C.F. of 777 and 1147.***

**Sol.**

```
777 ) 1147 ( 1
      777
      ----
      370 ) 777 ( 2
            740
            ----
             37 ) 370 ( 10
                  370
                  ---
                   ×
```

∴ *H.C.F. of 777 and 1147 is 37.*

**Ex. 2.** ***Show that 403 and 517 are co-prime.***

**Hint.** Show that H.C.F. of 403 and 517 is 1.

**Ex. 3.** ***Find the H.C.F. of 6851, 9061 and 18462.***

**Sol.** First consider any two of the numbers say 6851 and 9061.

```
6851) 9061 ( 1
      6851
      ----
      2210) 6851 ( 3
            6630
            ----
             221) 2210 ( 10
                  2210
                  ----
                    ×
```

∴ *H.C.F. of 6851 and 9061 is 221.*

Again, consider 221 and the third number 18462.

```
221) 18462 ( 83
     1768
     ----
      782
      663
      ---
      119) 221 ( 1
           119
           ---
           102) 119 ( 1
                102
                ---
                 17) 102 ( 6
                     102
                     ---
                      ×
```

∴ *H.C.F. of 221 and 18462 is 17.*

Hence, the required H.C.F. = 17.

**Ex. 4.** ***Express*** $\frac{87}{145}$ ***in simplest form.***

**Sol.** H.C.F. of 87 and 145 is 29.

Dividing numerator and denominator by 29 we get, $\frac{87}{145} = \frac{3}{5}$.

**Lowest Common Multiple (L.C.M.).** The least number which is exactly divisible by each one of the given numbers is called their L.C.M.

**Ex. 5.** ***Consider the numbers 12 and 18.***

**Sol.** Multiples of 12 are 12, 24, 36, 48, 60, ......

Multiples of 18 are 18, 36, 54, 72, .....

∴ *L.C.M. of 12 and 18 is 36.*

**Factorization Method.** *We may find the L.C.M. of given numbers by resolving each one of them into prime fators and then taking the product of highest powers of all the factors, that occur in these numbers.*

**Ex. 6.** ***Find the L.C.M. of 48, 108 and 140.***

**Sol.** $48 = 2 \times 2 \times 2 \times 2 \times 3 = 2^4 \times 3$

$108 = 2 \times 2 \times 3 \times 3 \times 3 = 2^2 \times 3^3$

$140 = 2 \times 2 \times 5 \times 7 = 2^2 \times 5 \times 7.$

$\therefore$ L.C.M. $= 2^4 \times 3^3 \times 5 \times 7 = 15120$.

**Ex. 7. (Short Cut Method)** ***Find the L.C.M. of 12, 15, 20, 27.***

**Sol.**

| | |
|---|---|
| 3 | 12 – 15 – 20 – 27 |
| 4 | 4 – 5 – 20 – 9 |
| 5 | 1 – 5 – 5 – 9 |
| | 1 – 1 – 1 – 9 |

$\therefore$ L.C.M. $= 3 \times 4 \times 5 \times 9 = 540$.

**Theorem.** *The product of two given numbers is equal to the product of their H.C.F. and L.C.M.*

**Proof.** Let $a$ and $b$ be two given numbers. Let their H.C.F. and L.C.M. be $h$ and $k$ respectively. On dividing $a$ and $b$ by $h$, let the quotients be $m$ and $n$ respectively, so that

$$a = m\,h \text{ and } b = n\,h$$

$\therefore$ L.C.M. of $a$ and $b$ is $m\,n\,h$ and therefore, $k = m\,n\,h$.

Now, $a \times b = m\,h \times n\,h = m\,n\,h \times h = k \times h$.

Thus,

**Product of Given Numbers = Product of thier H.C.F. & L.C.M.**

**Rule 2.** If two numbers are given, then, to find their H.C.F.

$$L.C.M. = \frac{\text{Product of these numbers}}{\text{H.C.F. of these numbers}}.$$

**If more than two numbers are given,** find L.C.M. of any two of these numbers. Then, L.C.M. of this L.C.M. and the third number gives the L.C.M. of these three numbers and so on.

**Ex. 8.** ***Find the L.C.M. of 852 and 1491.***

**Sol.**

```
852 ) 1491 ( 1
      852
      ----
      639 ) 852 ( 1
            639
            ---
            213 ) 639 ( 3
                  639
                  ---
                   ×
```

$\therefore$ H.C.F. of 852 and 1491 is 213.

Hence, L.C.M. $= \frac{852 \times 1491}{213} = 5964$.

**Ex. 9.** ***Find the L.C.M. of 572,462 and 187.***

**Sol.** First consider any two numbers, say 572, 462.

It is easy to find that H.C.F. of 572 and 462 is 22.

So, L.C.M. of 572 and 462 is $\frac{572 \times 462}{22} = 12012$.

Now, consider this L.C.M. and the third number 187.

It is easy to find that H.C.F. of 12012 and 187 is 11.

$\therefore$ L.C.M. of 12012 and 187 is $\dfrac{12012 \times 187}{11} = 204204.$

Hence, the required L.C.M. = 204204.

**H.C.F. and L.C.M. of Fractions :**

*If we are given some fractions, then their*

$$H.C.F. = \frac{H.C.F.\ of\ numerators}{L.C.M.\ of\ denominators}$$

*and*

$$L.C.M. = \frac{L.C.M.\ of\ numerators}{H.C.F.\ of\ denominators}.$$

**Ex.10.** ***Find the H.C.F. and L.C.M. of*** $\frac{3}{8}, \frac{5}{12}$ ***and*** $\frac{9}{14}$.

**Sol.** $\text{H.C.F.} = \dfrac{\text{H.C.F. of 3, 5 \& 9}}{\text{L.C.M. of 8, 12, 14}}$

$$\begin{array}{r|l} 2 & 8-12-14 \\ 2 & 4-6-7 \\ \hline & 2-3-7 \end{array}$$

$$= \frac{1}{2 \times 2 \times 2 \times 3 \times 7} = \frac{1}{168}.$$

$\text{L.C.M.} = \dfrac{\text{L.C.M. of 3, 5, 9}}{\text{H.C.F. of 8, 12, 14}}$

$$\begin{array}{r|l} 3 & 3-5-9 \\ \hline & 1-5-3 \end{array}$$

$$= \frac{3 \times 1 \times 5 \times 3}{2} = \frac{45}{2}.$$

$\therefore$ $\quad \text{H.C.F.} = \dfrac{1}{168}, \text{ L.C.M.} = \dfrac{45}{2}.$

**Ex.11.** ***Arrange the following fractions in descending order :—***

$$\frac{1}{2}, \frac{2}{3}, \frac{3}{5}, \frac{5}{7}$$

**Sol.** L.C.M. of denominators = 210.

Changing each fraction into an equivalent fraction with 210 as denominator, we have :—

$$\frac{1}{2} = \frac{1 \times 105}{2 \times 105} = \frac{105}{210}; \frac{2}{3} = \frac{2 \times 70}{3 \times 70} = \frac{140}{210};$$

$$\frac{3}{5} = \frac{3 \times 42}{5 \times 42} = \frac{126}{210}; \frac{5}{7} = \frac{5 \times 30}{7 \times 30} = \frac{150}{210}.$$

Clearly, $\dfrac{150}{210} > \dfrac{140}{210} > \dfrac{126}{210} > \dfrac{105}{210}$

Or $\dfrac{5}{7} > \dfrac{2}{3} > \dfrac{3}{5} > \dfrac{1}{2}.$

## SOLVED PROBLEMS ON H.C.F. AND L.C.M.

**Ex. 1.** ***Find the greatest number that will divide 804 and 1745 leaving remainders 5 and 6 respectively.***

**Sol.** Clearly, the required number is the H.C.F. of (804 – 5) *i.e.*, 799 and (1745 – 6) *i.e.* 1739.

Now H.C.F. of 799 and 1739 is 47.

*Hence, the required number is 47.*

**Ex. 2.** ***Find the greatest number that will divide 640, 710, and 1526 so as to leave 11, 7 and 9 as remainders respectively.***

**Sol.** Clearly, the required number is the H.C.F. of (640 – 11), (710 – 7) and (1526 – 9) *i.e.*, H.C.F. of 629, 703 and 1517, which is 37.

*Hence, the required number is 37.*

**Ex. 3.** ***Find the greatest number which will divide 590, 908 and 1014 so as to leave the same remainder in each case.***

**Sol.** Since the required number which divides 590, 908 and 1014 leaves the same remainder in each case, so this number will exactly divide the difference of any two numbers.

Consequently, the required number is H.C.F. of (908 – 590), (1014 – 908) and (1014 – 590). In other words, it is the H.C.F. of 318, 106 and 424, which is 53.

*Hence, the required number is 53.*

**Ex. 4.** ***The sum of two numbers is 721 and their H.C.F. is 103. Find all possible pairs of such numbers.***

**Sol.** Since the H.C.F. of numbers is 103, so they must be of the form 103 $a$ and 103 $b$, where $a$ and $b$ are co-primes.

Consequently, $103\,a + 103\,b = 721$ or $a + b = 7$.

But, the possible pairs of numbers, which are co-prime and whose sum is 7 are (1, 6), (2, 5) and (3, 4).

*Hence, the required numbers are*

$$\left.\begin{array}{ll} & (103\times 1, 103\times 6) \text{ or } (103, 618) \\ & (103\times 2, 103\times 5) \text{ or } (206, 515) \\ \text{and} & (103\times 3, 103\times 4) \text{ or } (309, 412) \end{array}\right\}$$

**Ex. 5.** ***The product of two numbers is 19712 and their H.C.F. is 16. Find all possible pairs of such numbers.***

**Sol.** Let the numbers be 16$a$ and 16$b$, where $a$ and $b$ are co-prime.

$\therefore\ 16\,a \times 16\,b = 19712$ or $a\,b = 77$.

But, the possible pairs of numbers, which are co-prime and whose product is 77 are (1, 77) and (7, 11).

*Hence, the required numbers are :*

$$\left.\begin{array}{ll} & (16\times 1, 16\times 77) \text{ or } (16, 1232) \\ \text{and} & (16\times 7, 16\times 11) \text{ or } (112, 176) \end{array}\right\}$$

**Ex. 6.** ***What is the greatest possible length which can be used to measure exactly the following lengths :***

*7 metres, 3 metres 85 cm and 12 metres 95 cm. ?*

**Sol.** The lengths to be measured are 700 cm, 385 cm and 1295 cm. The required length in cm. is the H.C.F. of 700, 385 and 1295, which is 35 cm.

**Ex. 7.** ***Find the least number which when divided by 15, 27, 35 and 42, leaves in each case a remainder 9 ?***

**Sol.** The least number which is exactly divisibe by each one of the numbers 15, 27, 35 and 42 is the L.C.M. of these numbers.

| | |
|---|---|
| 3 | 15 – 27 – 35 – 42 |
| 5 | 5 – 9 – 35 – 14 |
| 7 | 1 – 9 – 7 – 14 |
| | 1 – 9 – 1 – 2 |

Now L.C.M. of 15, 27, 35 & 42 is

$3 \times 5 \times 7 \times 9 \times 2 = 1890.$

$\therefore$ *The required number* $= (1890 + 9) = 1899.$

**Ex. 8.** ***Find the smallest number which when increased by 8 is exactly divisible by 30, 45, 65 and 78.***

**Sol.** The required number

= { (L.C.M. of 30, 45, 65, 78) – 8 }

Now, L.C.M. of 30, 45, 65 & 78 is

$= 5 \times 3 \times 13 \times 2 \times 3 = 1170.$

| | |
|---|---|
| 5 | 30 – 45 – 65 – 78 |
| 3 | 6 – 9 – 13 – 78 |
| 13 | 2 – 3 – 13 – 26 |
| 2 | 2 – 3 – 1 – 2 |
| | 1 – 3 – 1 – 1 |

Hence, the required number

= (1170 – 8) = 1162.

**Ex.9.** ***Find the smallest number which when decreased by 5, is exactly divisible by 20, 28, 35 and 105.***

**Sol.** Clearly, the required number when decreased by 5 is the L.C.M. of 20, 28, 35 and 105.

| | |
|---|---|
| 5 | 20 – 28 – 35 – 105 |
| 7 | 4 – 28 – 7 – 21 |
| 4 | 4 – 4 – 1 – 3 |
| | 1 – 1 – 1 – 3 |

$\therefore$ Required number

= (L.C.M. of 20, 28, 35, 105) + 5.

Now, L.C.M. of 20, 28, 35, 105 is $= 5 \times 7 \times 4 \times 3 = 420.$

*Hence, the required number = (420 + 5) = 425.*

**Ex.10.** ***Find the least number which when divided by 35, 45 and 55 leaves the remainders 18, 28 and 38 respectively.***

**Sol.** Note here that the difference between any divisor and the corresponding remainder is the same, which is 17.

| | |
|---|---|
| 5 | 35 – 45 – 55 |
| | 7 – 9 – 11 |

$\therefore$ *Required number* = (L.C.M. of 35, 45, 55) – 17

$= (5 \times 7 \times 9 \times 11) - 17 = (3465 - 17) = 3448.$

**Ex.11.** *Find the greatest number of 4 digits, which when divided by 12, 18, 21 and 28 leaves in each case a remainder 5.*

**Sol.** L.C.M. of 12, 18, 21 and 28 is

$= 3 \times 7 \times 2 \times 2 \times 3 = 252$

| | |
|---|---|
| 3 | 12 – 18 – 21 – 28 |
| 7 | 4 – 6 – 7 – 28 |
| 2 | 4 – 6 – 1 – 4 |
| 2 | 2 – 3 – 1 – 2 |
| | 1 – 3 – 1 – 1 |

Greatest number of 4 digits = 9999

On dividing 9999 by 252, the remainder = 171.

∴ Greatest number of 4 digits exactly divisible by 12, 18, 21 and 28 is (9999 – 171) = 9828.

∴ Required number = (9828 + 5) = 9833.

**Ex. 12.** *The H.C.F. of two numbers is 42 and their L.C.M. is 1260. If one of the numbers is 210, find the other.*

**Sol.** *The other number* $= \dfrac{42 \times 1260}{210} = 252.$

**Ex. 13.** *Find the least number which when divided by 5, 6, 7 and 8 leaves a remainder 3, but when divided by 9, leaves no remainder.*

| | |
|---|---|
| 2 | 5 – 6 – 7 – 8 |
| | 5 – 3 – 7 – 4 |

**Sol.** L.C.M. of 5, 6, 7 & 8

$= 2 \times 5 \times 3 \times 7 \times 4 = 840.$

The number is therefore of the form (840 $k$ + 3).

Now, we have to find the least value of $k$ for which (840 $k$ + 3) is divisible by 9.

$$9 \overline{)\,840k + 3\,}\; 93k$$
$$\underline{837k}$$
$$3k + 3$$

Putting $3k + 3 = 9$ we get $k = 2$.

*Hence, the required number*

*$= (840 \times 2 + 3) = 1683.$*

**Ex.14.** *Find the greatest number of 5 digits which when divided by 15, 18, 21 and 24 leaves 11, 14, 17 and 20 as remainders respectively.*

**Sol.** L.C.M. of 15, 18, 21 & 24

$= 3 \times 2 \times 5 \times 3 \times 7 \times 4 = 2520.$

| | |
|---|---|
| 3 | 15 – 18 – 21 – 24 |
| 2 | 5 – 6 – 7 – 8 |
| | 5 – 3 – 7 – 4 |

Now, greatest number of 5 digits = 99999.

But, on dividing 99999 by 2520 we get 1719 as remainder.

So, the greatest number of 5 digits exactly divisible by 2520 is

= (99999 – 1719) = 98280.

Now, the difference between any divisor and the corresponding remainder is the same, which is 4.

∴ *The required number = (98280 – 4) = 98276.*

**Ex.15.** *Find the least multiple of 23, which when divided by 18, 21 and 24 leaves the remainders 7, 10 and 13 respectively.*

| | |
|---|---|
| 3 | 18 – 21 – 24 |
| 2 | 6 – 7 – 8 |
| | 3 – 7 – 4 |

**Sol.** L.C.M. of 18, 21 and 24

$= 3 \times 2 \times 3 \times 7 \times 4 = 504.$

$$23\overline{)504k - 11}(21k$$
$$\underline{483k}$$
$$21k - 11$$

Moreover, the difference between each divisor and the corresponding remainder is the same, which is 11.

∴ Required number is of the form (504 $k$ – 11), which is divisible by 23 for the least value of $k$.

Now, on dividing (504 $k$ – 11) by 23, we get (21 $k$ – 11) as the remainder. We find the least positive number $k$ for which (21 $k$ – 11) is divisible by 23. By inspection, $k = 6$.

*Hence, the required number* $= (504 \times 6 - 11) = 3013$.

**Ex. 16.** *Find two numbers of 3 digits each, whose H.C.F. is 80 and L.C.M. is 5760.*

**Sol.** Let the numbers be $80a$ and $80b$, where $a$ and $b$ are co-prime.

Then $80a \times 80b = 80 \times 5760$ or $ab = 72$

Now, the possible pairs of $a$ and $b$, which are co-prime and whose product is 72 are (1, 72) and (8, 9).

∴ Possible pairs of numbers with given H.C.F. and L.C.M. are :—

$(80 \times 1, 80 \times 72)$ or $(80, 5760)$

and $(80 \times 8, 80 \times 9)$ or $(640, 720)$

∴ *Required numbers of 3 digits each, are 640 and 720.*

**Ex. 17.** *Find the least number of square tiles required to pave the floor of a room 15 metres 17 cm long and 9 metres 2 cm broad.*

**Sol.** Since the number of tiles is to be least, their size must be largest, which is clearly in cm, the H.C.F. of 1517 and 902.

But, H.C.F. of 1517 and 902 is 41.

∴ Each side of the largest square tile = 41 cm.

*Hence, the total number of such tiles* $= \dfrac{1517 \times 902}{41 \times 41} = 814.$

**Ex.18.** *Five bells begin to toll together and toll respectively at intervals of 6, 7, 8, 9 and 12 seconds. How many times, they will toll together in one hour excluding the one at the start ?*

| | |
|---|---|
| 3 | 6 – 7 – 8 – 9 – 12 |
| 2 | 2 – 7 – 8 – 3 – 4 |
| 2 | 1 – 7 – 4 – 3 – 2 |
| | 1 – 7 – 2 – 3 – 1 |

**Sol.** L.C.M. of 6, 7, 8, 9 & 12

$= 3 \times 2 \times 2 \times 7 \times 2 \times 3 = 504$

Bells will toll together after 504 seconds.

*In 1 hour, they will toll together.*

$= \dfrac{60 \times 60}{504}$ times $= \dfrac{50}{7}$ times *i.e.*, 7 times.

**Ex.19.** *The sum and difference of the L.C.M. and the H.C.F. of two numbers are 592 and 518 respectively. If the sum of two numbers be 296, find the numbers.*

**Sol.** Let the L.C.M. and H.C.F. be h and k respectively.

$\therefore\ h + k = 592$ and $h - k = 518$

Consequently, $h = \dfrac{592 + 518}{2} = 555$ & $k = \dfrac{592 - 518}{2} = 37.$

*i.e.*, L.C.M. = 555 and H.C.F. = 37.

Now, let the numbers be $37a$ and $37b$, where $a$ and $b$ are co-primes.

$\therefore\ 37a + 37b = 296$ or $a + b = 8$.

Possible pairs of co-primes, whose sum is 8 are (1, 7) & (3, 5).

$\therefore$ *Possible pairs of numbers are :*

$$\left.\begin{array}{ll} & (37\times 1, 37\times 7) \text{ or } (37, 259) \\ \text{and} & (37\times 3, 37\times 5) \text{ or } (111, 185) \end{array}\right\}$$

Now, H.C.F. × L.C.M. = $555 \times 37 = 20535$.

Also, $111 \times 185 = 20535$, while $37 \times 259 \neq 20535$.

*Hence, the required numbers are 111 and 185.*

**Ex. 20.** *What least number must be subtracted from 1837, so that the remainder when divided by 8, 12 and 15 will leave in each case the same remainder 5.*

**Sol.** Let the required number be r. Then (1837 – r – 5) or (1832 – r) must be exactly divisible by the L.C.M. of 8, 12 and 15.

Now, L.C.M. of 8, 12, 15 is

$= 2 \times 2 \times 3 \times 2 \times 5 = 120.$

| | |
|---|---|
| 2 | 8 – 12 – 15 |
| 2 | 4 – 6 – 15 |
| 3 | 2 – 3 – 15 |
| | 2 – 1 – 5 |

Now, (1832 – r) when divided by 120 leaves the remainder (32 – r) which must be 0.

Thus, (32 – r) = 0 or r = 32.

*Hence, the required number = 32.*

**Ex. 21.** *The circumferences of the fore and hind wheels of a carriage are $6\frac{3}{14}$ metres and $8\frac{1}{18}$ metres respectively. At any given moment, a chalk mark is put on the point of contact of each wheel with the ground. Find the distance travelled by the carriage so that both the chalk marks are again on the ground at the same time.*

**Sol.** *The required distance in metres*

$$= \text{L.C.M. of } \frac{87}{14} \text{ and } \frac{145}{18}$$

$$= \frac{\text{L.C.M. of 87 \& 145}}{\text{H.C.F. of 14 \& 18}} = \left(\frac{435}{2}\right) \text{m.} = 217.5 \text{ m.}$$

## EXERCISE 2A (*Subjective*)

1. *Find the H.C.F. of :*

(*i*) 598, 874 (*ii*) 32712, 4002 (*iii*) 2923, 3239

(*iv*) 42, 63, 40 (*v*) 78, 117, 156

(*vi*) 23562, 27846, 34034

**2.** *Find the L.C.M. of :*

(*i*) 15, 35, 42, 50 (*ii*) 8, 12, 27, 40 (*iii*) 12, 64, 112, 204

(*iv*) 22, 54, 108, 135, 198 (*v*) 36, 63, 77, 84

(*vi*) 7, 17, 51, 63

**3.** (*i*) Express $\frac{1271}{1395}$ in the simplest form.

**Hint.** *H.C.F. of 1271 and 1395 is 31. Divide numerator and denominator by 31.*

(*ii*) Reduce $\frac{481}{629}$ to lowest terms.

**4.** *Find the H.C.F. and L.C.M. of :*

(*i*) $\frac{2}{3}, \frac{4}{9}, \frac{6}{13}$ (*ii*) $\frac{1}{2}, \frac{1}{3}, \frac{3}{5}, \frac{7}{9}$ (*iii*) $\frac{2}{5}, \frac{8}{15}, \frac{16}{25}, 4$

(*iv*) $5, \frac{1}{3}, \frac{4}{7}, \frac{9}{14}$ (*v*) $5, 9\frac{1}{3}, \frac{5}{6}, \frac{2}{9}$ (*vi*) $\frac{3}{2}, \frac{27}{8}, \frac{15}{16}, 12$

**5.** The product of two numbers is 20535. If their H.C.F. is 37, find their L.C.M.

**6.** The L.C.M. of two numbers is 3855 and their H.C.F. is 257. If one number is 1285, find the other.

**7.** *Find the L.C.M. of :*

(*i*) 111, 592 (*ii*) 6851, 9061

(*iii*) 364, 2520 & 5265 (*iv*) 572, 462, 187

**8.** *Show that the following numbers are co-prime :*

(*i*) 1147 & 983 (*ii*) 1403 & 1139 (*iii*) 3567 & 5917.

**Hint.** *H.C.F. of co-primes is 1*

**9.** A merchant has three kinds of wine; of the first kind 403 gallons, of the second 527 gallons and of the third 589 gallons. What is the least number of full casks of equal size in which this can be stored without mixing ?

**Hint.** *Find H.C.F. of 403, 527 and 589.*

**10.** Find the least number of square tiles required for a terrace 15.17 m long and 9.02 m broad.

**Hint.** *Tiles are least, when size of each is largest. So, H.C.F. of 1517 cm and 902 cm gives each side of a tile, which is 41 cm.*

$\therefore$ *Number of tiles* $= \left(\frac{1517 \times 902}{41 \times 41}\right)$.

**11.** Three pieces of timber 24 meters, 28.8 meters nd 33.6 meters long have to be divided into planks of the same length. What is the greatest possible length of each plank ?

**Hint.** *Find H.C.F. of 2400 cm, 2880 cm and 3360 cm.*

**12.** Four bells toll at intervals of 6, 8, 12 and 18 minutes respectively. If they start tolling together at 12 a.m; find after what interval will they toll together and how many times will they toll together in 6 hours.

**Hints.** *L.C.M. of 6, 8, 12, 18 min. = 72 min. = 1 hr. 12 min.*

*So, they will toll together after 1 hr. 12 min.*

*In 6 hours, they will toll together*

$$= \left( \frac{360}{72} \textit{ times} + 1 \textit{ time at the start} \right).$$

**13.** Three persons A, B, C run along a circular path 12 km long. They start their race from the same point and at the same time with a speed of 3 km/hr, 7 km/hr and 13 km/hr respectively. After what time will they meet again ?

**Hint.** *Time taken by A, B, C to cover 12 km is 4 hours,* $\frac{12}{7}$ *hours and* $\frac{12}{13}$ *hours respectively.*

*L.C.M. of* $4, \frac{12}{7}$ *and* $\frac{12}{13} = 12$.

*So, they will meet again after 12 hours.*

**14.** The sum of two numbers is 1080 and their H.C.F. is 36. Find all possible pairs of such numbers.

**15.** The product of two numbers is 16464 and their H.C.F. is 14. Find all possible pairs of such numbers.

**16.** Find the greatest numer of four digits and least number of five digits that have 144 as their H.C.F.

**Hint.** *Greatest number of 4 digits exactly divisible by 144 is 9936 and the least number of 5 digits exactly divisible by 144 is 10080.*

**17.** Find the least number of five digits which when divided by 52, 56, 78 and 91 leaves no remainder.

**Hint.** *L.C.M. of 52, 56, 78, 91 is 2184.*

*Now, 10000 when divided by 2184 leaves 1264 as remainder. So, required number = [ (10000) + (2184 – 1264) ].*

**18.** Find the greatest number which divides 3453 and 9370, leaving 2 and 3 as remainders respectively.

**Hint.** *Find H.C.F. of (3453 – 2) and (9370 – 3).*

**19.** Find the greatest number which is such that when 76, 151 and 226 are divided by it, the remainders are all alike. Find also the remainder.

**Hint.** *Required number is the H.C.F. of :*

$$(151 - 76), (226 - 151) \text{ and } (226 - 76).$$

*For remainder, divide 76 by this number.*

**20.** Find the smallest number which when increased by 5 is divisible by 18, 21, 35 and 48.

**Hint.** *Required number* = [ *L.C.M. of* (18, 21, 35, 48) − 5 ].

**21.** Find the smallest number which when diminished by 8 is divisible by 21, 28, 36, 45.

**Hint.** *Required number* = [ (*L.C.M. of* 21, 28, 36, 45) + 8 ].

**22.** Find the least number which when divided by 16, 18 and 21 leaves the remainders 3, 5 and 8 respectively.

**Hint.** $(16 - 3) = 13, (18 - 5) = 13 \;\&\; (21 - 8) = 13.$

∴ *Required number* = [ (*L.C.M. of* 16, 18, 21) − 13 ].

**23.** Find the least number of five digits which when divided by 52, 56, 78 and 91 leaves 28, 32, 54 and 67 as remainders respectively.

**Hint.** $(52 - 28) = (56 - 32) = (78 - 54) = (91 - 67) = 24.$

*L.C.M. of 52, 56, 78, 91 is 2184.*

*Least number of 5 digits divisible by 2184 is 10920.*

∴ *Required number* = (10920 − 24).

**FILL IN THE BLANKS WITH CORRECT ANSWERS :—**

**24.** (*i*) The largest number that will divide 192, 236 and 280 leaving remainders 3, 5 and 7 respectively is (.........).

(*ii*) The largest number which divides 158, 282 and 344 so as to leave the same remainder in each case is (..........).

(*iii*) The greatest possible length which can be used to measure 2 metres 5 cms, 3 metres 69 cms and 8 metres 61 cms exactly is (.......) cms.

(*iv*) The greatest number of four digits which when divided by 6, 8, 14 and 21 leaves in each case a remainder 3 is (........).

(*v*) The greatest number of four digits which when divided by 7, 11, 14 22 leaves respectively 4, 8, 11, 19 as remainders, is (.....).

**25.** (*i*) The smallest number which when decreased by 5 is exactly divisible by 5, 6, 8 and 9 is (..........).

(*ii*) The least number which when divided by 5, 8, 12, and 15 leaves a remainder 2, 5, 9 and 12 respectively, is (........).

(*iii*) The least number which when divided by 2, 3, 4 and 5 leaves a remainder 1 but is exactly divisible by 11 is (.........).

(*iv*) The least even number, which when divided by 18, 24 and 32 leaves remainders 6, 12 and 20 respectively is (........).

(*v*) The least number that should be added to 363 to form a number which when divided by 6, 8 and 15 leaves in each case a remainder 5 is (.......).

**26.** (*i*) Two numbers are Co-prime, if their (......) is 1.

(*ii*) L.C.M. of two numbers = $\dfrac{\text{(Product of numbers)}}{(\ldots\ldots\ldots\ldots)}$.

(*iii*) H.C.F. of given fractions = $\dfrac{(\ldots\ldots\ldots\ldots)\text{ of numerators}}{(\ldots\ldots\ldots\ldots)\text{ of denominators}}$.

(*iv*) L.C.M. of given fractions = $\dfrac{(\ldots\ldots\ldots\ldots)\text{ of numerators}}{(\ldots\ldots\ldots\ldots)\text{ of denominators}}$.

(v) If the sum of two numbers is 99 and the sum and difference of their L.C.M. and H.C.F. are 231 and 209 respectively. Then the numbers are (..........) and (...........).

## ANSWERS (*EXERCISE 2A*)

**1.** (*i*) 46 (*ii*) 174 (*iii*) 79 (*iv*) 7
(*v*) 39 (*vi*) 238

**2.** (*i*) 1050 (*ii*) 1080 (*iii*) 22848 (*iv*) 5940
(*v*) 2772 (*vi*) 1071 **3.** (*i*) $\frac{41}{45}$ (*ii*) $\frac{13}{17}$

**4.** (*i*) H.C.F. = $\frac{2}{117}$, L.C.M. = 12 (*ii*) H.C.F. = $\frac{1}{90}$, L.C.M. = 21
(*iii*) H.C.F. = $\frac{2}{75}$, L.C.M. = $\frac{16}{5}$ (*iv*) H.C.F. = $\frac{1}{42}$, L.C.M. = 180
(*v*) H.C.F. = $\frac{1}{90}$, L.C.M. = 140 (*vi*) H.C.F. = $\frac{3}{16}$, L.C.M. = 270

**5.** 555 **6.** 771 **7.** (*i*) 1776 (*ii*) 280891 (*iii*) 294840
(*iv*) 204204 **9.** 31 gallons **10.** 814 **11.** 4.8 m **12.** 72 min, 6 times
**13.** 12 hours **14.** (36, 1044), (252, 828), (396, 684) & (468, 612)
**15.** (14, 1176), (42, 392), (56, 294), (98, 168)
**16.** 9936, 10080 **17.** 10920 **18.** 493 **19.** 75, 1
**20.** 5035 **21.** 1268 **22.** 995 **23.** 10896
**24.** (*i*) 21 (*ii*) 31 (*iii*) 41 (*iv*) 9918
(*v*) 9853 **25.** (*i*) 365 (*ii*) 117 (*iii*) 121
(*iv*) 276 (*v*) 2
**26.** (*i*) H.C.F. (*ii*) H.C.F. of numbers
(*iii*) H.C.F. in $N^r$ & L.C.M. in $D^r$.
(*iv*) L.C.M. in $N^r$ & H.C.F. in $D^r$. (*v*) 55, 44

## EXERCISE 2B (*OBJECTIVE TYPE QUESTIONS*)

**1.** H.C.F. of 2923, 3239 is :

(*a*) 37 (*b*) 73
(*c*) 79 (*d*) 47

**2.** H.C.F. of $2^3$, $3^2$, 4 and 15 is :

(*a*) $2^3$ (*b*) 1
(*c*) $3^2$ (*d*) 360

**3.** L.C.M. of 87 and 145 is :

(*a*) 870 (*b*) 1305
(*c*) 435 (*d*) 1740

**4.** L.C.M. of 33, 4, 42 and 3 is :

(*a*) 432 (*b*) 12
(*c*) 48 (*d*) none of these

**5.** $\frac{561}{748}$ when reduced to lowest terms is :

(*a*) $\frac{13}{14}$ (*b*) $\frac{3}{4}$
(*c*) $\frac{11}{14}$ (*d*) $\frac{23}{24}$

**6.** The H.C.F. of $\frac{1}{2}, \frac{3}{4}, \frac{5}{6}, \frac{7}{8}, \frac{9}{10}$ is :

(*a*) $\frac{1}{2}$ (*b*) $\frac{1}{10}$
(*c*) $\frac{9}{120}$ (*d*) $\frac{1}{120}$

**7.** The G.C.D. of $\frac{3}{16}, \frac{5}{12}, \frac{7}{8}$ is :

(*a*) $\frac{1}{4}$ (*b*) $\frac{1}{48}$
(*c*) $\frac{105}{4}$ (*d*) $\frac{105}{48}$

**8.** The L.C.M. of $\frac{1}{3}, \frac{5}{6}, \frac{5}{9}, \frac{10}{27}$ is :

(*a*) $\frac{5}{54}$ (*b*) $\frac{5}{27}$
(*c*) $\frac{10}{3}$ (*d*) $\frac{5}{3}$

**9.** Which of the following fractions is the greatest ?

$$\frac{7}{8}, \frac{6}{7}, \frac{4}{5}, \frac{5}{6}$$

(a) $\frac{6}{7}$ (b) $\frac{5}{6}$

(c) $\frac{4}{5}$ (d) $\frac{7}{8}$

**(Railways, 1991)**

**10.** Which of the following is in ascending order :

(a) $\frac{5}{7}, \frac{7}{8}, \frac{9}{11}$ (b) $\frac{5}{7}, \frac{9}{11}, \frac{7}{8}$

(c) $\frac{7}{8}, \frac{5}{7}, \frac{9}{11}$ (d) $\frac{9}{11}, \frac{7}{8}, \frac{5}{7}$

**(Bank P.O. 1988)**

**11.** Of the fractions $\frac{2}{3}, \frac{5}{7}, \frac{9}{13}, \frac{9}{14}, \frac{7}{4}$, the least is :

(a) $\frac{9}{14}$ (b) $\frac{2}{3}$

(c) $\frac{7}{4}$ (d) $\frac{5}{7}$

**12.** Of the fractions $\frac{11}{14}, \frac{14}{17}, \frac{17}{20}, \frac{23}{26}, \frac{29}{32}$, the largest is :

(a) $\frac{14}{17}$ (b) $\frac{23}{26}$

(c) $\frac{17}{20}$ (d) $\frac{29}{32}$

**13.** The product of two numbers is 2025 and their H.C.F. is 15. Their L.C.M. is :

(a) 30375 (b) 2040

(c) 135 (d) 2010

**14.** The L.C.M. of two numbers is 2310 and their H.C.F. is 30. If one number is 210, the other is :

(a) 16170 (b) 2100

(c) 1470 (d) 330

**(Clerks' Grade Exam. 1991)**

**15.** The product of two digit numbers is 2160 and their G.C.M. is 12. The numbers are :

(a) (12, 180) (b) (72, 30)

(c) (36, 60) (d) (96, 25)

**16.** The sum of two numbers is 216 and their H.C.F. is 27. The numbers are :

(a) (54, 162) (b) (108, 108)

(c) (27, 189) (d) none of these

**17.** The largest number which exactly divides 522, 1276 and 1624 is :

(a) 29 (b) 58

(c) 4 (d) none of these

**18.** The largest number which divides 77, 147, 252 to leave the same remainder in each case, is :

(*a*) 9 (*b*) 15
(*c*) 25 (*d*) 35

**19.** The greatest number which can divide 432, 534 and 398 leaving the same remainder 7 in each case, is :

(*a*) 17 (*b*) 208
(*c*) 34 (*d*) none of these

**20.** The least square number which is divisible by 6, 8 and 15 is :

(*a*) 2500 (*b*) 3600
(*c*) 4900 (*d*) 4500

**21.** The least number which when divided by 16, 18 and 21 leaves the remainders 3, 5 and 8 respectively, is :

(*a*) 893 (*b*) 995
(*c*) 1024 (*d*) 982

**22.** The least number which when divided by 15, 27, 35 and 42 leaves in each case a remainder 7, is :

(*a*) 1883 (*b*) 1897
(*c*) 1987 (*d*) 2007

**23.** The smallest number which when diminished by 3 is divisible by 21, 28, 36 and 45, is :

(*a*) 420 (*b*) 1257
(*c*) 1260 (*d*) 1263

**24.** The greatest number of 4 digits, which is divisible by each one of the numbers 12, 18, 21 and 28, is :

(*a*) 9848 (*b*) 9864
(*c*) 9828 (*d*) 9636

**25.** The greatest number by which if 373 and 813 are divided the remainders will be 8 and 10 respectively, is :

(*a*) 63 (*b*) 69
(*c*) 71 (*d*) 73

**26.** The smallest number which when divided by 20, 25, 35 and 40 leaves the remainders 14, 19, 24 and 34 respectively, is :

(*a*) 1394 (*b*) 1404
(*c*) 1664 (*d*) 1406

**27.** The least multiple of 7 which leaves a remainder 4 when divided by 6, 9, 15 and 18, is :

(*a*) 74 (*b*) 94
(*c*) 184 (*d*) 364

**28.** The ratio of two numbers is 13 : 15 and their L.C.M. is 39780. The numbers are :

(*a*) 2652, 3060 (*b*) 884, 1020

(c) 884, 1040 (d) 670, 1340

**29.** The ratio of three numbers is 35 : 55 : 77 and their H.C.F. is 24. The numbers are :

(a) 420, 660, 924 (b) 280, 440, 616
(c) 840, 1320, 1848 (d) 105, 165, 231

**30.** Three pieces of timber 42 metres, 49 metres and 63 metres long have to be divided into planks of the same length. The greatest possible length of each plank is :

(a) 63 metres (b) 42 metres
(c) 7 metres (d) 14 metres

**31.** Three different containers contain different qualities of mixtures of milk and water whose measurements are 1653 kg, 2261 kg and 2527 kg respectively. What biggest measure of milk must be there to measure all different quantities exactly ?

(a) 19 kg (b) 26 kg
(c) 29 kg (d) 31 kg

**32.** The greatest possible length which can be used to measure exactly the lengths :

7 m ; 3 m 85 cms; 12 m 95 cms. is :

(a) 15 cms (b) 25 cms
(c) 35 cms (d) 42 cms

**33.** Which of the following numbers are co-primes :

(a) (14, 35) (b) (18, 25)
(c) (31, 93) (d) (23, 62)

**34.** The least number of square tiles required to pave the ceiling of a room 15 m 17 cm long and 8 m 2 cm broad, is :

(a) 902 (b) 656
(c) 738 (d) 814

**35.** Five bells begin to toll together and toll respectively at intervals of 6, 7, 8, 9 and 12 seconds. After how many seconds will they toll together again ?

(a) 72 sec. (b) 612 sec.
(c) 504 sec. (d) 318 sec.

**36.** The H.C.F. and L.C.M. of two numbers are 44 and 264 respectively. If the first number is divided by 2, the quotient is 44. The other number is :

(a) 132 (b) 33
(c) 66 (d) 264 **(Astt. Grade, 1987)**

**37.** Out of $2^{2^2}$, $2^{22}$, 222, $(22)^2$ the largest one is :

(a) $2^{2^2}$ (b) $2^{22}$
(c) 222 (d) $(22)^2$ **(Astt. Grade, 1987)**

**38.** The number of prime factors in $2^{222} \times 3^{333} \times 5^{555}$ *is* :

(*a*) 1110 (*b*) 1107

(*c*) 3 (*d*) 1272 **(Astt. Grade, 1987)**

## HINTS & SOLUTIONS
## (*EXERCISE 2B*)

**1.**

```
2923 ) 3239 ( 1
       2923
       ----
        316 ) 2923 ( 9
              2844
              ----
                79 ) 316 ( 4
                     316
                     ---
                      ×
```

$\therefore$ H.C.F. of 2923, 3239 is 79.

**2.** Clearly, 1 is the H.C.F. of the given numbers.

**3.**

```
87 ) 145 ( 1
      87
      --
      58 ) 87 ( 1
           58
           --
           29 ) 58 ( 2
                58
                --
                 ×
```

$\therefore$ H.C.F. of given numbers is 29.

So, $\text{L.C.M.} = \frac{87 \times 145}{29} = 435.$

**4.** The given numbers are $3^3$, $2^2$, $2^4$ and 3.

$\therefore$ L.C.M. $= 3^3 \times 2^4 = 27 \times 16 = 432.$

**5.** H.C.F. of 561 and 748 is 187.

Dividing Nr. and Dr. by 187, the fraction is $\frac{3}{4}$.

**6.** H.C.F. of 1, 3, 5, 7, 9 is 1.

L.C.M. of 2, 4, 6, 8, 10 is 120.

$\therefore$ H.C.F. of given fractions $= \frac{1}{120}$.

**7.** $\text{G.C.D.} = \frac{\text{H.C.F. of } 3, 5, 7}{\text{L.C.M. of } 16, 12, 8} = \frac{1}{48}$.

**8.** $\text{L.C.M.} = \frac{\text{L.C.M. of } 1, 5, 5, 10}{\text{H.C.F. of } 3, 6, 9, 27} = \frac{10}{3}$.

**9.** $\frac{7}{8} = .875, \frac{6}{7} = .857, \frac{4}{5} = 0.8, \frac{5}{6} = 0.833.$

$\therefore \ \frac{7}{8}$ is the greatest of all.

**10.** $\frac{5}{7} = .714; \frac{7}{8} = .875, \frac{9}{11} = .818.$

Clearly, $\frac{5}{7} < \frac{9}{11} < \frac{7}{8}.$

**11.** $\frac{2}{3} = .666, \frac{5}{7} = .714, \frac{9}{13} = .692, \frac{9}{14} = .64, \frac{7}{4} = 1.75.$

Clearly, the least fraction is $\frac{9}{14}.$

**12.** The difference in the numerator and denominator of each of the given fractions is the same, namely 3. So, the one with largest numerator is the largest *i.e.* $\frac{29}{32}.$

**13.** $\text{L.C.M.} = \frac{\text{Product of numbers}}{\text{H.C.F.}} = \frac{2025}{15} = 135.$

**14.** The other number $= \frac{\text{L.C.M.} \times \text{H.C.F.}}{\text{given number}} = \frac{2310 \times 30}{210} = 330.$

**15.** Let the numbers be $12a$ and $12b$.

Then, $12a \times 12b = 2160$ or $ab = 15$.

$\therefore$ Values of co-primes $a$ and $b$ are (1, 15), (3, 5).

So, the two digit numbers are $(3 \times 12, 5 \times 12)$ *i.e.* (36, 60).

**16.** Let the numbers be $27a$ and $27b$. Then,

$$27\,a + 27\,b = 216 \text{ or } a + b = 8.$$

Values of coprimes $a$ and $b$ such that $a + b = 8$ are (1, 7), (3, 5).

$\therefore$ Numbers are $(27 \times 1, 27 \times 7)$ *i.e.* (27, 189).

**17.** The required number is the H.C.F. of 522, 1276 and 1624.

Now, $522 = 2 \times 3^2 \times 29$; $1276 = 2^2 \times 11 \times 29$ and

$1624 = 2^3 \times 7 \times 29$.

$\therefore$ H.C.F. $= 2 \times 29 = 58$.

**18.** Required number is the H.C.F. of (147 – 77), (252 – 147) and (252 – 77) *i.e.* H.C.F. of 70, 105 and 175.

Now, $70 = 2 \times 5 \times 7$, $105 = 3 \times 7 \times 5$, $175 = 5^2 \times 7$.

$\therefore$ Required number $= 5 \times 7 = 35$.

**19.** Required number is the H.C.F. of (432 – 7), (534 – 7) and (398 – 7) *i.e.* H.C.F. of 425, 527 & 391.

$\therefore$ Required number = 17.

**20.** The least number divisible by 6, 8, 15 is their L.C.M. which is 120.

Now, $120 = 2 \times 2 \times 2 \times 3 \times 5$.

To make it a perfect square, it must be multiplied by $2 \times 3 \times 5$.

Required number $= 120 \times 2 \times 3 \times 5 = 3600$

**21.** Here $(16 - 3) = 13$, $(18 - 5) = 13$, $(21 - 8) = 13$.

So, required number is [(L.C.M. of 16, 18, 21) – 13]

$= (1008 - 13) = 995.$

**22.** Required number $=$ (L.C.M. of 15, 27, 35, 42) + 7

$= 1890 + 7 = 1897.$

**23.** Required number $=$ (L.C.M. of 21, 28, 36, 45) + 3

$= 1260 + 3 = 1263.$

**24.** Required number is divisible by 12, 18, 21, 28 *i.e.* 252. Now, greatest number of 4 digits = 9999. On dividing 9999 by 252, the remainder is 171.

$\therefore$ Required number $= (9999 - 171) = 9828.$

**25.** Required number = H.C.F. of $(373 - 8)$ and $(813 - 10)$

= H.C.F. of 365 and 803 = 73.

**26.** Required number $=$ (L.C.M. of 20, 25, 35, 40) – 6

$= 1400 - 6 = 1394.$

**27.** L.C.M. of 6, 9, 15 and 18 is 90.

Let $x$ be the least multiple of 7, which when divided by 90 leaves the remainder 4. Then, $x$ is of the form $90k + 4$.

Now, minimum value of $k$ for which $90k + 4$ is divisible by 7 is 4.

$\therefore \; x = 4 \times 90 + 4 = 364.$

**28.** Let the numbers be $13x$ and $15x$.

Clearly, their L.C.M. is $(13 \times 15)\, x$.

Now, $(13 \times 15)\, x = 39780 \Rightarrow x = \dfrac{39780}{13 \times 15} = 204.$

$\therefore$ The numbers are $(13 \times 204, 15 \times 204)$ *i.e.* (2652, 3060).

**29.** Let the numbers be $35x$, $55x$ and $77x$.

Clearly, their H.C.F. is $x$. Now, $x = 24$.

$\therefore$ The numbers are $(35 \times 24, 55 \times 24, 77 \times 24)$

*i.e.* (840, 1320, 1848).

**30.** Required length = H.C.F. of 42m, 49m, 63m = 7m.

**31.** Biggest measure = H.C.F. of 1653 kg, 2261 kg, 2527 kg

= 19 kg.

**32.** Required length is the H.C.F. of 700, 385 and 1295 in cms. *i.e.* 35 cms.

**33.** H.C.F. of 18 and 25 is 1. So, 18 and 25 are co-primes.

**34.** Size of largest square tile

= H.C.F. of 1517 and 902 cms = 41 cms.

$\therefore$ Least number of tiles required $= \dfrac{1517 \times 902}{41 \times 41} = 814.$

**35.** L.C.M. of 6, 7, 8, 9 and 12 is 504.

$\therefore$ The bells will toll together again after 504 sec.

**36.** First number $= 2 \times 44 = 88.$

Second number $= \dfrac{44 \times 264}{88} = 132.$

**37.** $2^{2^2} = 2^4;\quad 222 < 2^8;\quad (22)^2 = (2 \times 11)^2\ 2^2 \times (11)^2 < 2^2 \times 2^8 = 2^{10}.$

$\therefore$ $2^{22}$ is the largest.

**38.** The only prime factors in the given product are 2, 3, 5.

## ANSWERS
## (*EXERCISE 2B*)

| | | | | | | |
|---|---|---|---|---|---|---|
| **1.** (*c*) | **2.** (*b*) | **3.** (*c*) | **4.** (*a*) | **5.** (*b*) | **6.** (*d*) | **7.** (*b*) |
| **8.** (*c*) | **9.** (*d*) | **10.** (*b*) | **11.** (*a*) | **12.** (*d*) | **13.** (*c*) | **14.** (*d*) |
| **15.** (*c*) | **16.** (*c*) | **17.** (*b*) | **18.** (*d*) | **19.** (*a*) | **20.** (*b*) | **21.** (*b*) |
| **22.** (*b*) | **23.** (*d*) | **24.** (*c*) | **25.** (*d*) | **26.** (*a*) | **27.** (*d*) | **28.** (*a*) |
| **29.** (*c*) | **30.** (*c*) | **31.** (*a*) | **32.** (*c*) | **33.** (*b*) | **34.** (*d*) | **35.** (*c*) |
| **36.** (*a*) | **37.** (*b*) | **38.** (*c*) | | | | |

# 3

# DECIMAL FRACTIONS

**Decimal Fractions.** *The fractions in which the denominators are powers of 10, are called decimal fractions such as* $\frac{3}{10}, \frac{7}{100}, \frac{9}{1000}$ *etc.*

We denote the tenth part of a unit by $\frac{1}{10}$; the hundredth part of a unit by $\frac{1}{100}$ and thousandth part of a unit by $\frac{1}{1000}$ and so on.

**Ex.** (*i*) $\frac{3}{10}$ *represents 3 tenths of a unit, denoted by* .3,

(read as *'point three'* or *'decimal three'*)

(*ii*) $\frac{7}{10}$ *represents 7 hundredths of a unit, denoted by* .07,

(read as *'point zero seven'*, or *'decimal zero seven'*).

(*iii*) 54.367 represents a number, read as *'fifty four point three six seven.*

It contains 5 tens, 4 units, 3 tenths, 6 hundredths and 7 thousandths.

$$54.367 = 50 + 4 + .3 + .06 + .007.$$

**To Convert a Decimal Fraction into a Vulgar Fraction.** *We put 1 under the decimal point in the denominator and annex as many zeros as is the number of digits after the decimal point. Now, we remove the decimal point and put the fraction in simplest form.*

For example :—

(*i*) $0.7 = \frac{7}{10}$  (*ii*) $1.56 = \frac{156}{100} = \frac{39}{25}$.

(*iii*) $134.9721 = \frac{1349721}{10000}$  (*iv*) $0.0531 = \frac{531}{10000}$.

**Remark :** Annexing zeros to the right of a decimal fraction does not change its value.

Thus, 0.7 = 0.70 = 0.700 = 0.7000 etc.

**Addition and Subtraction of Decimal Fractions.** (*Rule*). *The given numbers are so placed under each other that the decimal points lie in one column. Numbers so arranged can now be added and subtracted in a usual way as shown below.*

**Ex.1.** *(i) Add 6.209, .89, 12.7, 131.3719 and 0.6.*

*(ii) Subtract 23.3765 from 41.052.*

**Sol.** (i)

$$\begin{array}{r} 6.209 \\ .89 \\ 12.7 \\ 131.3719 \\ 0.6 \\ \hline 151.7709 \end{array}$$

(ii)

$$\begin{array}{r} 41.052 \\ -23.3765 \\ \hline 17.6755 \end{array}$$

## MULTIPLICATION OF DECIMAL FRACTIONS

### Multiplication by a Whole Number :

**Rule.** *The given decimal numeral considered as whole number* (removing the decimal point) *is multiplied by the given whole number. Now in this product, the decimal point is marked in such a way that this number has as many decimal places as are there in the multiplicand.*

**Ex.2.** *Multiply 43.5037 by 93.*

**Method :** First multiply 435037 by 93.

The product is 40458441.

The multiplicand contains 4 places of decimals. So mark off decimal point in the product taking 4 places from the right.

$$\begin{array}{r} 435037 \\ \times 93 \\ \hline 1305111 \\ 391533\times \\ \hline 40458441 \end{array}$$

Thus $43.5037 \times 93 = 4045.8441$.

### Multiplication by a Power of 10 :

**Rule.** *This is done by shifting the decimal point to the right by as many places as is the power of 10* (*i.e.* the number of zeros).

**Ex.3.** *Multiply :*

***(i) 563.7 by 10 (ii) 5.8691 by 100 (iii) 16.05 by 1000.***

**Sol.** (i) $563.7 \times 10 = 5637$ (*Shifting decimal 1 place to the right*)

(ii) $5.8691 \times 100 = 586.91$

(*Shifting decimal 2 places to the right*)

(iii) $16.05 \times 1000 = 16050.$

(*Shifting decimal 3 places to the right*)

### Multiplication of Two Decimal Numbers :

**Rule.** In this case, *we multiply the given numbers considering them without the decimal point. Now, in the product, the decimal point is marked off to obtain as many places of decimals as is the sum of the numbers of the decimal places in the multiplicand and the multiplier.*

**Ex. 4.** *Multiply 48.0387 by 8.039.*

**Method :** First multiply 480387 by 8039 and in the product, mark off (4 + 3) or 7 places of decimals from the right.

**Sol.**

```
     480387
    × 8039
   4323483
  1441161×
3843096×××
3861831093
```

$\therefore$ 480387 × 8039 = 3861831093.

So, 48.0387 × 8.039 = 386.1831093.

## DIVISION

**When The Divisor is a Counting Number :**

**Rule.** *We divide the given decimal numeral without considering the decimal point, by the given counting number. Now, in the quotient, we put the decimal point to give as many places of decimals as are there in the dividend.*

**Ex. 5.** ***Divide 30.27495 by 15.***

**Sol.** $\dfrac{3027495}{15} = 201833.$

$\therefore \quad \dfrac{30.27495}{15} = 2.01833.$

**Ex. 6.** ***Divide 0.05922 by 6.***

**Sol.** $\dfrac{5922}{6} = 987.$

$\therefore \quad \dfrac{0.05922}{6} = 0.00987.$

**When The Divisor as well as Dividend is a Decimal :**

**Rule.** *We multiply both the dividend and the divisor by a suitable multiple of 10 to make divisor a whole number. Now we proceed as above.*

**Ex. 7.** ***Divide : (i) 0.0182 by 0.014.***
***(ii) 63.5535 by 13.05.***

**Sol.** (*i*) $\dfrac{0.0182}{0.014} = \dfrac{0.0182 \times 1000}{0.014 \times 1000} = \dfrac{18.2}{14} = 1.3.$

(*ii*) $\dfrac{63.5535}{13.05} = \dfrac{63.5535 \times 100}{13.05 \times 100} = \dfrac{6355.35}{1305} = 4.87.$

**Division by Power of 10 :**

**Rule.** *We shift the decimal places to left by as many places as is the power of 10.*

**Ex. 8.** ***Divide : (i) 15.08 by 10 (ii) 16.97 by 1000.***

**Sol.** (*i*) $\dfrac{15.08}{10} = 1.508$ (*Shifting decimal 1 place to the left*) (*ii*) $\frac{16.97}{1000} = 0.01697$ (*Shifting decimal 3 places to the left*)

## H.C.F. AND L.C.M. OF DECIMALS

**Rule.** *In given numbers, make the same number of places of decimals by annexing zeros in some numbers, if necessary. Considering these numbers without decimal point, find H.C.F. or L.C.M. as the case may be. Now, in the result, mark off as many decimal places as are there in each of the numbers.*

**Ex. 9.** *Find the H.C.F. and L.C.M. of 1.8, 5.4 and 12.*

**Sol.** *The numbers may be written as 1.8, 5.4 and 12.0.*

H.C.F. of 18, 54 and 120 is 6.

L.C.M. of 18, 54, 120 is 1080.

∴ H.C.F. of given numbers = 0.6

L.C.M. of given numbers = 108.0 = 108.

**Recurring Decimals.** *If in a decimal, some figure or a set of figures is repeated continually, we call it a recurring decimal.*

The set of repeated figures is called period.

**Ex** (*i*) $\frac{2}{3} = 0.6666\ldots = 0.\dot{6}$.

(We write it by putting a dot on the repeated digit)

(*ii*) $\frac{22}{7} = 3.142857142857\ldots\ldots = 3.\overline{142857}$.

(We write it by putting a bar on the period)

**Pure & Mixed Recurring Decimals.** *A decimal in which all the figures after the decimal point are repeated, is known as a Pure recurring decimal. e.g.* $0.\dot{6}$, $3.\overline{142857}$, $1.\overline{16}$ etc.

**Ex. 9. *Convert the folowing numbers into recurring decimals :***

(*i*) $\frac{1}{15}$ (*ii*) $\frac{16}{99}$ (*iii*) $\frac{34}{275}$.

**Sol.** (*i*)

```
15) 100 ( .066....
     90
    ---
    100
     90
    ---
     10
```

∴ $\frac{1}{15} = 0.066\ldots = 0.0\dot{6}$

(*ii*)

```
99) 160 ( .1616...
     99
    ---
    610
    594
    ---
    160
     99
    ---
    610
    594
    ---
     16
```

∴ $\frac{16}{99} = 0.1616\ldots. = 0.\overline{16}$

(*iii*)

$$
\begin{array}{r}
275 \big)\ 340\ \big(\ .123636... \\
\underline{275\phantom{0}} \\
650 \\
\underline{550} \\
1000 \\
\underline{825} \\
1750 \\
\underline{1650} \\
100
\end{array}
$$

$\therefore \quad \frac{34}{275} = 0.123636.... = 0.12\overline{36}$

**Rule For Converting Pure Recurring Decimals into Vulgar Fractions.** *Write the repeated figures only once in the numerator and take as many nines in the denominator as is the number of repeating figures.*

**Ex. 10.** ***Express the following decimals as vulgar fractions.***

(*i*) **0.666.....** (*ii*) **0.161616....** (*iii*) $\mathbf{0.\overline{234}}$

(*iv*) **3.5555.....** (*v*) **2.035035....** (*v*) $\mathbf{6.\overline{14}}$

**Sol.** (*i*) $0.6666..... = 0.\dot{6} = \frac{6}{9} = \frac{2}{3}$.

(*ii*) $0.161616..... = 0.\overline{16} = \frac{16}{99}$.

(*iii*) $0.\overline{234} = \frac{234}{999}$.

(*iv*) $3.555.... = 3.\dot{5} = 3 + 0.\dot{5} = 3 + \frac{5}{9} = 3\frac{5}{9}$.

(*v*) $2.035035.... = 2.\overline{035} = 2 + 0.\overline{035} = 2 + \frac{35}{999} = 2\frac{35}{999}$.

(*vi*) $6.\overline{14} = 6 + 0.\overline{14} = 6 + \frac{14}{99} = 6\frac{14}{99}$.

**Rule For Converting A Mixed Recurring Decimal into A Vulgar Fraction.** *Form a fraction in which the numerator is the difference between the number formed by all the digits after the decimal point, taking the repeated digits only once and that formed by the digits, which are not repeated and the denominator is the number formed by as many nines as there are repeating digits followed by as many zeros as is the number of non-repeating digits.*

**Ex. 11.** ***Convert the following decimals as vulgar fractions :***

(*i*) ***0.1777 .....*** (*ii*) ***0.12545454....*** (*iii*) ***2.53666....***

**Sol.** (*i*) $0.1777.... = 0.1\dot{7} = \frac{17-1}{90} = \frac{16}{90} = \frac{8}{45}$.

(*ii*) $0.125454.... = 0.12\overline{54} = \frac{1254-12}{9900} = \frac{1242}{9900} = \frac{69}{550}$.

(*iii*) $2.536666..... = 2.53\dot{6} = 2 + \left(\frac{536-53}{900}\right) = 2\frac{161}{300}$.

**Ex.12.** ***Write the following fractions in descending order :***

$$\frac{7}{8}, \frac{6}{7}, \frac{4}{5}, \frac{5}{6}$$ **(Railways 1991)**

**Sol.** Expressing each of the given fractions in decimal form, we get :

$$\frac{7}{8} = 0.875; \frac{6}{7} = 0.857; \frac{4}{5} = 0.8; \frac{5}{6} = 0.833.$$

Clearly, $0.875 > 0.857 > 0.833 > 0.8$.

$\therefore \frac{7}{8} > \frac{6}{7} > \frac{5}{6} > \frac{4}{5}$.

**Ex.13.** ***Simplify :*** $\frac{(0.623)^3 + (0.377)^3}{(0.623)^2 - (0.623 \times 0.377) + (0.377)^2}$.

**(L.I.C. 1991)**

**Sol.** Given expression

$$= \frac{a^3 + b^3}{(a^2 - ab + b^2)} = (a+b), \text{ where } a = 0.623, b = 0.377$$

$$= (0.623 + 0.377) = 1.$$

## EXERCISE 3A

**1.** Convert the following fractions into decimals :

(*i*) $\frac{5}{8}$ (*ii*) $\frac{4}{9}$ (*iii*) $\frac{5}{12}$ (*iv*) $\frac{2}{13}$

(*v*) $\frac{13}{44}$ (*vi*) $5\frac{5}{24}$ (*vii*) $\frac{437}{999}$ (*viii*) $8\frac{9}{1100}$

**2.** Convert the following decimals into vulgar fractions :

(*i*) .00026 (*ii*) 5.6875 (*iii*) 2.064 (*iv*) 10.20105

**3.** Arrange in descending order of magnitude :

(*i*) 7.70, 0.77, 0.07, 0.007, 7.0007

(*ii*) $\frac{1}{2}, \frac{3}{7}, \frac{3}{5}, \frac{4}{9}$ **(Railways 1991)**

(*iii*) $\frac{4}{13}, \frac{7}{29}, \frac{13}{53}$, 0.2460

(*iv*) $\frac{2}{3}, \frac{5}{8}, \frac{7}{10}, \frac{13}{18}$ **(L.I.C. 1991)**

**Hint.** *Convert each fraction in decimal form.*

**4.** Find the largest fraction among the following :

(*i*) $\frac{1}{2}, \frac{3}{4}, \frac{5}{6}, \frac{7}{12}, \frac{2}{5}$ (*ii*) $\frac{7}{9}, \frac{6}{11}, \frac{16}{21}, \frac{12}{17}$

**Hint.** *Convert each fraction in decimal form.*

**5.** Find the H.C.F. and L.C.M. of :—

(*i*) 2.22, 33.3 and 37, (*ii*) 0.24, 1.8, 2.7 and 16,

(*iii*) 178.5 and 30.45, (*iv*) 2.5, .075 and 5.

**6.** Add :

(*i*) 3.5427, 0.79, 18.6, 110.035, 32

(*ii*) 5.7349, 0.09, 2.146, 0.5, 66, 109.003, 8.05712.

**7.** Subtract :

(*i*) 2.3756 from 4.2304 (*ii*) 9.8103 from 10.01

(*iii*) 1.1359 from 2 (*iv*) 1.0009 from 2.1

**8.** Multiply :

(*i*) 3.4972 by 24 (*ii*) 12.367 by 1.9

(*iii*) 4.00378 by 0.37 (*iv*) 3.579 by 0.3792

(*v*) .0036 by .0006 (*vi*) $1.3 \times .13 \times .013$

(*vii*) $.8 \times .016 \times .0035$ (*viii*) $8 \times .8 \times .08 \times 1.5$

**9.** Divide :

(*i*) 96.555 by 41 (*ii*) 169 by 1.3

(*iii*) 132 by 0.006 (*iv*) 0.00169 by 0.13

(*v*) .1 by 5000 (*vi*) 5.30123 by 3.15

**10.** Find the value of :

(*i*) $\frac{.2 \times .2 \times .2 + .02 \times .02 \times .02}{.6 \times .6 \times .6 + .06 \times .06 \times .06}$ (*ii*) $\frac{.0203 \times 2.92}{.0073 \times 14.5 \times .7}$

(*iii*) $\frac{0.35 \times 1.25 \times 0.075}{.328125}$ (*iv*) $\frac{3 \times 0.57 \times 0.133 \times 11.9}{1.9 \times 0.19 \times .017}$

**11.** Simplify :

(*i*) $16.579 \times 13.637 + 16.579 \times 6.363$

**Hint.** Use $a \times b + a \times c = a \times (b + c)$.

(*ii*) $\frac{0.25 \times 0.25 - 0.24 \times 0.24}{0.49}$ **(Railways 1991)**

**Hint.** *Use,* $a^2 - b^2 = (a - b)(a + b)$.

(*iii*) $\frac{0.5 \times 0.5 \times 0.5 + 0.6 \times 0.6 \times 0.6}{0.5 \times 0.5 - 0.3 + 0.6 \times 0.6}$ **(Hotel Management 1991)**

**Hint.** *Use,* $\frac{(a^3 + b^3)}{(a^2 - ab + b^2)} = (a + b)$, *where* $a = 0.5, b = 0.6$.

(*iv*) $\frac{0.86 \times 0.86 \times 0.86 - 0.14 \times 0.14 \times 0.14}{0.86 \times 0.86 + 0.86 \times 0.14 + 0.14 \times 0.14}$.

**Hint.** *Use,* $\dfrac{(a^3 - b^3)}{(a^2 + ab + b^2)} = (a - b)$, *where* $a = 0.86, b = 0.14$.

(*v*) $(0.98 \times 0.98 - 0.98 \times 1.52 + 0.76 \times 0.76)$

**Hint.** *Use,* $(a^2 - 2ab + b^2) = (a - b)^2$, *where* $a = 0.98, b = 0.76$.

**12.** Convert the following decimals into vulgar fractions :

(*i*) $0.\dot{7}$ (*ii*) $4.\dot{6}$ (*iii*) $0.\overline{35}$ (*iv*) $1.\overline{432}$

(*v*) $0.\overline{052}$ (*vi*) $2.\overline{13}$ (*vii*) $3.\overline{001}$ (*viii*) $3.\overline{142857}$

**13.** Convert the following decimals into vulgar fractions :

(*i*) $0.2\dot{5}$ (*ii*) $1.3\dot{7}$ (*iii*) $0.57\dot{9}$ (*iv*) $0.3\overline{15}$

(*v*) $0.57\overline{92}$ (*vi*) $3.05\overline{87}$ (*vii*) $1.005\dot{7}$

**Hint.** $1.005\dot{7} = \left(1 + \dfrac{57 - 5}{9000}\right)$.

**14.** Simplify :

(*i*) $0.\dot{6} + 0.\dot{7} + 0.\dot{2}$ **Hint.** *Exp.* $= \left(\dfrac{6}{9} + \dfrac{7}{9} + \dfrac{2}{9}\right)$.

(*ii*) $0.\overline{47} + 3.\overline{53} + 0.\overline{05}$

**Hint.** *Given Exp.* $= \left(2 + \dfrac{47}{99} + 3 + \dfrac{53}{99} + \dfrac{5}{99}\right)$.

(*iii*) $3.\overline{57} \times 0.4$ **Hint.** *Exp.* $= \left(3\dfrac{57}{99} \times \dfrac{4}{10}\right)$.

(*iv*) $29.\overline{55} \div 0.\dot{1}$ **Hint.** *Exp.* $= \left(29\dfrac{55}{99} \div \dfrac{1}{9}\right)$.

(*v*) $0.34\overline{67} + 0.1\overline{333}$

**Hint.** *Sum* $= \left(\dfrac{3467 - 34}{9900}\right) + \left(\dfrac{1333 - 13}{9900}\right)$.

## ANSWERS (EXERCISE 3A)

**1.** (*i*) 0.625 (*ii*) $0.\dot{4}$ (*iii*) $0.41\dot{6}$ (*iv*) $0.\overline{153846}$

(*v*) $0.29\overline{54}$ (*vi*) $5.208\dot{3}$ (*vii*) $\overline{.437}$ (*viii*) $8.00\overline{81}$

**2.** (*i*) $\dfrac{13}{50000}$ (*ii*) $5\dfrac{11}{16}$ (*iii*) $2\dfrac{8}{125}$ (*iv*) $10\dfrac{4021}{200000}$

**3.** (*i*) $7.70 > 7.0007 > 0.77 > 0.07 > 0.007$

(*ii*) $\dfrac{3}{5} > \dfrac{1}{2} > \dfrac{4}{9} > \dfrac{3}{7}$ (*iii*) $\dfrac{4}{13} > 0.2460 > \dfrac{13}{53} > \dfrac{7}{29}$

(*iv*) $\dfrac{13}{18} > \dfrac{7}{10} > \dfrac{2}{3} > \dfrac{5}{8}$ (*v*) $\dfrac{8}{15} > \dfrac{11}{23} > \dfrac{3}{8} > \dfrac{26}{81}$

**4.** (*i*) $\frac{5}{6}$ (*ii*) $\frac{7}{9}$

**5.** (*i*) 0.74, 333 (*ii*) 0.02, 432 (*iii*) 1.05, 5176.5 (*iv*) 0.025, 15

**6.** (*i*) 164.9677 (*ii*) 193.53102

**7.** (*i*) 1.8548 (*ii*) 0.1997 (*iii*) 0.8641 (*iv*) 1.0991

**8.** (*i*) 83.9328 (*ii*) 23.4973 (*iii*) 1.4813986 (*iv*) 1.3571568
(*v*) .00000216 (*vi*) .002197 (*vii*) .0000448 (*viii*) 0.7680

**9.** (*i*) 2.355 (*ii*) 130 (*iii*) 22000 (*iv*) .013
(*v*) 0.00002 (*vi*) 1.68293

**10.** (*i*) $.\overline{037}$ (*ii*) 0.8 (*iii*) 0.1 (*iv*) 441

**11.** (*i*) 331.58 (*ii*) 0.01 (*iii*) 1.1 (*iv*) 0.72
(*v*) .0484

**12.** (*i*) $\frac{7}{9}$ (*ii*) $4\frac{2}{3}$ (*iii*) $\frac{35}{99}$ (*iv*) $1\frac{432}{999}$
(*v*) $\frac{52}{999}$ (*vi*) $2\frac{13}{99}$ (*vii*) $3\frac{1}{999}$ (*viii*) $\frac{22}{7}$

**13.** (*i*) $\frac{23}{90}$ (*ii*) $1\frac{17}{45}$ (*iii*) $\frac{29}{50}$ (*iv*) $\frac{52}{165}$
(*v*) $\frac{1147}{1980}$ (*vi*) $3\frac{97}{1650}$ (*vii*) $\frac{2263}{2250}$

**14.** (*i*) $1.\dot{6}$ (*ii*) $6.\overline{06}$ (*iii*) $1.4\overline{303}$ (*iv*) 266
(*v*) $0.4\overline{801}$

## EXERCISE 3B (*DECIMAL FRACTIONS*)

## OBJECTIVE TYPE QUESTIONS

**1.** The value of (1 + .1 + .01 + .001) is :
(*a*) 1.003 (*b*) 1.111
(*c*) 1.011 (*d*) 1.001

**2.** 6202.5 + 620.25 + 62.2025 + 6.2025 + 0.62025 = ?
(*a*) 68915.9775 (*b*) 689.159775
(*c*) 6891.77525 (*d*) 689159.775

**3.** 32.6 + 12.6 – ? = 25
(*a*) 20.2 (*b*) 20
(*c*) 20.1 (*d*) 22

**4.** 3.006 × 3.2 – 3.54 × 0.13 – 8.049 = ?
(*a*) 83.541 (*b*) 1.09
(*c*) 1.22 (*d*) 1.11

**5.** 0.144 ÷ 0.012 = ?
(*a*) .12 (*b*) 1.2
(*c*) 12 (*d*) .012

6. $37.09 \times ? = (41 - .201)$
(a) 11 (b) 1.1
(c) 11.1 (d) 9.1

7. $50.8 \div 2540 = ?$
(a) 20 (b) 2.0
(c) 0.2 (d) 0.02

8. $172.23 - ? = 63.83 + 22$
(a) 130.4 (b) 86.40
(c) 108.18 (d) 85.83 **(Bank P.O., 1991)**

9. $15.60 \times 0.30 = ?$
(a) 4.68 (b) 0.458
(c) 0.468 (d) 0.0468 **(Bank P.O., 1991)**

10. $(.\overline{6} + .\overline{7} + .\overline{8} + .\overline{3})$ yields :
(a) $2\frac{3}{10}$ (b) $2\frac{33}{100}$
(c) $2\frac{2}{3}$ (d) $2.\overline{35}$

**(Income Tax & Central Excise, 1988)**

11. $0.07 \times 0.008 = ?$
(a) .056 (b) .0056
(c) .00056 (d) .56

12. The value of $(25 \div .0005)$ is :
(a) 50 (b) 500
(c) 5000 (d) None of these

13. $0.000033 \div 0.11 = ?$
(a) .003 (b) .03
(c) .0003 (d) 0.30

14. $0.001 \div ? = 0.01$ :
(a) 10 (b) .1
(c) .01 (d) .001

15. $? \div .0025 = 800$
(a) .2 (b) .02
(c) 2000 (d) 2

16. $7.83 - (3.79 - 2.56) = ?$
(a) 4.04 (b) 1.48
(c) 6.06 (d) 6.6 **(Bank P.O., 1991)**

17. ? % *of* 10.8 = 32.4
(a) 3 (b) 30
(c) 300 (d) .3

18. $\frac{3420}{19} = \frac{?}{0.01} \times 7$

(*a*) $\frac{35}{9}$ (*b*) $\frac{63}{5}$

(*c*) $\frac{18}{7}$ (*d*) None of these

**(Bank P.O. Exam., 1988)**

**19.** If $12276 \div 155 = 79.2$, the value of $122.76 \div 15.5$ is equal to :

(*a*) 7.092 (*b*) 7.92

(*c*) 79.02 (*d*) 79.2 **(CDS, 1991)**

**20.** If $\frac{1}{3.718} = .2689$, then the value of $\frac{1}{.0003718}$ is :

(*a*) 2689 (*b*) 2.689

(*c*) 26890 (*d*) .2689

**21.** $(.351 \times .897 + .351 \times .103) = ?$

(*a*) .897 (*b*) .351

(*c*) 3.518970 (*d*) 8.973510

**22.** $2 \times 0.2 \times 0.02 \times 0.002 \times 20 = ?$

(*a*) 0.000032 (*b*) 0.00032

(*c*) 0.00320 (*d*) 0.03200

**23.** $8.32 \times 0.999 = ?$

(*a*) 0.831168 (*b*) 8.31618

(*c*) 8.31168 (*d*) 8.31668

**24.** $\frac{17.28 \div ?}{3.6 \times 0.2} = 200$

(*a*) 120 (*b*) 1.20

(*c*) 12 (*d*) 0.12

**(Bank P.O. Exam., 1988)**

**25.** $80.40 \div 20 - (-4.2) = ?$

(*a*) 497.8 (*b*) 5.786

(*c*) 947.0 (*d*) 8.22

**(Bank P.O. Exam., 1986)**

**26.** $12 \div 0.09$ of $0.3 \times 2 = ?$

(*a*) 0.80 (*b*) 8.0

(*c*) 80 (*d*) None of these

**(Bank Clerical, 1988)**

**27.** $\frac{20 + 8 \times 0.5}{20 - ?} = 12$

(*a*) 8 (*b*) 18

(*c*) 2 (*d*) None of these

**(Bank Clerical, 1990)**

28. $\dfrac{3}{3+\dfrac{0.3-3.03}{3\times 0.91}}=?$

(a) 1.5
(b) 15
(c) .75
(d) 1.75

29. The local value of 7 in 0.03759 is :

(a) 7
(b) $\dfrac{7}{100}$
(c) $\dfrac{7}{1000}$
(d) $\dfrac{7}{10000}$

30. If $\sqrt{5}=2.24$, then the value of $\dfrac{3\sqrt{5}}{2\sqrt{5}-0.48}$ will be :

(a) 0.168
(b) 1.68
(c) 16.8
(d) 168

**(Central Excise & I. Tax 1988)**

31. Greatest fraction out of $\dfrac{2}{5}, \dfrac{5}{6}, \dfrac{11}{12}, \dfrac{7}{8}$ is :

(a) $\dfrac{7}{8}$
(b) $\dfrac{11}{12}$
(c) $\dfrac{5}{6}$
(d) $\dfrac{2}{5}$

**(BSRB Exam., 1989)**

32. Which part contains the fractions in ascending order ?

(a) $\dfrac{11}{14}, \dfrac{16}{19}, \dfrac{16}{21}$
(b) $\dfrac{16}{19}, \dfrac{11}{14}, \dfrac{16}{21}$
(c) $\dfrac{16}{21}, \dfrac{11}{14}, \dfrac{16}{19}$
(d) $\dfrac{16}{19}, \dfrac{16}{21}, \dfrac{11}{14}$

**(Bank P.O. Exam., 1987**

33. What should be less by the multiplication of 0.527 and 2.013 to get 1

(a) 0.06081
(b) 2.060851
(c) 0.939085
(d) 1.939085

34. $10 \div 0.5 - 10 \times 0.5 = ?$

(a) 1
(b) 0
(c) 10
(d) 15

35. $\dfrac{30+8\times 0.5}{30-?}=17$

(a) 13
(b) 28
(c) 2
(d) 11

36. $0.15 \div \dfrac{0.5}{15}=?$

(a) 4.5
(b) 45
(c) 0.03
(d) 0.45

**(Bank Clerical, 199**

**37.** 1 litre of water weighs 1 kilogram. How many cubic millimeters of water will weigh .1 gm ?

(*a*) 0.1 (*b*) 1
(*c*) 10 (*d*) 100

**38.** What decimal fraction is 20 mm. of a metre ?

(*a*) .2 (*b*) .02
(*c*) .05 (*d*) .5

**39.** It being given that $\sqrt{15} = 3.88$, the best approximation to $\sqrt{\frac{5}{3}}$ is :

(*a*) 0.43 (*b*) 1.89
(*c*) 1.29 (*d*) 1.63

**(S.B.I. P.O. Exam. 1987)**

**40.** $\left(\frac{6.5 \times 4.7 + 6.5 \times 5.3}{1.3 \times 7.9 - 1.3 \times 6.9}\right) = ?$

(*a*) 3.9 (*b*) 39
(*c*) 34.45 (*d*) 50

**41.** The value of $(13.065 \times 13.065 - 3.065 \times 3.065)$ is :

(*a*) 100 (*b*) 161.3
(*c*) 159.5 (*d*) 141.6

**42.** $6\frac{1}{4} \times 0.25 + 0.75 - 0.3125 = ?$

(*a*) 5.9375 (*b*) 4.2968
(*c*) 2.1250 (*d*) 2.0000

**(BSRB Exam., 1990)**

**43.** The value of $(.875 \times .875 + 2 \times .875 \times .225 + .225 \times .225)$ is :

(*a*) 2.175 (*b*) 7.275
(*c*) 1.21 (*d*) 6.31

**44.** The value of $(9.75 \times 9.75 - 2 \times 9.75 \times 5.75 + 5.75 \times 5.75)$ is :

(*a*) 13.25 (*b*) 3.625
(*c*) 4 (*d*) 16

**45.** $\left(\frac{5.7 \times 5.7 \times 5.7 + 2.3 \times 2.3 \times 2.3}{5.7 \times 5.7 + 2.3 \times 2.3 - 5.7 \times 2.3}\right) = ?$

(*a*) 2.3 (*b*) 3.4
(*c*) 5.7 (*d*) 8.0

**(CBI Exam., 1990)**

**46.** $\left(\frac{0.89 \times 0.89 \times 0.89 - 0.64 \times 0.64 \times 0.64}{0.89 \times 0.89 + 0.89 \times 0.64 + 0.64 \times 0.64}\right) = ?$

(*a*) 2.5 (*b*) 0.25
(*c*) 0.93 (*d*) 1.53

**47.** The simplification of $\frac{(0.87)^3 + (0.13)^3}{(0.87)^2 + (0.13)^2 - (0.87)(0.13)}$ yields the result :

(*a*) 0.13 (*b*) 0.74
(*c*) 0.87 (*d*) 1

**(I.Tax & Central Excise, 1988)**

**48.** $\left(\frac{0.03 \times 0.03 + 0.01 \times 0.01 - 0.02 \times 0.03}{0.02}\right) = ?$

(*a*) 0.2 (*b*) 0.04
(*c*) 0.5 (*d*) 0.02

**49.** $(2.03125 \times 5.936 + 8.125 \times 6.516) = ?$

(*a*) 64.5 (*b*) 68.725
(*c*) 65 (*d*) 65.125

**50.** $\frac{(0.05)^2 + (0.41)^2 + (0.073)^2}{(0.005)^2 + (0.041)^2 + (0.0073)^2} = ?$

(*a*) 0.1 (*b*) 10
(*c*) 100 (*d*) 1000

**51.** The value of $\left\{\frac{(2.3)^3 - .027}{(2.3)^2 + .69 + .09}\right\}$ is :

(*a*) 2.6 (*b*) 2
(*c*) 2.33 (*d*) 2.27

**52.** $\left(\frac{.538 \times .538 - .462 \times .462}{1 - .924}\right) = ?$

(*a*) .076 (*b*) 1.042
(*a*) .076 (*b*) 1.042
(*c*) 2 (*d*) 1

**53.** $\left(\frac{1.04 \times 1.04 + 1.04 \times 0.04 + 0.04 \times 0.04}{1.04 \times 1.04 \times 1.04 - 0.04 \times 0.04 \times 0.04}\right)$

(*a*) 0.001 (*b*) 0.1
(*c*) 1 (*d*) 0.01

**(Astt. Grade, 1987)**

**54.** $(0.6 \times 0.6 \times 0.6 + 0.4 \times 0.4 \times 0.4 + 3 \times 0.6 \times 0.4)$ is equal to :

(*a*) 21.736 (*b*) 2.1736
(*c*) 217.36 (*d*) 1

**55.** $(3.64 \times 3.64 + 7.28 \times .36 + .36 \times .36) = ?$

(*a*) 4 (*b*) 8
(*c*) 16 (*d*) 13.28

**56.** $\frac{(.728 \times .728 - .272 \times .272)}{.456} = ?$

(*a*) 3 (*b*) 4
(*c*) 2 (*d*) 1

**57.** $\frac{.125 + .027}{.5 \times .5 - .15 + .09}$ is equal to :

(*a*) .08 (*b*) 1

(*c*) 2 (*d*) .8

**58.** The value of :

$(0.58 \times 0.58 \times 0.58 - 3 \times 0.58 \times 0.42 \times 0.16 - 0.42 \times 0.42 \times 0.42)$ is :

(*a*) 0.16 (*b*) 1

(*c*) 0.004096 (*d*) 1.3976

**59.** $\dfrac{(3.537 - .948)^2 + (3.537 + .948)^2}{(3.537)^2 + (.948)^2} = ?$

(*a*) 4.485 (*b*) 2.589

(*c*) 4 (*d*) 2

**60.** The H.C.F. of 0.24, 1.8 and 2.7 is :

(*a*) 6 (*b*) .6

(*c*) .06 (*d*) .006

**61.** G.C.D. of 1.08, .36 and .9 is :

(*a*) .03 (*b*) .9

(*c*) .18 (*d*) .108

**62.** The L.C.M.of 3, 0.09 and 2.7 is :

(*a*) 2.7 (*b*) .27

(*c*) 0.027 (*d*) 27

**63.** The value of $0.\overline{43}$ is :

(*a*) $\dfrac{43}{100}$ (*b*) $\dfrac{43}{99}$

(*c*) $\dfrac{43}{90}$ (*d*) $\dfrac{4}{9}$

**64.** The value of $.1\overline{36}$ is :

(*a*) $\dfrac{136}{1000}$ (*b*) $\dfrac{136}{999}$

(*c*) $\dfrac{136}{990}$ (*d*) $\dfrac{3}{22}$

**65.** The value of $.53\dot{6}$ is :

(*a*) $\dfrac{536}{1000}$ (*b*) $\dfrac{536}{999}$

(*c*) $\dfrac{536}{990}$ (*d*) $\dfrac{161}{300}$

**66.** The value of $0.5\dot{7}$ is :

(*a*) $\dfrac{57}{1000}$ (*b*) $\dfrac{57}{99}$

(*c*) $\dfrac{26}{45}$ (*d*) $\dfrac{52}{9}$

**67.** $(0.\overline{63} + 0.\overline{37}) = ?$

(*a*) 1 (*b*) $1.\overline{01}$

(*c*) $.\overline{101}$ (*d*) 1.01

**68.** The value of $(0.34\overline{67} + 0.13\overline{33})$ is :

(*a*) 0.48 (*b*) $0.48\overline{01}$

(*c*) $0.\overline{48}$ (*d*) $0.4\overline{8}$

**69.** $(0.\overline{45} \times 0.1) = ?$

(*a*) 0.045 (*b*) $.0\overline{45}$

(*c*) 0.05 (*d*) $.0\dot{5}$

**70.** The value of $(0.\overline{55} \div 0.\dot{1})$ is :

(*a*) 5.5 (*b*) 1.8

(*c*) 5 (*d*) .55

**71.** If $\sqrt{4096} = 64$, then the value of

$\sqrt{40.96} + \sqrt{0.4096} + \sqrt{0.004096} + \sqrt{0.00004096}$

upto two places of decimals is :

(*a*) 7.09 (*b*) 7.10

(*c*) 7.1104 (*d*) 7.12

**72.** $\dfrac{0.25 \times 0.25 - 0.24 \times 0.24}{0.49} = ?$

(*a*) 0.0006 (*b*) 0.49

(*c*) 0.01 (*d*) 0.1

**(Railway Recruitment, 1991)**

**73.** $\sqrt{\dfrac{0.289}{0.00121}} = ?$

(*a*) $\dfrac{170}{11}$ (*b*) $\dfrac{17}{110}$

(*c*) $\dfrac{0.17}{11}$ (*d*) $\dfrac{17}{11}$

**(Railway Recruitment, 1991)**

**74.** The cube root of 5.832 is :

(*a*) .018 (*b*) 18

(*c*) .18 (*d*) 1.8

**75.** Fill in the blanks :

(*i*) $(\ldots\ldots\ldots) \div 0.05 = 5$

(*ii*) $(\ldots\ldots) \times 1000 = 34.7$

(*iii*) $2 \div 0.0025 = (\ldots\ldots\ldots)$

(*iv*) $(0.0001) \div (\ldots\ldots\ldots) = 0.01$

(*v*) $.5 \times .05 \times .005 \times .0005 = (\ldots\ldots\ldots)$

(*vi*) $0.\overline{63} + 0.\overline{37} = (\ldots\ldots\ldots)$

(*vii*) H.C.F. of .09, 2.7 and .243 is (.........)

(*viii*) L.C.M. of 1.1, .99 and 1 is (.........).

# HINTS & SOLUTIONS (*EXERCISE 3B*)

**1.**
$$\begin{array}{r} 1.000 \\ .1 \\ .01 \\ .001 \\ \hline 1.111 \end{array}$$

**2.**
$$\begin{array}{r} 6202.5 \\ 620.25 \\ 62.2025 \\ 6.2025 \\ 0.62025 \\ \hline 6891.77525 \end{array}$$

**3.** Let $32.6 + 12.6 - x = 25$.
Then, $x = (32.6 + 12.6) - 25 = (45.2 - 25) = 20.2$.

**4.** $3.006 \times 3.2 - 3.54 \times 0.13 - 8.049$
$= 9.6192 - 0.4602 - 8.049 = 1.11$.

**5.** $\frac{0.144}{0.012} = \frac{144}{12} = 12.$

**6.** Let $37.09 \times x = 41 - 0.201 = 40.799$
$x = \frac{40.799}{37.090} = 1.1.$

**7.** $\frac{50.8}{2540} = \frac{508}{25400} = \frac{1}{50} = 0.02.$

**8.** Let $172.23 - x = 63.83 + 22$
$\Rightarrow x = 172.23 - 85.83 \Rightarrow x = 86.40.$

**9.** $15.60 \times 0.30 = 4.6800 = 4.68.$

**10.** $.\bar{6} + .\bar{7} + .\bar{8} + .\bar{3} = \frac{6}{9} + \frac{7}{9} + \frac{8}{9} + \frac{3}{9} = \frac{24}{9} = \frac{8}{3} = 2\frac{2}{3}.$

**11.** $7 \times 8 = 56 \Rightarrow 0.07 \times 0.008 = 0.00056.$

**12.** $\frac{25}{0.0005} = \frac{25 \times 10000}{5} = 50000.$

**13.** $\frac{0.000033}{0.110000} = \frac{33}{110000} = \frac{3}{10000} = 0.0003.$

**14.** Let $\frac{0.001}{x} = 0.01$. Then, $x = \frac{0.001}{0.010} = \frac{1}{10} = 0.1.$

**15.** Let $\frac{x}{.0025} = 800$. Then, $x = \frac{800 \times 25}{10000} = 2.$

**16.** $7.83 - (3.79 - 2.56) = (7.83 - 1.23) = 6.6.$

**17.** Let $x$ % of $10.8 = 32.4$.
Then, $\frac{x}{100} \times 10.8 = 32.4 \Rightarrow x = \frac{32.4 \times 100}{10.8} = 300.$

**18.** $\frac{3420}{19} = \frac{x}{0.01} \times 7 \Rightarrow x = \frac{3420}{19} \times \frac{0.01}{7} = \frac{9}{35}.$

**19.** $\frac{122.76}{15.50}=\frac{12276}{1550}=\frac{12276}{155}\times\frac{1}{10}=\frac{79.2}{10}=7.92.$

**20.** $\frac{1}{.0003718}=\frac{10000}{3.718}=\left(10000\times\frac{1}{3.718}\right)$

$=(10000\times 0.2689)=2689.$

**21.** $(.351\times .897+.351\times .103)=.351\times(.897+.103)=.351\times 1=.351.$

**22.** $2\times 2\times 2\times 2\times 20=320.$

$\therefore\ 2\times 0.2\times 0.02\times 0.002\times 20=0.000320=0.00032.$

**23.** $8.32\times 0.999=8.32\times(1-0.001)=8.32-0.00832=8.31168.$

**24.** $17.28\div x=200\times 3.6\times 0.2=144.$

$\Rightarrow \frac{17.28}{x}=144$ or $x=\frac{17.28}{144}=0.12.$

**25.** $\frac{80.40}{20}+4.2=4.02+4.2=8.22.$

**26.** $12\div 0.09$ or $0.3\times 2=12\div 0.027\times 2=\frac{12000}{27}\times 2=\frac{8000}{9}.$

**27.** $\frac{20+8\times 0.5}{20-x}=12\Rightarrow 20+4=240-12x.$

$\therefore\ 12x=(240-24)=216$ or $x=18.$

**28.** $\frac{3}{3+\frac{0.3-3.03}{3\times 0.91}}=\frac{3}{3-\frac{2.73}{2.73}}=\frac{3}{3-1}=\frac{3}{2}=1.5.$

**29.** Local value of 7 in 0.03759 is $\frac{7}{1000}$.

**30.** Given expression $=\frac{3\times 2.24}{2\times 2.24-0.48}=\frac{6.72}{4}=1.68.$

**31.** $\frac{2}{5}=0.40,\ \frac{5}{6}=0.83,\ \frac{11}{12}=0.91\ \&\ \frac{7}{8}=0.87.$

$\therefore$ Greatest of given fractions is $\frac{11}{21}$.

**32.** $\frac{11}{14}=0.78,\ \frac{16}{19}=0.84,\ \frac{16}{21}=0.76$

Since $0.76<0.78<0.84$, so $\frac{16}{21}<\frac{11}{14}<\frac{16}{19}$.

**33.** $0.527\times 2.013=1.060851.$

$\therefore$ Required number $=0.060851.$

**34.** $10\div 0.5-10\times 0.5=\frac{1}{0.5}-5=20-5=15.$

**35.** $\dfrac{30+8\times 0.5}{30-x} = 17 \Rightarrow 34 = 510 - 17x.$

$\therefore \quad 17x = 510 - 34 = 476 \Rightarrow x = 28.$

**36.** $0.15 \div \dfrac{0.5}{15} = \dfrac{0.15\times 15}{0.50} = \dfrac{15\times 15}{50} = \dfrac{9}{2} = 4.5.$

**37.** 1000 gm is the weight of 1000 cu. cm.

1 gm is the weight of 1 cu. cm i.e. 1000 cu. mm.

$\dfrac{1}{10}$ gm is the weight of $\dfrac{1}{10}\times 1000 = 100$ cu. mm.

**38.** Required fraction $= \dfrac{20}{1000\times 10} = 0.020 = 0.02.$

**39.** $\sqrt{\dfrac{5}{3}} = \sqrt{\dfrac{5\times 3}{3\times 3}} = \dfrac{\sqrt{15}}{3} = \dfrac{3.88}{3} = 1.29.$

**40.** Given expression $= \dfrac{6.5\times(4.7+5.3)}{1.3\times(7.9-6.9)} = \dfrac{6.5\times 10}{1.3\times 1} = 50.$

**41.** Given expression $= [\,(13.065)^2 - (3.065)^2\,]$

$= (13.065 + 3.065)\times(13.065 - 3.065)$

$= 16.130 \times 10 = 161.30.$

**42.** Given expression $= 6.25\times 0.25 + 0.75 - 0.3125$

$= 1.5625 + 0.75 - 0.3125 = 2.$

**43.** Given expression $= (a^2 + 2ab + b^2)$, where $a = .875$, $b = .225$

$= (a+b)^2 = (.875 + .225)^2 = (1.1)^2 = 1.21.$

**44.** Given expression $= (a^2 - 2ab + b^2)$, where $a = 9.75$, $b = 5.75$

$= (a-b)^2 = (9.75 - 5.75)^2 = (4)^2 = 16.$

**45.** Given expression $= \left(\dfrac{a^3+b^3}{a^2-ab+b^2}\right)$, where $a = 5.7$, $b = 2.3$

$= (a+b) = (5.7 + 2.3) = 8.$

**46.** Given expression $= \left(\dfrac{a^3-b^3}{a^2+ab+b^2}\right)$, where $a = 0.89$, $b = 0.64$

$= (a-b) = (0.89 - 0.64) = 0.25.$

**47.** Given expression $= \left(\dfrac{a^3+b^3}{a^2+b^2-ab}\right)$, where $a = 0.87$, $b = 0.13$

$= (a+b) = (0.87 + 0.13) = 1.$

**48.** Given expression $= \dfrac{0.03\times 0.03 - 0.02\times 0.03 + 0.01\times 0.01}{0.02}$

$= \dfrac{(0.03 - 0.02)\times 0.03 + 0.01\times 0.01}{0.02} = \dfrac{0.01\times 0.03 + 0.01\times 0.01}{0.02}$

$$= \frac{0.01 \times (0.03 + 0.01)}{0.02} = \frac{0.01 \times 0.04}{0.02} = 0.02.$$

**49.** Given expression $= 2.03125 \times 5.936 + 4 \times 2.03125 \times 6.516$

$$= 2.03125 \times (5.936 + 26.064)$$
$$= 2.03125 \times 32 = 65.$$

**50.** Given expression

$$= \frac{a^2 + b^2 + c^2}{\left(\frac{a}{10}\right)^2 + \left(\frac{b}{10}\right)^2 + \left(\frac{c}{10}\right)^2} = \frac{100\,(a^2 + b^2 + c^2)}{(a^2 + b^2 + c^2)} = 100.$$

**51.** Given expression $= \left(\frac{a^3 - b^3}{a^2 + ab + b^2}\right)$, where $a = 2.3, b = 0.3$

$$= (a - b) = (2.3 - 0.3) = 2.$$

**52.** Given expression $= \frac{(.538)^2 - (0.462)^2}{(0.076)}$

$$= \frac{(.538 + .462)\,(.538 - .462)}{(0.076)} = \frac{1 \times 0.076}{0.076} = 1.$$

**53.** Given expression $= \left(\frac{a^2 + ab + b^2}{a^3 - b^3}\right)$, where $a = 1.04, b = 0.04$

$$= \left(\frac{1}{a - b}\right) = \left(\frac{1}{1.04 - 0.04}\right) = 1.$$

**54.** Given expression $= a^3 + b^3 + 3ab\,(a + b)$, where

$a = 0.6, b = 0.4, a + b = 1$

$$= (a + b)^3 = (0.6 + 0.4)^3 = 1.$$

**55.** Given expression $= (3.64)^2 + 2 \times (3.64) \times 0.36 + (0.36)^2$

$$= (3.64 + 0.36)^2 = (4)^2 = 16.$$

**56.** Given expression $= \frac{(0.728)^2 - (0.272)^2}{0.456}$

$$= \frac{(0.728 + 0.272)\,(0.728 - 0.272)}{0.456} = \frac{1 \times 0.456}{0.456} = 1.$$

**57.** Given expression $= \frac{(.5)^3 + (.3)^3}{(.5)^2 - (.5 \times .3) + (.3)^2}$

$$= (.5 + .3) = (.8).$$

**58.** Given expression $= [\, a^3 - 3ab\,(a - b) - b^3 \,]$, where

$a = 0.58, b = 0.42$

$$= (a - b)^3 = (0.58 - 0.42)^3 = (0.16)^3$$
$$= 0.004096.$$

**59.** Given expression $= \dfrac{(a-b)^2 + (a+b)^2}{(a^2+b^2)} = \dfrac{2(a^2+b^2)}{(a^2+b^2)} = 2.$

**60.** 0.24 = 0.24, 1.8 = 1.80, 2.7 = 2.70.

H.C.F. of 24, 180 and 270 is 6.

$\therefore$ H.C.F. of 0.24, 1.80 and 2.70 is 0.06.

**61.** The given numbers are 1.08, 0.36 and 0.90.

H.C.F. of 108, 36 and 90 is 18.

$\therefore$ H.C.F. of given numbers = 0.18.

**62.** The given numbers are 3.00, 0.09 and 2.70.

L.C.M. of 300, 9 and 270 is 2700.

$\therefore$ L.C.M. of given numbers = 27.00 i.e. 27.

**63.** $0.\overline{43} = \dfrac{43}{99}$.

**64.** $0.1\overline{36} = \dfrac{136-1}{990} = \dfrac{135}{990} = \dfrac{3}{22}$.

**65.** $0.53\dot{6} = \dfrac{536-53}{900} = \dfrac{483}{900} = \dfrac{161}{300}$.

**66.** $0.5\dot{7} = \dfrac{57-5}{90} = \dfrac{52}{90} = \dfrac{26}{45}$.

**67.** $0.\overline{63} + 0.\overline{37} = \dfrac{63}{99} + \dfrac{37}{99} = \dfrac{100}{99} = 1.\overline{01}$.

**68.** $0.34\overline{67} + 0.13\overline{33} = \dfrac{3467-34}{9900} + \dfrac{1333-13}{9900}$

$= \dfrac{3433+1320}{9900} = \dfrac{4753}{9900} = 0.48\overline{01}$.

**69.** $0.\overline{45} \times 0.1 = \dfrac{45}{99} \times 0.1 = \dfrac{45}{99} \times \dfrac{1}{10} = 0.0\overline{45}$

**70.** $0.\overline{55} \div 0.\dot{1} = \dfrac{55}{99} \div \dfrac{1}{9} = \dfrac{55}{99} \times \dfrac{9}{1} = 5.$

**71.** Given expression

$= \sqrt{\dfrac{4096}{100}} + \sqrt{\dfrac{4096}{10000}} + \sqrt{\dfrac{4096}{1000000}} + \sqrt{\dfrac{4096}{100000000}}$

$= \dfrac{64}{10} + \dfrac{64}{100} + \dfrac{64}{1000} + \dfrac{64}{10000}$

$= 6.4 + 0.64 + 0.064 + 0.0064 = 7.1104.$

**72.** Given expression $= \dfrac{(0.25)^2 - (0.24)^2}{0.49}$

$= \dfrac{(0.25+0.24) \times (0.25-0.24)}{0.49} = \dfrac{0.49 \times 0.01}{0.49} = 0.01.$

**73.** $\sqrt{\dfrac{0.289}{0.00121}} = \sqrt{\dfrac{0.28900}{0.00121}} = \sqrt{\dfrac{28900}{121}} = \dfrac{170}{11}$.

**74.** $(5.832)^{1/3} = \left(\dfrac{5832}{1000}\right)^{1/3} = \left(\dfrac{18\times18\times18}{10\times10\times10}\right)^{1/3} = \dfrac{18}{10} = 1.8.$

**75.** (*i*) $\dfrac{x}{0.05} = 5 \Rightarrow x = 0.25.$

(*ii*) $x \times 1000 = 34.7 \Rightarrow x = \dfrac{34.7}{1000} = 0.0347.$

(*iii*) $\dfrac{2}{0.0025} = \dfrac{2\times10000}{25} = 800.$

(*iv*) $\dfrac{0.0001}{x} = 0.01 \Rightarrow x = \dfrac{0.0001}{0.0100} = \dfrac{1}{100} = 0.01.$

(*v*) $5\times5\times5\times5 = 625.$

$\therefore \quad .5\times.05\times.005\times.0005 = 0.0000000625.$

(*vi*) $0.\overline{63} + 0.\overline{37} = \dfrac{63}{99} + \dfrac{37}{99} = \dfrac{100}{99} = 1.\overline{01}.$

(*vii*) Given numbers are 0.090, 2.700 and 0.243.

H.C.F. of 90, 2700 and 243 is 9.

$\therefore$ H.C.F. of given numbers = 0.009.

(*viii*) Given numbers are 1.10, 0.99 and 1.00.

L.C.M. of 110, 99 and 100 is 9900.

$\therefore$ L.C.M. of given numbers = 99.00 *i.e.* 99.

## ANSWERS (*EXERCISE 3B*) [*DECIMAL FRACTIONS*]

| | | | | | | |
|---|---|---|---|---|---|---|
| **1.** (b) | **2.** (c) | **3.** (a) | **4.** (d) | **5.** (c) | **6.** (b) | **7.** (d) |
| **8.** (b) | **9.** (a) | **10.** (c) | **11.** (c) | **12.** (d) | **13.** (c) | **14.** (b) |
| **15.** (d) | **16.** (d) | **17.** (c) | **18.** (d) | **19.** (b) | **20.** (a) | **21.** (b) |
| **22.** (b) | **23.** (c) | **24.** (d) | **25.** (d) | **26.** (d) | **27.** (b) | **28.** (a) |
| **29.** (c) | **30.** (b) | **31.** (b) | **32.** (c) | **33.** (b) | **34.** (d) | **35.** (b) |
| **36.** (a) | **37.** (d) | **38.** (b) | **39.** (c) | **40.** (d) | **41.** (b) | **42.** (d) |
| **43.** (c) | **44.** (d) | **45.** (d) | **46.** (b) | **47.** (d) | **48.** (d) | **49.** (c) |
| **50.** (c) | **51.** (b) | **52.** (d) | **53.** (c) | **54.** (d) | **55.** (c) | **56.** (d) |
| **57.** (d) | **58.** (c) | **59.** (d) | **60.** (c) | **61.** (c) | **62.** (d) | **63.** (b) |
| **64.** (d) | **65.** (d) | **66.** (c) | **67.** (b) | **68.** (b) | **69.** (b) | **70.** (c) |
| **71.** (c) | **72.** (c) | **73.** (a) | **74.** (d) | **75.** (i) 0.25 | (ii) 0.0347 | |

(*iii*) 800 (*iv*) 0.01 (*v*) 0.0000000625 (*vi*) $1.\overline{01}$

(*vii*) 0.009 (*viii*) 99.

# 4

# SIMPLIFICATION

*In simplifying an expression containing fractions, the order of various operations must be strictly maintained in the manner given below :—*

I. **Vinculum or Bar.** For example, $-\overline{3-5} = -(-2) = 2$,

whereas, $-3-5=-8$.

II. **Removal of Brackets in the order ( ), { }, & [ ].** For example,

$[4-\{3-(2-1)\}] = [4-\{3-1\}] = [4-2] = 2$.

III Of

IV Division

V Multiplication

VI Addition

VII Subtraction

> *Remember*— **VBODMAS,**
> *Where V, B, O, D, M, A, S stand for Vinculum, Bracket, of, Division, Multiplication, Addition & subtraction respectively.*

## ILLUSTRATIVE EXAMPLES

**Ex. 1.** *Simplify :—*

$$7\frac{1}{2}-\left[5\frac{1}{2}-\left\{4\frac{1}{4}-\left(3\frac{1}{2}-\overline{2\frac{1}{2}-\frac{1}{2}}\right)\right\}\right]$$

**Sol.** The given expression

$$=\frac{15}{2}-\left[\frac{11}{2}-\left\{\frac{17}{4}-\left(\frac{7}{2}-\overline{\frac{5}{2}-\frac{1}{2}}\right)\right\}\right]$$

$$=\frac{15}{2}-\left[\frac{11}{2}-\left\{\frac{17}{4}-\left(\frac{7}{2}-2\right)\right\}\right] \quad \text{(Removing Vinculum)}$$

$$=\frac{15}{2}-\left[\frac{11}{2}-\left\{\frac{17}{4}-\frac{3}{2}\right\}\right]$$

$$=\frac{15}{2}-\left[\frac{11}{2}-\frac{11}{4}\right]=\frac{15}{2}-\frac{11}{4}=\frac{(30-11)}{4}$$

$$=\frac{19}{4}=4\frac{3}{4}.$$

**Ex. 2.** *Simplify :*

$$1\div[1+1\div\{1+1\div(1+1\div3)\}]$$

**Sol.** The given expression is

$$= 1 \div \left[ 1 + 1 \div \left\{ 1 + 1 \div \left( 1 + \frac{1}{3} \right) \right\} \right]$$

$$= 1 \div \left[ 1 + 1 \div \left\{ 1 + 1 \div \frac{4}{3} \right\} \right]$$

$$= 1 \div \left[ 1 + 1 \div \left\{ 1 + 1 \times \frac{3}{4} \right\} \right]$$

$$= 1 \div \left[ 1 + 1 \div \left\{ 1 + \frac{3}{4} \right\} \right]$$

$$= 1 \div \left[ 1 + 1 \div \frac{7}{4} \right]$$

$$= 1 \div \left[ 1 + 1 \times \frac{4}{7} \right]$$

$$= 1 \div \left[ 1 + \frac{4}{7} \right]$$

$$= 1 \div \frac{11}{7} = 1 \times \frac{7}{11} = \frac{7}{11}.$$

**Ex. 3.** ***Simplify :***

$$\frac{3}{11} + \frac{4}{3} \div \frac{10}{11} \times \frac{5}{11} - \frac{14}{33} \text{ of } 1\frac{2}{7}.$$

**Sol.** The given expression

$$= \frac{3}{11} + \frac{4}{3} \div \frac{10}{11} \times \frac{5}{11} - \frac{14}{33} \text{ of } \frac{9}{7}$$

$$= \frac{3}{11} + \frac{4}{3} \div \frac{10}{11} \times \frac{5}{11} - \frac{6}{11}$$

$$= \frac{3}{11} + \frac{4}{3} \times \frac{11}{10} \times \frac{5}{11} - \frac{6}{11}$$

$$= \frac{3}{11} + \frac{2}{3} - \frac{6}{11}$$

$$= \frac{9 + 22 - 18}{33} = \frac{13}{33}.$$

**Ex. 4.** ***Simplify :***

$$4.59 \times 1.8 \div 3.6 + 5.4 \text{ of } \frac{1}{9} - \frac{1}{5}.$$

**Sol.** The given expression

$$= 4.59 \times 1.8 \div 3.6 + .6 - .2$$

$= 4.59 \times 1.8 \times \frac{1}{3.6} + .6 - .2$

$= 2.295 + .6 - .2$

$= 2.895 - .2 = 2.695.$

**Ex. 5.** ***Find the value of :***

$$\cfrac{1}{2+\cfrac{1}{2+\cfrac{1}{2-\cfrac{1}{2}}}}$$

**Sol.** The given expression

$$= \cfrac{1}{2+\cfrac{1}{2+\cfrac{1}{\left(\frac{3}{2}\right)}}} = \cfrac{1}{2+\cfrac{1}{2+\frac{2}{3}}} = \cfrac{1}{2+\cfrac{1}{\left(\frac{8}{3}\right)}}$$

$$= \frac{1}{2+\frac{3}{8}} = \frac{1}{\frac{19}{8}} = \frac{8}{19}.$$

**Ex. 6.** ***Find the value of the expression given below upto 3 places of decimals :***

$$3.5 \div .7 \text{ of } 7 + .5 \times .3 - .1$$

**Sol.** The given expression

$= 3.5 \div 4.9 + .5 \times .3 - .1$

$= \frac{3.5}{4.9} + .5 \times .3 - .1$

$= \frac{5}{7} + .15 - .1$

$= \frac{5}{7} + \frac{3}{20} - \frac{1}{10} = \frac{100 + 21 - 14}{140} = \frac{107}{140} = .764.$

**Ex. 7.** ***Find the value of :***

$$\frac{0.47 \times 0.47 + 0.35 \times 0.35 - 2 \times 0.47 \times 0.35}{0.12}$$

**Sol.** The given expression

$= \frac{0.47 \times 0.47 + 0.35 \times 0.35 - 2 \times 0.47 \times 0.35}{0.47 - 0.35}$

$= \frac{a^2 + b^2 - 2ab}{a - b} = \frac{(a-b)^2}{a-b}$, where $a = 0.47$ & $b = 0.35$

$= a - b = 0.47 - 0.35 = 0.12$

**Ex. 8.** ***Find the value of :***

$$\frac{6.431 \times 6.431 \times 6.431 + .569 \times .569 \times .569}{6.431 \times 6.431 - 6.431 \times .569 + .569 \times .569}$$

**Sol.** The given expression

$$= \frac{a^3 + b^3}{a^2 - ab + b^2} = a + b, \text{ where } a = 6.431 \ \& \ b = .569$$

$$= 6.431 + .569 = 7.$$

**Ex. 9.** ***Find the value of :***

$$(0.6)^3 + (0.4)^3 + 3 \times 0.6 \times 0.4$$

**Sol.** The given expression

$$= (0.6)^3 + (0.4)^3 + 3 \times 0.6 \times 0.4 \times 1$$

$$= (0.6)^3 + (0.4)^3 + 3 \times 0.6 \times 0.4\ (0.6 + 0.4)$$

$$= a^3 + b^3 + 3ab\,(a + b) = (a + b)^3, \text{ where } a = 0.6 \ \& \ b = 0.4$$

$$= (0.6 + 0.4)^3 = (1)^3 = 1.$$

**Ex. 10.** ***Simplify :***

$$\frac{.0016 \times .025}{.325 \times .05} + \frac{.1216 \times .105 \times .002}{.08512 \times .625 \times .039}$$

**Sol.** The given expression is

$$= \frac{16 \times 25}{325 \times 5 \times 100} + \frac{1216 \times 105 \times 2 \times 10}{8512 \times 625 \times 39}$$

$$= \frac{4}{65 \times 25} + \frac{4}{25 \times 13} = \frac{4}{65 \times 25} \times \frac{25 \times 13}{4} = \frac{1}{5} = 0.2.$$

**Ex. 11.** ***Simplify :***

$$\frac{.538 \times .538 - .462 \times .462}{1 - .924}$$

**Sol.** The given expression

$$= \frac{(.538)^2 - (.462)^2}{.076} = \frac{(.538)^2 - (.462)^2}{(.538 - .462)}$$

$$= \frac{a^2 - b^2}{a - b}, \text{ where } a = .538 \ \& \ b = .462$$

$$= (a + b) = .538 + .462 = 1.$$

**Ex. 12.** ***Simplify :*** $\dfrac{(2.3)^3 - .027}{(2.3)^2 + .69 + .09}$.

**Sol.** The given expression

$$= \frac{(2.3)^3 - (.3)^3}{(2.3)^2 + (2.3 \times .3) + (.3)^2} = \frac{(a^3 - b^3)}{(a^2 + ab + b^2)} = (a - b),$$

where $a = 2.3$ & $b = .3$.

$$= 2.3 - .3 = 2.$$

## EXERCISE 4 A

*Find the values of :*

1. $\left(1\frac{3}{5}-\frac{2}{3}+\frac{12}{13}+\frac{7}{5}\times\frac{1}{3}\right)$.

2. $2-[3-\{6-(5-\overline{4-3})\}]$.

3. $\frac{1}{3}\div\frac{5}{3}+\frac{1}{4}\times\frac{3}{5}-\frac{2}{5}$ of $\frac{5}{7}$.

4. $\frac{1}{5}+\frac{1}{3}\left\{\frac{1}{2}-\frac{1}{3}+3\left(\frac{1}{2}-\overline{\frac{1}{4}+\frac{1}{5}}\right)\right\}$.

5. $7\frac{1}{2}+\frac{1}{2}\div\frac{1}{2}$ of $\frac{1}{4}-\frac{2}{5}\times2\frac{1}{3}\div1\frac{7}{8}$ of $\left(1\frac{2}{5}-1\frac{1}{3}\right)$.

6. $\frac{5+5\times5}{5\times5+5}\times\frac{\frac{1}{5}\div\frac{1}{5}\text{ of }\frac{1}{5}}{\frac{1}{5}\text{ of }\frac{1}{5}\div\frac{1}{5}}\times\left(5-\frac{1}{5}\right)\times\frac{1}{\left\{\frac{46}{5}-\frac{3}{\left(1-\frac{2}{3}\right)}\right\}}$.

7. $\frac{7}{5-\frac{8}{3}}\div\frac{3-\frac{3}{2}}{4-\frac{3}{2}}-\frac{5}{7}$ of $\left\{\frac{1}{1\frac{3}{7}}\div\frac{6}{5}\text{ of }\frac{3\frac{1}{3}-2\frac{1}{2}}{\frac{47}{21}-2}\right\}$.

8. $\frac{7.85\times7.85-2.15\times2.15}{7.85-2.15}-\frac{\frac{1}{3}+\frac{1}{5}+\frac{1}{7}}{\frac{1}{9}\div\left(\frac{1}{2}+\frac{1}{13}\right)}$.

9. $\frac{4\frac{1}{7}-2\frac{1}{4}}{3\frac{1}{2}+1\frac{1}{7}}\div\frac{1}{2+\frac{1}{2+\frac{1}{5-\frac{1}{5}}}}$.

10. (*i*) $1+\frac{1}{2+\frac{1}{1-\frac{1}{3}}}$ **(Excise & I.T. 1989)**

(*ii*) $1+\frac{1}{1+\frac{1}{1+\frac{1}{9}}}$ **(Assistant Grade 1990)**

11. $\frac{75983\times75983-45983\times45983}{75983+45983}$ **(Clerk's Grade 1991)**

12. $\frac{885\times885\times885+115\times115\times115}{885\times885+115\times115-885\times115}$ **(Clerk's Grade 1991)**

**Hint.** $\left(\dfrac{a^3+b^3}{a^2+b^2-ab}\right)=(a+b)=(885+115)=1000.$

**13.** $\dfrac{.7541\times.7541\times.7541-.2459\times.2459\times.2459}{.7541\times.7541+.7541\times.2459+.2459\times.2459}.$

**Hint.** $\left(\dfrac{a^3-b^3}{a^2+ab+b^2}\right)=(a-b)=(.7541-.2459).$

**14.** $\dfrac{1.75\times1.75\times1.75+1.25\times1.25\times1.25}{1.75\times1.75+1.25\times1.25-1.75\times1.25}.$

**Hint.** $\left(\dfrac{a^3+b^3}{a^2+b^2-ab}\right)=(a+b)=(1.75+1.25)=3.$

**15.** $\dfrac{0.125+0.027}{(.5)^2-.15+(.3)^2}.$ **Hint.** $\left(\dfrac{a^3+b^3}{a^2-ab+b^2}\right)=(a+b)=(.5+.3)=.8.$

**16.** $\dfrac{.527\times.527-2\times.527\times.495+.495\times.495}{.032}.$

**Hint.** Given Exp. $=\left(\dfrac{a^2-2ab+b^2}{a-b}\right)=(a-b),$

where $a=.527$ & $b=.495.$

**17.** $\dfrac{.896\times.752+.896\times.248}{.7\times.034+.7\times.966}.$

**Hint.** Given Exp. $=\dfrac{.896\times(.752+.248)}{.7\times(.034+.966)}=\dfrac{.896}{.700}=1.28.$

**18.** $.7\times.7\times.7+.3\times.3\times.3+3\times.7\times.3.$

**Hint.** Given exp. $=a^3+b^3+3ab\,(a+b)=(a+b)^3=(.7+.3)^3.$

**19.** $\dfrac{.356\times.356-2\times.356\times.106+.106\times.106}{.632\times.632+2\times.632\times.368+.368\times.368}.$

**Hint.** Given Exp. $=\left(\dfrac{a^2-2ab+b^2}{c^2+2cd+d^2}\right)=\dfrac{(a-b)^2}{(c+d)^2}=\dfrac{(.356-.106)^2}{(.632+.368)^2}.$

## ANSWERS (*EXERCISE 4 A*)

**1.** $1\frac{31}{90}$ **2.** 1 **3.** $\frac{9}{140}$ **4.** $\frac{11}{36}$ **5.** $4\frac{1}{30}$ **6.** 600 **7.** $1\frac{1}{2}$

**8.** 1 **9.** 1 **10.** (*i*) $1\frac{2}{7}$ (*ii*) $1\frac{10}{19}$ **11.** 30000 **12.** 1000

**13.** 0.5082 **14.** 3 **15.** 0.8 **16.** 0.032 **17.** 1.28 **18.** 1 **19.** .0625

## EXERCISE 4 B
## (*OBJECTIVE TYPE QUESTIONS*)

**1.** $5+5+5\times5=?$

(*a*) $\frac{2}{5}$ (*b*) 10

(*c*) $5\frac{1}{5}$ (*d*) 30

**2.** $171 \div 19 \times 9 = ?$

(*a*) 1 (*b*) 18

(*c*) 81 (d) 0

**(S.B.I. P.O. Exam. 1987)**

**3.** $5005 - 5000 \div 10.00 = ?$

(*a*) 0.5 (*b*) 50

(*c*) 5000 (*d*) 4505

**(Bank P.O. Exam. 1988)**

**4.** $30 - 5 \times 2 + 3 - 8 \div 2 = ?$

(*a*) $22\frac{1}{2}$ (*b*) 19

(*c*) $7\frac{1}{2}$ (*d*) 25

**5.** $8 \div 4\,(3 - 2) \times 4 + 3 - 7 = ?$

(*a*) – 3 (*b*) 5

(*c*) 4 (*d*) – 4 **(Railways 1991)**

**6.** $\frac{10}{3} \times \frac{12}{5} \times \frac{?}{4} = 16$

(*a*) 6 (*b*) 2

(*c*) 8 (*d*) 4 **(Railways 1991)**

**7.** $\frac{1}{7} + \left[\frac{7}{9} - \left(\frac{3}{9} + \frac{2}{9}\right) - \frac{2}{9}\right] = ?$

(*a*) $\frac{1}{9}$ (*b*) $\frac{3}{7}$

(*c*) $\frac{2}{9}$ (*d*) $\frac{1}{7}$ **(Clerical Grade 1991)**

**8.** $\frac{12 \times 8 - 19 \times 2}{6 \times 7 - 6.5 \times 2} = ?$

(*a*) 2 (*b*) $\frac{1}{2}$

(*c*) $\frac{29}{3}$ (*d*) none of these

**(BSRB Exam 1990)**

**9.** $\frac{\frac{1}{2} + 4 + 20}{\frac{1}{2} \times 4 + 20} = ?$

(*a*) $\frac{81}{88}$ (*b*) $2\frac{3}{11}$

(*c*) $\frac{161}{176}$ (*d*) 1 **(Bank P.O. Exam. 1990)**

**10.** $\dfrac{\frac{1}{5} \div \frac{1}{5} \text{ of } \frac{1}{5}}{\frac{1}{5} \text{ of } \frac{1}{5} \div \frac{1}{5}} = ?$

(*a*) 1 (*b*) 5

(*c*) $\frac{1}{5}$ (*d*) 25 **(Railways 1991)**

**11.** $(20 \div 5) \div 2 + (16 \div 8) \times 2 + (10 \div 5) \times (3 \div 2) = ?$

(*a*) 9 (*b*) 12

(*c*) 15 (*d*) 18 **(Clerical Grade 1991)**

**12.** $\frac{31}{10} \times \frac{3}{10} + \frac{7}{5} \div 20 = ?$

(*a*) 0 (*b*) 1

(*c*) 100 (*d*) $\frac{107}{200}$ **(Bank P.O. 1988)**

**13.** $\dfrac{16 - 6 \times 2 + 3}{23 - 3 \times 2} = ?$

(*a*) $\frac{7}{17}$ (*b*) $\frac{23}{40}$

(*c*) $\frac{14}{23}$ (*d*) $\frac{2}{17}$ **(Bank P.O. 1988)**

**14.** $7\frac{1}{2} - \left[2\frac{1}{4} \div \left\{1\frac{1}{4} - \frac{1}{2}\left(1\frac{1}{2} - \frac{1}{3} - \frac{1}{6}\right)\right\}\right] = ?$

(*a*) $\frac{2}{9}$ (*b*) 1

(*c*) $4\frac{1}{2}$ (*d*) $1\frac{77}{228}$

**(Central Excise 1989)**

**15.** $.6 \times .6 + .6 \div .6 = ?$

(*a*) .16 (*b*) .46

(*c*) .37 (*d*) .42

**16.** $\dfrac{(7 + 7 + 7) \div 7}{5 + 5 + 5 \div 5} = ?$

(*a*) 1 (*b*) $\frac{1}{5}$

(*c*) $\frac{15}{11}$ (*d*) $\frac{3}{11}$

**17.** $4 - 3.6 \div 4 + 0.2 \times 0.5 = ?$

(*a*) 3.2 (*b*) .2

(*c*) 1.65 (*d*) .15

**18.** $2 \text{ of } \frac{3}{4} \div \frac{3}{4} + \frac{1}{4} = ?$

(*a*) $\frac{3}{2}$ (*b*) $\frac{9}{4}$

(*c*) $\frac{5}{2}$ (*d*) $\frac{8}{3}$

**19.** $.05 \times 5 - .005 \times 5 = ?$

(*a*) 2.25 (*b*) .225

(*c*) 2.025 (*d*) .29875

**20.** $.7 \div .2 \times .5 + .5 = ?$

(*a*) 2.25 (*b*) 3.5

(*c*) 1.80 (*d*) 2.05

**21.** $.01 \times .3 \div .4 \times .5 = ?$

(*a*) .015 (*b*) .0375

(*c*) .00375 (*d*) .1

**22.** $3 \div \left[ (8-5) \div \left\{ (4-2) \div \left( 2 + \frac{8}{13} \right) \right\} \right] = ?$

(*a*) $\frac{13}{17}$ (*b*) $\frac{17}{13}$

(*c*) $\frac{68}{13}$ (*d*) $\frac{13}{68}$

**(Hotel Management 1991)**

**23.** $\dfrac{\frac{1}{4} \div \frac{1}{4} + 1\frac{1}{4}}{\frac{1}{4} \times \frac{1}{4} + 2\frac{1}{4}} = ?$

(*a*) $\frac{16}{25}$ (*b*) $\frac{32}{185}$

(*c*) $\frac{36}{185}$ (*d*) None of these

**24.** $\frac{3}{4} \div 2\frac{1}{4} \text{ of } \frac{2}{3} - \dfrac{\frac{1}{2} - \frac{1}{3}}{\frac{1}{2} + \frac{1}{3}} \times 3\frac{1}{3} + \frac{5}{6} = ?$

(*a*) $\frac{7}{18}$ (*b*) $\frac{49}{54}$

(*c*) $\frac{2}{3}$ (*d*) $\frac{1}{6}$

**25.** $\left\{7\frac{1}{2} \div \frac{1}{2} \div \frac{1}{2} \text{ of } \frac{1}{4} - \frac{2}{5} \times 2\frac{1}{3} \div 1\frac{7}{8} \text{ of } \left(1\frac{2}{5} - 1\frac{1}{3}\right)\right\} = ?$

(*a*) $3\frac{1}{5}$  (*b*) $2\frac{1}{24}$

(*c*) $4\frac{1}{30}$  (*d*) none of these

**26.** The value of $1 + \cfrac{1}{1 + \cfrac{1}{1 + \cfrac{1}{9}}}$ is :

(*a*) $\frac{29}{19}$  (*b*) $\frac{10}{19}$

(*c*) $\frac{29}{10}$  (*d*) $\frac{10}{9}$

**(Hotel Management 1991)**

**27.** $\frac{17.28 \div ?}{3.6 \times 0.2} = 2$

(*a*) 120  (*b*) 1.20

(*c*) 12  (*d*) 0.12

**(S.B.I. P.O. Exam. 1988)**

**28.** $108 \div 36 \text{ of } \frac{1}{3} + \frac{2}{5} \times 3\frac{3}{4} = ?$

(*a*) $8\frac{3}{4}$  (*b*) $6\frac{1}{4}$

(*c*) $2\frac{1}{2}$  (*d*) $10\frac{1}{2}$

**29.** $\left(4.59 \times 1.8 \div 3.6 + 5.4 \text{ of } \frac{1}{9} - \frac{1}{5}\right) = ?$

(*a*) 3.015  (*b*) 2.705

(*c*) 2.695  (*d*) none of these

**30.** $\frac{5}{6} \div \frac{6}{7} \times ? - \frac{8}{9} \div 1\frac{3}{5} + \frac{3}{4} \times 3\frac{1}{3} = 2\frac{7}{9}$.

(*a*) $\frac{7}{6}$  (*b*) $\frac{6}{7}$

(*c*) 1  (*d*) none of these

**(G.I.C.A.A.O. Exam 1988)**

**31.** $1 + 1 \div \left\{1 + 1 \div \left(1 + \frac{1}{3}\right)\right\} = ?$

(*a*) $1\frac{1}{3}$  (*b*) $1\frac{4}{7}$

(*c*) $1\frac{1}{8}$ (*d*) $1\frac{2}{3}$

**32.** $15\frac{2}{3} \times 3\frac{1}{6} = 11\frac{7}{8} + ?$

(*a*) $39\frac{5}{9}$ (*b*) $137\frac{4}{9}$

(*c*) $29\frac{7}{9}$ (*d*) none of these

**33.** $4\frac{1}{2} + 3\frac{1}{6} + ? + 2\frac{1}{3} = 13\frac{2}{5}$.

(*a*) $3\frac{2}{5}$ (*b*) $1\frac{2}{5}$

(*c*) $4\frac{1}{5}$ (*d*) $4\frac{1}{6}$

**(S.B.I. P.O. Exam. 1987)**

**34.** The value of $\dfrac{1}{3 + \dfrac{2}{2 + \dfrac{1}{2}}}$ is :

(*a*) $\frac{4}{5}$ (*b*) $\frac{5}{4}$

(*c*) $\frac{5}{19}$ (*d*) $\frac{19}{5}$ **(S.S.C. Exam. 1987)**

**35.** How many $\frac{1}{8}$'s are there in $37\frac{1}{2}$ ?

(*a*) 300 (*b*) 400

(*c*) 500 (*d*) cannot be determined

**(S.B.I.P.O. Exam. 1988)**

**36.** In a college, $\frac{1}{5}$ of the girls and $\frac{1}{8}$ of the boys took part in a social camp. What of the total number of students in the college took part in the camp ?

(*a*) $\frac{13}{40}$ (*b*) $\frac{13}{80}$

(*c*) $\frac{2}{13}$ (*d*) Data inadequate

**(S.B.I.P.O. Exam. 1988)**

**37.** A boy was asked to find $\frac{7}{9}$ of a fraction. He made a mistake of dividing the fraction by $\frac{7}{9}$ and so got an answer which exceeded the correct answer by $\frac{8}{21}$. The correct answer is

(*a*) $\frac{3}{7}$ (*b*) $\frac{7}{12}$

(*d*) $\frac{2}{21}$ (*d*) $\frac{1}{3}$

**38.** If we multiply a fraction by itself and divide the product by its reciprocal, the fraction thus obtained is $18\frac{26}{27}$. The original fraction is :

(*a*) $\frac{8}{27}$ (*b*) $2\frac{2}{3}$

(*c*) $1\frac{1}{3}$ (*d*) none of these

**(L.I.C.A.A.O. Exam. 1988)**

**39.** What fraction must be subtracted from the sum of $\frac{1}{4}$ and $\frac{1}{6}$ to have an average of $\frac{1}{12}$ of all the three fractions ?

(*a*) $\frac{1}{2}$ (*b*) $\frac{1}{3}$

(*c*) $\frac{1}{4}$ (*d*) $\frac{1}{6}$ **(S.B.I.P.O. Exam. 1988)**

**40.** The highest score in an inning was $\frac{3}{11}$ of the total and the next highest was $\frac{3}{11}$ of the remainder. If these scores differed by 9, the total score is :

(*a*) 121 (*b*) 99

(*c*) 110 (*d*) 132

**41.** In a certain office (1/ 3) of the workers are women, (1/ 2) of the women are married and (1/ 3) of the married women have children. If (3/ 4) of the men are married and (2/ 3) of the married men have children, what part of workers are without children ?

(*a*) $\frac{5}{18}$ (*b*) $\frac{4}{9}$

(*c*) $\frac{11}{18}$ (*d*) $\frac{17}{36}$ **(S.B.I.P.O. Exam. 1987)**

## HINTS & SOLUTIONS (*EXERCISE 4B*)

**1.** $5 + 5 \div 5 \times 5 = 5 + 5 \times \frac{1}{5} \times 5 = 10.$

**2.** $171 \div 19 \times 9 = 171 \times \frac{1}{19} \times 9 = 81.$

**3.** Given expression $= 5005 - \frac{5000}{10} = 5005 - 500 = 4505.$

**4.** $30 - 5 \times 2 + 3 - 8 \div 2 = 30 - 5 \times 2 + 3 - 4$

$= 30 - 10 + 3 - 4 = 19.$

**5.** $8 \div 4\ (3 - 2) \times 4 + 3 - 7 = 8 \div 4 \times 1 \times 4 + 3 - 7$

$= 2 \times 1 \times 4 + 3 - 7 = 8 + 3 - 7 = 4$

**6.** $\frac{10}{3} \times \frac{12}{5} \times \frac{x}{4} = 16 \Leftrightarrow x = \frac{16 \times 3 \times 5 \times 4}{10 \times 12} = 8.$

**7.** Given expression $= \frac{1}{7} + \left[\frac{7}{9} - \frac{5}{9} - \frac{2}{9}\right] = \frac{1}{7}.$

**8.** Given expression $= \frac{96 - 38}{42 - 13} = \frac{58}{29} = 2.$

**9.** Given expression $= \frac{\frac{1}{8} + 20}{2 + 20} = \frac{161}{8} \times \frac{1}{22} = \frac{161}{176}.$

**10.** Given expression $= \frac{\frac{1}{5} + \frac{1}{25}}{\frac{1}{25} + \frac{1}{5}} = \frac{5}{\left(\frac{1}{5}\right)} = 5 \times 5 = 25.$

**11.** Given expression $= 4 \div 2 + 2 \times 2 + 2 \times \frac{3}{2} = 2 + 4 + 3 = 9.$

**12.** Given expression $= \frac{31}{10} \times \frac{3}{10} + \frac{7}{5} \times \frac{1}{20}$

$= \frac{93}{100} + \frac{7}{100} = \frac{100}{100} = 1.$

**13.** Given expression $= \frac{16 - 12 + 3}{23 - 6} = \frac{7}{17}.$

**14.** Given expression $= \frac{15}{2} - \left[\frac{9}{4} \div \left\{\frac{5}{4} - \frac{1}{2}\left(\frac{3}{2} - \frac{1}{3} - \frac{1}{6}\right)\right\}\right]$

$= \frac{15}{2} - \left[\frac{9}{4} \div \left\{\frac{5}{4} - \frac{1}{2} \times 1\right\}\right]$

$= \frac{15}{2} - \left[\frac{9}{4} \times \frac{4}{3}\right] = \frac{9}{2} = 4\frac{1}{2}.$

**15.** $.6 \times .6 + .6 \div .6 = .6 \times .6 + .1 = .36 + .1 = .46.$

**16.** Given expression $= \dfrac{21 \div 7}{5+5+1} = \dfrac{3}{11}.$

**17.** Given expression $= 4 - 3.6 \div 4 + .2 \times .5$

$$= 4 - \frac{3.6}{4} + .2 \times .5 = 4 - .9 + .1$$

$$= 4.1 - .9 = 3.2.$$

**18.** 2 of $\dfrac{3}{4} \div \dfrac{3}{4} + \dfrac{1}{4}$

$$= \frac{3}{2} \div \frac{3}{4} + \frac{1}{4} = \frac{3}{2} \times \frac{4}{3} + \frac{1}{4} = 2 + \frac{1}{4} = \frac{9}{4}.$$

**19.** $.05 \times 5 - .005 \times 5 = .25 - .025 = .225.$

**20.** Given expression $= \dfrac{.7}{.2} \times .5 + .5 = 3.5 \times .5 + .5$

$$= 1.75 + .5 = 2.25.$$

**21.** $.01 \times .3 \div .4 \times 5 = .01 \times \dfrac{.3}{.4} \times .5 = .01 \times .75 \times .5$

$$= 0.00375.$$

**22.** Given expression $= 3 \div \left[3 \div \left\{2 \div \dfrac{34}{13}\right\}\right]$

$$= 3 \div \left[3 \div 2 \times \frac{13}{34}\right] = 3 \div \frac{3 \times 34}{26}$$

$$= \frac{3 \times 26}{3 \times 34} = \frac{13}{17}.$$

**23.** Given expression $= \dfrac{\frac{1}{4} + \frac{1}{4} \times \frac{4}{5}}{\frac{1}{16} + \frac{9}{4}} = \dfrac{\frac{1}{4} + \frac{1}{5}}{\frac{1+36}{16}}$

$$= \frac{9}{20} \times \frac{16}{37} = \frac{36}{185}.$$

**24.** Given expression $= \dfrac{3}{4} \div \dfrac{9}{4}$ of $\dfrac{2}{3} - \dfrac{\frac{1}{2} - \frac{1}{3}}{\frac{1}{2} + \frac{1}{3}} \times \dfrac{10}{3} + \dfrac{5}{6}.$

$$= \frac{3}{4} \div \frac{3}{2} - \left(\frac{1}{6} \times \frac{6}{5}\right) \times \frac{10}{3} + \frac{5}{6}$$

$$= \frac{3}{4} \times \frac{2}{3} - \frac{1}{5} \times \frac{10}{3} + \frac{5}{6}$$

$$= \frac{1}{2} - \frac{2}{3} + \frac{5}{6} = \frac{3-4+5}{6} = \frac{4}{6} = \frac{2}{3}.$$

**25.** Given expression $= \frac{15}{2} + \frac{1}{2} \div \frac{1}{8} - \frac{2}{5} \times \frac{7}{3} \div \frac{15}{8} \text{ of } \frac{1}{15}$

$$= \frac{15}{2} + \frac{1}{2} \times \frac{8}{1} - \frac{2}{5} \times \frac{7}{3} \div \frac{1}{8}$$

$$= \frac{15}{2} + \frac{1}{2} \times \frac{8}{1} - \frac{2}{5} \times \frac{7}{3} \times \frac{8}{1}$$

$$= \frac{15}{2} + 4 - \frac{112}{15} = \frac{121}{30} = 4\frac{1}{30}.$$

**26.** Given expression $= 1 + \cfrac{1}{1 + \cfrac{1}{\frac{10}{9}}} = 1 + \cfrac{1}{1 + \frac{9}{10}}$

$$= 1 + \frac{10}{19} = \frac{29}{19}.$$

**27.** Let $\frac{17.28 \div x}{3.6 \times 0.2} = 2$. Then, $\frac{17.28}{x} = 2 \times 3.6 \times 0.2$

$$\Leftrightarrow \frac{17.28}{x} = 1.44 \Leftrightarrow x = \frac{17.28}{1.44} = 12.$$

**28.** Given expression $= 108 \div 12 + \frac{2}{5} \times \frac{15}{4}$

$$= \frac{108}{12} + \frac{2}{5} \times \frac{15}{4} = 9 + \frac{3}{2} = 10\frac{1}{2}.$$

**29.** Given expression $= 4.59 \times \frac{1.8}{3.6} + .6 - .2$

$$= 2.295 + .6 - .2 = 2.695.$$

**30.** Let $\frac{5}{6} \div \frac{6}{7} \times x - \frac{8}{9} \div \frac{8}{5} + \frac{3}{4} \times \frac{10}{3} = \frac{25}{9}$. Then,

$$\frac{5}{6} \times \frac{7}{6} \times x - \frac{8}{9} \times \frac{5}{8} + \frac{3}{4} \times \frac{10}{3} = \frac{25}{9}.$$

Or $\frac{35}{36} \times x - \frac{5}{9} + \frac{5}{2} = \frac{25}{9}$

Or $\frac{35}{36} x = \frac{25}{9} + \frac{5}{9} - \frac{5}{2} = \frac{50 + 10 - 45}{18} = \frac{15}{18}.$

$\therefore \quad x = \frac{15}{18} \times \frac{36}{35} = \frac{6}{7}.$

**31.** Given expression

$$= 1 + 1 \div \left\{ 1 + 1 \div \frac{4}{3} \right\} = 1 + 1 \div \left\{ 1 + 1 \times \frac{3}{4} \right\}$$

$$= 1 + 1 \div \left\{ 1 + \frac{3}{4} \right\} = 1 + 1 \div \frac{7}{4} = 1 + 1 \times \frac{4}{7} = \frac{11}{7} = 1\frac{4}{7}.$$

**32.** Let $\frac{47}{3} \times \frac{19}{6} + \frac{19}{3} = \frac{205}{18} + x$. Then,

$$\frac{893}{18} + \frac{19}{3} = \frac{205}{18} + x \quad \text{or} \quad \frac{893 + 114}{18} = \frac{205}{18} + x.$$

$$\text{Or} \quad \frac{1007}{18} = \frac{205}{18} + x \quad \text{or} \quad x = \left( \frac{1007}{18} - \frac{205}{18} \right) = \frac{802}{18} = 44\frac{5}{9}.$$

**33.** Let $\frac{9}{2} + \frac{19}{6} + x + \frac{7}{3} = \frac{67}{5}$. Then,

$$x = \frac{67}{5} - \left( \frac{9}{2} + \frac{19}{6} + \frac{7}{3} \right) = \frac{67}{5} - 10 = \frac{17}{5} = 3\frac{2}{5}.$$

**34.** Given expression $= \dfrac{1}{3 + \frac{4}{5}} = \dfrac{5}{19}.$

**35.** $37\frac{1}{2} = \frac{75}{2} = \frac{1}{8} \times \left( \frac{75 \times 8}{2} \right) = \frac{1}{8} \times 300.$

Thus, the number of $\frac{1}{8}$'s in $37\frac{1}{2}$ is 300.

**36.** Out of 5 girls, 1 took part in the camp and out of 8 boys, 1 took part in the camp. Thus out of 13 students, 2 took part in the camp. *i.e.* $\frac{2}{13}$ of total students joined the camp.

**37.** Let $x$ be the fraction. Then,

$$\frac{9}{7}x - \frac{7}{9}x = \frac{8}{21} \quad \text{or} \quad \frac{32x}{63} = \frac{8}{21}$$

$$\text{Or} \quad x = \frac{8 \times 63}{21 \times 32} = \frac{3}{4}.$$

$\therefore$ Correct answer $= \frac{7}{9}$ of $\frac{3}{4} = \frac{7}{12}$.

**38.** Let the fraction be $\left( \frac{a}{b} \right)$. Then, $\left( \frac{a}{b} \times \frac{a}{b} \right) \div \frac{b}{a} = 18\frac{26}{27} = \frac{512}{27}.$

$$\text{Or} \quad \frac{a}{b} \times \frac{a}{b} \times \frac{a}{b} = \frac{512}{27} \quad \text{or} \quad \left( \frac{a}{b} \right)^3 = \left( \frac{8}{3} \right)^3$$

$\therefore \ \frac{a}{b}=\frac{8}{3}=2\frac{2}{3}$.

**39.** Let $\frac{1}{4}+\frac{1}{6}-x=3\times\frac{1}{12}$. Then,

$\frac{1}{4}+\frac{1}{6}-x=\frac{1}{4}$ or $x=\frac{1}{6}$.

**40.** Let the total score be $x$.

Then, the highest score is $\frac{3x}{11}$.

Remainder $=\left(x-\frac{3x}{11}\right)=\frac{8x}{11}$.

Next highest score $=\frac{3}{11}$ of $\frac{8x}{11}=\frac{24x}{121}$.

Now, $\frac{3x}{11}-\frac{24x}{121}=9 \Rightarrow \frac{9x}{121}=9$ or $x=121$.

**41.** Let the total number of workers be $x$.

Then, number of women $=\frac{1}{3}x$;

And, number of men $=\frac{2}{3}x$.

Number of women having children $=\frac{1}{3}$ of $\frac{1}{2}$ of $\frac{1}{3}x=\frac{1}{18}x$.

Number of men having children $=\frac{2}{3}$ of $\frac{3}{4}$ of $\frac{2}{3}x=\frac{1}{3}x$.

Number of workers having children $=\frac{1}{18}x+\frac{1}{3}x=\frac{7}{18}x$.

Number of workers having no children $=\left(x-\frac{7}{18}x\right)=\frac{11}{18}x$.

## ANSWERS (*EXERCISE 4 B*)

| | | | | | | |
|---|---|---|---|---|---|---|
| **1.** (*b*) | **2.** (*c*) | **3.** (*d*) | **4.** (*b*) | **5.** (*c*) | **6.** (*c*) | **7.** (*d*) |
| **8.** (*a*) | **9.** (*c*) | **10.** (*d*) | **11.** (*a*) | **12.** (*b*) | **13.** (*a*) | **14.** (*c*) |
| **15.** (*b*) | **16.** (*d*) | **17.** (*a*) | **18.** (*b*) | **19.** (*b*) | **20.** (*a*) | **21.** (*c*) |
| **22.** (*a*) | **23.** (*c*) | **24.** (*c*) | **25.** (*c*) | **26.** (*a*) | **27.** (*c*) | **28.** (*d*) |
| **29.** (*c*) | **30.** (*b*) | **31.** (*b*) | **32.** (*d*) | **33.** (*a*) | **34.** (*c*) | **35.** (*a*) |
| **36.** (*c*) | **37.** (*b*) | **38.** (*b*) | **39.** (*d*) | **40.** (*a*) | **41.** (*c*) | |

# 5

# SQUARE ROOT & CUBE ROOT

**Square Root.** *Square root of a given number is that number, the product of which by itself is equal to the given number.*

We use radical sign $\sqrt{x}$ to denote square root of a number $x$.

Thus, $\sqrt{9} = 3$ ; $\sqrt{25} = 5$ ; $\sqrt{144} = 12$, $\sqrt{.16} = 0.4$.

**Square Root by Means of Factors.** *Resolve the given number into prime factors and take the product of prime factors choosing one out of every pair.*

**Ex. 1.** ***Find the square root of 1156.***

**Sol.** Resolving 1156 into prime factors, we get

| | |
|---|---|
| 2 | 1156 |
| 2 | 578 |
| 17 | 289 |
| 17 | 17 |
| | 1 |

$1156 = 2 \times 2 \times 17 \times 17 = 2^2 \times 17^2$.

$\therefore \quad \sqrt{1156} = (2 \times 17) = 34$.

**Ex. 2.** ***By what least number should 1440 be multiplied or divided in order to make it a perfect square ? Find these perfect square numbers.***

**Sol.** Resolving 1440 into prime factors, we have

| | |
|---|---|
| 2 | 1440 |
| 2 | 720 |
| 2 | 360 |
| 2 | 180 |
| 2 | 90 |
| 3 | 45 |
| 3 | 15 |
| 5 | 5 |
| | 1 |

$1440 = 2 \times 2 \times 2 \times 2 \times 2 \times 3 \times 3 \times 5$

$= 2^2 \times 2^2 \times 2 \times 3^2 \times 5$.

So, 1440 must be multiplied or divided by $2 \times 5$ *i.e.* 10 to make it a perfect square.

$\therefore$ *Perfect square numbers* $= 1440 \times 10$ & $1440 \div 10$

$= 14400$ & $144$.

**Ex. 3.** ***Find the least square number which is exactly divisible by 16, 36, 44 and 60.***

**Sol.** The least number exactly divisible by each one of the given numbers is their L.C.M.

| | |
|---|---|
| 4 | 16 – 36 – 44 – 60 |
| 3 | 4 – 9 – 11 – 15 |
| | 4 – 3 – 11 – 5 |

Now, *L.C.M. of given numbers*

$= (4 \times 3 \times 4 \times 3 \times 11 \times 5) = 7920$.

Also, $7920 = (2^2 \times 3^2 \times 2^2 \times 11 \times 5)$.

To make it a perfect square it must be multiplied by $11 \times 5$ or 55.

$\therefore$ *Required number* $= (7920 \times 55) = 435600$.

**General Method.** *In given number*, mark *off the digits in pairs* (each pair is called a period), from the right and then find the square root as shown in the examples given below.

**Ex. 1.** ***Find the square root of 134689.*** **(L.I.C. 1991)**

**Sol.**

| | |
|---|---|
| 3 | $\overline{13}\,\overline{46}\,\overline{89}$ ( 367 |
| | 9 |
| 66 | 446 |
| | 396 |
| 727 | 5089 |
| | 5089 |
| | × |

$\therefore \quad \sqrt{134689} = 367.$

**Ex. 2.** ***Find the square root of 1734489.***

**Sol.**

| | |
|---|---|
| 1 | $\overline{1}\,\overline{73}\,\overline{44}\,\overline{89}$ ( 1 3 1 7 |
| | 1 |
| 23 | 73 |
| | 69 |
| 261 | 444 |
| | 261 |
| 2627 | 18389 |
| | 18389 |
| | × |

$\therefore \quad \sqrt{1734489} = 1317.$

**Square Root of Decimal Fractions.** In the given number, make even number of decimal places by affixing a zero, if necessary. Mark off the periods and extract the square roots as shown below.

**Ex. 3.** ***Find the square root of 176.252176.***

**Sol.**

| | |
|---|---|
| 1 | $\overline{1}\,\overline{76}\,.\,\overline{25}\,\overline{21}\,\overline{76}$ ( 13.276 |
| | 1 |
| 23 | 76 |
| | 69 |
| 262 | 725 |
| | 524 |
| 2647 | 20121 |
| | 18529 |
| 26546 | 159276 |
| | 159276 |
| | × |

$\therefore \quad \sqrt{176.252176} = 13.276.$

**Ex. 4.** ***Extract the square root of 0.56423 upto four places of decimal.***

**Sol.** In the given number, the number of decimal places is odd. We make it even, by affixing a zero at the end. Mark off periods and find the square root as given below.

| | |
|---|---|
| 7 | $0.\overline{56}\,\overline{42}\,\overline{30}$ ( .7511 |
| | 49 |
| 145 | 742 |
| | 725 |
| 1501 | 1730 |
| | 1501 |
| 15021 | 22900 |
| | 15021 |

$\therefore \quad \sqrt{0.564230} = 0.7511.$

**Ex. 5.** ***Find the value of*** $\sqrt{0.4}$ ***upto three places of decimal.***

**Sol.** 0.4 = 0.40.

$\therefore$ Its square root can be obtained as under.

| | |
|---|---|
| 6 | $0.\overline{40}$ ( .632 |
| | 36 |
| 123 | 400 |
| | 369 |
| 1262 | 3100 |
| | 2524 |

$\therefore \quad \sqrt{0.4} = 0.632.$

**Ex. 6.** ***Find the square root of 0.00059049.***

**Sol.**

| | |
|---|---|
| 2 | $0.\overline{00}\,\overline{05}\,\overline{90}\,\overline{49}$ ( .0243 |
| | 4 |
| 44 | 190 |
| | 176 |
| 483 | 1449 |
| | 1449 |
| | × |

$\sqrt{0.00059049} = 0.0243.$

**Square Root of Vulgar Fraction :**

(i) If $a$ and $b$ are perfect squares, then

$$\sqrt{\frac{a}{b}} = \frac{\sqrt{a}}{\sqrt{b}}.$$

(ii) If $b$ is not a perfect square, then

$$\sqrt{\frac{a}{b}} = \sqrt{\frac{a}{b} \times \frac{b}{b}} = \sqrt{\frac{ab}{b^2}} = \frac{\sqrt{ab}}{b}.$$

**Ex. 7.** ***Evaluate upto three places of decimals :***

**(i)** $\sqrt{\dfrac{2}{7}}$ **(ii)** $\sqrt{3\dfrac{5}{12}}$ **(iii)** $\dfrac{1}{\sqrt{5}}$.

**Sol. (i)** $\sqrt{\dfrac{2}{7}} = \sqrt{\dfrac{2 \times 7}{7 \times 7}} = \dfrac{\sqrt{14}}{\sqrt{7 \times 7}} = \dfrac{\sqrt{14}}{7}$.

| | |
|---|---|
| 3 | 14 ( 3.741 |
| | 9 |
| 67 | 500 |
| | 469 |
| 744 | 3100 |
| | 2976 |
| 7481 | 12400 |
| | 7481 |

$\therefore \quad \dfrac{\sqrt{14}}{7} = \dfrac{3.741}{7} = 0.534.$

**(ii)** $\sqrt{3\dfrac{5}{12}} = \sqrt{\dfrac{41}{12}} = \sqrt{\dfrac{41 \times 12}{12 \times 12}} = \dfrac{\sqrt{492}}{12}$

$= \dfrac{22.181}{12} = 1.848. \quad [\because \sqrt{492} = 22.181]$

**(iii)** $\dfrac{1}{\sqrt{5}} = \dfrac{1}{\sqrt{5}} \times \dfrac{\sqrt{5}}{\sqrt{5}} = \dfrac{\sqrt{5}}{5}$.

We find the value of $\sqrt{5}$ as under

| | |
|---|---|
| 2 | 5 ( 2.235 |
| | 4 |
| 42 | 100 |
| | 84 |
| 443 | 1600 |
| | 1329 |
| 4465 | 27100 |
| | 22325 |

$\therefore \quad \sqrt{5} = 2.235.$

Hence, $\dfrac{1}{\sqrt{5}} = \dfrac{\sqrt{5}}{5} = \dfrac{2.235}{5} = 0.447.$

**Ex. 8.** ***Find the value of*** $\left(\dfrac{\sqrt{5}+\sqrt{3}}{\sqrt{5}-\sqrt{3}}\right)$.

**Sol.** $\left(\dfrac{\sqrt{5}+\sqrt{3}}{\sqrt{5}-\sqrt{3}}\right) = \dfrac{(\sqrt{5}+\sqrt{3})}{(\sqrt{5}-\sqrt{3})} \times \dfrac{(\sqrt{5}+\sqrt{3})}{(\sqrt{5}+\sqrt{3})} = \dfrac{(\sqrt{5}+\sqrt{3})^2}{[(\sqrt{5})^2 - (\sqrt{3})^2]}$

$= \dfrac{5+3+2\sqrt{15}}{(5-3)} = \dfrac{8+2\sqrt{15}}{2}$

$$= 4 + \sqrt{15} = 4 + 3.8729 = 7.8729.$$

$$[\because \sqrt{15} = 3.8729]$$

**Ex. 9.** ***Find the square root of*** $\left(\dfrac{\sqrt{3}+\sqrt{2}}{\sqrt{3}-\sqrt{2}}\right)$.

**Sol.** $\left(\dfrac{\sqrt{3}+\sqrt{2}}{\sqrt{3}-\sqrt{2}}\right) = \dfrac{(\sqrt{3}+\sqrt{2})}{(\sqrt{3}-\sqrt{2})} \times \dfrac{(\sqrt{3}+\sqrt{2})}{(\sqrt{3}+\sqrt{2})} = \dfrac{(\sqrt{3}+\sqrt{2})^2}{(3-2)} = (\sqrt{3}+\sqrt{2})^2$

$$\therefore \sqrt{\left(\frac{\sqrt{3}+\sqrt{2}}{\sqrt{3}-\sqrt{2}}\right)} = (\sqrt{3}+\sqrt{2}) = 1.732 + 1.414 = 3.146.$$

[The values of $\sqrt{3}$ and $\sqrt{2}$ may be evaluated as $\sqrt{3} = 1.732$ and $\sqrt{2} = 1.414$.]

**Ex. 10.** ***Find the least number of six digits which is a perfect square.***

**Sol.** The least number of six digits is 100000.

| | |
|---|---|
| 3 | $\overline{10}\,\overline{00}\,\overline{00}$ (316 |
| | 9 |
| 61 | 100 |
| | 61 |
| 626 | 3900 |
| | 3756 |
| | 144 |

Thus, $(316)^2 < 100000 < (317)^2$.

$\therefore$ *The least number of six digits which is a perfect square is* $(317)^2 = 100489$.

**Ex. 11.** ***Find the greatest number of four digits, which is a perfect square.***

**Sol.** The greatest number of 4 digits = 9999.

Extracting the square root of 9999 we find that :

| | |
|---|---|
| 9 | 99 99 (99 |
| | 81 |
| 189 | 18 99 |
| | 17 01 |
| | 198 |

$\therefore$ $(99)^2 < 9999$ by 198.

*Hence, the required number* = $9999 - 198 = 9801$ *or* $(99)^2$.

**Ex. 12.** ***What least number must be added to 594 to make the sum a perfect square.***

**Sol.** While extracting the square root of 594, we observe that

$(24)^2 < 594 < (25)^2$.

| | | |
|---|---|---|
| 2 | $\overline{5}\,\overline{94}$ | (24 |
| | 4 | |
| 44 | 194 | |
| | 176 | |
| | 18 | |

$\therefore$ *The number to be added* $= (25)^2 - 594 = 625 - 594 = 31.$

**Ex. 13.** ***Find the cost of errecting a fence round a square field of area 10 hectares at 35 paise per metre.***

**Sol.** Area of the field = $(10 \times 10000)$ sq. metres.

$\therefore$ Side of the field = $\sqrt{100000}$ metres = 316.227 metres.

So, the perimeter of the field = $316.227 \times 4 = 1264.908$ metres.

*Hence, the cost of fencing* $= Rs. \dfrac{1264.908 \times 35}{100} = Rs. 442.72.$

**Ex. 14.** ***A general wishing to daw up his 64025 men in the form of a solid square, found that he had 16 men over. Find the number of men in the front row.***

**Sol.** Number of men arranged in a solid square

$$= (64025 - 16) = 64009.$$

$\therefore$ *Number of men in the front row* $= \sqrt{64009} = 253.$

**Cube Root.** *Cube root of a given number x is the number whose cube is x. The cube root of x is written as* $\sqrt[3]{x}$.

Thus, $\sqrt[3]{8} = \sqrt[3]{2 \times 2 \times 2} = 2$ and $\sqrt[3]{27} = \sqrt[3]{3 \times 3 \times 3} = 3.$

**Rule.** Resovle the given number into prime factors and take the product of prime numbers choosing one each from three of the same type.

**Ex. 15.** ***Find the cube root of 9261.***

**Sol.** Resolving 9261 into prime factors, we get

| | |
|---|---|
| 3 | 9261 |
| 3 | 3087 |
| 3 | 1029 |
| 7 | 343 |
| 7 | 49 |
| | 7 |

$$9261 = 3 \times 3 \times 3 \times 7 \times 7 \times 7$$

$$\therefore \quad \sqrt[3]{9261} = \sqrt[3]{(3 \times 3 \times 3 \times 7 \times 7 \times 7)}$$

$$= 3 \times 7 = 21.$$

**Ex. 16.** ***By what least number must 21600 be multiplied to make it a perfect cube ?***

***Also find the perfect cube number & its cube-root.***

**Sol.** Resolving 21600 into prime factors, we get :

$$21600 = 2 \times 2 \times 2 \times 2 \times 2 \times 3 \times 3 \times 3 \times 5 \times 5$$

$$= 2^3 \times 2^2 \times 3^3 \times 5^2.$$

Clearly, the number with which 21600 be multiplied to make a perfect cube is $2 \times 5 = 10$.

| | |
|---|---|
| 2 | 21600 |
| 2 | 10800 |
| 2 | 5400 |
| 2 | 2700 |
| 2 | 1350 |
| 3 | 675 |
| 3 | 225 |
| 3 | 75 |
| 5 | 25 |
| | 5 |

Also the perfect cube number

$= 21600 \times 10 = 216000 = 2^3 \times 2^3 \times 3^3 \times 5^3$.

*Cube root of this number* $= 2 \times 2 \times 3 \times 5 = 60$.

## EXERCISE 5 A

**1.** Find the square root by factorization method :

(*i*) 1225 (*ii*) 1296 (*iii*) 1764

(*iv*) 4096 (*v*) 7744 (*vi*) 9216

(*vii*) 18496 (*viii*) 11025

**2.** Find the square root of :

(*i*) 286225 (*ii*) 298116 (*iii*) 226576

(*iii*) 1471369 (*v*) 63409369 (*vi*) 78765625

(*vii*) 100489 (*viii*) 1000014129

**3.** Find the square root of :

(*i*) 432.64 (*ii*) 133.6336 (*iii*) .327184

(*iv*) .00367236 (*v*) .00038809 (*vi*) .00099856

**4.** Find the square root of each of the following upto three decimal places :

(*i*) 427 (*ii*) 7536 (*iii*) 0.5

(*iv*) 0.121 (*v*) 321.73025 (*vi*) 0.000945

**5.** Find the least number which is a perfect square and which is divisible by each of the numbers 9, 12 and 32.

**Hint.** L.C.M. of 9, 12, 32 is $3^2 \times 4^2 \times 2 = 288$.

$\therefore$ Required number $= (3^2 \times 4^2 \times 2^2)$.

**6.** What must be subtracted from 16160 to get a perfect square number ? What is this perfect square number ? Also, find is square root.

**Hint.** While extracting the square root of 16160, we find that $(127)^2 > 16160$ by 31. Hence 31 must be subtracted from given number.

**7.** What must be added to 2582415 to make the sum a perfect square ? What is this perfect square number ? Also, find its square root.

**Hint.** We find that $(1606)^2 < 2582415 < (1607)^2$.

$\therefore$ Number to be added $= (1607)^2 - 2582415 = 34$.

**8.** Find the least number of four digits which is a perfect square.

**9.** Find the greatest number of four digits which is a perfect square.

**10.** A certain number of persons agree to subscribe as many paise as there are subscribers. The whole subscription being Rs. 25824.49, find the number of subscribers.

**11.** Find the value of each of the following upto four decimal places :

(i) $\frac{1}{\sqrt{2}}$ (ii) $\frac{1}{\sqrt{8}}$ (iii) $\sqrt{\frac{2}{3}}$ (iv) $\sqrt{\frac{7}{8}}$

**12.** Find the value of $\frac{7}{3+\sqrt{2}}$ upto four places or decimal.

**Hint.** $\frac{7}{3+\sqrt{2}} = \frac{7}{(3+\sqrt{2})} \times \frac{(3-\sqrt{2})}{(3-\sqrt{2})} = \frac{7\times(3-\sqrt{2})}{(9-2)} = (3-\sqrt{2})$.

**13.** Simplify : $\left(\frac{\sqrt{5}+\sqrt{3}}{\sqrt{5}-\sqrt{3}} + \frac{\sqrt{5}-\sqrt{3}}{\sqrt{5}+\sqrt{3}}\right)$. **(L.I.C. 1991)**

**Hint.** Given expression $= \frac{(\sqrt{5}+\sqrt{3})^2 + (\sqrt{5}-\sqrt{3})^2}{(\sqrt{5}-\sqrt{3})(\sqrt{5}+\sqrt{3})}$

$= \frac{2(5+3)}{(5-3)} = 8.$

**14.** Find the square root of $\left(\frac{\sqrt{2}-1}{\sqrt{2}+1}\right)$ upto four decimal places.

**15.** Find the least square number which is exactly divisible by 8, 12, 15 and 20. **(Clerk's Grade 1991)**

**16.** Find the value of $\left(\frac{1}{\sqrt{9}-\sqrt{8}}\right)$ upto 3 places of decimal. **(C.B.I. 1990)**

**Hint.** $\frac{1}{\sqrt{9}-\sqrt{8}} = \frac{1}{\sqrt{9}-\sqrt{8}} \times \frac{\sqrt{9}+\sqrt{8}}{\sqrt{9}+\sqrt{8}} = (\sqrt{9}+\sqrt{8}) = (3+2\sqrt{2})$.

**17.** Which is greater 1.1375 or $\sqrt{\frac{9}{7}}$ and by how much ?

**18.** Find the value of $\sqrt{8+2\sqrt{15}}$ upto 3 decimal places.

**Hint.** $8+2\sqrt{15} = (5+3+2\sqrt{5\times3}) = (\sqrt{5}+\sqrt{3})^2$.

**19.** Find the number whose square is equal to the difference of the squares of 75.15 and 60.12.

**Hint.** Required number $= \sqrt{(75.15)^2 - (60.12)^2}$.

**20.** If the sum of two numbers be multiplied by each separately, the products so obtained are 2418 and 3666. Find the numbers.

**Hint.** Let the numbers be $a$ and $b$.

Then $a(a+b) = 2418$ & $b(a+b) = 3666$

Adding we get, $(a+b)^2 = 6084$ or $a+b = \sqrt{6084} = 78.$

Subtracting, we get, $a^2 - b^2 = 1248.$

$\therefore$ $(a+b)(a-b) = 1248$ or $(a-b) = 16.$

Solve $a+b=78$ and $a-b=16$ for $a$ and $b$.

**21.** Find the value of $\sqrt[3]{\left(1-\frac{91}{216}\right)}$. **(Clerk's Grade 1991)**

**22.** Find the cube root of :

(*i*) 512 (*ii*) 2197 (*iii*) 0.000125

**23.** By what smallest number must 21600 be divided to make it a perfect cube ?

**24.** The volume of a cubical box is 32.768 cu. metres. Find the length of each size of the box.

**Hint.** $\sqrt[3]{32.768} = \sqrt[3]{\frac{32768}{1000}} = \frac{\sqrt[3]{32768}}{10} = \frac{32}{10} = 3.2.$

**25.** Fill in the blanks :

(*i*) The least number by which 176 must be multiplied to make the result a perect square is (.......).

(*ii*) The least number which must be added to 6156 to make it a perfect square is (..........).

(*iii*) The least number which must be subtracted from 1200 to make the result a perfect square is (........).

(*iv*) The least square number divisible by each one of 8, 9 and 10 is (.........).

## ANSWERS (*EXERCISE 5 A*)

**1.** (*i*) 35 (*ii*) 36 (*iii*) 42 (*iv*) 64
(*v*) 88 (*vi*) 96 (*vii*) 136 (*viii*) 105

**2.** (*i*) 535 (*ii*) 546 (*iii*) 476 (*iv*) 1213
(*v*) 7963 (*vi*) 8875 (*vii*) 317 (*viii*) 31623

**3.** (*i*) 20.8 (*ii*) 11.56 (*iii*) 0.572 (*iv*) 0.0606
(*v*) 0.0197 (*vi*) 0.0316

**4.** (*i*) 20.663 (*ii*) 86.811 (*iii*) 0.707 (*iv*) 0.348
(*v*) 17.936 (*vi*) 0.030

**5.** 576 **6.** 31, 16129, 127 **7.** 34, 2582449, 1607

**8.** 1024 **9.** 9801 **10.** 1607

**11.** (*i*) 0.7071 (*ii*) 0.353 (*iii*) 0.816 (*iv*) 0.935

**12.** 1.5858 **13.** 8 **14.** 0.4142 **15.** 3600

**16.** 5.828 **17.** 1.1375, 0.0037 **18.** 3.968 **19.** 45.09

**20.** 47, 31 **21.** $\frac{5}{6}$ **22.** (*i*) 8 (*ii*) 13 (*iii*) 0.05

**23.** 100. **24.** 3.2

**25.** (*i*) 11 (*ii*) 85 (*iii*) 44 (*iv*) 3600.

## EXERCISE 5 B
## (*OBJECTIVE TYPE QUESTIONS*)

**1.** $\sqrt{64009} = ?$

(*a*) 803 (*b*) 363

(*c*) 253 (*d*) 347

**2.** $\sqrt{67621} = ?$

(*a*) 320.04 (*b*) 260.04

(*c*) 260.40 (*d*) 280.04

**3.** $\dfrac{\sqrt{3125}}{\sqrt{5}} = ?$

(*a*) 24.25 (*b*) 35

(*c*) 25 (*d*) 27.25

**4.** $\dfrac{?}{\sqrt{64}} = 8$

(*a*) 1 (*b*) 64

(*c*) $\sqrt{8}$ (*d*) none of these

**5.** $\dfrac{\sqrt{288}}{\sqrt{128}} = ?$

(*a*) $\dfrac{\sqrt{3}}{2}$ (*b*) $\dfrac{3}{\sqrt{2}}$

(*c*) $\dfrac{3}{2}$ (*d*) $\sqrt{\dfrac{3}{2}}$

**6.** $\sqrt{5} \times \sqrt{500} = ?$

(*a*) 46.95 (*b*) 43.75

(*c*) 50.25 (*d*) 50

**7.** $\dfrac{34}{\sqrt{289}} \times \dfrac{\sqrt{196}}{70} = ?$

(*a*) $\dfrac{16}{35}$ (*b*) $\dfrac{2}{5}$

(*c*) $\dfrac{5}{2}$ (*d*) $\dfrac{1}{5}$

**8.** If $\dfrac{250}{\sqrt{x}} = 10$, then the value of $x$ is :

(*a*) 25 (*b*) 250

(*c*) 625 (*d*) 2500

**9.** $\dfrac{112}{\sqrt{196}} \times \dfrac{\sqrt{576}}{12} \times \dfrac{\sqrt{256}}{8} = ?$

(*a*) 8 (*b*) 16

(*c*) 32 (*d*) 12

**10.** $\frac{1872}{\sqrt{?}} = 234$

(*a*) 324 (*b*) 64

(*c*) 8 (*d*) 256

**11.** If $\sqrt{\frac{x}{144}} = \frac{21}{36}$, then the value of $x$ is :

(*a*) 1296 (*b*) 441

(*c*) 196 (*d*) 49 **(Bank Clerical 1991)**

**12.** $\sqrt{248 + \sqrt{52 + \sqrt{144}}} = ?$

(*a*) 14 (*b*) 16

(*c*) 18.8 (*d*) 16.6

**13.** $\sqrt{\frac{?}{3136}} = \frac{28}{56}$

(*a*) 1568 (*b*) 28

(*c*) 784 (*d*) 56

**14.** $\frac{\sqrt{?}}{200} = 0.02$

(*a*) 0.4 (*b*) 4

(*c*) 16 (*d*) 1.6 **(Bank Clerical 1990)**

**15.** $\sqrt{?} - 3 = 19$

(*a*) 529 (*b*) 22

(*c*) 441 (*d*) none of these

**16.** $\sqrt{\frac{1694}{?}} + 14 = 25$

(*a*) 14 (*b*) 12

(*c*) 22 (*d*) 11

**17.** $\sqrt{\frac{25}{15625}} = \sqrt{\frac{?}{30625}}$

(*a*) 2 (*b*) 35

(*c*) 49 (*d*) 1225

**18.** The value of $\sqrt{0.9}$ is

(*a*) 0.3 (*b*) 0.03

(*c*) 0.94 (*d*) 0.33

**19.** $\sqrt{.009} = ?$

(*a*) .03 (*b*) 0.3

(*c*) .094 (*d*) none of these

**20.** $\frac{?}{20} = \frac{45}{?}$

(*a*) 25 (*b*) 30

(*c*) 35 (*d*) none of these

21. $\sqrt{4\frac{2}{3}} = ?$

(*a*) 2.34 (*b*) 1.967

(*c*) 2.16 (*d*) 2.061

22. $\sqrt{\frac{36.1}{102.4}} = ?$

(*a*) $\frac{19}{28}$ (*b*) $\frac{19}{32}$

(*c*) $\frac{2.9}{32}$ (*d*) none of these

23. $\frac{\sqrt{7.84}}{\sqrt{1.96}} = ?$

(*a*) 2 (*b*) 4

(*c*) 5.6 (*d*) 20

24. $\sqrt{0.121} = ?$

(*a*) .11 (*b*) 1.1

(*c*) .347 (*d*) none of these

25. $\sqrt{0.064} = ?$

(*a*) 0.8 (*b*) 0.08

(*c*) 0.008 (*d*) 0.252

26. $\sqrt{\frac{0.16}{0.4}} = ?$

(*a*) 0.2 (*b*) 0.02

(*c*) 0.63 (*d*) $\frac{2\sqrt{5}}{5}$

27. $\sqrt{0.0009} + \sqrt{0.01} = ?$

(*a*) 3 (*b*) 0.3

(*c*) $\frac{1}{3}$ (*d*) none of these

**(Bank Clerical 1991)**

28. $\sqrt{0.01 + \sqrt{0.0064}} = ?$

(*a*) 0.3 (*b*) 0.03

(*c*) $\sqrt{0.18}$ (*d*) none of these

**(Bank Clerical 1991)**

29. $\frac{\sqrt{?}}{.25} = 100$

(*a*) 25 (*b*) 625

(*c*) 3125 (*d*) 5

30. $\sqrt{(75.24 + ?)} = 8.71$

(*a*) 0.6241 (*b*) 6.241
(*c*) 62.41 (*d*) none of these

**31.** $\frac{1}{\sqrt{3}} = ?$

(*a*) .377 (*b*) .477
(*c*) .577 (*d*) .677

**32.** $\sqrt{\frac{2}{9}} = ?$

(*a*) .232 (*b*) .317
(*c*) .471 (*d*) .417

**33.** $\sqrt{\frac{7}{8}} = ?$

(*a*) .715 (*b*) .805
(*c*) .935 (*d*) .917

**34.** If $\sqrt{\left(1 + \frac{27}{169}\right)} = \left(1 + \frac{x}{13}\right)$, then $x$ equals :

(*a*) 1 (*b*) 3
(*c*) 5 (*d*) 7

**35.** $\frac{\sqrt{32} + \sqrt{48}}{\sqrt{8} + \sqrt{12}} = ?$

(*a*) $\sqrt{2}$ (*b*) 2
(*c*) 4 (*d*) 8

**36.** $\frac{7}{3 + \sqrt{2}} = ?$

(*a*) 1.5858 (*b*) 1.6136
(*c*) 1.7316 (*d*) 1.5947

**37.** If $\sqrt{2401} = \sqrt{(7)^x}$, then the value of $x$ is :

(*a*) 2 (*b*) 3
(*c*) 4 (*d*) 5

**38.** If $\sqrt{15625} = 125$, then the value of
$\sqrt{15625} + \sqrt{156.25} + \sqrt{1.5625} = ?$

(*a*) 1.3875 (*b*) 13.875
(*c*) 138.75 (*d*) 156.25

**(Assistant Grade 1990)**

**39.** $\left(\frac{4 + \sqrt{2}}{\sqrt{2} + 1}\right) = ?$

(*a*) 2.242 (*b*) 2.4136
(*c*) 2.3216 (*d*) 2.4136

**40.** $\sqrt{2\sqrt{2\sqrt{2\sqrt{2\sqrt{2}}}}} = ?$

(*a*) 0 (*b*) 1
(*c*) 2 (*d*) $2^{31/32}$

**41.** If $\sqrt{2^n} = 64$, then the value of *n* is :

(*a*) 2 (*b*) 4
(*c*) 6 (*d*) 12 **(Assistant Grade 1990)**

**42.** $\sqrt{\dfrac{\sqrt{2}-1}{\sqrt{2}+1}} = ?$

(*a*) 0.732 (*b*) 1.3142
(*c*) 0.4142 (*d*) 0.3652

**43.** If $\sqrt{0.04 \times 0.4 \times a} = 0.4 \times 0.04 \times \sqrt{b}$, then the value of $\dfrac{a}{b}$ is :

(*a*) 0.016 (*b*) 1.6
(*c*) 0.16 (*d*) none of these

**44.** $\dfrac{\sqrt{2}}{2+\sqrt{2}} = ?$

(*a*) .4142 (*b*) .2071
(*c*) .4713 (*d*) .828

**45.** The value of $\sqrt{\dfrac{1.21 \times 0.9}{1.1 \times 0.11}}$ is

(*a*) 2 (*b*) 3
(*c*) 9 (*d*) 11 **(C.D.S. 1991)**

**46.** The value of $(\sqrt{8})^{1/3}$ is

(*a*) 2 (*b*) 4
(*c*) $\sqrt{2}$ (*d*) 8 **(C.D.S. 1991)**

**47.** The product of two numbers is 120 and the sum of their squares is 289. The sum of the two numbers is :

(*a*) 23 (*b*) $\sqrt{409}$
(*c*) 20 (*d*) 169 **(Bank Clerical 1991)**

**48.** The sum of squares of two numbers is 80 and the square of their difference is 36. The product of the two numbers is

(*a*) 22 (*b*) 44
(*c*) 58 (*d*) 116

**49.** The largest number of five digits, which is a perfect square is :

(*a*) 99999 (*b*) 99764
(*c*) 99976 (*d*) 99856

**50.** The smallest number of four digits, which is a perfect square is :

(*a*) 1000 (*b*) 1016
(*c*) 1024 (*d*) 1036

51. What smallest number must be added to 269 to make it a perfect square :

(a) 31 (b) 16
(c) 7 (d) 20

52. What least number must be subtracted from 16800 to make it a perfect square :

(a) 249 (b) 159
(c) 169 (d) 219

53. $\sqrt{.00059049} = ?$

(a) .243 (b) .00243
(c) .0243 (d) .000243 **(Railways 1988)**

54. The least number by which 176 should be multiplied to make the result a perfect square is :

(a) 10 (b) 9
(c) 11 (d) 8

55. The least number by which 216 must be divided to make the result a perfect square is :

(a) 6 (b) 4
(c) 9 (d) 3

56. The cube root of .000027 is :

(a) .3 (b) .03
(c) .003 (d) none of these

57. $\sqrt[3]{4\frac{12}{125}} = ?$

(a) $1\frac{3}{5}$ (b) $1\frac{2}{5}$
(c) $2\frac{2}{5}$ (d) $1\frac{4}{5}$

58. $\sqrt[3]{1 - \frac{91}{216}} = ?$

(a) $1 - \frac{5}{6}$ (b) $\frac{5}{6}$
(c) $1 - \frac{\sqrt[3]{91}}{6}$ (d) none of these

59. By what least number must 21600 be multiplied to make it a perfect cube ?

(a) 6 (b) 10
(c) 30 (d) 60

60. What is the smallest number by which 3600 be divided to make it a perfect cube ?

(a) 9 (b) 50
(c) 300 (d) 450

61. The length of diagonal of a square is 8 cm. The length of the side of the square is :

(*a*) 2 cm (*b*) 2.8 cm
(*c*) 1.414 cm (*d*) 5.64 cm

62. $\frac{\sqrt{324}}{1.5} = \frac{?}{\sqrt{256}}$

(*a*) 192 (*b*) 432
(*c*) 288 (*d*) 122 **(Bank P.O. 1991)**

63. $\frac{1}{\sqrt{9} - \sqrt{8}}$ is equal to :

(*a*) $\frac{1}{2}(3 - 2\sqrt{2})$ (*b*) $\frac{1}{3 + 2\sqrt{2}}$
(*c*) $(3 - 2\sqrt{2})$ (*d*) $(3 + 2\sqrt{2})$ **(C.B.I. 1990)**

64. $\sqrt{\frac{0.324 \times 0.081 \times 4.624}{1.5625 \times 0.0289 \times 72.9 \times 64}} = ?$

(*a*) 24 (*b*) 2.40
(*c*) 0.024 (*d*) none of these

**(Assistant Grade 1990)**

65. If $a = \frac{\sqrt{5} + 1}{\sqrt{5} - 1}$ and $b = \frac{\sqrt{5} - 1}{\sqrt{5} + 1}$, then the value of $\left(\frac{a^2 + ab + b^2}{a^2 - ab + b^2}\right)$ is

(*a*) $\frac{3}{4}$ (*b*) $\frac{4}{3}$
(*c*) $\frac{3}{5}$ (*d*) $\frac{5}{3}$ **(C.B.I. 1990)**

66. The sum $3 + \frac{1}{\sqrt{3}} + \frac{1}{3 + \sqrt{3}} + \frac{1}{\sqrt{3} - 3}$ equals :

(*a*) 0 (*b*) 1
(*c*) 3 (*d*) $3 + \sqrt{3}$

67. $\sqrt{\frac{?}{169}} = \frac{54}{39}$

(*a*) 108 (*b*) 324
(*c*) 2916 (*d*) 4800

**(Railway Recruitment, 1991)**

68. $\frac{\sqrt{24} + \sqrt{216}}{\sqrt{96}} = ?$

(*a*) $2\sqrt{6}$ (*b*) $6\sqrt{2}$
(*c*) 2 (*d*) $\frac{2}{\sqrt{6}}$

**(Railway Recruitment, 1991)**

# HINTS AND SOLUTIONS

1.

| | |
|---|---|
| 2 | $\overline{6}\ \overline{40}\ \overline{09}$ (253 |
| | 4 |
| 45 | 240 |
| | 225 |
| 503 | 1509 |
| | 1509 |
| | × |

$\therefore \quad \sqrt{64009} = 253.$

2.

| | |
|---|---|
| 2 | $\overline{6}\ \overline{76}\ \overline{21}$ (260.04 |
| | 4 |
| 46 | 276 |
| | 276 |
| 520 | 21 |
| | 00 |
| 5200 | 2100 |
| | 0000 |
| 52004 | 210000 |
| | 208016 |

$\therefore \quad \sqrt{67621} = 260.04.$

3. $\dfrac{\sqrt{3125}}{\sqrt{5}} = \sqrt{\dfrac{3125}{5}} = \sqrt{625} = 25.$

4. $\dfrac{x}{\sqrt{64}} = 8 \Rightarrow \dfrac{x}{8} = 8 \Rightarrow x = 64.$

5. $\dfrac{\sqrt{288}}{\sqrt{128}} = \sqrt{\dfrac{288}{128}} = \sqrt{\dfrac{9}{4}} = \dfrac{3}{2}.$

6. $\sqrt{5} \times \sqrt{500} = \sqrt{5 \times 500} = \sqrt{2500} = 50.$

7. $\dfrac{34}{\sqrt{289}} \times \dfrac{\sqrt{196}}{70} = \dfrac{34}{17} \times \dfrac{14}{70} = \dfrac{2}{5}.$

8. $\dfrac{250}{\sqrt{x}} = 10 \Rightarrow \sqrt{x} = \dfrac{250}{10} = 25 \Rightarrow x = (25)^2 = 625.$

9. $\dfrac{112}{\sqrt{196}} \times \dfrac{\sqrt{576}}{12} \times \dfrac{\sqrt{256}}{8} = \dfrac{112}{14} \times \dfrac{24}{12} \times \dfrac{16}{8} = 32.$

10. $\dfrac{1872}{\sqrt{x}} = 234 \Rightarrow \sqrt{x} = \dfrac{1872}{234} = 8 \Rightarrow x = (8)^2 = 64.$

11. $\sqrt{\dfrac{x}{144}} = \dfrac{21}{36} \Rightarrow \dfrac{\sqrt{x}}{\sqrt{144}} = \dfrac{21}{36} \Rightarrow \dfrac{\sqrt{x}}{12} = \dfrac{21}{36}$

$\therefore \quad \sqrt{x} = \left(\dfrac{21}{36} \times 12\right) = 7$ and so $x = (7)^2 = 49.$

12. $\sqrt{248 + \sqrt{52 + \sqrt{144}}} = \sqrt{248 + \sqrt{52 + 12}} = \sqrt{248 + \sqrt{64}}$

$= \sqrt{248 + 8} = \sqrt{256} = 16.$

13. $\sqrt{\dfrac{x}{3136}} = \dfrac{28}{56} = \dfrac{1}{2} \Rightarrow \dfrac{\sqrt{x}}{\sqrt{3136}} = \dfrac{1}{2} \Rightarrow \dfrac{\sqrt{x}}{56} = \dfrac{1}{2}.$

$\therefore \sqrt{x} = 56 \times \dfrac{1}{2} = 28.$ So, $x = (28)^2 = 784.$

**14.** $\frac{\sqrt{x}}{200} = 0.02 \Rightarrow \sqrt{x} = 200 \times 0.02 = 4 \Rightarrow x = (4)^2 = 16.$

**15.** Let $\sqrt{x} - 3 = 19$. Then, $\sqrt{x} = 22$. So, $x = (22)^2 = 484$.

**16.** $\sqrt{\frac{1694}{x}} + 14 = 25 \Rightarrow \sqrt{\frac{1694}{x}} = 11$

$\therefore \quad \frac{1694}{x} = 121 \Rightarrow x = \frac{1694}{121} = 14.$

**17.** $\sqrt{\frac{25}{15625}} = \sqrt{\frac{x}{30625}} \Rightarrow \frac{25}{15625} = \frac{x}{30625}.$

$\therefore \quad x = \frac{25 \times 30625}{15625} = 49.$

**18.** $\sqrt{0.90} = \sqrt{\frac{90}{100}} = \frac{\sqrt{90}}{10} = \frac{9.4}{10} = 0.94.$

**19.** $\sqrt{0.0090} = \sqrt{\frac{90}{10000}} = \frac{\sqrt{90}}{100} = \frac{9.4}{100} = 0.094.$

**20.** Let $\frac{x}{20} = \frac{45}{x}$. Then, $x^2 = 45 \times 20 = 900$.

$\therefore \quad x = \sqrt{900} = 30.$

**21.** $\sqrt{4\frac{2}{3}} = \sqrt{\frac{14}{3} \times \frac{3}{3}} = \frac{\sqrt{42}}{3} = \frac{6.48}{3} = 2.16.$

**22.** $\sqrt{\frac{36.1}{102.4}} = \sqrt{\frac{361}{1024}} = \frac{\sqrt{361}}{\sqrt{1024}} = \frac{19}{32}.$

**23.** $\frac{\sqrt{7.84}}{\sqrt{1.96}} = \sqrt{\frac{7.84}{1.96}} = \sqrt{\frac{784}{196}} = \sqrt{4} = 2.$

**24.** $\sqrt{0.121} = \sqrt{0.1210} = \sqrt{\frac{1210}{10000}} = \frac{\sqrt{1210}}{100} = \frac{34.7}{100} = 0.347.$

**25.** $\sqrt{0.064} = \sqrt{\frac{640}{10000}} = \frac{\sqrt{640}}{100} = \frac{25.2}{100} = 0.252.$

**26.** $\sqrt{\frac{0.16}{0.4}} = \sqrt{\frac{16}{40}} = \sqrt{\frac{160}{400}} = \frac{\sqrt{160}}{\sqrt{400}} = \frac{\sqrt{160}}{20} = \frac{12.64}{20} = 0.63.$

**27.** $\frac{\sqrt{0.0009}}{\sqrt{0.01}} = \sqrt{\frac{0.0009}{0.0100}} = \sqrt{\frac{9}{100}} = \frac{3}{10} = 0.3.$

**28.** $\sqrt{0.01 + \sqrt{0.0064}} = \sqrt{0.01 + 0.08} = \sqrt{0.09} = 0.3.$

**29.** Let $\frac{\sqrt{x}}{0.25} = 100$. Then, $\sqrt{x} = 0.25 \times 100 = 25$.

$\therefore \quad x = (25)^2 = 625.$

**30.** Let $\sqrt{75.24 + x} = 8.71$. Then, $75.24 + x = (8.71)^2$.

$\therefore \quad x = (75.8641 - 75.24) = 0.6241.$

**31.** $\dfrac{1}{\sqrt{3}} = \dfrac{1}{\sqrt{3}} \times \dfrac{\sqrt{3}}{3} = \dfrac{1.732}{3} = 0.577.$

**32.** $\sqrt{\dfrac{2}{9}} = \dfrac{\sqrt{2}}{3} = \dfrac{1.414}{3} = 0.471.$

**33.** $\sqrt{\dfrac{7}{8}} = \sqrt{\dfrac{7}{8} \times \dfrac{2}{2}} = \dfrac{\sqrt{14}}{\sqrt{16}} = \dfrac{\sqrt{14}}{4} = \dfrac{3.74}{4} = 0.935.$

**34.** $\sqrt{\left(1 + \dfrac{27}{169}\right)} = \left(1 + \dfrac{x}{13}\right) \Rightarrow 1 + \dfrac{x}{13} = \sqrt{\dfrac{196}{169}} = \dfrac{14}{13}.$

$\therefore \quad \dfrac{x}{13} = \dfrac{14}{13} - 1 = \dfrac{1}{13}.$ So, $x = 1$.

**35.** $\dfrac{\sqrt{32} + \sqrt{48}}{\sqrt{8} + \sqrt{12}} = \dfrac{\sqrt{16 \times 2} + \sqrt{16 \times 3}}{\sqrt{4 \times 2} + \sqrt{4 \times 3}} = \dfrac{4\sqrt{2} + 4\sqrt{3}}{2\sqrt{2} + 2\sqrt{3}}$

$= \dfrac{4(\sqrt{2} + \sqrt{3})}{2(\sqrt{2} + \sqrt{3})} = 2.$

**36.** $\dfrac{7}{3 + \sqrt{2}} = \dfrac{7}{3 + \sqrt{2}} \times \dfrac{3 - \sqrt{2}}{3 - \sqrt{2}} = \dfrac{21 - 7\sqrt{2}}{9 - 2} = \dfrac{7(3 - \sqrt{2})}{7}$

$= (3 - \sqrt{2}) = (3 - 1.4142) = 1.5858.$

**37.** $\sqrt{2401} = \sqrt{7^x} \Rightarrow 7^x = 2401 = 7^4 \Rightarrow x = 4.$

**38.** $\sqrt{15625} + \sqrt{156.25} + \sqrt{1.5625}$

$= 125 + \sqrt{\dfrac{15625}{100}} + \sqrt{\dfrac{15625}{10000}}$

$= 125 + \dfrac{\sqrt{15625}}{10} + \dfrac{\sqrt{15625}}{100} = 125 + \dfrac{125}{10} + \dfrac{125}{100}$

$= 125 + 12.5 + 1.25 = 138.75.$

**39.** $\dfrac{4 + \sqrt{2}}{\sqrt{2} + 1} = \dfrac{(4 + \sqrt{2})}{(\sqrt{2} + 1)} \times \dfrac{(\sqrt{2} - 1)}{(\sqrt{2} - 1)} = \dfrac{3\sqrt{2} - 2}{(2 - 1)}$

$= (3\sqrt{2} - 2) = (3 \times 1.414 - 2) = 2.242.$

**40.** $\sqrt{2\sqrt{2\sqrt{2\sqrt{2\sqrt{2}}}}} = \sqrt{2\sqrt{2\sqrt{2\sqrt{(2 \times 2^{1/2})}}}}$

$= \sqrt{2\sqrt{2\sqrt{2 \times (2^{3/2})^{1/2}}}} = \sqrt{2\sqrt{2\sqrt{2^{(1 + 3/4)}}}}$

$= \sqrt{2\sqrt{2 \times (2^{7/4})^{1/2}}} = \sqrt{2\sqrt{2^{(1 + 7/8)}}} = \sqrt{2 \times (2^{15/8})^{1/2}}$

$= \sqrt{2 \times 2^{15/16}} = \left(2^{1 + 15/16}\right)^{1/2} = 2^{31/32}.$

**41.** $\sqrt{2^n} = 64 = 2^6 \Rightarrow 2^{n/2} = 2^6 \Rightarrow \dfrac{n}{2} = 6 \Rightarrow n = 12.$

**42.** $\dfrac{\sqrt{2} - 1}{\sqrt{2} + 1} = \dfrac{\sqrt{2} - 1}{\sqrt{2} + 1} \times \dfrac{\sqrt{2} - 1}{\sqrt{2} - 1} = \dfrac{(\sqrt{2} - 1)^2}{(2 - 1)} = (\sqrt{2} - 1)^2.$

$\therefore \quad \sqrt{\left(\dfrac{\sqrt{2}-1}{\sqrt{2}+1}\right)} = (\sqrt{2}-1) = (1.4142-1) = 0.4142.$

**43.** $\sqrt{0.04 \times 0.4 \times a} = 0.4 \times 0.04 \times \sqrt{b}$

$\Rightarrow 0.04 \times 0.4 \times a = 0.4 \times 0.4 \times 0.04 \times 0.04 \times b$

$\Rightarrow \dfrac{a}{b} = \dfrac{0.4 \times 0.4 \times 0.04 \times 0.04}{0.04 \times 0.4} = 0.016.$

**44.** $\dfrac{\sqrt{2}}{2+\sqrt{2}} = \dfrac{\sqrt{2}}{2+\sqrt{2}} \times \dfrac{2-\sqrt{2}}{2-\sqrt{2}} = \dfrac{2\sqrt{2}-2}{4-2} = \dfrac{2(\sqrt{2}-1)}{2}$

$= (\sqrt{2}-1) = (1.4142-1) = 0.4142.$

**45.** $\sqrt{\dfrac{1.21 \times 0.9}{1.1 \times 0.11}} = \sqrt{\dfrac{121 \times 9}{11 \times 11}} = \sqrt{9} = 3.$

**46.** $(\sqrt{8})^{1/3} = \left\{(2^3)^{1/2}\right\}^{1/3} = (2^{3/2})^{1/3} = 2^{(3/2 \times 1/3)} = 2^{1/2} = \sqrt{2}.$

**47.** $a^2 + b^2 = 289$ and $ab = 120$.

$\therefore \quad (a+b)^2 = (a^2 + b^2 + 2ab) = (289 + 240) = (529).$

So, $(a+b) = \sqrt{529} = 23.$

**48.** $a^2 + b^2 = 80$ and $(a-b)^2 = 36$.

$(a-b)^2 = 36 \Rightarrow a^2 + b^2 - 2ab = 36$

$\Rightarrow 80 - 2ab = 36$

$\Rightarrow 2ab = 80 - 36 = 44$

$\Rightarrow ab = 22.$

**49.** The largest number of five digits is 99999.

| | |
|---|---|
| 3 | $\overline{9}\,\overline{99}\,\overline{99}$ (316 |
| | 9 |
| 61 | 99 |
| | 61 |
| 626 | 3899 |
| | 3756 |
| | 143 |

$\therefore$ Required number $= (99999 - 143) = 99856.$

**50.** The smallest number of four digits is 1000.

| | |
|---|---|
| 3 | 1000 (31 |
| | 9 |
| 61 | 100 |
| | 61 |
| | 39 |

$\therefore$ Required number $= (32)^2 = 1024.$

**51.**
$$\begin{array}{r|l} 1 & \overline{26}\,9\ (\ 16 \\ & 1 \\ \hline 26 & 169 \\ & 156 \\ \hline & 13 \end{array}$$

$\therefore$ Required numbe to be added $= (17)^2 - 269 = 20.$

**52.**
$$\begin{array}{r|l} 1 & 1\ \overline{68}\ \overline{00}\ (\ 129 \\ & 1 \\ \hline 22 & 68 \\ & 44 \\ \hline 249 & 2400 \\ & 2241 \\ \hline & 159 \end{array}$$

$\therefore$ Required number to be subtracted = 159.

**53.** $\sqrt{.00059049} = \sqrt{\dfrac{59049}{100000000}} = \dfrac{\sqrt{59049}}{10000} = \dfrac{243}{10000} = .0243.$

**54.** $176 = 2 \times 2 \times 2 \times 2 \times 11.$

$\therefore$ To make it a perfect square, the given number should be multiplied by 11.

**55.** $216 = 2 \times 2 \times 2 \times 3 \times 3 \times 3.$

$\therefore$ To make it a perfect square, 216 must be divided by 6.

**56.** $(.000027)^{1/3} = \left(\dfrac{27}{1000000}\right)^{1/3} = \dfrac{(27)^{1/3}}{(10^6)^{1/3}} = \dfrac{3}{100} = .03.$

**57.** $\left(4\dfrac{12}{125}\right)^{1/3} = \left(\dfrac{512}{125}\right)^{1/3} = \dfrac{(8 \times 8 \times 8)^{1/3}}{(5 \times 5 \times 5)^{1/3}} = \dfrac{8}{5} = 1\dfrac{3}{5}.$

**58.** $\left(1 - \dfrac{91}{216}\right)^{1/3} = \left(\dfrac{125}{216}\right)^{1/3} = \left(\dfrac{5 \times 5 \times 5}{6 \times 6 \times 6}\right)^{1/3} = \dfrac{5}{6}.$

**59.** $21600 = 6 \times 6 \times 6 \times 10 \times 10.$

To make the given number a perfect cube, it must be multiplied by 10.

**60.** $3600 = 2 \times 2 \times 2 \times 2 \times 3 \times 3 \times 5 \times 5.$

So, it make it a perfect cube, the given number must be divided by $2 \times 3 \times 3 \times 5 \times 5 = 450.$

**61.** $a^2 + a^2 = (\text{diagonal})^2 \Rightarrow 2a^2 = (8)^2 = 64 \Rightarrow a^2 = 32$

$\Rightarrow a = \sqrt{32} = 4\sqrt{2} = 4 \times 1.41 = 5.64$ cm.

**62.** $\dfrac{\sqrt{324}}{1.5} = \dfrac{x}{\sqrt{256}} \Rightarrow \dfrac{18}{1.5} = \dfrac{x}{16} \Rightarrow x = \dfrac{16 \times 18}{1.5} = 192.$

**63.** $\frac{1}{\sqrt{9}-\sqrt{8}} = \frac{1}{(\sqrt{9}-\sqrt{8})} \times \frac{(\sqrt{9}+\sqrt{8})}{(\sqrt{9}+\sqrt{8})} = \frac{\sqrt{9}+\sqrt{8}}{(9-8)} = (\sqrt{9}+\sqrt{8})$

$= (3 + 2\sqrt{2})$.

**64.** Given expression $= \sqrt{\frac{324 \times 81 \times 4624}{15625 \times 289 \times 729 \times 64}}$

$= \frac{18 \times 9 \times 68}{125 \times 17 \times 27 \times 8} = \frac{3}{125} = 0.024.$

**65.** $a = \frac{\sqrt{5}+1}{\sqrt{5}-1} \times \frac{\sqrt{5}+1}{\sqrt{5}+1} = \frac{(\sqrt{5}+1)^2}{(5-1)} = \frac{6+2\sqrt{5}}{4}$;

$b = \frac{\sqrt{5}-1}{\sqrt{5}+1} \times \frac{\sqrt{5}-1}{\sqrt{5}-1} = \frac{(\sqrt{5}-1)^2}{(5-1)} = \frac{6-2\sqrt{5}}{4}$.

$\therefore\ ab = \frac{(6+2\sqrt{5})}{4} \times \frac{(6-2\sqrt{5})}{4} = \frac{36-20}{16} = 1.$

$a + b = \frac{6+2\sqrt{5}}{4} + \frac{6-2\sqrt{5}}{4} = \frac{12}{4} = 3$;

$a - b = \frac{6+2\sqrt{5}}{4} - \frac{6-2\sqrt{5}}{4} = \frac{4\sqrt{5}}{4} = \sqrt{5}.$

$\therefore\ \frac{a^2+ab+b^2}{a^2-ab+b^2} = \frac{(a+b)^2-ab}{(a-b)^2+ab} = \frac{3^2-1}{(\sqrt{5})^2+1} = \frac{8}{6} = \frac{4}{3}$.

**66.** Given expression

$= 3 + \frac{1}{\sqrt{3}} \times \frac{\sqrt{3}}{\sqrt{3}} + \frac{1}{3+\sqrt{3}} \times \frac{3-\sqrt{3}}{3-\sqrt{3}} + \frac{1}{\sqrt{3}-3} \times \frac{\sqrt{3}+3}{\sqrt{3}+3}$

$= 3 + \frac{\sqrt{3}}{3} + \frac{3-\sqrt{3}}{6} + \frac{\sqrt{3}+3}{-6} = \frac{18+2\sqrt{3}+3-\sqrt{3}-\sqrt{3}-3}{6} = 3.$

## ANSWERS

| | | | | | | |
|---|---|---|---|---|---|---|
| **1.** (*c*) | **2.** (*b*) | **3.** (*c*) | **4.** (*b*) | **5.** (*c*) | **6.** (*d*) | **7.** (*b*) |
| **8.** (*c*) | **9.** (*c*) | **10.** (*b*) | **11.** (*d*) | **12.** (*b*) | **13.** (*c*) | **14.** (*c*) |
| **15.** (*d*) | **16.** (*a*) | **17.** (*c*) | **18.** (*c*) | **19.** (*c*) | **20.** (*b*) | **21.** (*c*) |
| **22.** (*b*) | **23.** (*a*) | **24.** (*c*) | **25.** (*d*) | **26.** (*c*) | **27.** (*b*) | **28.** (*a*) |
| **29.** (*b*) | **30.** (*a*) | **31.** (*c*) | **32.** (*c*) | **33.** (*c*) | **34.** (*a*) | **35.** (b) |
| **36.** (*a*) | **37.** (*c*) | **38.** (*c*) | **39.** (*a*) | **40.** (*d*) | **41.** (*d*) | **42.** (*c*) |
| **43.** (*a*) | **44.** (*a*) | **45.** (*b*) | **46.** (*c*) | **47.** (*a*) | **48.** (*a*) | **49.** (*d*) |
| **50.** (*c*) | **51.** (*d*) | **52.** (*b*) | **53.** (*c*) | **54.** (*c*) | **55.** (*a*) | **56.** (*b*) |
| **57.** (*a*) | **58.** (*b*) | **59.** (*b*) | **60.** (*d*) | **61.** (*d*) | **62.** (*a*) | **63.** (*d*) |
| **64.** (*c*) | **65.** (*b*) | **66.** (*c*) | | | | |

# 6

# PERCENTAGE

***By a certain per cent, we mean that many hundredths.***

Thus, $x$ percent means $x$ hundredths, denoted by $x\%$.

Also, a percentage is a fraction whose denominator is 100.

Thus, $9\% = \frac{9}{100}$ ; $75\% = \frac{75}{100} = \frac{3}{4}$ ; $0.5\% = \frac{0.5}{100} = .005$ .

**Ex. 1.** ***Express each of the following as a fraction :***

(*i*) 36% (*ii*) 3.6%

(*iii*) 150% (*iv*) $12\frac{1}{2}\%$

**Sol.** (*i*) $36\% = \frac{36}{100} = \frac{9}{25}$. (*ii*) $3.6\% = \frac{3.6}{100} = \frac{36}{1000} = \frac{9}{250}$.

(*iii*) $150\% = \frac{150}{100} = \frac{3}{2}$. (*iv*) $12\frac{1}{2}\% = \frac{25}{2 \times 100} = \frac{1}{8}$.

**Ex. 2.** ***Express each of the following as a decimal :***

(*i*) 65% (*ii*) 6%.

(*iii*) 120%. (*iv*) $\frac{1}{2}\%$.

**Sol.** (*i*) $65\% = \frac{65}{100} = .65$. (*ii*) $6\% = \frac{6}{100} = .06$.

(*iii*) $120\% = \frac{120}{100} = 1.20$. (*iv*) $\frac{1}{2}\% = \frac{1}{2 \times 100} = .005$.

**Ex. 3.** ***Express as rate per cent :***

(*i*) $\frac{3}{4}$ (*ii*) $\frac{7}{8}$

(*iii*) $\frac{1}{15}$ (*iv*) 0.025

**Sol.** (*i*) $\frac{3}{4} = \left(\frac{3}{4} \times 100\right)\% = 75\%$. (*ii*) $\frac{7}{8} = \left(\frac{7}{8} \times 100\right)\% = 87\frac{1}{2}\%$.

(*iii*) $\frac{1}{15} = \left(\frac{1}{15} \times 100\right)\% = 6\frac{2}{3}\%$. (*iv*) $0.025 = \left(\frac{25}{1000} \times 100\right)\% = 2.5\%$

**Ex. 4.** ***Find :***

(*i*) $37\frac{1}{2}\%$ of Rs. 450. (*ii*) 0.5% of Rs. 8 in paise

(*iii*) 2% of 1.5 kg in gms (*iv*) 1.2% of 5 litres in $cm^3$.

**Sol.** (*i*) $37\frac{1}{2}\%$ of Rs. 450 = Rs. $\left(\frac{75}{2 \times 100} \times 450\right)$ = Rs.168.75.

(*ii*) 0.5% of Rs. 8 = 0.5% of (800 paise)

$= \left(\frac{0.5}{100} \times 800\right)$ paise = 4 paise.

(*iii*) 2% of 1.5 kg = $\frac{2}{100}$ of 1500 gms = 30 gm.

(*iv*) 1.2% of 5 litres = $\frac{1.2}{100} \times 5000 \text{ cm}^3 = 60 \text{ cm}^3$.

**Ex. 5.** ***(i) What per cent of 80 is 18 ?***

***(ii) 15 is what per cent of 90 ?***

**Sol.** (*i*) Let $x$ % of 80 = 18.

Then, $\frac{x}{100} \times 80 = 18$ or $x = \frac{18 \times 100}{80} = 22\frac{1}{2}$.

**Alternative method :**

Fraction = $\frac{18}{80} = \left(\frac{18}{80} \times 100\right)\% = 22\frac{1}{2}\%$.

(*ii*) Fraction = $\frac{15}{90} = \left(\frac{15}{90} \times 100\right)\% = 16\frac{2}{3}\%$.

**Ex. 6.** ***A man saves 16% of his total income of Rs 5000 per month. How much does he spend ?***

**Sol.** Money spent = (100 – 16) % of Rs 5000

= (84 % of Rs 5000) = Rs $\left(\frac{84}{100} \times 5000\right)$ = Rs 4200.

**Ex. 7.** ***In an examination, 36% are pass marks. If an examinee gets 17 marks and fails by 10 marks, what are the maximum marks ?***

**Sol.** Total pass marks = (17 + 10) = 27.

Let maximum marks be $x$.

Then, 36% of $x$ = 27 or $\frac{36}{100} \times x = 27$.

$\therefore x = \left(\frac{27 \times 100}{36}\right) = 75$.

*Hence, maximum marks* = 75.

**Ex. 8.** ***Madan's salary is 25% of Ram's salary and Ram's salary is 40% of Sudin's salary. If the total salary of all the three for a month is Rs 12000, how much did Madan earn that month?*** **(Bank P.O. 1991)**

**Sol.** Let Sudin's salary = Rs $x$. Then,

Ram's salary = $\left(\frac{40}{100}x\right)$ = Rs $\frac{2x}{5}$

and Madan's salary = Rs $\left(\frac{25}{100} \times \frac{2x}{5}\right)$ = Rs $\left(\frac{x}{10}\right)$.

$$\therefore x + \frac{2x}{5} + \frac{x}{10} = 12000 \Rightarrow x = 8000.$$

$$\text{so, Madan's salary} = \text{Rs}\left(\frac{8000}{10}\right) = \text{Rs } 800.$$

**Ex. 9.** ***A candidate scoring 25% in an examination fails by 30 marks while another candidate who scores 50% marks gets 20 marks more than the minimum required for a pass. Find the minimum pass percentage.***

**(Hotel Management 1991)**

**Sol.** Let the maximum pass marks be $x$.

Then, (25% of $x$) + 30 = (50% of $x$) – 20

or $\frac{x}{4} + 30 = \frac{x}{2} - 20$ or $\frac{x}{2} - \frac{x}{4} = 30 + 20$

or $\frac{x}{4} = 50$ or $x = 200$.

So, maximum marks = 200.

$$\text{Minimum pass marks} = \left(\frac{200}{4} + 30\right) = 80.$$

$$\therefore \text{Minimum pass percentage} = \left(\frac{80}{200} \times 100\right)\% = 40\%.$$

**Ex. 10.** ***A man donated 5% of his income to a charitable trust and deposited 20% of the remainder in a bank. If he now has, Rs 1919 left, what is his income ?***

**Sol.** Suppose income = Rs $x$.

Then, donated money = (5% of Rs $x$) = Rs $\frac{x}{20}$.

$$\text{Remainder} = \text{Rs}\left(x - \frac{x}{20}\right) = \text{Rs } \frac{19x}{20}.$$

$$\text{Money deposited in the bank} = 20\% \text{ of Rs}\left(\frac{19x}{20}\right)$$

$$= \text{Rs}\left(\frac{20}{100} \times \frac{19x}{20}\right) = \text{Rs } \frac{19x}{100}.$$

$$\text{Balance} = \text{Rs}\left(\frac{19x}{20} - \frac{19x}{100}\right) = \text{Rs } \frac{76}{100}x.$$

$$\therefore \frac{76}{100}x = 1919 \text{ or } x = \frac{1919 \times 100}{76} = 2525.$$

*Hence the man's income* = Rs *2525* .

**Ex. 11.** ***In an examination, 42% students failed in Hindi and 52% failed in English. If 17% students failed in both the subjects, find the percentage of those students who passed in both the subjects.***

**Sol.** Let the number of students appeared be 100.

Number of students who failed in Hindi only = (42 – 17) = 25 .

Number of students who failed in English only = (52 – 17) = 35 .
Number of students who failed in at least one of the subjects
= (25 + 35 + 17) = 77 .
*Number of students who passed in both the subjects*
= (100 – 77) = 23% .

**Ex. 12.** ***If A's income is 25% more than that of B, how much per cent B's income is less than that of A ?***

**Sol.** Let *B*'s income be Rs 100.
Then, *A*'s income = Rs 125.
If *A*'s income is Rs 125, *B*'s income = Rs 100.
If A's income is Rs 100, *B*'s income = Rs $\left(\frac{100}{125} \times 100\right)$ = Rs 80.
∴ *B's income is less than A's income by (100– 80)% = 20%* .

**Ex. 13.** ***Two numbers are 30% and 37% less than a third number. How much per cent is the second number less than the first ?***

**Sol.** Let the third number be 100.
Then, 1st number = 70; 2nd number = 63.
Difference of 1st and 2nd numbers = (70 – 63) =7.
∴ *Required percentage* = $\left(\frac{7}{70} \times 100\right)\%$ = 10% .

**Ex. 14.** ***From the salary of an officer 10% is deducted as house rent; 15% of the rest he spends on children's education; 10% of the balance he spends on clothes. After this expenditure, he is left with Rs. 2754. Find his salary.***

**Sol.** Let the salary be Rs. 100. Then,
House rent = Rs. 10; Balance = Rs. (100 – 10) = Rs. 90.
Expenditure on children's education = 15% of Rs. 90 = Rs. $\left(\frac{27}{2}\right)$.
Balance now = Rs. $\left(90 - \frac{27}{2}\right)$ = Rs. $\frac{153}{2}$ .
Expenditure on clothes = $\left(10\% \text{ of Rs. } \frac{153}{2}\right)$ = Rs. $\left(\frac{153}{20}\right)$ .
Now, balance = Rs. $\left(\frac{153}{2} - \frac{153}{20}\right)$ = Rs. $\frac{1377}{20}$ .
If last balance is Rs. $\frac{1377}{20}$ , then salary = Rs. 100.
If last balance is Rs. 2754, then salary = Rs. $\left(\frac{100 \times 20}{1377} \times 2754\right)$
= Rs. 4000.

**Ex.15.** ***If the price of a commodity be raised by 20%, find by how much per cent must a house holder reduce the consumption, so that expenditure***

***on it remains unchanged.***

**Sol.** Let commodity used be 1 unit and its cost be Re. 1.

New cost of 1 unit = (120% of Re. 1) = Rs. $\left(\frac{6}{5}\right)$.

Now, Rs. $\left(\frac{6}{5}\right)$ yield 1 unit.

So, Re 1 will yield = $\left(1 \times \frac{5}{6}\right) = \frac{5}{6}$ unit.

Reduction in consumption = $\left(1 - \frac{5}{6}\right) = \frac{1}{6}$.

$\therefore$ *Reduction percent* = $\left(\frac{1}{6} \times \frac{1}{1} \times 100\right)\% = 16\frac{2}{3}\%$.

**Short Cut Method :**

(i) If price is increased by r%, then

reduction in consumption = $\left[\frac{r}{(100+r)} \times 100\right]\%$.

(ii) If price is decreased by r%, then

increase in consumption = $\left[\frac{r}{(100-r)} \times 100\right]\%$.

**Ex. 16.** ***If the price of sugar be decreased by 25%, find by how much per cent must its consumption be increased, to keep the expenditure fixed on sugar.***

**Sol.** Increase in consumption = $\left[\frac{25}{(100-25)} \times 100\right]\%$

$= \left(\frac{25}{75} \times 100\right)\% = 33\frac{1}{3}\%$.

**Ex. 17.** ***In an election, one of the two candidates gets 42% of total votes and still loses by 368 votes. Find the total number of votes.***

**Sol.** Let the total number of votes be 100.

Votes polled by defeated candidate = 42.

Votes polled by winning candidate = (100 – 42) = 58.

Difference of votes polled = (58 – 42) = 16.

If the difference of votes is 16, total votes = 100.

If the difference of votes is 368, total votes

$= \left(\frac{100}{16} \times 368\right) = 2300$.

**Ex. 18.** ***In an election contested by two candidates, 5% of the voters did not cast their votes. The successful candidate won by 4900 votes, securing 51% of the total votes. How many votes were cast for each candidate ?***

**Sol.** Let the total votes be 100.

Votes cast = (100 – 5) = 95.
Votes polled by successful candidate = 51.
Votes polled by defeated candidate = (95 – 51) = 44.
Difference of votes = (51 – 44) = 7.
If difference is 7, votes polled by winner = 51.
If difference is 4900, votes polled by winner

$$= \left(\frac{51}{7} \times 4900\right) = 35700.$$

*Votes polled by defeated candidate = (35700 – 4900) = 30800.*

**Ex. 19.** ***The tax on an article decreases by 10% and its consumption increases by 10%. Find the effect per cent on its revenue.***

**Sol.** Let the original consumption be 1 unit & tax on it be Rs. 100.
So, revenue = Rs. (100 × 1) = Rs. 100.

New consumption = $\left(\frac{110}{100} \times 1\right) = \frac{11}{10}$ units.

Now, tax on 1 unit = Rs. 90

Tax on $\frac{11}{10}$ units = Rs. $\left(90 \times \frac{11}{10}\right)$ = Rs. 99.

∴ *Decrease in revenue = 1%.*

**Ex. 20.** ***What is 80% of a number whose 200% is 90 ?***

**Sol.** Let the number be $x$.

Then, 200% of $x = 90 \Rightarrow \frac{200}{100} \times x = 90$ or $x = 45$.

∴ *80% of 45* = $\left(\frac{80}{100} \times 45\right) = 36$.

**Ex. 21.** ***If the numerator of a fraction is increased by 20% and its denominator is decreased by 10%, the fraction becomes*** $\frac{16}{21}$. ***Find the original fraction.***

**Sol.** Let the original fraction be $\frac{x}{y}$.

$$\text{New fraction} = \frac{120\% \text{ of } x}{90\% \text{ of } y} = \frac{\frac{120}{100} \times x}{\frac{90}{100} \times y} = \frac{4x}{3y}.$$

$$\text{Now } \frac{4}{3} \cdot \frac{x}{y} = \frac{16}{21} \Rightarrow \frac{x}{y} = \left(\frac{16}{21} \times \frac{3}{4}\right) = \frac{4}{7}.$$

∴ *original fraction* = $\frac{4}{7}$.

**Ex. 22.** ***Due to a fall of 10% in the rate of tea, 500 gm. more tea can be purchased for Rs. 140. Find the original rate and the reduced rate.***

**Sol.** Money spent originally = Rs. 140.

Less money to be spent now = (10% of Rs. 140) = Rs. 14.

∴ Rs. 14, now yields 500 gm. tea.

∴ Present rate of tea = Rs. 28 per kg.

If present value is Rs.90, then original value = Rs. 100

If present value is Rs. 28, then original value

$$= \text{Rs.} \left(\frac{100}{90} \times 28\right) = \text{Rs. } 31.11.$$

**Problems based on population :—**

**Formulae :**

(i) If the population of a town (or the length of a tree in length units) is P and the annual increase is r%, then the population (or length of tree) after t years is $= P\left(1 + \frac{r}{100}\right)^t$.

(ii) If the population of a town (or value of a machine in Rupees) is P and the annual decrease (or depreciation) is r%, then the population (or value of machine) after t years is $= P\left(1 - \frac{r}{100}\right)^t$.

**Ex. 23.** ***The population of a town increases 10% annually. If its present population be 60000, find its population after 3 years.***

**Sol.** Population after 3 years $= 60000 \times \left(1 + \frac{10}{100}\right)^3$

$$= \left(60000 \times \frac{11}{10} \times \frac{11}{10} \times \frac{11}{10}\right) = 79860.$$

**Ex. 24.** ***The value of a car depreciates at 12% annually. If its present price is 156250, what will be its value after 3 years ?***

**Sol.** Depreciated value after 3 years

$$= \text{Rs.} \left[156250 \times \left(1 - \frac{12}{100}\right)^3\right] = \text{Rs.} \left(156250 \times \frac{22}{25} \times \frac{22}{25} \times \frac{22}{25}\right)$$

= *Rs. 106480.*

**Ex. 25.** ***The population of a town increases by 5% annually. If it is 15435 now, what it was 2 years ago ?***

**Sol.** Let the population 2 years ago be $P$. Then,

$$P\left(1 + \frac{5}{100}\right)^2 = 15435 \text{ or } P \times \frac{21}{20} \times \frac{21}{20} = 15435.$$

$$\therefore \text{P} = \frac{15435 \times 20 \times 20}{21 \times 21} = 14000.$$

*Hence the population 2 years ago was 14000.*

**Ex. 26.** ***The value of a machinery depreciates at the rate of 10% after each year. Find, its purchase price, if at the end of 3 years its value is only Rs. 8748.***

**Sol.** Let the purchase price be Rs. $P$.

Then, $P\left(1-\frac{10}{100}\right)^3 = 8748$ or $P \times \frac{9}{10} \times \frac{9}{10} \times \frac{9}{10} = 8748$.

$\therefore P = \frac{8748 \times 10 \times 10 \times 10}{9 \times 9 \times 9} = 12000$.

*Hence, the purchase price of the machine is Rs. 12000.*

## EXERCISE 6A

**1.** Express each of the following as a fraction :

(a) 15% (b) $16\frac{2}{3}$% (c) 20.5% (d) 0.3%

**2.** Express each one of the following as decimal :

(a) 18% (b) 45% (c) 2.6% (d) 0.1%

**3.** Express each of the following as a per cent :

(a) $\frac{1}{8}$ (b) $\frac{4}{25}$ (c) $\frac{5}{3}$ (d) $\frac{1}{125}$

(e) 0.47 (f) 0.15 (g) 0.005 (h) 0.01

**4.** Find :

(i) 36% of Rs. 5 (ii) 8% of 1.25 kg.

(iii) 45% of 6 litres (iv) 10% of 2.5 metres

**5.** (i) What per cent of 60 is 72 ?

**Hint.** *Let x% of 60 = 72. Then,* $x = \left(\frac{72 \times 100}{60}\right)$.

(ii) What per cent of Rs. 11.40 is 38 paise ?

**Hint.** *Let x% of 1140 = 38. Then,* $x = \left(\frac{38 \times 100}{1140}\right)$.

(iii) What per cent of 7.4 kg is 37 gms. ?

**Hint.** *Let x% of 7400 = 37. Then,* $x = \left(\frac{37 \times 100}{7400}\right)$.

(iv) What per cent is 7 gm. of a kilogram ?

**Hint.** *Let x% of 1000 = 7. Then,* $x = \left(\frac{7 \times 100}{1000}\right)$.

(v) What is the number whose 40% is 30 ?

**6.** A number increased by $37\frac{1}{2}$% gives 33. Find the number.

**Hint.** *Let the number be x.*

Then, $\left(x + \frac{75 \times x}{2 \times 100}\right) = 33$.

**7.** A candidate has to obtain 33% of the total marks to pass. He obtains 262 marks and fails by 200 marks. Find the maximum marks.

**8.** Out of an earning of Rs. 6250, Mr. Verma spends 86%. What are his savings ?

**9.** If $37\frac{1}{2}$ % of a number is 900, what is $62\frac{1}{2}$ % of the number ?

**10.** A peon saves 20% of his monthly income. Due to a 20% increase in dearness, he can save Rs. 60 only. Find his monthly income.

**Hint.** *Let his income be Rs. 100. Then, expenditure = Rs. 80.*

*Now, expenditure = (120% of Rs. 80) = Rs. 96.*
*Saving now = Rs. 4.*
*If present saving is Rs. 4, income = Rs. 100*
*If present saving is Rs. 60, income = Rs.* $\left(\frac{100}{4} \times 60\right)$.

**11.** In an examination, A gets 28% of total marks and failed by 55 marks, while B gets 37% of the total marks and gets 17 more than the pass marks. Find the maximum marks and those necessary for passing.

**Hint.** *Let maximum marks be x. Then,*

$\frac{28}{100}x + 55 = \frac{37}{100}x - 17$. Find $x$ & $\left(\frac{28}{100}x + 55\right)$.

**12.** In an examination 60% of the candidates pass in English and 52% pass in Mathematics, while 32% failed in both the subjects. If 220 candidates passed, find how many candidates appeared in all.

**13.** A man had a certain sum of money. He gave 20% of it to his eldest son, 30% of the remaining he gave to his younger son. 10% of the remaining he donated to a school. If the man still had Rs. 4032, find his total sum.

**Hint.** *Suppose he had Rs 100.*

*Given to eldest son = Rs 20; Balance = Rs 80;*
*Given to younger son = Rs 24; Balance = Rs 56:*
*Donated to school = Rs 5.60.*
*Final balance = Rs (56 – 5.60) = Rs 50.40.*
*If final balance is Rs 50.40, he had Rs 100*
*If final balance is Rs 4032, he had Rs* $\left(\frac{100}{50.40} \times 4032\right)$.

**14.** An army lost 10% of its men in war. 10% of the remaining died due to disease and 10% of the rest were declared disabled. Thus, the strength of the Army was reduced to 729000 active men. Find out the original strength of the army.

**15.** **If *A*'s income is $26\frac{2}{3}$ % more than that of *B*, find how much per cent is *B*'s income less than that of A.**

**16.** **If *A*'s income is 20% less than that of *B*, find how much per cent is *B*'s income more than that of *A*.**

**17.** **Mustard oil is now being sold at Rs. 15 per kg. During last month its rate was Rs. 13 per kg. Find, by how much percent must a family reduce its consumption to keep the expenditure fixed.**

**Hint.** *Let there be a consumption of 1 kg originally.*

*Then, expenditure = Rs. 13*

*Now Rs. 15 fetch = 1 kg.*

*So, Rs. 13 fetch* $= \left(\frac{1}{15} \times 13\right)$ *kg.*

*Reduction in consumption* $= \left(1 - \frac{13}{15}\right) = \frac{2}{15}$.

$\therefore$ *Reduction %* $= \left(\frac{2}{15 \times 1} \times 100\right) \% = 13\frac{1}{3} \%$.

**18.** **Due to a 20% fall in the price of eggs, a man purchases 15 eggs more for Rs 22.50. Find the original and reduced prices.**

**Hint.** *Present price of 15 eggs* = **(Rs 22.50 – 20% of Rs 22.50)**

$\therefore$ *Present price per dozen* $= \left(\frac{18}{15} \times 12\right) =$ *Rs 14.40.*

*Original price per dozen* = *Rs* $\left(\frac{100}{80} \times 14.40\right)$.

**19.** **A reduction of 25% in the price of sugar enables me to get 50 kg more for Rs. 500. What is the reduced price per kg ?**

**Hints.** *Let the original rate be Rs. x per kg.*

*Reduced rate* = Rs. $\left(\frac{75}{100} \times x\right)$ *per kg* = Rs. $\left(\frac{3x}{4}\right)$ *per kg.*

$\therefore \frac{500}{\frac{3x}{4}} = \left(\frac{500}{x} + 50\right)$. *Find x.*

**20.** **A man spends 75% of his income. His income increased by 20% and he increased his expenditure by 10%. Find by how much per cent his savings are increased.**

**21.** **The population of a town is 8000. It increases by 10% during first year and by 20% during second year. Find the population after 2 years.**

**Hint.** *Population after 2 years*

$= 8000 \times \left(1 + \frac{10}{100}\right) \times \left(1 + \frac{20}{100}\right)$.

**22.** The population of a village increases by 5% annually. If the present population is 4410, what it was 2 years ago ? **(L.I.C. 1991)**

**Hint.** *Let it be P.*

*Then,* $P\left(1+\frac{5}{100}\right)^2 = 4410$. *Find P.*

**23.** If the annual decrease in the population of a town is 5% and the present population is 68590, what was it 3 years ago ?

**Hint.** *Let it be P. Then,*

$$P\left(1-\frac{5}{100}\right)^3 = 68590. \text{ Find P.}$$

**24.** The population of a town increases by 12% during first year and decreases by 10% during second year. If the present population is 50400, what it was 2 years ago ? **(L.I.C. 1991)**

**Hint.** *Let it be P. Then,*

$$P\left(1+\frac{12}{100}\right)\left(1-\frac{10}{100}\right) = 50400.$$

**25.** A papaya tree was planted 2 years ago. It increases at the rate of 20% per annum. If at present, the height of the tree is 540 cm, what it was when the tree was planted ?

**26.** In an examination 65% of the total examinees passed. If the number of failures is 420, find the total number of examinees.

**27.** In an election contested by two candidates 4% of the votes cast are invalid. A candidate gets 55% of the total votes and wins the election by 280 valid votes. Find the total number of votes cast.

**Hints.** *Let the total votes be 100.*

*Valid votes = 96*

*Votes polled by winning candidate =55*

*Votes polled by defeated candidate = (96 – 55) = 41*

*If he wins by 14 votes, number of votes = 100*

*If he wins by 280 votes, number of votes =* $\left(\frac{100}{14}\times 280\right)$.

## ANSWERS (Exercise 6A)

| | | | |
|---|---|---|---|
| **1.** (*a*) $\frac{3}{20}$ | (*b*) $\frac{1}{6}$ | (*c*) $\frac{41}{200}$ | (*d*) $\frac{3}{1000}$ |
| **2.** (*a*) 0.18 | (*b*) 0.45 | (*c*) 0.026 | (*d*) 0.001 |
| **3.** (*a*) $12\frac{1}{2}\%$ | (*b*) 16% | (*c*) $166\frac{2}{3}\%$ | (*d*) 0.8 % |
| (*e*) 47% | (*f*) 15% | (*g*) 0.5% | (*h*) 1% |
| **4.** (*i*) Rs 1.80 | (*ii*) 100 gm | (*iii*) 2.7 litres | (*iv*) 25 cm |

**5.** (*i*) 120% (*ii*) $3\frac{1}{3}$ % (*iii*) $\frac{1}{2}$ % (*iv*) 0.7%

(*v*) 75 **6.** 24 **7.** 1400 **8.** Rs 875

**9.** 1500 **10.** Rs 1500 **11.** 800, 279 **12.** 500

**13.** Rs 8000 **14.** 1000000 **15.** 21.05% **16.** 25%

**17.** $13\frac{1}{3}$% **18.** Rs 18 per dozen & Rs 14.40 per dozen

**19.** Rs 2.50 per kg **20.** 50% **21.** 10560 **22.** 4000

**23.** 80000 **24.** 50000 **25.** 375 cm **26.** 1200

**27.** 2000

## EXERCISE 6B (Objective Type Questions)

**1.** 4% of 400 – 2% of 800 = ?

(*a*) 2 (*b*) –4
(*c*) 0 (*d*) 16

**2.** ? % of 130 = 11.7

(*a*) 90 (*b*) 9
(*c*) 0.9 (*d*) 0.09 **(S.B.I.P.O. Exam, 1987)**

**3.** 40% of 70 = 4 × ?

(*a*) 28 (*b*) 280
(*c*) 7 (*d*) 70 **(Bank clerical, 1990)**

**4.** The fraction equivalent to $\frac{2}{5}$% is :

(*a*) $\frac{1}{40}$ (*b*) $\frac{1}{125}$
(*c*) $\frac{1}{250}$ (*d*) $\frac{1}{500}$

**5.** 0.2% of ? = 0.03

(*a*) 20 (*b*) 2.5
(*c*) 15 (*d*) 1.5

**6.** 0.025 in terms of rate percent is

(*a*) $\frac{1}{4}$% (*b*) 25%
(*c*) 2.5% (*d*) $37\frac{1}{2}$%

**7.** 5% of 10% of Rs 175 = ?

(*a*) Rs 8.75 (*b*) Re 0.50
(*c*) Re 0.875 (*d*) Rs 17.50

**8.** What is 25% of 25% equal to ?

(*a*) 6.25 (*b*) .625
(*c*) .0625 (*d*) .00625 **(Astt. Grade 1987)**

**9.** If 40% of 40% of $x = 40$, then the value of $x$ is :

(*a*) 100 (*b*) 400

(*c*) 250 (*d*) 1000

**10.** What percent of 7.2 kg is 18 gms ?

(*a*) 25% (*b*) 2.5%

(*c*) .25% (*d*) .025%

**11.** 5 out of 2250 parts of earth is sulphur. What is the percentage of sulphur in earth ?

(*a*) $\frac{1}{45}$ (*b*) $\frac{2}{9}$

(*c*) $\frac{11}{50}$ (*d*) $\frac{2}{45}$ **(Hotel Management 1991)**

**12.** $\frac{30\ \%\ \text{of}\ 80}{?} = 24$

(*a*) $\frac{3}{10}$ (*b*) 1

(*c*) $\frac{3}{17}$ (*d*) 2 **(Delhi Police 1989)**

**13.** $45 \times ? = 25\%$ of 900

(*a*) 16.20 (*b*) 500

(*c*) 4 (*d*) 5

**14.** $8\frac{1}{3}\%$ expressed as a fraction is

(*a*) $\frac{25}{3}$ (*b*) $\frac{3}{25}$

(*c*) $\frac{1}{12}$ (*d*) $\frac{1}{4}$ **(C.B.I. Exam 1989)**

**15.** If $37\frac{1}{2}\%$ of a number is 900, then $62\frac{1}{2}\%$ of the number will be :

(*a*) 1200 (*b*) 1350

(*c*) 1500 (*d*) 540

**16.** 75% of a number when added to 75 becomes the number itself. The number is :

(*a*) 150 (*b*) 200

(*c*) 225 (*d*) 300

**(Railway Recruitment 1991)**

**17.** What percent of $\frac{2}{7}$ is $\frac{1}{35}$ ?

(*a*) 25% (*b*) 2.5%

(*c*) 1000% (*d*) 10%.

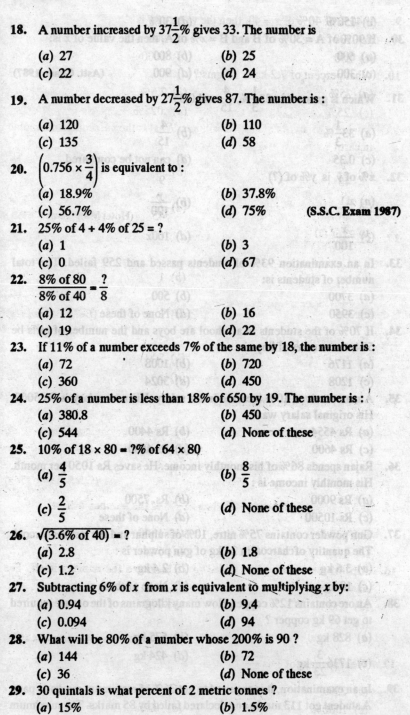

**18.** A number increased by $37\frac{1}{2}\%$ gives 33. The number is

(*a*) 27 (*b*) 25
(*c*) 22 (*d*) 24

**19.** A number decreased by $27\frac{1}{2}\%$ gives 87. The number is :

(*a*) 120 (*b*) 110
(*c*) 135 (*d*) 58

**20.** $\left(0.756 \times \frac{3}{4}\right)$ is equivalent to :

(*a*) 18.9% (*b*) 37.8%
(*c*) 56.7% (*d*) 75% **(S.S.C. Exam 1987)**

**21.** 25% of 4 + 4% of 25 = ?

(*a*) 1 (*b*) 3
(*c*) 0 (*d*) 67

**22.** $\frac{8\% \text{ of } 80}{8\% \text{ of } 40} = \frac{?}{8}$

(*a*) 12 (*b*) 16
(*c*) 19 (*d*) 22

**23.** If 11% of a number exceeds 7% of the same by 18, the number is :

(*a*) 72 (*b*) 720
(*c*) 360 (*d*) 450

**24.** 25% of a number is less than 18% of 650 by 19. The number is :

(*a*) 380.8 (*b*) 450
(*c*) 544 (*d*) None of these

**25.** 10% of 18 × 80 = ?% of 64 × 80

(*a*) $\frac{4}{5}$ (*b*) $\frac{8}{5}$
(*c*) $\frac{2}{5}$ (*d*) None of these

**26.** $\sqrt{(3.6\% \text{ of } 40)} = ?$

(*a*) 2.8 (*b*) 1.8
(*c*) 1.2 (*d*) None of these

**27.** Subtracting 6% of $x$ from $x$ is equivalent to multiplying $x$ by:

(*a*) 0.94 (*b*) 9.4
(*c*) 0.094 (*d*) 94

**28.** What will be 80% of a number whose 200% is 90 ?

(*a*) 144 (*b*) 72
(*c*) 36 (*d*) None of these

**29.** 30 quintals is what percent of 2 metric tonnes ?

(*a*) 15% (*b*) 1.5%

(c) 150% (d) 30%

30. If 90% of A = 30% of B and B = $x$% of A, then the value of $x$ is:
(a) 600 (b) 800
(c) 300 (d) 900 **(Astt. Grade 1987)**

31. Which is greatest in $33\frac{1}{3}\%$, $\frac{4}{15}$ and 0.35 ?
(a) $33\frac{1}{3}\%$ (b) $\frac{4}{15}$
(c) 0.35 (d) can not be compared

32. $x$% of $y$ is $y$% of (?)
(a) $x$ (b) $\frac{y}{100}$
(c) $\frac{x}{100}$ (d) $100x$

33. In an examination 93% of students passed and 259 failed. The total number of students is:
(a) 3700 (b) 500
(c) 3950 (d) None of these

34. If 70% of the students in a school are boys and the number of girls be 504, the number of boys is:
(a) 1176 (b) 1008
(c) 1208 (d) 3024

35. A man received 10% increase in his salary. His new salary is Rs 5060. His original salary was :
(a) Rs 4554 (b) Rs 4400
(c) Rs 4600 (d) Rs 4560

36. Rajan spends 86% of his monthly income. He saves Rs 1050 per month. His monthly income is :
(a) Rs 9000 (b) Rs. 7500
(c) Rs 10500 (d) None of these

37. Gun powder contains 75% nitre, 10% of sulphur and the rest is charcoal. The quantity ofcharcoalin 24 kg of gun powder is :
(a) 3.6 kg (b) 2.4 kg
(c) 18 kg (d) None of these

38. An ore contains 12% copper, How many kilograms of the ore are required to get 69 kg copper ?
(a) 828 kg (b) 575 kg
(c) $1736\frac{3}{23}$ kg (d) 424 kg

39. In an examination, it is required to get 36% of maximum marks to pass. A student got 113 marks and declared failed by 85 marks. The maximum

marks are :

(*a*) 500 (*b*) 550

(*c*) 640 (*d*) 1008

**40.** In an examination, 52% of the candidates failed in English, 42% failed in Mathematics and 17% failed in both. The number of those who have passed in both the subjects is :

(*a*) 23% (*b*) 35%

(*c*) 25% (*d*) 40% **(C.D.S. 1991)**

**41.** After deducting a commission of 5% a T.V. set costs Rs 7600. Its gross value is :

(*a*) Rs 7980 (*b*) Rs 8000

(*c*) Rs 8200 (*d*) None of these

**42.** The marked price is 10% higher than the cost price. A discount of 10% is given on the marked price. In this kind of sale, the seller

(*a*) Bears no loss, no gain (*b*) Gains 1%

(*c*) Loses 10% (*d*) Loses 1% **(C.D.S. 1991)**

**43.** A man's wages were decreased by 50%. Again the reduced wages were increased by 50%. He has a loss of :

(*a*) 0% (*b*) 2.5%

(*c*) 0.25% (*d*) 25%

**44.** A sweet seller declares that he sells sweets at the cost price. However, he uses a weight of 450 gm instead of 500 gm. His percentage profit is

(*a*) 10 (*b*) $11\frac{1}{9}$

(*c*) 12 (*d*) $12\frac{2}{9}$

**45.** In an examination, there were 2000 candidates out of which 900 candidates were boys and rest are girls. If 32% of the boys and 38% of the girls passed, then the total percentage of failed candidates is :

(*a*) 35.3 % (*b*) 70%

(*c*) 64.7% (*d*) 68.5%

**46.** If *x* is 90% of *y*, what percent of *x* is *y* ?

(*a*) 90 (*b*) 190

(*c*) 101.1 (*d*) 111.1 **(Assistant Grade 1990)**

**47.** The price of an article has been reduced by 25%. In order to restore the original price, the new price must be increased by

(*a*) $33\frac{1}{3}\%$ (*b*) $9\frac{1}{11}\%$

(*c*) $11\frac{1}{9}\%$ (*d*) $66\frac{2}{3}\%$

**48.** The price of sugar is increased by 20%. If the expenditure is not allowed to increase, the ratio between the reduction in consumption and original consumption is :

(a) $\frac{1}{3}$ (b) $\frac{1}{4}$

(c) $\frac{1}{6}$ (d) $\frac{1}{8}$

**49.** The price of cooking oil has increased by 25%. The percentage of reduction that a family should effect in the use of cooking oil so as not to increase the expenditure on this account is :

(a) 15% (b) 20%

(c) 25% (d) 30%

(Central Excise & I. Tax 1988)

**50.** In a vocational course in a college 15% seats increase annually. If there were 800 students in 1991 how many students will be there in 1993 ?

(a) 920 (b) 1058

(c) 1040 (d) 1178

**51.** When the price of fans was reduced by 20%, the number of fans sold increased by 40%. What was the effect on the revenue received by the shopkeeper ?

(a) 12% decrease (b) 12% increase

(c) 40% increase (d) 30% increase

**52.** Two numbers are less than a third number by 30% and 37% respectively. How much percent is the second number less than the first ?

(a) 7% (b) 10%

(c) 4% (d) 3%

**53.** If the price of tea is increased by 1%, how much percent must a householder reduce her consumption of tea to have no extra expenditure ?

(a) $\frac{1}{101}\%$ (b) $\frac{101}{100}\%$

(c) $\frac{100}{101}\%$ (d) 1%

**54.** 1 litre of water is added to 5 litres of a 20% solution of alcohol in water. The strength of alcohol is now:

(a) $12\frac{1}{2}\%$ (b) $16\frac{2}{3}\%$

(c) 24% (d) 16%

**55.** The value of a machinery depreciates 10% annually. What will be its value 2 years hence if the present value is Rs 4000 ?

(a) Rs 3200 (b) Rs 3240

(c) Rs 3260 (d) Rs 3280

**56.** The value of a T.V. set depreciates at the rate of 10% after each year. It was purchased 3years ago. If its present value is Rs 8748, its purchase price was :

(*a*) Rs 10000 (*b*) Rs 11372.40
(*c*) Rs 12000 (*d*) None of these

**57.** The population of a town is 8000. It increases by 10% during first year and by 20% during second year. The population after 2 years will be :

(*a*) 10400 (*b*) 10560
(*c*) 10620 (*d*) None of these

**58.** The population of a town increases by 5% annually. If it is 15435 now, its population 2 years ago was :

(*a*) 14000 (*b*) 15000
(*c*) 13700 (*d*) 14800

**59.** There were 600 students in a college. If 75% offered English and 45% Hindi, how many offered both ?

(*a*) 48 (*b*) 60
(*c*) 80 (*d*) 120

**60.** In an election between two candidates, the candidate who gets 30% of the votes polled is defeated by 15000 votes. The number of votes polled by the winning candidate is :

(*a*) 11250 (*b*) 15000
(*c*) 26250 (*d*) 37500

**61.** At an election, a candidate secures 40% of the votes but is defeated by the other candidate by a majority of 298 votes. The total number of votes recorded were :

(*a*) 1360 (*b*) 1490
(*c*) 1520 (*d*) 1602

**62.** If the numerator of a fraction is increased by 20% and its denominator be diminished by 10%, the value of the fraction is $\frac{16}{21}$. The original fraction is :

(*a*) $\frac{3}{5}$ (*b*) $\frac{4}{7}$
(*c*) $\frac{2}{3}$ (*d*) $\frac{5}{7}$

**63.** A man spends 75% of his income. His income is increased by 20% and he increased his expenditure by 10%. His savings are increased by :

(*a*) 10% (*b*) 25%
(*c*) $37\frac{1}{2}\%$ (*d*) 50%

**64.** The boys and girls in a college are in the ratio 3:2. If 20% of the boys, and 25% of the girls are adults, the percentage of students who are not adults is :

(*a*) 67.5% (*b*) 82.5%
(*c*) 78% (*d*) 58%

**65.** If the side of a square is increased by 25%, then how much percent does its area get increased ?

(*a*) 56.25 (*b*) 50
(*c*) 125 (*d*) 156.25

**66.** In measuring the side of a square, an error of 5% in excess was made. The error percent in the calculated area is :

(*a*) 10% (*b*) 25%
(*c*) 10.25% (*d*) 10.5%

**67.** The length of a rectangle is increased by 10% and breadth decreased by 10%. Then the area of new rectangle :

(*a*) is neither decreased nor increased
(*b*) is increased by 1%
(*c*) is decreased by 1%
(*d*) is decreased by 2%

**68.** If the diameter of a circle is increased by 100%, its area is increased by

(*a*) 300% (*b*) 200%
(*c*) 100% (*d*) 400%

**69.** The length and breadth of a square are increased by 30% and 20% respectively. The area of the rectangle so formed exceeds the area of the square by :

(*a*) 20% (*b*) 36%
(*c*) 50% (*d*) 56%

**70.** The radius of a circle is increased by 1%. What is the increase percent in its area ?

(*a*) 1.1% (*b*) 1%
(*c*) 2% (*d*) 2.01%

**71.** If all the edges of a cube are doubled, its volume is increased by:

(*a*) 300% (*b*) 400%
(*c*) 700% (*d*) 800%

**72.** If the diameter of a sphere is doubled, the increase in its surface area will be

(*a*) 100% (*b*) 200%
(*c*) 300% (*d*) 400%

**73.** For a sphere of radius 10 cms, the numerical value of surface area is how many percent of the numerical value of its volume ?

(*a*) 26.5% (*b*) 30%

(*c*) 24% (*d*) 45%

**74.** A mixture of 40 litres of milk and water contains 10% water. How much water should be added to this so that water may be 20% in the new mixture ?

(*a*) 5 litres (*b*) 4 litres

(*c*) 6.5 litres (*d*) 7.5 litres

**75.** By how much percent is four- fifth of 70 lesser than five- seventh of 122 ?

(*a*) 42% (*b*) 30%

(*c*) 24% (*d*) 36%

**76.** One litre of water is evaporated from 6 litres of a solution containing 5% salt. The percentage of salt in the remaining solution is

(*a*) 6% (*b*) $5\frac{5}{7}\%$

(*c*) $4\frac{4}{9}\%$ (*d*) 5%

**77.** After spending 40% in machinery 25% in building, 15% in raw material and 5% on furniture, Harilal had a balance of Rs. 1305. Total money with him was :

(*a*) Rs 6500 (*b*) Rs 7225

(*c*) Rs 8700 (*d*) Rs 1390

**78.** Which number is 60% less than 80?

(*a*) 48 (*b*) 42

(*c*) 32 (*d*) 12 **(Assistant Grade 1990)**

**79.** If the base of a rectangle is increased by 10% and the area is unchanged, then its corresponding altitude must be decreased by :

(*a*) $11\frac{1}{9}\%$ (*b*) $9\frac{1}{11}\%$

(*c*) 11% (*d*) 10% **(C.B.I. Exam, 1990)**

**80.** The price of an article was increased by *p*%. Later the new price was decreased by *p*%. If the latest price was Re. 1, the original price was :

(*a*) Re 1 (*b*) Rs $\frac{1-p^2}{100}$

(*c*) Rs $\frac{\sqrt{1-p^2}}{100}$ (*d*) Rs $\left(\frac{10000}{10000-p^2}\right)$

**(C.P.O. Exam 1990)**

**81.** If 10% of m is the same as 20% of n then m : n is equal to :

(*a*) 1 : 2 (*b*) 2 : 1
(*c*) 5 :1 (*d*) 10 :1 **(C.B.I. Exam 1990)**

**82.** The cost price of an article is 40% of the selling price. The percent that the selling price is of cost price is
(*a*) 40 (*b*) 60
(*c*) 240 (*d*) 250

**83.** A owns a house worth Rs 10000. he sells it to B at a profit of 10% based on the worth of the house. B sells the house back to A at a loss of 10%. In this tranaction , A gets
(*a*) a profit of Rs 2000 (*b*) a profit of Rs 1,100
(*c*) a profit of Rs 1,000 (*d*) no profit no loss
**(C.B.I. Exam 1990)**

**84.** The price of an article is cut by 10%. To restore it to the former value, the new price must be increased by
(*a*) 10% (*b*) $9\frac{1}{11}\%$
(*c*) $11\frac{1}{9}\%$ (*d*) 11% **(C.P.O. Exam 1990)**

**85.** *p* is six times as large as *q*. The percent that *q* is less than *p*, is :
(*a*) $16\frac{2}{3}$ (*b*) 60
(*c*) 90 (*d*) $83\frac{1}{3}$ **(C.P.O. Exam 1990)**

**86.** In an examination 70% candidates passed in English and 65% in Mathematics. If 27% candidates failed in both the subjects and 248 passed the examination, the total number of candidates was :
(*a*) 400 (*b*) 348
(*c*) 420 (*d*) 484 **(Bank Clerical 1991)**

## HINTS & SOLUTIONS

## (EXERCISE 6B)

**1.** (4% of 400) – (2% of 800) = $\left(\frac{4}{100} \times 400\right) - \left(\frac{2}{100} \times 800\right) = 16 - 16 = 0.$

**2.** Let $x$% of 130 = 11.7.

Then, $\frac{x}{100} \times 130 = 11.7 \Rightarrow x = \frac{11.7 \times 100}{130} = 9.$

**3.** Let 40% of 70 = $4 \times x$.

Then, $\frac{40}{100} \times 70 = 4 \times x \Rightarrow x = \left(\frac{40}{100} \times 70 \times \frac{1}{4}\right) = 7.$

**4.** $\frac{2}{5}\% = \frac{2}{5} \times \frac{1}{100} = \frac{1}{250}$.

**5.** $0.2\%$ of $x = 0.03 \Rightarrow \frac{0.2}{100} \times x = 0.03$.

$\therefore x = \frac{0.03 \times 100}{0.20} = 15$.

**6.** $0.025 = \frac{25}{1000} = \frac{1}{40} = \left(\frac{1}{40} \times 100\right)\% = 2.5\%$.

**7.** 5% of 10% of Rs 175 = Rs $\left(\frac{5}{100} \times \frac{10}{100} \times 175\right)$ = Re 0.875.

**8.** $25\%$ of $25\% = \left(\frac{25}{100} \times \frac{25}{100}\right) = \frac{625}{10000} = 0.0625$.

**9.** $40\%$ of $40\%$ of $x = 40$.

$\therefore \frac{40}{100} \times \frac{40}{100} \times x = 40$ or $x = \frac{40 \times 100 \times 100}{40 \times 40} = 250$.

**10.** Required percentage $= \left(\frac{18}{7.2 \times 1000} \times 100\right)\% = 0.25\%$.

**11.** Percentage of sulphur $= \left(\frac{5}{2250} \times 100\right)\% = \frac{2}{9}\%$.

**12.** $\frac{30\% \text{ of } 80}{x} = 24 \Rightarrow 24x = \left(\frac{30}{100} \times 80\right) = 24 \Rightarrow x = 1$.

**13.** Let $45 \times x = 25\%$ of 900. Then, $45x = \frac{25}{100} \times 900 = 225$.

$\therefore x = \frac{225}{45} = 5$.

**14.** $8\frac{1}{3}\% = \frac{25}{3 \times 100} = \frac{1}{12}$.

**15.** $37\frac{1}{2}\%$ of $x = 900 \Rightarrow \frac{75}{2 \times 100} \times x = 900$.

$\therefore x = \frac{2 \times 100 \times 900}{75} = 2400$.

Now, $62\frac{1}{2}\%$ of $2400 = \left(\frac{125}{2} \times \frac{1}{100} \times 2400\right) = 1500$.

**16.** $75 + (75\%$ of $x) = x$. So, $75 + \left(\frac{75}{100}\right)x = x$.

or $75 + \frac{3}{4}x = x$ or $x - \frac{3}{4}x = 75$.

$\therefore \frac{1}{4}x = 75$. So, $x = 75 \times 4 = 300$.

**17.** Let $x\%$ of $\frac{2}{7} = \frac{1}{35}$.

Then, $\frac{x}{100} \times \frac{2}{7} = \frac{1}{35} \Rightarrow x = \frac{1}{35} \times \frac{100 \times 7}{2} = 10$.

**18.** $x + 37\frac{1}{2}\%$ of $x = 33$.

$$x + \frac{75}{2 \times 100} x = 33 \text{ or } x + \frac{3x}{8} = 33$$

$$\text{or } \frac{11x}{8} = 33. \text{ So, } x = \frac{33 \times 8}{11} = 24.$$

**19.** $x - 27\frac{1}{2}\%$ of $x = 87$.

$$\text{or } x - \frac{55}{2 \times 100} \times x = 87 \text{ or } x - \frac{11}{40} x = 87$$

$$\text{or } \frac{29}{40} x = 87 \text{ or } x = \frac{87 \times 40}{29} = 120.$$

**20.** $0.756 \times \frac{3}{4} = \left(\frac{756}{1000} \times \frac{3}{4}\right) = \left(\frac{756 \times 3}{1000 \times 4} \times 100\right)\% = 56.7\%$

**21.** $\frac{25\% \text{ of } 4}{4\% \text{ of } 25} = \frac{\frac{25}{100} \times 4}{\frac{4}{100} \times 25} = 1$.

**22.** $\frac{8\% \text{ of } 80}{8\% \text{ of } 40} = \frac{x}{8}$ or $\frac{\frac{8}{100} \times 80}{\frac{8}{100} \times 40} = \frac{x}{8}$ or $x = 16$.

**23.** $11\%$ of $x - 7\%$ of $x = 18 \Rightarrow 4\%$ of $x = 18$.

$$\therefore \frac{4}{100} \times x = 18 \text{ or } x = \frac{18 \times 100}{4} = 450.$$

**24.** $(18\% \text{ of } 650) - (25\% \text{ of } x) = 19$.

$$\therefore \frac{18}{100} \times 650 - \frac{25}{100} \times x = 19 \text{ or } \frac{x}{4} = 117 - 19 = 98$$

or $x = (98 \times 4) = 392$.

**25.** $10\%$ of $18 \times 80 = x\%$ of $64 \times 80$.

$$\therefore \frac{10}{100} \times 18 \times 80 = \frac{x}{100} \times 64 \times 80$$

$$\text{So, } 180 = 64x \text{ or } x = \frac{180}{64} = \frac{45}{16}.$$

**26.** $\sqrt{(3.6\% \text{ of } 40)} = \sqrt{\frac{3.6}{100} \times 40} = \sqrt{\frac{36 \times 40}{1000}} = \frac{6 \times 2}{10} = 1.2.$

**27.** $(x - 6\% \text{ of } x) = \left(x - \frac{6}{100}x\right) = \frac{94}{100} x = 0.94x.$

$\therefore$ Required number = 0.94.

**28.** $200\%$ of $x = 90 \Rightarrow \frac{200}{100} \times x = 90 \Rightarrow x = 45.$

Now, 80% of 45 = $\left(\frac{80}{100} \times 45\right) = 36.$

**29.** Required percentage = $\left(\frac{3}{2} \times 100\right)\% = 150\%$.

Note that 1 metric tonne = 1000 kg = 10 quintals.

**30.** $\frac{90}{100}A = \frac{30}{100}B = \frac{30}{100} \times \frac{x}{100}A$

$\therefore x = \frac{90}{100} \times \frac{100 \times 100}{30} = 300.$

**31.** $33\frac{1}{3}\% = \frac{100}{3} \times \frac{1}{100} = \frac{1}{3} = 0.33; \frac{4}{15} = 0.266.$

$\therefore$ out of $33\frac{1}{3}\%, \frac{4}{15}$ and .35, the greatest number is 0.35.

**32.** $x\%$ of $y = y\%$ of $z$

$\Rightarrow \frac{x}{100} \times y = \frac{y}{100} \times z$ or $z = \frac{xy}{100} \times \frac{100}{y} = x.$

**33.** $(93\%$ of $x) + 259 = x \Rightarrow \frac{93}{100}x + 259 = x.$

$\therefore x - \frac{93}{100}x = 259$ or $\frac{7}{100}x = 259$

or $x = \frac{259 \times 100}{7} = 3700.$

**34.** If number of girls is 30, the number of boys = 70.

If number of girls is 504, the number of boys = $\left(\frac{70}{30} \times 504\right) = 1176.$

**35.** If new salary is Rs.110, original salary = 100.
If new salary is Rs.5060, original salary

$= \left(\frac{100}{110} \times 5060\right) =$ Rs.4600.

**36.** Let Rajan's monthly income be Rs. $x$.

Then, 14% of $x = 1050 \Rightarrow \frac{14}{100}x = 1050$

or $x = \frac{1050 \times 100}{14} = 7500.$

**37.** Charcoal = $[100 - (75 + 10)]\% = 15\%$.
$\therefore$ quantity of charcoal = 15% of 24 kg.

$= \left(\frac{15}{100} \times 24\right)$ kg. = 3.6 kg.

**38.** For 12 kg. copper, ore required = 100 kg.

For 69 kg. copper, ore required = $\left(\frac{100}{12} \times 69\right)$ kg. = 575 kg.

**39.** 36% of $x = 113 + 85 = 198.$

$\therefore \frac{36}{100} \times x = 198$ or $x = \frac{198 \times 100}{36} = 550.$

**40.** Failed in English only = (52 – 17)% = 35%;
Failed in Mathematics only = (42 – 17)% = 25%;
Failed in both = 17%
Failed = (35 + 25 + 17)% = 77%.
∴ Passed in both the subjects = (100 – 77)% = 23%.

**41.** If net value is Rs.95, gross value = Rs.100.

If net value is Rs.7600, gross value = Rs. $\left(\frac{100}{95} \times 7600\right)$ = Rs. 8000.

**42.** Let C.P. = Rs.100. Then, marked price = Rs.110.

Discount = (10% of Rs.110) = $\left(\frac{10}{100}\right.$ of Rs.110$\left.\right)$ = Rs.11.

∴ S.P. = Rs. (110 – 11) = Rs.99.
So, the seller loses 1% .

**43.** Let original wages = Rs.100.
Decreased wages = Rs.50.

Increased wages = Rs. $\left(\frac{150}{100} \times 50\right)$ = Rs.75.

Loss = 25%.

**44.** Let C.P. of each gm. = Re.1.
C.P. of 450 gm. = Rs.450.
S.P. of 450 gm. = Rs.500.

$\therefore$ Profit = $\left(\frac{50}{450} \times 100\right)\% = 11\frac{1}{9}\%.$

**45.** Students passed = (32% of 900 + 38% of 1100)

$= \left(\frac{32}{100} \times 900 + \frac{38}{100} \times 1100\right) = (288 + 418) = 706.$

∴ Students failed= (2000 – 706) = Rs.1294.

Percentage of failure = $\left(\frac{1294}{2000} \times 100\right)\% = 64.7\%.$

**46.** $x = 90\%$ of $y = \frac{90}{100}y = \frac{9y}{10} \Rightarrow \frac{y}{x} = \frac{10}{9}$

Let $y = z\%$ of $x = \frac{z}{100}x \Rightarrow \frac{y}{x} = \frac{z}{100}.$

$\therefore \frac{z}{100} = \frac{10}{9} \Rightarrow z = \frac{10 \times 100}{9} = 111.1\%.$

**47.** Let original price = Rs.100.
Reduced price = Rs.75.
Increase on Rs.75 = Rs.25

Increase on Rs.100 = $\left(\frac{25}{75} \times 100\right)\% = 33\frac{1}{3}\%.$

**48.** Reduction in consumption $= \left[\frac{r}{100+r} \times 100\right]\%$

$= \left(\frac{20}{120} \times 100\right)\% = \frac{50}{3}\%.$

$\therefore \frac{\text{Reduction in consumption}}{\text{Original consumption}} = \frac{50}{3 \times 100} = \frac{1}{6}.$

**49.** Reduction in consumption $= \left(\frac{25}{125} \times 100\right)\% = 20\%.$

**50.** Number of students in 1993 $= 800 \times \left(1 + \frac{15}{100}\right)^2 = 1058.$

**51.** Let the number of fans originally sold be 100 and let the price of each fan be Rs.100.

Then revenue = Rs. (100 × 100) = Rs.10000.

New revenue = Rs. (80 × 140) = Rs.11200.

$\therefore$ Revenue increased $= \left(\frac{1200}{10000} \times 100\right)\% = 12\%.$

**52.** Let third number = 100. Then,

First number = 70, second number = 63.

$\therefore$ Second number is less than the first by $\left(\frac{7}{70} \times 100\right) = 10\%.$

**53.** Reduction in consumption $= \left(\frac{1}{101} \times 100\right)\% = \frac{100}{101}\%.$

**54.** Alcohol in 5 litres $= \left(\frac{20}{100} \times 5\right)$ litres = 1 litre.

Strength of new solution $= \left(\frac{1}{6} \times 100\right)\% = 16\frac{2}{3}\%.$

**55.** Value after 2 years = Rs. $\left[4000 \times \left(1 - \frac{10}{100}\right)^2\right]$ = Rs.3240.

**56.** Let purchased price of the T.V. set be Rs.$x$.

Then, $x\left(1 - \frac{10}{100}\right)^3 = 8748 \Rightarrow x = \left(8748 \times \frac{10}{9} \times \frac{10}{9} \times \frac{10}{9}\right)$ = Rs.12000.

**57.** Population after 2 years $= 8000 \times \left(1 + \frac{20}{100}\right) \times \left(1 + \frac{20}{100}\right) = 10560.$

**58.** Let the population two years ago be $x$.

Then, $x\left(1 + \frac{5}{100}\right)^2 = 15435$. So, $x = \left(15435 \times \frac{20}{21} \times \frac{20}{21}\right) = 14000.$

**59.** Let $x\%$ offered both.

Then, $75 + 45 - x = 100$ or $x = 20$.

$\therefore$ Number of students offering both the subjects

$= 20\% \text{ of } 600 = \left(\frac{20}{100} \times 600\right) = 120.$

**60.** $70\% \text{ of } x - 30\% x = 15000 \Rightarrow 40\% \text{ of } x = 15000.$

$\therefore$ Total votes polled $= x = \frac{15000 \times 100}{40} = 37500.$

Votes polled by winning candidate $= \left(\frac{70}{100} \times 37500\right) = 26250.$

**61.** $60\% \text{ of } x - 40\% \text{ of } x = 298 \Rightarrow 20\% \text{ of } x = 298.$

$\therefore \frac{20}{100} \times x = 298$ or $x = \frac{298 \times 100}{20} = 1490.$

**62.** Let the fraction be $\frac{x}{y}$.

Then, $\frac{120\% \text{ of } x}{90\% \text{ of } y} = \frac{16}{21} \Rightarrow \frac{\frac{120}{100}x}{\frac{90}{100}y} = \frac{16}{21}$

or $\frac{4x}{3y} = \frac{16}{21}$ or $\frac{x}{y} = \frac{16}{21} \times \frac{3}{4} = \frac{4}{7}.$

**63.** Let income = Rs.100.

Then, expenditure = Rs.75, savings = Rs. 25.

New income = Rs. 120

New expenditure = Rs. $\left(\frac{110}{100} \times 75\right)$ = Rs. $\frac{165}{2}$.

New savings = Rs. $\left(120 - \frac{165}{2}\right)$ = Rs. $\frac{75}{2}$.

Increase in savings $= \left(\frac{25}{2} \times \frac{1}{25} \times 100\right)\% = 50\%.$

**64.** Let the number of boys and girls be 300 and 200 respectively. Then, the number of students who are not adults

$= (80\% \text{ of } 300) + (75\% + 200) = (240 + 150) = 390.$

$\therefore$ Required percentage $= \left(\frac{390}{500} \times 100\right) = 78\%.$

**65.** Let original side = 100 units. Then, new side = 125 units.

Increase in area

$= \left[(125)^2 - (100)^2\right] = [(125 + 100)(125 - 100)] = 5625.$

$\therefore$ Percentage increase $= \left(\frac{5625}{100 \times 100} \times 100\right)\% = 56.25\%.$

**66.** Let actual side be 100 m.

Measured side = 105 m.

Error in area

$= \left[(105)^2 - (100)^2\right] = (105 + 100) \times (105 - 100) = 1025 \text{ m}^2.$

Required percentage $= \left(\frac{1025}{100 \times 100} \times 100\right)\% = 10.25\%.$

**67.** Let length = 100 m & breadth = 100 m.

New length = 110 m, New breadth = 90 m.

Area decreased $= (100 \times 100 - 110 \times 90) = 100 \text{ m}^2$.

Decrease percent $= \left(\frac{100}{100 \times 100} \times 100\right)\% = 1\%.$

**68.** Let diameter = 100 m. New diameter = 200 m.

Change in area

$= \left[\pi \times (100)^2 - \pi \times (50)^2\right] = \pi\left[(100)^2 - (50)^2\right] = 7500\,\pi.$

$\therefore$ Percentage increase $= \left(\frac{7500\,\pi}{\pi \times 50 \times 50} \times 100\right)\% = 300\%.$

**69.** Let the original length = m.

New length = 130 m & New breadth = 120 m.

Increase in area $= (130 \times 120 - 100 \times 100) = 5600 \text{ m}^2$.

Increase percent $= \left(\frac{5600}{100 \times 100} \times 100\right)\% = 56\%.$

**70.** Let the original radius be 100 m.

Then, new radius = 101 m.

$\therefore$ Increase in area

$= \pi \times \left[(101)^2 - (100)^2\right] = \pi \times (101 + 100)(101 - 100) = 201\,\pi.$

Increase percent $= \left(\frac{201\,\pi}{100 \times 100 \times \pi} \times 100\right)\% = 2.01\%.$

**71.** Let the original length of the edge = 100 m.

New length = 200 m.

Increase in volume $= \left[(200)^3 - (100)^3\right] = 7000000.$

Increase percent $= \left(\frac{7000000}{1000000} \times 100\right)\% = 700\%.$

**72.** Let original length of diameter = 100 m.

New diameter = 200 m.

Then, increase in surface area $= \pi \times \left[(200)^2 - (100)^2\right]$

$= \pi \times (200 + 100) \times (200 - 100) = 30000\,\pi$

$\therefore$ Increase percent $= \left(\frac{30000\,\pi}{\pi \times 100 \times 100} \times 100\right)\% = 300\%.$

**73.** Surface area $= 4\pi \times 10 \times 10 = 400\pi$.

Volume $= \frac{4}{3}\pi \times 10 \times 10 \times 10 = \frac{4000}{3}\pi.$

Required percentage $= \left(400\pi \times \frac{3}{4000\pi} \times 100\right)\% = 30\%.$

**74.** Milk in 40 litres = 90% of 40 = $\left(\frac{90}{100} \times 40\right)$ = 36 litres.

Let $x$ litres of water be added to it.

Then, $\frac{4+x}{40+x} \times 100 = 20 \Rightarrow \frac{4+x}{40+x} = \frac{1}{5} \Rightarrow x = 5$ litres.

**75.** $\left(\frac{5}{7} \times 122 - \frac{4}{5} \times 70\right) = (80 - 56) = 24.$

$\therefore$ Required percentage = $\left(\frac{24}{80} \times 100\right)\% = 30\%$.

**76.** Salt in 5 litres = $\left(\frac{5}{100} \times 6\right) = \frac{3}{10}$ litres.

Salt in 100 litres = $\left(\frac{3}{10} \times \frac{1}{5} \times 100\right)$ = 6 litres = 6%.

**77.** $[100 - (40 + 25 + 15 + 5)]\%$ of $x = 1305$.

$\Rightarrow 15\%$ of $x = 1305 \Rightarrow \frac{15}{100} \times x = 1305.$

$\therefore\ x = \frac{1305 \times 100}{15} = 8700.$

**78.** $(80 - 60\%\text{ of } 80) = \left(80 - \frac{60}{100} \times 80\right) = (80 - 48) = 32.$

**79.** Let length = 100 m & height = $x$ m.
Area = $(100\,x)$.
New length = 110 m & let new height = $(x - y\%\text{ of }x)$

Then, $110 \times \left(x - \frac{y}{100}x\right) = 100 \times x$

or $110 \times \left(1 - \frac{y}{100}\right) = 100$ or $1 - \frac{y}{100} = \frac{100}{110}$

or $\frac{y}{100} = 1 - \frac{100}{110} = \frac{10}{110} = \frac{1}{11}$.

$\therefore\ y = \frac{100}{11} = 9\frac{1}{11}\%.$

**80.** Let original price = Rs. $x$.

$\left(\frac{100-p}{100}\right)$ of $\left(\frac{100+p}{100}\right)$ of $x = 1$

$\therefore\ x = \frac{100 \times 100}{(100-p)(100+p)} = \frac{10000}{(10000 - p^2)}.$

**81.** $\frac{10}{100}m = \frac{20}{100}n \Rightarrow \frac{m}{n} = \left(\frac{20}{100} \times \frac{100}{10}\right) = \frac{2}{1}$. $\therefore\ m : n = 2 : 1.$

**82.** Let S.P. = Rs.100. Then, C.P. = Rs.40.
If C.P. is Rs.40, S.P. = Rs.100

If C.P. is Rs.100, S.P. = $\left(\frac{100}{40} \times 100\right)\% = 250\%$.

**83.** Price paid by $B$ = Rs. $\left(\frac{110}{100} \times 10000\right)$ = Rs.11000.

Price paid by $A$ = Rs. $\left(\frac{90}{100} \times 11000\right)$ = Rs.9900.

Thus profit made by $A$ in two transactions
= Rs. (1000 + 100) = Rs.1100.

**84.** Let the original price = Rs.100.
Reduced price = Rs 90.
Increment on Rs 90 = Rs 10
Increment on Rs 100 = $\left(\frac{10}{90} \times 100\right)\% = 11\frac{1}{9}\%$.

**85.** $p = 6q \Rightarrow q = \frac{1}{6}p$.

$\therefore$ q is less than p by $\left(p - \frac{1}{6}p\right) = \frac{5}{6}p$.

$\therefore$ Required percentage = $\left(\frac{5}{6}p \times \frac{1}{p} \times 100\right)\% = 83\frac{1}{3}\%$.

**86.** Failed in English only = (30 – 27) = 3% ;
Failed in Mathematics only = (35 – 27) = 8% ;
Failed in both the subjects = 27% .
Failed in one or both of the subjects = (3 + 8 + 27)% = 38% .

$\therefore$ 62% of $x = 248 \Rightarrow \frac{62}{100} \times x = 248$.

$\therefore x = \frac{248 \times 100}{62} = 400$.

## ANSWERS (EXERCISE 6 B)

| | | | | | |
|---|---|---|---|---|---|
| **1.** (c) | **2.** (b) | **3.** (c) | **4.** (c) | **5.** (c) | **6.** (c) |
| **7.** (c) | **8.** (c) | **9.** (c) | **10.** (c) | **11.** (b) | **12.** (b) |
| **13.** (d) | **14.** (c) | **15.** (c) | **16.** (d) | **17.** (d) | **18.** (d) |
| **19.** (a) | **20.** (c) | **21.** (a) | **22.** (b) | **23.** (d) | **24.** (d) |
| **25.** (d) | **26.** (c) | **27.** (a) | **28.** (c) | **29.** (c) | **30.** (c) |
| **31.** (c) | **32.** (a) | **33.** (a) | **34.** (a) | **35.** (c) | **36.** (b) |
| **37.** (a) | **38.** (b) | **39.** (b) | **40.** (a) | **41.** (b) | **42.** (d) |
| **43.** (d) | **44.** (b) | **45.** (c) | **46.** (d) | **47.** (a) | **48.** (c) |
| **49.** (b) | **50.** (b) | **51.** (b) | **52.** (b) | **53.** (c) | **54.** (b) |
| **55.** (b) | **56.** (c) | **57.** (b) | **58.** (a) | **59.** (d) | **60.** (c) |
| **61.** (b) | **62.** (b) | **63.** (d) | **64.** (c) | **65.** (a) | **66.** (c) |
| **67.** (c) | **68.** (a) | **69.** (d) | **70.** (d) | **71.** (c) | **72.** (c) |
| **73.** (b) | **74.** (a) | **75.** (b) | **76.** (a) | **77.** (c) | **78.** (c) |
| **79.** (b) | **80.** (d) | **81.** (b) | **82.** (d) | **83.** (b) | **84.** (c) |
| **85.** (d) | **86.** (a) | | | | |

# 7

# AVERAGE

$$\text{Average} = \frac{\text{Sum of Quantities}}{\text{Number of Quantities}}.$$

**Ex. 1.** ***The marks obtained by a student in English, Mathematics and Hindi are 65, 87 and 31 respectively. Find his average score.***

**Sol.** *Average Score* $= \left(\frac{65 + 87 + 31}{3}\right) = 61.$

**Ex. 2.** ***The daily earnings of a rickshaw puller during a week are Rs. 31.60, Rs. 29.40, Rs. 40, Rs. 10.90, Rs. 27, Rs. 29.80 and Rs. 27.30 respectively. Find his average daily earning for the week.***

**Sol.** Total earning during 7 days

= Rs. (31.60 + 29.40 + 40 + 10.90 + 27 + 29.80 + 27.30)

= Rs.196.

$\therefore$ *Average daily earning* = *Rs.* $\left(\frac{196}{7}\right)$ = *Rs. 28.*

**Ex. 3.** ***13 Sheep and 9 pigs were bought for Rs. 1291.85. If the average price of a sheep be Rs. 74, what is the average price of a pig ?***

**Sol.** Average price of a sheep = Rs.74

$\therefore$ Total price of 13 sheep = Rs. (74 × 13) = Rs. 962

But, total price of 13 sheep & 9 pigs = Rs. 1291.85

Total price of 9 pigs = Rs.(1291.85 – 962) = Rs. 329.85

*Hence, the average price of a pig* = *Rs.* $\left(\frac{329.85}{9}\right)$ = *Rs. 36.65.*

**Ex. 4.** ***The weights of 12 students (in kgs.) are :***

48.2, 50, 44.5, 49.3, 50.4, 45, 51, 42, 46.8, 48.4, 52, 50.8

**Find** : (i) the average weight of the students;

(ii) the average weight of all students together with that of the teacher whose weight is 62.5 kg;

(iii) the average weight of 11 students, if the one whose weight is 48.2 kg. goes away. (L.I.C. 1991)

**Sol.** Total weight of 12 students = 578.4 kg.

$\therefore$ (i) Average weight of 12 students = $\left(\frac{578.4}{12}\right)$ kg.

= 48.2 kg.

(ii) Total weight of 12 students and 1 teacher

$= (578.4 + 62.5)$ kg. $= 640.9$ kg.

*Average weight of these 13 persons* $= \left(\frac{640.9}{13}\right)$ *kg.* $= 49.3$ *kg.*

(iii) Weight of 11 students $= (578.4 - 48.2)$ kg. $= 530.2$ kg.

*Average weight of these 11 students* $= \left(\frac{530.2}{11}\right)$ kg. $= 48.2$ kg.

**Ex. 5.** ***The average monthly expenditure of a family was Rs. 1050 during first 3 months, Rs. 1260 during next 4 months and Rs. 1326 during last 5 months of the year. If the total saving during the year be Rs. 720, find average monthly income.***

**Sol.** Total income during the year

$=$ Rs. $[(1050 \times 3 + 1260 \times 4 + 1326 \times 5) + 720] =$ Rs.15540.

$\therefore$ Average monthly income $=$ Rs.$\frac{15540}{12} =$ Rs.1295.

**Ex. 6.** ***The average age of 30 children in a class is 9 years. If the teacher's age be included, the average age becomes 10 years. Find the teacher's age.*** **(Bank P.O. 1991)**

**Sol.** Average age of 30 children $= 9$ years.

Total age of 30 children $= (9 \times 30)$ years $= 270$ years.

Average age of 30 children & 1 teacher $= 10$ years.

Sum of their ages $= (10 \times 31)$ years $= 310$ years.

Teacher's age $= (310 - 270)$ years $= 40$ years.

**Ex. 7.** ***The average temperature for Monday, Tuesday and Wednesday was 40° C. The average temperature for Tuesday, Wednesday and Thursday was 41° C. If the temperature on Thursday be 42° C, what was the temperature on Monday ?***

**Sol.** Sum of temperatures on Monday; Tuesday & Wednesday

$= 40 \times 3 = 120°\ C$ ...(i)

Sum of temperatures on Tuesday; Wednesday &Thursday

$= 41 \times 3 = 123°\ C$ ...(ii)

Also, temperature on Thursday = 42° C ...(iii)

$\therefore$ Temperature on Monday $= \{(i) + (iii) - (ii)\}$

$= (120 + 42 - 123) = 39^0 C.$

**Ex. 8.** ***Out of 24 students of a class 6 are 1 m 15 cm in height; 8 are 1 m 5 cm and the rest are 1 m 11 cm. What is the average height of the students ?***

**Sol.** Sum of the heights of all the 24 students

$= (6 \times 115 + 8 \times 105 + 10 \times 111)$ cm. $= 2640$ cm.

$\therefore$ *Average height* $= \frac{2640}{24} = 110$ *cm.* $= 1$ *m* $10$ *cm.*

**Ex. 9.** ***13 chairs and 5 tables were bought for Rs. 8280. If the average price of a table be Rs.1227, find the average price of a chair.***

**Sol.** Total price of 5 tables = Rs. (1227 × 5) = Rs.6135.

Total price of 13 chairs and 5 tables = Rs.8280.

Total price of 13 chairs = Rs. (8280 – 6135) = 2145.

*Average price of a chair* = *Rs.* $\left(\frac{2145}{13}\right)$ = *Rs.165.*

**Ex. 10.** ***The average age of the husband and wife who were married 7 years ago was 25 years then. The average age of the family including the husband, wife and the child born during the interval is 22 years, now. How old is the child now ?*** **(Railways 1991)**

**Sol.** Total age of husband & wife, 7 years ago

= (2 × 25) years = 50 years.

Total age of husband & wife, now = (50 + 7 + 7) = 64 years

Total age of husband, wife and child, now

= (3 × 22) years = 66 years.

*Age of the child, now = (66 – 64) years = 2 years.*

**Ex. 11.** ***A train runs between two stations A & B. While going from A to B, its average speed is 100 km/hr. and while coming back from B to A, its average speed is 150 km/hr. Find the average speed during the whole journey.***

**Sol.** In such questions, the average speed is obtained by using the formula;

$$\text{Average speed} = \frac{2xy}{x+y}$$

$$= \left(\frac{2 \times 100 \times 150}{100 + 150}\right) km/hr. = 120\ km/hr.$$

**Ex. 12.** ***A batsman in his 12th inning makes a score of 63 runs and thereby increases his average score by 2. What is his average after the 12th inning ? He had never been 'not out'.***

**Sol.** Average score after 11th inning = 63 – (2 × 12) = 39

∴ *Average score after 12th inning = 39 + 2 = 41.*

**Ex. 13.** ***A batsman has a certain average runs for 16 innings. In the 17th inning he made a score of 87 runs thereby increasing his average by 3. What is his average after 17th inning ?***

**Sol.** Let the average for 16 innings be $x$ runs.

Total runs in 16 innings = $16x$

Total runs in 17 innings = $16x + 87$.

$$\text{Average of 17 innings} = \frac{16x + 87}{17}.$$

$$\therefore \frac{16x + 87}{17} = x + 3 \text{ or } x = 36.$$

*Thus, average of 17 innings = 36 + 3 = 39.*

**Ex. 14.** ***The average weight of 10 oarsman in a boat is increased by 1.5 kg. when one of the crew, who weighs 68 kg. is replaced by a new man. Find the weight of the new man.***

**Sol.** Total weight increased = (10 × 1.5) kg. = 15 kg.

$\therefore$ *Weight of the new man = (68 + 15) kg. = 83 kg.*

**Ex. 15.** ***The average age of 8 persons in a committee is increased by 2 years when two men aged 35 years and 45 years are substituted by two women. Find the average age of these two women.***

**Sol.** Total sum of ages increased = 2 × 8 = 16 years.

Total sum of ages of persons to be replaced = 35 + 45 = 80 years.

Total sum of ages of two women = 80 + 16 = 96 years.

$\therefore$ *Average age of women* $= \left(\frac{96}{2}\right)$ *years = 48 years.*

**Ex. 16.** ***The average age of a family of 6 members is 22 years. If the age of the youngest member be 7 years, then what was the average age of the family at the birth of the youngest member ?***

**Sol.** Total sum of ages of all members = 6 × 22 = 132 years.

Total sum of ages of all members 7 years ago

= 132 – (6 × 7) = 90 years.

But at that time there were 5 members in the family.

$\therefore$ *Average age at the birth of youngest* $= \left(\frac{90}{5}\right)$ *= 18 years.*

**Ex. 17.** ***The average of 11 results is 50. If the average of first six results is 49 and that of last six is 52, find the sixth result.***

**Sol.** The total sum of 11 results = 11 × 50 = 550.

The total sum of first 6 results = 6 × 49 = 294.

The total sum of last 6 results = 6 × 52 = 312.

*Sixth result = 294 + 312 – 550 = 56.*

**Ex. 18.** ***The average of 6 numbers is 8. What is the 7th number so that average becomes 10 ?*** **(L.I.C. 1991)**

**Sol.** Let 7th number be $x$.

Sum of given 6 numbers = (6 × 8) = 48.

$\therefore$ Average of 7 numbers $= \frac{48 + x}{7}$.

$\therefore \frac{48 + x}{7} = 10$ or $48 + x = 70$ or $x = 22$.

*Hence, the 7th number is 22.*

**Ex. 19.** ***5 years ago, the average of Ram and Shyam's ages was 20 years. Now, the average age of Ram, Shyam and Mohan is 30 years. What will be Mohan's age 10 years hence ?*** **(L.I.C. 1991)**

**Sol.** Total age of Ram & Shyam 5 years ago = $(2 \times 20) = 40$ years.

$\therefore$ Total age of Ram & Shyam now= $(40 + 5 + 5) = 50$ years.

Total age of Ram, Shyam & Mohan now = $(3 \times 30) = 90$ years.

Mohan's age now = $(90 - 50)$ years = 40 years.

*Mohan's age 10 years hence = (40 + 10) years = 50 years.*

## EXERCISE 7A (Subjective)

1. The population of three towns is 46975, 28735 and 30352. Find their average population.
2. The average income of four earning members of a family is Rs. 2350. After the death of one of the earning members, the average income becomes Rs.2340. Find the income of the deceased.

   **Hint.** *Income of the deceased = Rs.* $[(2350 \times 4) - (2340 \times 3)]$.
3. 30 cows are bought for Rs.15,000. The average cost of 18 of them is Rs. 450. What is the average cost of the others ?
4. The average weight of 19 students is 15 kg. By the admission of a new student, the average comes down to 14.8 kg. Find the weight of the new student.
5. Of the three numbers, the first is twice the second and three times the third. If the average of all the three numbers is 88, find the numbers.

   **Hint.** *Let third number = x. Then, first number = 3x and second number =* $\frac{3x}{2}$.

   $$\therefore \frac{1}{3}\left(3x + \frac{3x}{2} + x\right) = 88. \text{ Find } x.$$
6. The average score of a cricketer for 10 matches is 38.9 runs. If the average for the first 6 matches is 42, find the average for the last 4 matches.
7. The average weight of a group of boys and girls is 38 kg. The average weight of the boys is 42 kg. and that of the girls is 33 kg. If the number of boys is 25, find the number of girls.

   **Hint.** *Let the number of girls be x.*

   $$\text{Then, } \frac{(25 \times 42) + 33x}{(25 + x)} = 38 \text{ or } x = 20.$$
8. A man covers a distance of 6 km. by walking at the rate of 2.4 km. per hour. He retraces his path by walking at the rate of 3.6 km. per hour. Find his average speed during whole journey.

   **Hint.** *Average speed* $= \frac{2xy}{x + y} = \left(\frac{2 \times 2.4 \times 3.6}{2.4 + 3.6}\right)$ *km/hr.*
9. The average salary per head of the entire staff of an office including the officers and clerks is Rs. 2650. The average salary per officer is Rs. 4200

and that of the clerks is Rs.2030. If the number of officers is 8, find the number of clerks.

**Hint.** *Let the number of clerks be x.*

*Then,* $\frac{8 \times 4200 + x \times 2030}{8 + x} = 2650.$

*Find x.*

**10.** Visitors to a show were charged Rs.15 each on the first day; Rs.7.50 on the second day and Rs.2.50 on the third day and the total attendance on the three days were in the ratio 2 : 5 : 13. Find the average charge per person for the whole show.

**Hint.** *Let the visitors on the first, second and third day be 2x, 5x and 13x respectively.*

*Then, Average charges* $= \left(\frac{15 \times 2x + 7.50 \times 5x + 2.50 \times 13x}{20x}\right).$

**11.** The average daily temperature from 9th January to 16th January (both days inclusive) was 38.6° C and that from 10th to 17th (both days inclusive) was 39.2° C. The temperature on 9th was 34.6° C. What was the temperature on 17th January ?

**Hint.** *Reqd. Temp.* $= [34.6 + (8 \times 39.2) - (8 \times 38.6)]°$ *C.*

**12.** The average of 8 numbers is 21. If each of the numbers is multiplied by 8, find the average of new set of numbers. **(Central Excise 1989)**

**Hint.** *Sum of eight numbers* $= (8 \times 21) = 168.$

*Sum of new eight numbers* $= (168 \times 8) = 1344.$

$\therefore$ *Average of new set of numbers* $= \left(\frac{1344}{8}\right) = 168.$

**13.** The average age of a committee of 7 trustees is the same as it was 5 years ago, a younger man having been substituted for one of them. How much younger was he than the trustee whose place he took ?

**Hint.** *If the new member would have not been substituted, then the increased age* $= (5 \times 7) = 35$ *years. So, the new member is 35 years younger than the trustee whose place he took.*

**14.** The average height of 40 students is 163 cm. On a particular day, three students *A, B, C* were absent and the average of the remaining 37 students was found to be 162 cm. If *A, B* have equal heights and the height of *C* be 2 cm. less than that of *A*, find the heights of *A, B* and *C*. **(L.I.C. 1991)**

**Hint.** *Let the heights of A, B, C be x cm., x cm. & (x – 2) cm.*

*Then,* $x + x + (x - 2) = (163 \times 40 - 162 \times 37)$. *Find x.*

**15.** The average age of a husband and a wife was 23 years when they were married 5 years ago. The average age of the husband, the wife and a child, who was born during the interval, is 20 years now. How old is the child now ?

## ANSWERS (Exercise 7 A)

**1.** 35354 **2.** Rs.2380 **3.** Rs.575 **4.** 11 kg. **5.** 144, 72, 48
**6.** 34.25 runs **7.** 20 **8.** 2.88 km/hr. **9.** 20 **10.** Rs.5
**11.** 39.4° C **12.** 168 **13.** 35 years **14.** 176 cm, 176 cm., 74 cm.
**15.** 4 years.

## EXERCISE 7 B (Objective Type Questions)

**1.** The average of first five multiples of 3 is :

(*a*) 3 (*b*) 9
(*c*) 12 (*d*) 15

**(Central Excise & I. Tax, 1988)**

**2.** If *a, b, c, d, e* are five consecutive odd numbers, their average is :

(*a*) $5(a + 4)$ (*b*) $\frac{abcde}{5}$
(*c*) $5(a + b + c + d + e)$ (*d*) none of these

**(Bank P.O. Exam. 1989)**

**3.** The average of 30 results is 20 and the average of other 20 results is 30. The average of all the results is :

(*a*) 25 (*b*) 24
(*c*) 50 (*d*) 48

**4.** The average age of three boys is 15 years. If their ages are in ratio 3 : 5 : 7, the age of the youngest boy is :

(*a*) 21 years (*b*) 18 years
(*c*) 15 years (*d*) 9 years

**(S.B.I. P.O. Exam. 1987)**

**5.** The average of 3 numbers is 17 and that of the first two is 16. The third number is :

(*a*) 1 (*b*) 16
(*c*) 17 (*d*) 19

**6.** The average of 25 results is 18; that of first 12 is 14 and of the last 12 is 17. Thirteenth result is :

(*a*) 78 (*b*) 85
(*c*) 28 (*d*) 72 **(C.B.I. Exam. 1990)**

**7.** Out of three numbers, the first is twice the second and is half of the third. If the average of the three numbers is 56, the three numbers in order are :

(*a*) 48, 96, 24 (*b*) 48, 24, 96
(*c*) 96, 24, 48 (*d*) 96, 48, 24

**(Central Excise & I. Tax 1988)**

**8.** Of the three numbers, second is twice the first and is also thrice the third. If the average of the three numbers is 44, the largest number is :

(*a*) 24 (*b*) 36

(*c*) 72 (*d*) 108

**(Central Excise & I. Tax 1989)**

**9.** The average of 50 numbers is 38. If two numbers namely, 45 and 55 are discarded, the average of remaining numbers is :

(*a*) 36.50 (*b*) 37.00

(*c*) 37.50 (*d*) 37.52 **(C.B.I. Exam. 1990)**

**10.** The average height of 30 girls out of a class of 40 is 160 cms. and that of the remaining girls is 156 cms. The average height of the whole class is :

(*a*) 158 cms. (*b*) 158.5 cms.

(*c*) 159 cms. (*d*) 159.5 cms.

**(Central Excise & I. Tax 1988)**

**11.** In a certain primary school, there are 60 boys of age 12 each, 40 of age 13 each, 50 of age 14 each and 50 of age 15 each. The average age (in years) of the boys of the school is :

(*a*) 13.50 (*b*) 13

(*c*) 13.45 (*d*) 14 **(Clerical Grade Exam. 1991)**

**12.** The sum of three numbers is 98. If the ratio between first and second be 2 : 3 and that between second and third be 5 : 8, then the second number is :

(*a*) 30 (*b*) 20

(*c*) 58 (*d*) 48 **(S.S.C. Exam. 1986)**

**13.** The average weight of 8 persons is increased by 2.5 kg. when one of them whose weight is 56 kg. is replaced by a new man. The weight of the new man is :

(*a*) 66 kg. (*b*) 75 kg.

(*c*) 76 kg. (*d*) 86 kg.

**(Central Excise & I. Tax. 1988)**

**14.** The average age of 11 players of a cricket team is decreased by 2 months when two of them aged 17 years and 20 years are replaced by two reserves. The average age of the reserves is :

(*a*) 18 years 3 months (*b*) 17 years 1 month

(*c*) 17 years 7 months (*d*) 17 years 11 months

**15.** The average age of a committee of seven trustees is the same as it was 5 years ago; a young man having been substituted for one of them. The new man compared to the replaced old man, is younger in age by :

(*a*) 5 years (*b*) 7 years

(*c*) 12 years (*d*) 35 years

**(Clerical Grade Exam. 1991)**

**16.** Average monthly income of a family of four earning members was Rs.735. One of the earning members died and therefore the average income came down to Rs.650. The income of the deceased was :

(*a*) Rs. 820 (*b*) Rs. 990
(*c*) Rs. 692.50 (*d*) Rs. 1385

**17.** The average age of an adult class is 40 years. 12 new students with an average age of 32 years join the class, thereby decreasing the average of the class by 4 years. The original strength of the class was :
(*a*) 10 (*b*) 11
(*c*) 12 (*d*) 15
**(Central Excise & I. Tax 1989)**

**18.** A ship sails out to a mark at the rate of 15 km/hr. and sails back at the rate of 10 km/hr. The average rate of sailing is :
(*a*) 12.5 km/hr. (*b*) 12 km/hr.
(*c*) 25 km/hr. (*d*) 5 km/hr.

**19.** On a journey across Delhi a taxi averages 30 km/hr. for 60% of the distance, 20 km/hr. for 20% of it and 10 km/hr. for the remainder. The average speed for the whole journey is :
(*a*) 20 km/hr. (*b*) 22.5 km/hr.
(*c*) 25 km/hr. (*d*) 24.625 km/hr.

**20.** The average consumption of petrol for a car for seven months is 110 litres and for next five months it is 86 litres. The average monthly consumption is :
(*a*) 98 litres (*b*) 100 litres
(*c*) 102 litres (*d*) 96 litres

**21.** The average weight of *A, B, C* is 45 kg. If the average weight of *A* and *B* be 40 kg. and that of *B* and *C* be 43 kg., then the weight of *B* is :
(*a*) 17 kg. (*b*) 20 kg.
(*c*) 26 kg. (*d*) 31 kg.

**22.** The average of the daily income of *A*, *B* and *C* is Rs.60. If *B* earns Rs.20 more than *C* and *A* earns double of what *C* earns, the daily income of *C* is :
(*a*) Rs. 40 (*b*) Rs. 60
(*c*) Rs. 75 (*d*) none of these

**23.** Average earning of a labourer for first four days of a week is Rs.18 and for the last four days is Rs.22. If he earns Rs. 21 on the fourth day, his average earning for the whole week is :
(*a*) Rs. 18.95 (*b*) Rs. 19.65
(*c*) Rs. 19.85 (*d*) Rs. 20.70

**24.** The average expenditure of a man for the first five months is Rs.120 and for the next seven months is Rs.130. His monthly average income if he saves Rs.290 in that year, is :
(*a*) Rs. 160 (*b*) Rs. 170
(*c*) Rs. 150 (*d*) Rs. 140

(Railways 1991)

**25.** Average temperature of first four days of a week is 38.6° C and that of the last 4 days is 40.3° C. If the average temperature of the week be 39.1°C, the temperature on 4th day is :

(*a*) 39.8° C (*b*) 41.9° C
(*c*) 36.7° C (*d*) 38.6° C

**26.** The average temperature of first 3 days is 27° and of the next 3 days is 29°. If the average of the whole week is 28.5° C, the temperature of the last day is :

(*a*) 31.5° (*b*) 10.5°
(*c*) 21° (*d*) 42°

(Railways, 1991)

**27.** The mean temperature of Monday to Wednesday was 37° C and of Tuesday to Thursday was 34° C. If the temperature on Thursday was $\frac{4}{5}$th that of Monday, the temperature on Thursday was :

(*a*) 34° C (*b*) 35.5° C
(*c*) 36° C (*d*) 36.5° C

**28.** A batsman has a certain average runs for 11 innings. In the 12th inning he made a score of 90 runs and thereby decreases his average by 5. His average after 12th inning is :

(*a*) 127 (*b*) 150
(*c*) 145 (*d*) 217

**29.** A cricketer scored 180 runs in the first test and 258 runs in the second. How many runs should he score in the third test so that his average score in the three tests would be 230 runs ?

(*a*) 219 (*b*) 242
(*c*) 334 (*d*) none of these

(Bank P.O. 1991)

**30.** The average salary of 20 workers in an office is Rs.1900 per month. If the manager's salary is added, the average becomes Rs.2000 p.m. The manager's salary is :

(*a*) Rs. 24,000 (*b*) Rs. 25,200
(*c*) Rs. 45,600 (*d*) none of these

(S.B.I. P.O. Exam 1988)

**31.** With an average speed of 40 km/hr., a train reaches its destination in time. If it goes with an average speed of 35 km/hr., it is late by 15 minutes. The total journey is :

(*a*) 40 km. (*b*) 70 km.
(*c*) 30 km. (*d*) 80 km.

**32.** In an examination, a student scores 4 marks for every correct answer and loses 1 mark for every wrong answer. A student attempts all the 100 questions and scores 210 marks. The number of questions, he answered correctly was :
(*a*) 70 (*b*) 62
(*c*) 64 (*d*) 68

**33.** The average weight of 3 men *A*, *B* and *C* is 84 kg. Another man *D* joins the group and the average now becomes 80 kg. If another man *E*, whose weight is 3 kg. more than that of *D*, replaces *A*, then average weight of *B*, *C*, *D* and *E* becomes 79 kg. The weight of *A* is :
(*a*) 70 kg. (*b*) 72 kg.
(*c*) 75 kg. (*d*) 80 kg. **(Bank P.O. 1989)**

**34.** 10 years ago, the average age of a family of 4 members was 24 years. 2 children having been born, the average age of the family is same today. The present age of youngest child if they differ in age by 2 years, is :
(*a*) 1 year (*b*) 2 years
(*c*) 3 years (*d*) 5 years

**35.** The average age of *A*, *B*, *C*, *D* 5 years ago was 45 years. By including *X*, the present average of all the five is 49 years. The present age of *X* is :
(*a*) 64 years (*b*) 48 years
(*c*) 45 years (*d*) 40 years **(Bank P.O., 1988)**

**36.** The average height of 30 boys, out of a class of 50, is 160 cm. If the average height of the remaining boys is 165 cm., the average height of the whole class (in cm.) is :
(*a*) 161 (*b*) 162
(*c*) 163 (*d*) 164
**(Clerical Grade Exam, 1989)**

**37.** The average of first five prime numbers is :
(*a*) 5.0 (*b*) 5.2
(*c*) 5.6 (*d*) 6.0
**(Clerical Grade Exam, 1989)**

**38.** The average score of a cricketer for 10 matches is 43.9 runs. If the average for the first six matches is 53, the average for the last 4 matches is :
(*a*) 30.25 (*b*) 29.75
(*c*) 31 (*d*) 17.15

**39.** There were 35 students in a hostel. If the number of students increased by 7, the expenses of the mess were increased by Rs.42 per day, while the average expenditure per head diminished by Re.1. The original expenditure of the mess was :

(*a*) Rs. 400 (*b*) Rs. 420
(*c*) Rs. 432 (*d*) Rs. 442

**40.** The average salary per head of all the workers in a workshop is Rs.850. If the average salary per head of 7 technicians is Rs.1000 and the average salary per head of the rest is Rs.780. The total number of workers in the workshop is :

(*a*) 18 (*b*) 20
(*c*) 22 (*d*) 24

## HINTS & SOLUTIONS (Exercise 7B)

**1.** Average $= \frac{3(1+2+3+4+5)}{5} = \left(\frac{3 \times 15}{5}\right) = 9.$

**2.** Average $= \frac{a+a+2+a+4+a+6+a+8}{5} = (a+4)$

**3.** Total of 50 results $= (30 \times 20 + 20 \times 30) = 1200.$

$\therefore$ Average $= \frac{1200}{50} = 24.$

**4.** Let their ages be $3x$, $5x$ and $7x$ years.

Then, $3x + 5x + 7x = 3 \times 15$ or $15x = 45$ or $x = 3$.

$\therefore$ Age of youngest boy $= (3 \times 3)$ years $= 9$ years.

**5.** Third number $= [(3 \times 17) - (2 \times 16)] = (51 - 32) = 19.$

**6.** Thirteenth result $= [(25 \times 18) - (12 \times 14 + 12 \times 17)]$

$= [450 - (168 + 204)] = 450 - 372 = 78.$

**7.** Let the numbers be $2x$, $x$ and $4x$.

Average $= \frac{2x + x + 4x}{3} \Rightarrow \frac{7x}{3} = 56$

$\therefore\ x = \frac{3 \times 56}{7} = 24.$

Hence, the numbers in order are 48, 24 and 96.

**8.** Let the numbers be $x$, $2x$ and $\frac{2}{3}x$.

Average $= \frac{x + 2x + \frac{2}{3}x}{3} \Rightarrow \frac{11x}{9} = 44$

$\therefore\ x = \frac{44 \times 9}{11} = 36.$

So, the numbers are 36, 72 and 24.

Hence, the largest one is 72.

**9.** Total of 50 numbers $= 50 \times 38 = 1900.$

Average of 48 numbers $= \frac{1900 - (45 + 55)}{48}$

$= \frac{1800}{48} = 37.5.$

**10.** The average of the whole class

$= \frac{(30 \times 160 + 10 \times 156)}{40} = 159$ cms.

**11.** Average $= \frac{60 \times 12 + 40 \times 13 + 50 \times 14 + 50 \times 15}{60 + 40 + 50 + 50}$

$= \frac{720 + 520 + 700 + 750}{200} = \frac{2690}{200} = 13.45$

**12.** Let the numbers be $x, y, z$. Then,

$x + y + z = 98, \frac{x}{y} = \frac{2}{3}$ and $\frac{y}{z} = \frac{5}{8}$.

$\therefore x = \frac{2y}{3}$ and $z = \frac{8y}{5}$.

So, $\frac{2y}{3} + y + \frac{8y}{5} = 98$ or $\frac{49y}{15} = 98$ or $y = 30$.

**13.** Total weight increased = (2.5 × 8) kg.= 20 kg.
Weight of the new man = (56 + 20) kg. = 76 kg.

**14.** Total decrease = (11 × 2) months = 1 yr 10 mths.
∴ Total age of reserves = (17 + 20) yrs. – (1 yr 10 mths.)
= 35 years 2 months.
∴ Average age of reserves = 17 years 7 months.

**15.** During these 5 years, the total age would have increased by (7 × 5) = 35 years.
But, it remains the same as it was 5 years ago .
∴ The new man is younger than the replaced old man by 35 years.

**16.** The income of the deceased
= Rs. (735 × 4 – 650 × 3) = Rs. 990.

**17.** $40x + 12 \times 32 = (12 + x) \times 36 \Rightarrow x = 12$.

**18.** **Formula :** If the body covers equal distances at $x$ km/hr. and $y$ km/hr., then average speed $= \frac{2xy}{x+y}$

∴ Average speed $= \left(\frac{2 \times 15 \times 10}{15 + 10}\right)$ km/hr. = 12 km/hr.

**19.** Let the total journey be $x$ km.

Time taken $= \left(\frac{60x}{100 \times 30} + \frac{20x}{100 \times 20} + \frac{20x}{100 \times 10}\right)$ hrs. $= \frac{x}{20}$ hrs.

Average speed $= \left(x \times \frac{20}{x}\right)$ km/hr. = 20 km/hr.

**20.** Total consumption of petrol in 12 months
= (110 × 7 + 86 × 5) litres = 1200 litres

Average consumption $= \left(\frac{1200}{12}\right)$ litres = 100 litres.

**21.** Total weight of $A, B, C = (45 \times 3)$ kg. = 135 kg.
Total weight of $A$ and $B = (40 \times 2)$ kg. = 80 kg.
Total weight of $B$ and $C = (43 \times 2)$ kg. = 86 kg.
Total weight of $(A + 2B + C) = (80 + 86)$ kg. = 166 kg.
$\therefore$ Weight of $B = (166 - 135)$ kg. = 31 kg.

**22.** Suppose $C$ earns Rs.$x$. Then,
$x + (x + 20) + 2x = 3 \times 60$ or $x = 40$.
$\therefore$ Daily income of $C$ is Rs. 40.

**23.** Income for 7 days
= Rs. $[18 \times 4 + 22 \times 4 - 21]$ = Rs.139.
$\therefore$ Average income = Rs. $\left(\frac{139}{7}\right)$ = Rs.19.85.

**24.** Total income for 12 months = Rs. $(120 \times 5 + 130 \times 7 + 290)$
= Rs.1800.
$\therefore$ Average monthly income = Rs. $\left(\frac{1800}{12}\right)$ = Rs.150.

**25.** Temperature on 4th day
$= (38.6 \times 4 + 40.3 \times 4 - 39.1 \times 7)^\circ C = 41.9^\circ C$.

**26.** Temperature of the last day
$= [28.5 \times 7 - (27 \times 3) - (29 \times 3)]^\circ = 31.5^\circ$.

**27.** Sum of temperatures on $(M + T + W)$
$= (37 \times 3)^\circ = 111^\circ$ ...(i)
Sum of temperatures on $(T + W + Th) = (34 \times 3)^\circ = 102^\circ$
Sum of temperatures on $\left(T + W + \frac{4}{5}M\right) = 102^\circ$ ...(ii)
$\therefore$ (i) – (ii) gives :
$\frac{1}{5}$th of temperature on Monday $= (111 - 102) = 9^\circ$.
or Temperature on Monday = $45^\circ$
$\therefore$ Temperature on Thursday $= \frac{4}{5} \times 45^\circ = 36^\circ C$.

**28.** Let the average for 11 innings be $x$. Then,
$\frac{11x + 90}{12} = (x - 5)$ or $x = 150$.
$\therefore$ Average for 12 innings $= (150 - 5) = 145$.

**29.** Let the runs he should score in third test be $x$. Then,
$\frac{180 + 258 + x}{3} = 230 \Rightarrow x = 252$.

**30.** Total monthly salary of 21 personals
= Rs. $(21 \times 2000)$ = Rs.42000.
Total monthly salary of 20 personals
= Rs. $(20 \times 1900)$ = Rs.38000.
Monthly salary of the manager = Rs.4000.
Annual salary of the manager = Rs.48000.

**31.** Let the required distance be $x$ km. Then,
$\frac{x}{35} - \frac{x}{40} = \frac{1}{4}$ or $\frac{x}{280} = \frac{1}{4}$ or $x = \frac{280}{4} = 70$ km.

**32.** Suppose he answered $x$ questions correctly.
Number of wrong answers = $(100 - x)$
$\therefore 4x - (100 - x) = 210$ or $x = 62$.

**33.** Weight of $D = (80 \times 4 - 84 \times 3)$ kg. = 68 kg.
Weight of $E = (68 + 3)$ kg. = 71 kg.
$(B + C + D + E)$'s weight = $(79 \times 4)$ kg. = 316 kg.
$\therefore (B + C)$'s weight = $[316 - (68 + 71)]$ kg. = 177 kg.
Hence, $A$'s weight = $[(84 \times 3) - 177]$ kg. = 75 kg.

**34.** Total age now = $(24 \times 6)$ years = 144 years.
Total age of 4 members 10 years ago
= $(24 \times 4)$ years = 96 years.
Total age of these 4 members now = $(96 + 40)$ years
= 136 years
Sum of the ages of two children now
= $(144 - 136)$ years = 8 years
$\therefore x + x + 2 = 8$ or $x = 3$.

**35.** Total age of $A, B, C, D$ 5 years ago
= $(45 \times 4)$ years = 180 years.
Total present age of $A, B, C, D$ and $X$
= $(49 \times 5)$ years = 245 years
Present age of $A, B, C$ and $D = (180 + 5 \times 4)$ years
= 200 years.
$\therefore$ Present age of $x$ = 45 years.

**36.** Total height of 30 boys = $30 \times 160 = 4800$.
Total height of 20 boys = $20 \times 165 = 3300$.
Total height of 50 boys = 8100.
Average height of 50 boys = $\frac{8100}{50} = 162$.

**37.** Average = $\frac{2 + 3 + 5 + 7 + 11}{5} = \frac{28}{5} = 5.6$.

**38.** Total of last 4 matches = $(43.9 \times 10 - 53 \times 6) = 121$.

Average of these matches $= \frac{121}{4} = 30.25$.

**39.** Let the original expenditure per head be Rs. $x$.
Then, $35x + 42 = (x - 1) \times 42$ or $x = 12$.
$\therefore$ Original expenditure = Rs. $(35 \times 12)$ = Rs.420.

**40.** Let the number of all workers be $x$.
Then, $850x = (7 \times 1000) + (x - 7) \times 780$ or $x = 22$.

## ANSWERS (Exercise 7B)

| | | | | | |
|---|---|---|---|---|---|
| **1.** (*b*) | **2.** (*d*) | **3.** (*b*) | **4.** (*d*) | **5.** (*d*) | **6.** (*a*) |
| **7.** (*b*) | **8.** (*c*) | **9.** (*c*) | **10.** (*c*) | **11.** (*c*) | **12.** (*a*) |
| **13.** (*c*) | **14.** (*c*) | **15.** (*d*) | **16.** (*b*) | **17.** (*c*) | **18.** (*b*) |
| **19.** (*a*) | **20.** (*b*) | **21.** (*d*) | **22.** (*a*) | **23.** (*c*) | **24.** (*c*) |
| **25.** (*b*) | **26.** (*a*) | **27.** (*c*) | **28.** (*c*) | **29.** (*d*) | **30.** (*d*) |
| **31.** (*b*) | **32.** (*b*) | **33.** (*c*) | **34.** (*c*) | **35.** (*c*) | **36.** (*b*) |
| **37.** (*c*) | **38.** (*a*) | **39.** (*b*) | **40.** (*c*) | | |

# 8

# Ratio & Proportion

**Ratio.** The ratio of two quantities (in same units) is the fraction that one quantity is of the other. Thus, the ratio 2 to 3 is $\frac{2}{3}$, written as 2 :3. The first term of a ratio is called antecedent and the second term is known as consequent.

It may be noted that $10 : 15 = \frac{10}{15} = \frac{2}{3} = 2 : 3$ and therefore multiplication of each term of a ratio by the same number does not affect the ratio.

**Proportion.** The equality of two ratios is called Proportion.

Since 2 : 3=4 : 6, we write 2 : 3 : : 4 : 6 and say that 2, 3, 4 & 6 are in proportion. We call 2, 3, 4 & 6 as the 1st, 2nd, 3rd & 4th proportional respectively. The first and the fourth proportionals are called the extreme terms while the second & the third proportionals are called mean terms.

Also, Product of means = Product of extremes.

**Ex. 1.** ***A school boy 1.4 metres tall casts a shadow 1.2 metres long at the same time when the school building casts a shadow 5.4 metres long. Find the height of the building.***

**Sol.** Let the height of the building be $x$ metres. Then,

$$\frac{1.4}{1.2} = \frac{x}{5.4} \text{ or } x = \frac{1.4 \times 5.4}{1.2} = 6.3.$$

Hence, the height of the building is 6.3 metres.

**Ex. 2.** ***Divide Rs. 81 in the ratio 2 : 7.***

**Sol.** Sum of ratios = 2+7 = 9.

$$\therefore \text{ First part} = \text{Rs. } \frac{81 \times 2}{9} = \text{Rs. } 18.$$

$$\text{And, Second part} = \text{Rs. } \frac{81 \times 7}{9} = \text{Rs. } 63.$$

**Ex. 3.** ***Divide 94 into two parts in such a way that fifth part of the first and eighth part of the second are in the ratio 3:4.***

**Sol.** Let these parts be $x$ & $y$. Then,

$$\frac{x}{5} : \frac{y}{8} = 3 : 4 \text{ or } 8x : 5y = 3 : 4.$$

$$\therefore \frac{8x}{5y} = \frac{3}{4} \text{ or } \frac{x}{y} = \frac{5 \times 3}{4 \times 8} = \frac{15}{32}.$$

Thus, $x : y = 15 : 32$.

Now, sum of ratios = 15+32 = 47.

$\therefore$ First part = $\frac{94 \times 15}{47} = 30$, second part = $\frac{94 \times 32}{47} = 64$.

**Ex. 4.** ***Compare the ratios 2 : 3 & 4 : 7.***

**Sol.** $2:3 = \frac{2}{3}$ & $4:7 = \frac{4}{7}$.

Now, L.C.M. of denominators of $\frac{2}{3}$ & $\frac{4}{7}$ is 21.

$\therefore \frac{2}{3} = \frac{2 \times 7}{3 \times 7} = \frac{14}{21}$ & $\frac{4}{7} = \frac{4 \times 3}{7 \times 3} = \frac{12}{21}$.

Clearly, $\frac{14}{21} > \frac{12}{21}$ or $(2:3) > (4:7)$.

**Ex. 5.** ***If a : b = 4 : 5 & b : c = 6 : 7, find a : c.***

**Sol.** $\frac{a}{b} = \frac{4}{5}$ & $\frac{b}{c} = \frac{6}{7}$.

$\therefore \frac{a}{c} = \frac{a}{b} \times \frac{b}{c} = \frac{4}{5} \times \frac{6}{7} = \frac{24}{35}$.

Hence, $a : c = 24 : 35$.

**Ex. 6.** ***In a ratio which is equal to 7 : 8, if the antecedent is 35, what is the consequent ?***

**Sol.** If antecedent is 7, then consequent = 8.

If antecedent is 35, then consequent = $\frac{8}{7} \times 35 = 40$.

**Ex. 7.** ***The ratio between two numbers is 2 : 3. If each number is increased by 4, the ratio between them becomes 5 : 7. What are the numbers ?***

**Sol.** Let the numbers be $2x$ and $3x$. Then,

$\frac{2x+4}{3x+4} = \frac{5}{7}$ or $14x + 28 = 15x + 20$ or $x = 8$.

$\therefore$ The numbers are 16 and 24.

**Ex. 8.** ***What must be added to each term of the ratio 7 : 13 so that the ratio becomes 2 : 3 ?***

**Sol.** Let $x$ be added. Then

$\frac{7+x}{13+x} = \frac{2}{3}$ or $21 + 3x = 26 + 2x$ or $x = 5$.

$\therefore$ 5 must be added to each term.

**Ex. 9.** ***The ratio between two numbers is 3 : 4. If their L.C.M. is 180, find the numbers.***

**Sol.** Let the numbers be $3x$ and $4x$. Then,

L.C.M. of $3x$ & $4x = 3 \times 4 \times x = 12x$.

$\therefore 12x = 180$ or $x = 15$.

So, the numbers are 45 and 60.

**Ex. 10.** ***(i) Find the fourth proportional to 3, 5 & 21.***

***(ii) Find the mean proportional between 64 & 81.***

***(iii) Find the third proportional to 9 & 12.***

**Sol.** (i) Let $3:5::21:x$.

Then, $\frac{3}{5}=\frac{21}{x}$ or $x=\frac{5\times 21}{3}=35$.

(ii) Let $64:x::x:81$.

Then, $\frac{64}{x}=\frac{x}{81}$ or $x=\sqrt{64\times 81}=72$.

(iii) Since third proportional to 9 & 12 is the same as fourth proportional to 9, 12, 12.

Let $9:12::12:x$.

Then, $\frac{9}{12}=\frac{12}{x}$ or $x=\frac{12\times 12}{9}=16$.

**Ex. 11.** ***Find a fraction which bears the same ratio to*** $\frac{1}{27}$ ***that*** $\frac{3}{7}$ ***does to*** $\frac{5}{9}$.

**Sol.** Let the number be $x$. Then,

$x:\frac{1}{27}=\frac{3}{7}:\frac{5}{9}$ or $27x:1=27:35$

or $27x\times 35=1\times 27$ or $x=\frac{1\times 27}{27\times 35}=\frac{1}{35}$.

Hence, the required fraction $=\frac{1}{35}$.

**Ex. 12.** ***Find three numbers in the ratio of 2 : 3 : 5 such that the sum of their squares is 608.***

**Sol.** Let the numbers be $2x$, $3x$ & $5x$. Then,

$4x^2+9x^2+25x^2=608$ or $38x^2=608$.

$\therefore x^2=16$ or $x=4$.

So, the numbers are 8, 12 & 20.

**Ex. 13.** ***Divide 1162 into three parts such that 4 times the first may be equal to 5 times the second and 7 times the third.***

**Sol.** $4\times$ (1st part) $=5\times$ (2nd part) $=7\times$ (3rd part) $=x$ (say)

Then, 1st part $=\frac{x}{4}$, 2nd part $=\frac{x}{5}$; 3rd part $=\frac{x}{7}$.

$\therefore$ Ratio of divisions $=\frac{x}{4}:\frac{x}{5}:\frac{x}{7}$

$= \frac{1}{4} : \frac{1}{5} : \frac{1}{7} = 35 : 28 : 20.$

Sum of ratios $= 35 + 28 + 20 = 83.$

$\therefore$ 1st part $= \frac{1162 \times 35}{83} = 490.$

2nd part $= \frac{1162 \times 28}{83} = 392.$

3rd part $= \frac{1162 \times 20}{83} = 280.$

**Ex. 14.** ***Divide Rs. 680 among A, B & C such that A gets $\frac{2}{3}$ of what B gets and B gets $\frac{1}{4}$th of what C gets.***

**Sol.** Suppose $C$ gets Re. 1. Then, $B$ gets Re. $\frac{1}{4}$ & $A$ gets Rs. $\left(\frac{2}{3} \text{ of } \frac{1}{4}\right)$ or Rs. $\frac{1}{6}$.

Ratios of $A$, $B$ & $C$'s shares $= \frac{1}{6} : \frac{1}{4} : 1 = 2 : 3 : 12.$

So, $A$'s share = Rs. $\frac{680 \times 2}{17}$ = Rs 80.

$B$'s share = Rs. $\frac{680 \times 3}{17}$ = Rs 120.

$C$'s share = Rs. $\frac{680 \times 12}{17}$ = Rs 480.

**Ex. 15.** ***Divide Rs. 1870 in three parts in such a way that half of the first part, one–third of the second part and one-sixth of the third part are equal.***

**Sol.** Let $\frac{1}{2}A = \frac{1}{3}B = \frac{1}{6}C = x$ (say).

Then, $A = 2x$, $B = 3x$ and $C = 6x$.

$\therefore 2x + 3x + 6x = 1870 \Rightarrow x = 170.$

Hence, $A = 2 \times x = 2 \times 170$ = Rs 340;

$B = 3 \times x = 3 \times 170$ = Rs 510;

$C = 6 \times x = 6 \times 170$ = Rs 1020.

**Ex. 16.** ***Divide Rs. 12540 among A, B, C in such a way that A gets $\frac{3}{7}$ of what B and C together receive and B gets $\frac{2}{9}$ of what A and C together receive.***

**Sol.** ($A$'s share) : ($B$ & $C$'s share) = 3 : 7

Sum of these ratios = 3 + 7 = 10

$\therefore$ *A*'s share = Rs. $\left(\frac{12540 \times 3}{10}\right)$ = Rs. 3762.

Again, (*B*'s Share) : (*A* & *C*'s share) = 2 : 9.

sum of these ratios = 2 + 9 = 11.

$\therefore$ *B*'s share = Rs. $\left(\frac{12540 \times 2}{11}\right)$ = Rs. 2280.

So, *C*'s Share = Rs. [12540 – (3762 + 2280)] = Rs. 6498.

**Ex. 17.** ***A sum of Rs. 430 has been distributed among 45 people consisting of men, women and children. The amounts given to men, women and children are in the ratio 12 : 15 : 16. But, the amounts received by each man, woman and child are in the ratio 6 : 5 : 4. Find, what each man, each woman and each child receives.***

**Sol.** Ratio of personal shares = 6 : 5 : 4.

Ratio of the amounts = 12 : 15 : 16.

Ratio of men, women & children = $\frac{12}{6} : \frac{15}{5} : \frac{16}{4}$ = 2 : 3 : 4.

Sum of these ratios = (2 + 3 + 4) = 9.

Number of men = $\left(\frac{45 \times 2}{9}\right)$ = 10,

Number of women = $\left(\frac{45 \times 3}{9}\right)$ = 15,

and, the number of children = 45 – (10 + 15) = 20.

Now, dividing Rs. 430 in the ratio 12 : 15 : 16.

Total amount of men's share = Rs. $\left(\frac{430 \times 12}{43}\right)$ = Rs. 120.

Total amount of women's share = Rs. $\left(\frac{430 \times 15}{43}\right)$ = Rs. 150.

Total amount of children's share = Rs. [430 – (120 + 150)] = Rs. 160.

$\therefore$ Each man's share = Rs. $\left(\frac{120}{10}\right)$ = Rs. 12;

Each woman's share = Rs. $\left(\frac{150}{15}\right)$ = Rs. 10;

Each child's share = Rs. $\left(\frac{160}{20}\right)$ = Rs. 8.

**Ex. 18.** ***A bag contains rupees, fifty paise & twenty-five paise coins in the proportion 5 : 6 : 8. If the total amount is Rs.210, find the number of coins of each kind.***

**Sol.** Let there be 5 rupee coins, 6 fifty paise coins & 8 twenty–five paise coins.

Then, value of 6 fifty paise coins = Rs. 3.

Value of 8 twenty–five paise coins = Rs. 2.

∴ Values of these coins are in the ratio 5 : 3 : 2.

So, value of all rupee coins = Rs. $\left(\frac{210 \times 5}{10}\right)$ = Rs. 105.

Value of all 50 paise coins = Rs. $\left(\frac{210 \times 3}{10}\right)$ = Rs. 63.

Value of all 25 paise coins = Rs. $\left(\frac{210 \times 2}{10}\right)$ = Rs. 42.

Number of rupee coins = (105 × 1) = 105.

Number of 50-paise coins = (63 × 2) = 126.

Number of 25-paise coins = (42 × 4) = 168.

**Ex. 19.** ***An employer reduces the number of his employees in the ratio of 9 : 8 and increases their wages in the ratio 14 : 15. In what ratio the wages bill is increased or decreased ?***

**Sol.** Initially, let the number of employees be 9 and wages per head be Rs.14. Then,

Total wages bill = Rs. (9 × 14) = Rs.126.

Further, reduced number of employees is 8 and increased wages per head is Rs. 15.

∴ New wages bill = Rs. (8 × 15) = Rs.120.

Thus, the ratio of wages bills = 126 : 120 = 21 : 20,

i.e., wages bill is decreased in the ratio 21 : 20.

**Ex. 20.** ***In a mixture of 35 litres, the ratio of milk and water is 4 : 1. Another 7 litres of water is added to the mixture. Find the ratio of milk and water in the resulting mixture.***

**Sol.** Quantity of milk in 35 litres of mix. = $\frac{35 \times 4}{5}$ = 28 litres.

Quantity of water in 35 litres of mix. = 35–28 = 7 litres.

Quantity of new mix. = 35 + 7 = 42 litres.

Quantity of milk in new mix. = 28 litres.

Quantity of water in new mix.= 7 + 7 = 14 litres.

Ratio of milk & water in new mixture = 28 : 14 = 2 : 1.

**Ex. 21.** ***20 litres of a mixture contains milk and water in the ratio 3 : 1. How much milk must be added to this mixture to have a mixture containing milk and water in the ratio 4 : 1 ?***

**Sol.** Quantity of milk in the given mixture = $\left(\frac{20 \times 3}{4}\right)$ = 15 litres.

Quantity of water in this mixture = (20–15) = 5 litres.

Let $x$ litres of milk be added to given mixture to have the requisite ratio of milk and water.

Then, $\left(\frac{15+x}{5}\right)=\frac{4}{1}$ or $15+x=20$ or $x=5$.

$\therefore$ Milk to be added = 5 litres.

**Ex. 22.** ***A and B are two alloys of gold and copper prepared by mixing metals in the ratio 7 : 2 and 7 : 11 respectively. If equal quantities of alloys are melted to form a third alloy C, find the ratio of gold and copper in C.***

**Sol.** Gold in 1 gm of $A=\frac{7}{9}$ gm.

Gold in 1 gm. of $B=\frac{7}{18}$ gm.

Gold in 2 gm. of $C=\left(\frac{7}{9}+\frac{7}{18}\right)=\left(\frac{21}{18}\right)$ gm $=\frac{7}{6}$ gm.

Copper in 1 gm. of $A=\frac{2}{9}$ gm.

Copper in 1 gm. of $B=\frac{11}{18}$ gm.

Copper in 2 gm. of $C=\left(\frac{2}{9}+\frac{11}{18}\right)$ gm. $=\left(\frac{15}{18}\right)$ gm. $=\frac{5}{6}$ gm.

$\therefore$ Ratio of gold and copper in $C=\frac{7}{6}:\frac{5}{6}=7:5$.

## EXERCISE 8A (Subjective)

1. Fill in the blanks :
   (i) 2 : 3 : : 45 : (....)
   (ii) 1 : 7 : : (....) : 3
   (iii) (....) : 0.5 : : 7 : 14
   (iv) $3^3 : 15^3$ : : 3 : (....)
2. A ratio is equal to 2 : 5. If its antecedent is 9, find its consequent.
3. If $A:B=3:4$ and $B:C=5:6$, find $A:C$.
4. Compare the ratios (2 : 3) and (5 : 6).
5. Find the fourth proportional to :
   (i) 2, 3, 42
   (ii) 6, 5, 36
   (iii) 0.2, 0.7, 1.8
6. Find the third proportional to :
   (i) 6, 8
   (ii) 2.7, 6.3
   (iii) 0.8, 0.02
7. Find the mean proportional between :

(i) 8 and 128

(ii) 7 and 28

(iii) 0.6 and 0.15

**8.** Two sums of money are proportional to 3 : 7. If the first sum is Rs. 71.40, find the other.

**Hint.** *Let 3 : 7 = 71.40 : x. Find x.*

**9.** Two numbers are in the ratio 9 : 4. If 7 is added to each of the numbers the ratio becomes 5 : 3. Find the numbers. **(L.I.C. 1991)**

**10.** Monthly salaries of *A* and *B* are in the ratio 3 : 5. If the salary of each is increased by Rs. 20, their ratio becomes 13 : 21. Find their original salaries. **(L.I.C. 1991)**

**Hint.** *Let their original salaries be 3x and 5x. Then,* $\frac{3x+20}{5x+20}=\frac{13}{21}$.

*Find x.*

**11.** Divide Rs. 2873 in the ratio 9 : 4.

**12.** Divide Rs. 777 among *A, B, C* in such a way that *A* gets half of what *B* gets and *B* gets half of what *C* gets.

**Hint.** *Suppose C gets Rs. x. Then, B gets Rs.* $\frac{1}{2}x$ *and A gets Rs.* $\frac{1}{4}x$.

$\therefore\ A:B:C=\frac{1}{4}x:\frac{1}{2}x:x=1:2:4.$

*Now, Divide Rs.777 in the ratio 1 : 2 : 4.*

**13.** Divide Rs. 3870 among *A, B, C* so that if their shares be diminished by Rs. 5, Rs. 9 and Rs. 16 respectively, the remainder shall be in the ratio 3 : 4 : 5.

**Hint.** *First divide Rs. [3870 – (5 + 9 + 16)] or Rs.3840 in the ratio 3 : 4 : 5. Now add Rs. 5, Rs. 9 and Rs. 16 to these shares.*

**14.** Divide Rs. 5083 among 4 men, 3 women and 5 children so that each man gets (3/2) times of what a woman gets and each woman gets twice of what a child gets.

**Hint.** *Let the child get Re. 1. Then, a woman gets Rs. 2 and a man gets Rs. 3.*

∴ *Ratio of shares of 1 man, 1 woman & 1 child = 3 : 2 : 1.*

*Ratio of shares of 4 men, 3 women & 5 children = 12 : 6 : 5.*

*Now, divide Rs. 5083 in the ratio 12 : 6 : 5.*

**15.** If Rs. 5000 be divided among 10 men, 5 women and 10 boys so that 2 men may have as much as 5 boys and each woman thrice as much as 3 boys, find how much each man, each woman and each boy received.

**Hint.** *Suppose each boy gets Re. 1. Then, each woman gets Rs. 3 and each man gets Rs. (5/2).*

∴ *Ratio of shares of 10 men, 5 women & 10 boys*

$= \left(10 \times \frac{5}{2}\right) : (5 \times 3) : (10 \times 1) = 25 : 15 : 10.$

**16.** In a mixture of 40 litres, the ratio of milk and water is 4 : 1. How much water must be added to this mixture so that the ratio of milk and water becomes 2 : 3.

**Hint.** *Milk* = $\left(40 \times \frac{4}{5}\right)$ = *32 litres, water = 8 litres.*

*Let x litres of water may be added.*

*Then,* $\frac{32}{x+8} = \frac{2}{3}$. *Find x.*

**17.** Milk and water in one pot are in the ratio 1 : 2, while in another pot they are in the ratio 3 : 4. A mixture is made by taking 1 kg. of mixture from each one of the two pots. What is the ratio of milk and water in this new mixture ?

**Hint.** *Milk in final mixture* $= \left(\frac{1}{3} + \frac{3}{7}\right)$ *kg.* $= \frac{16}{21}$ *kg.*

*Water in final mixture* $= \left(\frac{2}{3} + \frac{4}{7}\right)$ *kg.* $= \frac{26}{21}$ *kg.*

**18.** A man leaves Rs. 8600 to be divided among 5 sons, 4 daughters and 2 nephews. If each daughter receives 4 times as much as each nephew and each son five times as much as each nephew, how much does each daughter receive ?

**19.** Divide Rs. 1800 among *A*, *B*, *C* and *D* so that *A* and *B* together get thrice as much as *C* and *D* together, *B* may get four times of what *C* gets and *C* may get $1\frac{1}{2}$ times as much as *D*.

**Hint.** *Suppose D gets Rs. x.*

*Then, C gets Rs.* $\frac{3x}{2}$ *and B gets Rs.* 6 *x*.

*Let A get Rs. y.*

*Then,* $y + 6x = 3\left(\frac{3x}{2} + x\right) \Rightarrow y = \left(\frac{15x}{2} - 6x\right) = \frac{3x}{2}$.

$\therefore A : B : C : D = \frac{3x}{2} : 6x : \frac{3x}{2} : x = 3 : 12 : 3 : 2.$

**20.** A factory employs skilled workers, unskilled workers and clerks in the proportion 8 : 5 : 1 and the wages of a skilled worker, an unskilled worker and a clerk are in the ratio 5 : 2 : 3. When 20 unskilled workers are employed, the total daily wages of all amount to Rs. 3180. Find the daily wages paid to each category of employees.

**Hint.** *Skilled workers* $= \left(\frac{8}{5} \times 20\right) = 32.$

*Number of clerks* $= \left(\frac{1}{5} \times 20\right) = 4.$

*Ratio of amount to 32 skilled workers, 20 unskilled workers and 4 clerks* $= 5 \times 32 : 2 \times 20 : 3 \times 4 = 160 : 40 : 12$ *or* $40 : 10 : 3$.

*Now, divide Rs. 3180 in the ratio 40 : 10 : 3.*

**21.** In the journey of 168 km. performed by taxi, rail and bus, the distance travelled by three ways in that order are in the ratio 8 : 1 : 3 and the charges per km. are in the ratio 8 : 1 : 4. The taxi charges being Rs.3.60 per km., find the total charges of the journey.

**Hint.** *Divide 168 km. in the ratio 8 : 1 : 3 to get 112 km., 14 km. and 42 km.*

*Now, taxi charges = Rs. 3.60 per km;*

*rail charges* $=$ Rs. $\left(\frac{1}{8} \times 3.60\right) =$ *Rs. 0.45/km;*

*bus charges* $=$ Rs. $\left(\frac{4}{8} \times 3.60\right) =$ *Rs. 1.80/km.*

$\therefore$ *Total charges* $=$ Rs. $(112 \times 3.60 + 14 \times 0.45 + 42 \times 1.80)$.

## ANSWERS (EXERCISE 8A)

**1.** (i) 67.5 (ii) $\frac{3}{7}$ (iii) 0.25 (iv) 375

**2.** $\frac{45}{2}$ **3.** 5 : 8 **4.** (2 : 3) < (5 : 6)

**5.** (i) 63 (ii) 30 (iii) 6.3

**6.** (i) $\frac{32}{3}$ (ii) 14.7 (iii) 0.0005

**7.** (i) 32 (ii) 14 (iii) 0.3

**8.** Rs. 166.60 **9.** 18, 8 **10.** Rs. 240, Rs. 400

**11.** Rs. 1989, Rs. 884 **12.** Rs. 111, Rs. 222, Rs. 444

**13.** Rs. 965, Rs. 1289, Rs. 1616 **14.** Rs. 2652, Rs. 1326, Rs. 1105

**15.** Rs. 2500, Rs. 1500, Rs. 1000 **16.** 40 litres

**17.** 8 : 13 **18.** Rs. 800

**19.** Rs. 270, Rs. 1080, Rs. 270, Rs. 180 **20.** Rs. 2400, Rs. 600, Rs. 180

**21.** Rs. 485.10

## EXERCISE 8B (Objective Type Questions)

**1.** If 10% of $m$ is the same as 20% of $n$, then $m : n$ is equal to :

(*a*) 1 : 2 (*b*) 2 : 1

(*c*) 5 : 1 (*d*) 10 : 1 **(C.B.I. Exam. 1990)**

**2.** 0.5 of a number equals 0.07 of another. Then the ratio of the two numbers is :

(*a*) 7 : 50 (*b*) 7 : 25

(*c*) 7 : 5 (*d*) 1 : 14

**3.** The ratio which $\left(\frac{1}{3} \text{ of Rs. } 9.30\right)$ bears to (0.6 of Rs. 1.55) is :

(*a*) 1 : 3 (*b*) 10 : 3

(*c*) 3 : 10 (*d*) 3 : 1

**4.** The ratio of 20 miles to 25 km. is

(*a*) 6 : 5 (*b*) 4 : 5

(*c*) 32 : 25 (*d*) 33 : 50

**5.** There are 3 numbers $A$, $B$, $C$ given in such a way that one-third of $A$, one-fourth of $B$ and one-fifth of $C$ are equal. Then, ratio between $A$, $B$ and $C$ is :

(*a*) 3 : 4 : 5 (*b*) 4 : 3 : 5

(*c*) 5 : 4 : 3 (*d*) $\frac{1}{3} : \frac{1}{4} : \frac{1}{5}$

**6.** If $A : B = 2 : 3$ and $B : C = 4 : 5$, then $C : A$ is equal to :

(*a*) 15 : 8 (*b*) 12 : 10

(*c*) 8 : 5 (*d*) 8 : 15

**(Railway Recruitment, 1991)**

**7.** There are 3 numbers $A$, $B$, $C$ such that twice $A$ is equal to thrice $B$ and four times $B$ is equal to five times $C$. The ratio between $A$ and $C$ is :

(*a*) 3 : 4 (*b*) 8 : 15

(*c*) 15 : 8 (*d*) 4 : 3

**8.** The ratio between two numbers is 3 : 4. If each number be increased by 6, the ratio becomes 4 : 5. The difference between the numbers is :

(*a*) 1 (*b*) 3

(*c*) 6 (*d*) 8

**9.** Two numbers are in ratio 7 : 9. If 12 is subtracted from each of them, the ratio becomes 3 : 5. The product of the numbers is :

(*a*) 432 (*b*) 567

(*c*) 1575 (*d*) none of these

**10.** The sum of the cubes of two numbers in the ratio 3 : 2 is 4375. The sum of the numbers is :

(*a*) $(4375)^{1/3}$ (b) 5

(c) 15 (d) 25

**11.** What number should be subtracted from each of the numbers 54, 71, 75 and 99 so that the remainders may be proportional ?

(*a*) 1 (*b*) 2
(*c*) 3 (*d*) 6

**12.** What must be added to each term of the ratio 7 : 13 so that the ratio becomes 2 : 3 ?

(*a*) 1 (*b*) 2
(*c*) 3 (*d*) 5

**13.** What number should be added to 6, 14, 18, 38 to make it equally proportionate ?

(*a*) 1 (*b*) 2
(*c*) 3 (*d*) 4

**14.** A fraction bears the same ratio to $\frac{1}{27}$ as $\frac{3}{7}$ does to $\frac{5}{9}$. The fraction is

(*a*) $\frac{7}{45}$ (*b*) $\frac{1}{35}$
(*c*) $\frac{45}{7}$ (*d*) $\frac{5}{21}$

**15.** The fourth proportional to numbers 8, 12 and 18 is :

(*a*) 22 (*b*) 24
(*c*) 27 (*d*) 30

**16.** The third proportion to numbers 0.8 and 0.2 is :

(*a*) 0.05 (*b*) 0.8
(*c*) 0.032 (*d*) none of these

**17.** The mean proportional between the numbers 8 and 18 is :

(*a*) 12 (*b*) 13
(*c*) $\frac{32}{9}$ (d) none of these

**18.** What number must be added to the numbers 3, 7 and 13, so that they are in a continued proportion ?

(*a*) 9 (*b*) 8
(*c*) 6 (*d*) 5

**19.** Which number should replace both the question marks in $\frac{36}{?} = \frac{?}{81}$.

(*a*) 48 (*b*) 54
(*c*) 56 (*d*) 64

**20.** If $\frac{1}{5} : \frac{1}{x} = \frac{1}{x} : \frac{1}{1.25}$, the value of $x$ is nearly

(*a*) 1.25 (*b*) 1.50
(*c*) 2.5 (*d*) 2.25

**21.** Out of ratios 7 : 15, 15 : 23, 17: 25, 21 : 29, the smallest one is :

(*a*) 17 : 25 (*b*) 7 : 15
(*c*) 15 : 23 (*d*) 21 : 29

**22.** 94 is divided into two parts in such a way that fifth part of the first and eighth part of the second are in ratio 3 : 4. The first part is :
(*a*) 27 (*b*) 30
(*c*) 36 (*d*) 48

**23.** If $A : B = 7 : 9$ and $B : C = 3 : 5$, then $A : B : C$ is :
(*a*) 7 : 9 : 5 (*b*) 21 : 35 : 45
(*c*) 7 : 9 : 15 (*d*) 7 : 3 : 15

**(Astt. Grade Exam. 1990)**

**24.** A profit of Rs.1581 is to be divided amongst three partners *A*, *B* and *C* in ratio $\frac{1}{3} : \frac{1}{2} : \frac{1}{5}$. The share of *C* will be :
(*a*) Rs. 255 (*b*) Rs. 306
(*c*) Rs. 408 (*d*) none of these

**25.** Rs.68000 is divided amongst Ashish, Kapil and Sumit in the ratio $\frac{1}{2} : \frac{1}{4} : \frac{5}{16}$. The difference between largest and smallest share is :
(*a*) Rs.6,000 (*b*) Rs.9,200
(*c*) Rs.14,440 (*d*) Rs.16,000

**26.** 6 men, 8 women and 6 children complete a job for a sum of Rs. 950. If their individual wages are in the ratio 4 : 3 : 2, the total money earned by the children is :
(*a*) Rs. 230 (*b*) Rs. 215
(*c*) Rs. 195 (*d*) Rs. 190

**27.** Rs.680 has been divided among *A, B, C* such that *A* gets (2/3) of what *B* gets and *B* gets (1/4) of what *C* gets. Then, *B*'s share is :
(*a*) Rs. 60 (*b*) Rs. 80
(*c*) Rs. 120 (*d*) Rs. 160

**28.** Rs.1870 has been divided into 3 parts in such a way that half of first part, one third of second part and one-sixth of third part are equal. The third part is :
(*a*) Rs. 510 (*b*) Rs. 680
(*c*) Rs. 1020 (*d*) Rs. 850

**29.** Rs. 600 has been divided among *A, B* and *C* in such a way that Rs.40 more than (2/5) of *A*'s share, Rs.20 more than (2/7) of B's share, Rs.10 more than (9/17) of *C*'s share, are all equal. *A*'s share is :
(*a*) Rs. 280 (*b*) Rs. 170
(*c*) Rs. 150 (*d*) Rs. 200

**(Railway Recruitment, 1991)**

**30.** Rs.777 is divided among *A, B* and *C* in such a way that *A* gets half of what *B* gets and *B* gets half of whatever *C* gets. Then, *B*'s share is :

(*a*) Rs. 111 (*b*) Rs. 388.50
(*c*) Rs. 194.25 (*d*) Rs. 222

**31.** Rs. 385 has been divided among *A, B, C* in such a way that *A* receives (2/9)th of what *B* and *C* together receive. Then *A*'s share is :
(*a*) Rs. 70 (*b*) Rs. 77
(*c*) Rs. 82.50 (*d*) Rs. 85

**32.** Some money is divided among three persons *A, B* and *C* in such a way that 5 times *A*'s share, 3 times *B*'s share and 2 times *C*'s share are all equal. The ratio between *A, B* and *C*'s shares is :
(*a*) 5 : 3 : 2 (*b*) 2 : 2 : 5
(*c*) 15 : 10 : 6 (*d*) 6 : 10 : 15

**33.** Rs.53 is divided among *A, B* and *C* in such a way that *A* gets Rs.7 more than what *B* gets and *B* gets Rs.8 more than what *C* gets. Then, the ratio of their shares is :
(*a*) 16 : 9 : 18 (*b*) 25 : 18 : 10
(*c*) 18 : 25 : 10 (*d*) 15 : 8 : 30

**34.** Rs.2430 has been divided among three persons *A, B* and *C* in such a way that if their shares be diminished by Rs. 5, Rs. 10 and Rs. 15 respectively the remainders are in the ratio 3 : 4 : 5. Then, *A*'s share is :
(*a*) Rs. 800 (*b*) Rs. 600
(*c*) Rs. 595 (*d*) Rs. 605

**35.** A sum of Rs.1300 is divided between *A, B, C* and *D*. What is *A*'s share, if :

$$\frac{A\text{'s share}}{B\text{'s share}} = \frac{B\text{'s share}}{C\text{'s share}} = \frac{C\text{'s share}}{D\text{'s share}} = \frac{2}{3}$$

(*a*) Rs. 140 (*b*) Rs. 160
(*c*) Rs. 240 (*d*) Rs. 320

**36.** If $x : y = 2 : 3$ and $2 : x = 1 : 2$, the value of $y$ is :
(*a*) $\frac{1}{3}$ (*b*) $\frac{3}{2}$
(*c*) 6 (*d*) 4

**37.** A sum of money is divided among *A, B* and *C*, so that to each rupee *A* gets, *B* gets 65 paise and *C* gets 35 paise. If *C*'s share is Rs.28, the sum is :
(*a*) Rs. 120 (*b*) Rs. 140
(*c*) Rs. 160 (*d*) Rs. 180

**38.** A bag contains Rs.102 in the form of rupee, 50 paise and 10 paise coins in ratio 3 : 4 : 10. The number of 10 paise coins is :
(*a*) 60 (*b*) 80
(*c*) 170 (*d*) 340

**39.** A bag contains Rs. 5, Rs. 2 and Re.1 coins. Rs. 5 and Rs. 2 coins are in ratio 10 : 7 and Rs. 2 and Re.1 coins are in ratio 1 : 3. If the total amount in the bag is Rs. 340, number of Rs. 2 coins is :

(*a*) 28 (*b*) 36
(*c*) 40 (*d*) 48

**40.** The cost of 3 books is in the ratio 4 : 5 : 7. The difference between the highest and the lowest cost is Rs.60. The cost of the modest book is :

(*a*) Rs. 80 (*b*) Rs. 90
(*c*) Rs.100 (*d*) Rs. 120

**41.** If five chairs and four tables cost as much as three chairs and seven tables, the ratio of the cost of one chair to cost of one table is :

(*a*) 3 : 2 (*b*) 4 : 3
(*c*) 3 : 4 (*d*) 1 : 3

**42.** *A*, *B* and *C* play a cricket match. The ratio of the runs scored by them in the match are as follows : $A : B = 2 : 3$ and $B : C = 2 : 5$. If the total runs scored by all of them are 75, runs scored by B are :

(*a*) 15 (*b*) 18
(*c*) 21 (*d*) 24

**43.** Two whole numbers whose sum is 64, cannot be in the ratio :

(*a*) 5 : 3 (*b*) 7 : 1
(*c*) 3 : 4 (*d*) 9 : 7

**44.** The ratio between two numbers is 3 : 4 and their L.C.M. is 180. The first number is :

(*a*) 60 (*b*) 45
(*c*) 15 (*d*) 20

**45.** A man has some hens and cows. If the number of heads be 48 and number of feet equals 140, the number of hens will be :

(*a*) 26 (*b*) 22
(*c*) 23 (*d*) 24

**46.** The ratio of number of boys and girls in a school of 720 students is 7 : 5. How many more girls should be admitted to make the ratio 1 : 1 ?

(*a*) 90 (*b*) 120
(*c*) 220 (*d*) 240

**47.** 729 ml of a mixture contains milk and water in the ratio 7 : 2. How much more water is to be added to get a new mixture containing milk and water in the ratio 7 : 3 ?

(*a*) 600 ml. (*b*) 710 ml.
(*c*) 520 ml. (*d*) none of these

**(Railway Recruitment, 1991)**

**48.** Gold is 19 times as heavy as water and copper 9 times as heavy as water. The ratio in which these two metals be mixed so that the mixture is 15 times as heavy as water, is :

(*a*) 1 : 2 (*b*) 2 : 3
(*c*) 3 : 2 (*d*) 19 : 135

**(Delhi Police & C.B.I. 1990)**

**49.** If $A$ is $\frac{1}{3}$ of $B$ and $B$ is $\frac{1}{2}$ of $C$, then $A : B : C$ is :
(*a*) 1 : 3 : 6 (*b*) 2 : 3 : 6
(*c*) 3 : 2 : 6 (*d*) 3 : 1 : 2

**(Police Inspector Exam. 1988)**

**50.** A certain amount was divided between Kavita and Reena in the ratio 4 : 3. If Reena's share was Rs.2400, the amount is :
(*a*) Rs. 5,600 (*b*) Rs. 3,200
(*c*) Rs. 9,600 (*d*) none of these

**(S.B.I.P.O. Exam. 1988)**

**51.** The prices of a scooter and a television set are in the ratio 3 : 2. If a scooter costs Rs.6,000 more than the television set, the price of the television set is :
(*a*) Rs. 18,000 (*b*) Rs. 12,000
(*c*) Rs. 10,000 (*d*) Rs. 6,000

**(Bank P.O. Exam. 1989)**

**52.** If a carton containing a dozen mirrors is dropped, which of the following cannot be the ratio of broken mirrors to unbroken mirrors ?
(*a*) 2 : 1 (*b*) 3 : 1
(*c*) 3 : 2 (*d*) 7 : 5 **(S.B.I. P.O. Exam. 1987)**

**53.** A man spends Rs.8100/- in buying tables at Rs.1200/- each and chairs at Rs.300/- each. The ratio of chairs to tables when the maximum number of tables is purchased, is
(*a*) 1 : 4 (*b*) 5 : 7
(*c*) 1 : 2 (*d*) 2 : 1 **(S.B.I. P.O. Exam. 1988)**

**54.** In a mixture of 60 litres, the ratio of milk and water is 2 : 1. If the ratio of milk and water is to be 1 : 2, then the amount of water to be further added is :
(*a*) 42 litres (*b*) 56 litres
(*c*) 60 litres (*d*) 77 litres **(N.D.A. Exam. 1990)**

**55.** 20 litres of a mixture contains milk and water in the ratio 5 : 3. If 4 litres of this mixture are replaced by 4 litres of milk, the ratio of milk to water in the new mixture will become :
(*a*) 2 : 1 (*b*) 6 : 3
(*c*) 7 : 3 (*d*) 8 : 3

**56.** A mixture contains milk and water in the ratio 5 : 1. On adding 5 litres of water the ratio of milk to water becomes 5 : 2. The quantity of milk in the mixture is :
(*a*) 16 litres (*b*) 25 litres

(c) 32.5 litres (d) 22.75 litres

**57.** The ratio of milk and water in 85 kg. of adulterated milk is 27 : 7. The amount of water which must be added to make the ratio 3 : 1, is :

(a) 5 kg. (b) 6.5 kg.
(c) 7.25 kg. (d) 8 kg.

**58.** The proportion of zinc and copper in a brass piece is 13 : 7. How much zinc will be there in 100 kg. of such a piece ?

(a) 20 kg. (b) 35 kg.
(c) 55 kg. (d) 65 kg.

**59.** *A* and *B* are two alloys of gold and copper prepared by mixing metals in proportions 7 : 2 and 7 : 11 respectively. If equal quantities of alloys are melted to form a third alloy *C*, the proportion of gold and copper in *C* will be :

(a) 5 : 9 (b) 5 : 7
(c) 7 : 5 (d) 9 : 5 **(C.D.S. Exam. 1989)**

**60.** In what ratio must 25% alcohol be mixed with 60% alcohol to get a mixture of 40% alcohol strength ?

(a) 1 : 2 (b) 2 : 1
(c) 2 : 3 (d) 3 : 2

**61.** A sum of Rs. 86700 is to be divided among *A*, *B* and *C* in such a manner that for every rupee that *A* gets, *B* gets 90 paise and for every rupee that *B* gets, *C* gets 110 paise. *B*'s share is :

(a) Rs. 26,010 (b) Rs. 27,000
(c) Rs. 30,000 (d) none of these

**(L.I.C. A.A.O. Exam. 1988)**

**62.** Rs.5625 is to be divided among *A*, *B* and *C*, so that A may receive (1/2) as much as *B* and *C* together receive and *B* receives (1/4) of what *A* and *C* together receive. The share of *A* is more than that of *B* by :

(a) Rs. 750 (b) Rs. 775
(c) Rs. 1500 (d) Rs. 1600

**(Excise and I.Tax Exam. 1988)**

**63.** If the weight of a 13 metres long iron rod be 23.4 kg., the weight of 6 metres long of such rod will be :

(a) 7.2 kg. (b) 12.4 kg.
(c) 10.8 kg. (d) 18 kg.

**(Bank P.O. Exam. 1986)**

**64.** The ratio of the money with Ram and Gopal is 7 : 17 and that with Gopal and Krishan is 7 : 17. If Ram has Rs.490, Krishan has :

(a) Rs. 2890 (b) Rs. 2330
(c) Rs. 1190 (d) Rs. 2680

**(Astt. Grade Exam. 1987)**

**65.** The average age of 3 girls is 20 years and their ages are in the proportion

3 : 5 : 7. The age of youngest girl is :

(*a*) 4 years (*b*) 6 years 8 months

(*c*) 8 years 3 months (*d*) 12 years

**66.** A father's age was 5 times his son's age 5 years ago and will be 3 times the son's age after 2 years. The ratio of their present ages is :

(*a*) 5 : 2 (*b*) 5 : 3

(*c*) 10 : 3 (*d*) 11 : 5

**67.** The ratio between the ages of Gyatri and Savitri is 6 : 5 and the sum of their ages is 44 years. The ratio of their ages after 8 years will be :

(*a*) 5 : 6 (*b*) 7 : 8

(*c*) 8 : 7 (*d*) 14 : 13

**(Bank P.O. Exam. 1987)**

**68.** The ratio of the father's age to son's age is 4 : 1. The product of their ages is 196. The ratio of their ages after 5 years will be :

(*a*) 3 : 1 (*b*) 10 : 3

(*c*) 11 : 4 (*d*) 14 : 5

**69.** The ratio between Sumit's and Prakash's ages at present is 2 : 3. Sumit is 6 years younger than Prakash. The ratio of Sumit's age to Prakash's age after 6 years will be :

(*a*) 2 : 3 (*b*) 3 : 4

(*c*) 1 : 2 (*d*) 3 : 8

**(Railway Recruitment, 1991)**

**70.** The ages of Manoj and Amit are in the ratio 2 : 3. After 12 years, their ages will be in the ratio 11 : 15. The age of Amit is :

(*a*) 32 years (*b*) 40 years

(*c*) 48 years (*d*) 56 years

**71.** The speeds of three cars are in the ratio 2 : 3 : 4. The ratio between the times taken by these cars to travel the same distance is :

(*a*) 2 : 3 : 4 (*b*) 4 : 3 : 2

(*c*) 4 : 3 : 6 (*d*) 6 : 4 : 3

**72.** The income of A and B are in the ratio 3 : 2 and their expenditures in the ratio 5 : 3. If each saves Rs.1000, A's income is :

(*a*) Rs. 3,000 (*b*) Rs. 4,000

(*c*) Rs. 6,000 (*d*) Rs. 9,000

**73.** Two numbers are such that the ratio between them is 3 : 5 but if each is increased by 10, the ratio between them becomes 5 : 7. The numbers are :

(*a*) 3, 5 (*b*) 7, 9

(*c*) 13, 22 (*d*) 15, 25

**(R.R.B. Exam. 1989)**

74. Two equal glasses are respectively $\frac{1}{3}$ and $\frac{1}{4}$ full of milk. They are then filled up with water and the contents mixed in a tumbler. The ratio of milk and water in the tumbler is :

(a) 7 : 5 (b) 7 : 17
(c) 9 : 21 (d) 11 : 23

75. The ratio of the first and second class fares between two stations is 4 : 1 and that of the number of passengers travelling by first and second class is 1 : 40. If Rs.1100 is collected as fare, the amount collected from first class passengers is :

(a) Rs. 275 (b) Rs. 315
(c) Rs. 137.50 (d) Rs. 100

76. The students in three classes are in the ratio 2 : 3 : 5. If 20 students are increased in each class, the ratio changes to 4 : 5 : 7. The total number of students in the three classes before the increase were :

(a) 10 (b) 90
(c) 100 (d) none of these

**(L.I.C. A.A.O. Exam. 1988)**

77. One year ago the ratio between Laxman's and Gopal's salary was 3 : 4. The individual ratios between their last year's and this year's salaries are 4 : 5 and 2 : 3 respectively. At present the total of their salary is Rs.4160. The salary of Laxman now, is :

(a) Rs. 1600 (b) Rs. 2560
(c) Rs. 1040 (d) Rs. 3120

**(S.B.I. P.O. Exam. 1987)**

78. The cost of making an article is divided between materials, labour and overheads in the ratio of 3 : 4 : 1. If the materials cost Rs.67.50, the cost of article is :

(a) Rs. 180 (b) Rs. 122.50
(c) Rs. 380 (d) Rs. 540

79. The areas of two spheres are in the ratio 1 : 4. The ratio of their volumes is :

(a) 1 : 2 (b) 1 : 4
(c) 1 : 8 (d) 1 : 64

80. The sides of a triangle are in the ratio of $\frac{1}{3} : \frac{1}{4} : \frac{1}{5}$. If the perimeter is 94 cms., the length of smallest side is :

(a) 18.8 cm. (b) 23.5 cm.
(c) 24 cm. (d) 31.3 cm.

81. In a factory the ratio of male workers to female workers was 5 : 3. If the number of female workers was less by 40, the total number of workers in the factory was :

(*a*) 100 (*b*) 500
(*c*) 160 (*d*) 200 **(Bank P.O. Exam. 1988)**

**82.** The mean proportional of 0.32 and 0.02 is :
(*a*) 0.34 (*b*) 0.30
(*c*) 0.16 (*d*) 0.08
**(Central Excise & I.Tax 1988)**

**83.** If $a : b = 2 : 3$, $b : c = 4 : 5$ and $c : d = 6 : 7$, then $a : d$ is equal to :
(*a*) 2 : 7 (*b*) 7 : 8
(*c*) 16 : 35 (*d*) 4 : 13
**(Clerk's Grade Exam. 1991)**

**84.** The monthly salary of *A*, *B* and *C* is in the proportion 2 : 3 : 5. If *C*'s monthly salary is Rs.1200 more than *A*'s monthly salary, B's annual salary is :
(*a*) Rs. 14,400 (*b*) Rs. 24,000
(*c*) Rs. 1,200 (*d*) Rs. 2,000 **(P.O. Exam. 1990)**

**85.** A shopkeeper mixes two kinds of flour, one costing Rs.3.50 per kg. and the other Rs.2.25 per kg., so that the price of the mixture is Rs.2.75 per kg. The ratio of first kind of flour to that of the second must be :
(*a*) 3 : 4 (*b*) 4 : 3
(*c*) 3 : 2 (*d*) 2 : 3 **(C.B.I. Exam. 1990)**

## HINTS & SOLUTIONS (Exercise 8B)

**1.** 10% of $m$ = 20% of $n \Rightarrow \frac{10}{100} m = \frac{20}{100} n$

$\therefore \frac{m}{n} = \left(\frac{20}{100} \times \frac{100}{10}\right) = \frac{2}{1}$.

**2.** 0.5 of $x$ = 0.07 of $y \Rightarrow \frac{5}{10} x = \frac{7}{100} y \Rightarrow \frac{x}{y} = \frac{7}{100} \times \frac{10}{5} = \frac{7}{50}$.

**3.** Requisite Ratio $= \dfrac{\left(\frac{1}{3} \text{ of } 930\right)}{\left(\frac{6}{10} \text{ of } 155\right)} = \frac{310}{93} = \frac{10}{3}$. i.e., 10 : 3.

**4.** 1 mile = 1.6 km.

$\therefore \frac{20 \text{ miles}}{25 \text{ km.}} = \frac{20 \times 1.6}{25} = \frac{20 \times 16}{250} = \frac{32}{25}$.

**5.** $\frac{A}{3} = \frac{B}{4} = \frac{C}{5} = x$ (say).

Then, $A = 3x$; $B = 4x$ and $C = 5x$.

$\therefore A : B : C = 3 : 4 : 5$.

**6.** $\frac{A}{B} = \frac{2}{3}, \frac{B}{C} = \frac{4}{5} \Rightarrow \frac{A}{B} \times \frac{B}{C} = \frac{2}{3} \times \frac{4}{5} \Rightarrow \frac{A}{C} = \frac{8}{15} \Rightarrow \frac{C}{A} = \frac{15}{8}$.

**7.** $2A = 3B$ and $4B = 5C$.

$\therefore \frac{A}{B} = \frac{3}{2}$ and $\frac{B}{C} = \frac{5}{4}$.

Hence, $\frac{A}{C} = \frac{A}{B} \times \frac{B}{C} = \frac{3}{2} \times \frac{5}{4} = \frac{15}{8}$.

**8.** Let the numbers be $3x$ and $4x$. Then,

$\frac{3x+6}{4x+6} = \frac{4}{5} \Rightarrow 5(3x+6) = 4(4x+6)$

$\therefore 15x + 30 = 16x + 24$ or $x = 6$.

Hence, the numbers are 18 and 24.

$\therefore$ Their difference = $(24 - 18) = 6$.

**9.** Let the numbers be $7x$ and $9x$.

$\frac{7x-12}{9x-12} = \frac{3}{5} \Rightarrow 5(7x-12) = 3(9x-12)$.

$\therefore 35x - 60 = 27x - 36$ or $8x = 24$ or $x = 3$.

Hence, the numbers are 21 and 27.

So, the product of numbers = $(21 \times 27) = 567$.

**10.** Let the numbers be $3x$ and $2x$.

Then, $(3x)^3 + (2x)^3 = 4375 \Rightarrow 35x^3 = 4375$.

$\therefore x^3 = 125$ or $x = 5$. So, the numbers are 15 and 10.

Sum of the numbers = $(15 + 10) = 25$.

**11.** Let $x$ be subtracted. Then,

$\frac{54-x}{71-x} = \frac{75-x}{99-x}$ or $(99-x)(54-x) = (71-x)(75-x)$

or $5346 - 153x + x^2 = 5325 + x^2 - 146x$ or $x = 3$.

**12.** Let $x$ be added. Then,

$\frac{7+x}{13+x} = \frac{2}{3} \Rightarrow 3(7+x) = 2(13+x)$ or $x = 5$.

**13.** Let $x$ be added. Then,

$\frac{6+x}{14+x} = \frac{18+x}{38+x}$ or $(6+x)(38+x) = (14+x)(18+x)$

or $x^2 + 44x + 228 = x^2 + 32x + 252$ or $x = 2$.

**14.** Let the fraction be $\frac{x}{y}$.

Then, $\frac{x}{y} : \frac{1}{27} = \frac{3}{7} : \frac{5}{9} \Rightarrow \frac{x}{y} \times \frac{5}{9} = \frac{1}{27} \times \frac{3}{7}$

or $\frac{x}{y} = \left(\frac{1}{27} \times \frac{3}{7} \times \frac{9}{5}\right) = \frac{1}{35}$.

**15.** Let, $8 : 12 :: 18 : x$.

Then, $8x = 12 \times 18 \Rightarrow x = \frac{12 \times 18}{8} = 27.$

**16.** $0.8 : 0.2 :: 0.2 : x.$

So, $0.8x = 0.2 \times 0.2$ or $x = \frac{0.2 \times 0.2}{0.8} = 0.05.$

**17.** Mean proportional $= \sqrt{8 \times 18} = \sqrt{144} = 12.$

**18.** Let $x$ be added.

Then, $\frac{3+x}{7+x} = \frac{7+x}{13+x}$ or $(3+x)(13+x) = (7+x)^2$.

or $x^2 + 16x + 39 = x^2 + 14x + 49$ or $x = 5.$

**19.** Let $\frac{36}{x} = \frac{x}{81}$. Then, $x^2 = 36 \times 81$ or $x = \sqrt{36 \times 81} = 6 \times 9 = 54.$

**20.** $\frac{1}{5} : \frac{1}{x} = \frac{1}{x} : \frac{100}{125} \Rightarrow \left(\frac{1}{x}\right)^2 = \frac{1}{5} \times \frac{100}{125} = \frac{4}{25} = \left(\frac{2}{5}\right)^2.$

$\therefore \frac{1}{x} = \frac{2}{5}$ or $x = \frac{5}{2} = 2.5.$

**21.** $7 : 15 = \frac{7}{15} = 0.46;\ 15 : 23 = \frac{15}{23} = 0.65;$

$17 : 25 = \frac{17}{25} = 0.68$ and $21 : 29 = \frac{21}{29} = 0.72$.

Hence, the smallest ratio is 7 : 15.

**22.** Let the two parts be $x$ and $y$. Then, $\frac{x}{5} : \frac{y}{8} = 3 : 4$

$\therefore \frac{x}{5} \times 4 = \frac{y}{8} \times 3$ or $\frac{4x}{5} = \frac{3y}{8}$ or $\frac{x}{y} = \left(\frac{3}{8} \times \frac{5}{4}\right) = 15 : 32$

Hence, first part = Rs. $\left(94 \times \frac{15}{47}\right)$ = Rs.30.

**23.** $\frac{A}{B} = \frac{7}{9}$ and $\frac{B}{C} = \frac{3}{5} = \frac{9}{15}$.

$\therefore A : B : C = 7 : 9 : 15.$

**24.** $\frac{1}{3} : \frac{1}{2} : \frac{1}{5} = 10 : 15 : 6.$

$\therefore$ *C*'s share = Rs. $\left(1581 \times \frac{6}{31}\right)$ = Rs. 306.

**25.** $\frac{1}{2} : \frac{1}{4} : \frac{5}{16} = 8 : 4 : 5.$

Ashish's share = Rs. $\left(68000 \times \frac{8}{17}\right)$ = Rs. 32000;

Kapil's share = Rs. $\left(68000 \times \frac{4}{17}\right)$ = Rs. 16000;

Sumit's share = Rs. [68000 – (32000 + 16000)] = Rs.20000.

Difference between largest and smallest shares

= Rs. (32000 – 16000) = Rs.16000.

**26.** Ratio of their shares = (6 × 4 : 8 × 3 : 6 × 2) or (24 : 24 : 12) or 2 : 2 : 1

$\therefore$ Children's share = Rs. $\left(950 \times \frac{1}{5}\right)$ = Rs. 190.

**27.** Suppose $C$ gets Rs. $x$.

Then, $B$ gets $\frac{1}{4}x$ and $A$ gets $\left(\frac{2}{3} \text{ of } \frac{1}{4}x\right)$ i.e., $\frac{1}{6}x$.

$\therefore A : B : C = \frac{1}{6}x : \frac{1}{4}x : x$ or $2 : 3 : 12$

$\therefore$ $B$'s share = Rs. $\left(680 \times \frac{3}{17}\right)$ = Rs. 120.

**28.** $\frac{1}{2}A = \frac{1}{3}B = \frac{1}{6}C = x$ (say)

$\therefore A = 2x, B = 3x$ and $C = 6x$.

So, $A : B : C = 2 : 3 : 6$.

$\therefore$ Third part = Rs. $\left(1870 \times \frac{6}{11}\right)$ = Rs.1020.

**29.** $\frac{2}{5}A + 40 = \frac{2}{7}B + 20 = \frac{9}{17}C + 10 = x.$

$\therefore A = \frac{5}{2}(x - 40) = \frac{5}{2}x - 100; B = \frac{7}{2}(x - 20) = \frac{7}{2}x - 70$

and $C = \frac{17}{9}(x - 10) = \frac{17}{9}x - \frac{170}{9}$.

$\therefore \frac{5}{2}x - 100 + \frac{7}{2}x - 70 + \frac{17}{9}x - \frac{170}{9} = 600$

or $\frac{71}{9}x = \frac{7100}{9} \Rightarrow x = 100.$

$\therefore$ A's share = $\left(\frac{5}{2} \times 100 - 100\right)$ = Rs 150.

**30.** Suppose $C$ gets Rs.$x$. Then $B$ gets $\frac{1}{2}x$ & $A$ gets $\frac{1}{4}x$.

$\therefore A : B : x = \frac{1}{4}x : \frac{1}{2}x : x$ or $1 : 2 : 4$.

$B$'s share = Rs. $\left(777 \times \frac{2}{7}\right)$ = Rs.222.

**31.** Let B and C together receive Rs.$x$. Then, A receives Rs. $\frac{2}{9}x$.

$\therefore A : (B + C) = \frac{2}{9}x : x = 2 : 9.$

$\therefore$ A's share = Rs. $\left(385 \times \frac{2}{11}\right)$ = Rs. 70.

**32.** $5A = 3B = 2C = x$ (say).

Then, $A = \frac{x}{5}$; $B = \frac{x}{3}$ and $C = \frac{x}{2}$.

$\therefore A : B : C = \frac{x}{5} : \frac{x}{3} : \frac{x}{2} = 6 : 10 : 15.$

**33.** Suppose $C$ gets Rs. $x$.

Then, B gets Rs. $(x + 8)$ and A gets Rs. $(x + 15)$.

$\therefore x + 15 + x + 8 + x = 53 \Rightarrow x = 10.$

$\therefore A : B : C = (10 + 15) : (10 + 8) : 10$ or $25 : 18 : 10$.

**34.** Divide Rs. [2430 – (5 + 10 + 15)] = Rs. 2400 in the ratio 3 : 4 : 5.

Then, first part = Rs. $\left(2400 \times \frac{3}{12}\right)$ = Rs.600.

$\therefore$ A's share = Rs. (600 + 5) = Rs. 605.

**35.** $\frac{A}{B} = \frac{2}{3}, \frac{B}{C} = \frac{2}{3}$ and $\frac{C}{D} = \frac{2}{3}$

$$\Rightarrow \frac{A}{B} = \frac{2}{3}, \frac{B}{C} = \frac{2 \times \frac{3}{2}}{3 \times \frac{3}{2}} = \frac{3}{\frac{9}{2}} \;\&\; \frac{C}{D} = \frac{2 \times \frac{9}{4}}{3 \times \frac{9}{4}} = \frac{\frac{9}{2}}{\frac{27}{4}}$$

$\Rightarrow A : B : C : D = 2 : 3 : \frac{9}{2} : \frac{27}{4}$ or $8 : 12 : 18 : 27$.

$\therefore$ A's share = Rs. $\left(1300 \times \frac{8}{65}\right)$ = Rs. 160.

**36.** $\frac{2}{x} = \frac{1}{2}$ and $\frac{x}{y} = \frac{2}{3}$.

$\therefore \frac{2}{x} \times \frac{x}{y} = \frac{1}{2} \times \frac{2}{3} \Rightarrow \frac{2}{y} = \frac{1}{3}$ or $y = 6$.

**37.** $A : B : C = 100 : 65 : 35$ or $20 : 13 : 7$.

If C's share is Rs.7, total sum = Rs.40

If C's share is Rs.28, total sum = Rs. $\left(\frac{40}{7} \times 28\right)$ = Rs. 160.

**38.** Ratio of coins = 3 : 4 : 10.

3 rupee coins = Rs. 3; 4 fifty paise coins = Rs.2

and 10 ten paise coins = Re.1.

$\therefore$ Ratio of their values = 3 : 2 : 1.

$\therefore$ Value of 10 paise coins = Rs. $\left(102 \times \frac{1}{6}\right)$ = Rs. 17.

Number of 10 paise coins = 17 × 10 = 170.

**39.** Ratio of Rs. 5, Rs. 2 & Re.1 coins = 10 : 7 : 21.

Values of these coins = 50 : 14 : 21.

Value of Rs. 2 coins = Rs. $\left(340 \times \frac{14}{85}\right)$ = Rs. 56.

$\therefore$ Number of Rs. 2 coins = $\frac{56}{2} = 28$.

**40.** Let their costs be $4x$, $5x$ and $7x$ respectively.
Then, $(7x - 4x) = 60 \Rightarrow x = 20$.
Cost of the modest book = Rs. $5x$ = Rs. 100.

**41.** Let cost of 1 chair and 1 table be Rs. $x$ and Rs. $y$ respectively.

Then, $5x + 4y = 3x + 7y \Rightarrow 2x = 3y \Rightarrow \frac{x}{y} = \frac{3}{2}$.

**42.** $\frac{A}{B} = \frac{2}{3}$ and $\frac{B}{C} = \frac{2}{5} = \frac{2 \times \frac{3}{2}}{5 \times \frac{3}{2}} = \frac{3}{\frac{15}{2}}$.

$\therefore A : B : C = 2 : 3 : \frac{15}{2} = 4 : 6 : 15$.

Runs scored by $B = \left(75 \times \frac{6}{25}\right) = 18$.

**43.** Sum of the ratios must divide 64.
So, they can not be in the ratio 3 : 4.

**44.** Let the numbers be $3x$ and $4x$.
Then, their L.C.M. = $12x$
$\therefore 12x = 180$ or $x = 15$
So, the first number is 45.

**45.** Let there be $x$ hens and $y$ cows. Then,
$x + y = 48$ and $2x + 4y = 140$.
Solving these equations, we get $x = 26$.

**46.** Number of boys = $\left(720 \times \frac{7}{12}\right) = 420$.
Number of girls = $(720 - 420) = 300$.
So, to make the ratio 1 : 1, the number of more girls to be added = $(420 - 300) = 120$.

**47.** Quantity of milk = $\left(729 \times \frac{7}{9}\right)$ ml. = 567 ml.

Quantity of water = $(729 - 567)$ ml. = 162 ml.

Now, $\frac{567}{162 + x} = \frac{7}{3} \Rightarrow 567 \times 3 = 7(162 + x)$

$\Rightarrow 1701 = 1134 + 7x \Rightarrow x = 81$.

**48.** Let $x$ gm. of water be taken.

Then, gold = $19x$ gm. and copper= $9x$ gm.

Let 1 gm. of gold be mixed with $y$ gm. of copper.

Then, $19x + 9xy = 15x(1 + y) \Rightarrow y = \frac{2}{3}$..

**49.** $B = \frac{1}{2}C, A = \frac{1}{3}B, = \frac{1}{3}\left(\frac{1}{2}C\right) = \frac{1}{6}C$.

$\therefore A : B : C = \frac{1}{6}C : \frac{1}{2}C : C = \frac{1}{6} : \frac{1}{2} : 1$ or $1 : 3 : 6$.

**50.** Let the amount be Rs. $x$.

Then, Reena's share = Rs. $\left(x \times \frac{3}{7}\right)$.

$\therefore \frac{3x}{7} = 2400$ or $x = \left(\frac{2400 \times 7}{3}\right)$ = Rs. 5600.

**51.** Let the prices of a scooter and a television be Rs. $3x$ and Rs. $2x$ respectively. Then,

$3x - 2x = 6000$ or $x = 6000$.

$\therefore$ Cost of a television set = Rs.12000.

**52.** Sum of the ratios must divide 12.

Since $3 + 2 = 5$ does not divide 12, so it can not be $3 : 2$.

**53.** Maximum number of tables purchased at Rs.1200 per table spending within Rs. 8100 is clearly 6.

Remaining amount = Rs. $(8100 - 1200 \times 6)$ = Rs. 900.

Number of chairs for Rs.900 = $(900 \div 300) = 3$.

$\therefore$ Number of chairs : Number of tables = $3 : 6$ or $1 : 2$.

**54.** Milk in given mixture = $\left(60 \times \frac{2}{3}\right)$ litres = 40 litres.

Let $x$ litres of water be added to it.

Then, $\frac{40}{20 + x} = \frac{1}{2} \Rightarrow 20 + x = 80 \Rightarrow x = 60$.

**55.** 16 litres of mixture contains milk = $\left(16 \times \frac{5}{8}\right)$ = 10 litres

and water = 6 litres.

New mixture contains milk = 14 litres & water = 6 litres.

$\therefore$ Ratio of milk and water in the mix. = $14 : 6$ or $7 : 3$.

**56.** Let milk = $5x$ litres & water = $x$ litres.

Now, $\frac{5x}{x + 5} = \frac{5}{2} \Rightarrow 10x = 5x + 25 \Rightarrow x = 5$.

$\therefore$ Quantity of milk = 25 litres.

**57.** Milk in 85 kg. = $\left(85 \times \frac{27}{34}\right) = \frac{135}{2} = 67.5$ kg.

and water in it = $(85 - 67.5) = 17.5$ kg.

Let $x$ kg. of water be added to it.

Then, $\frac{67.5}{17.5+x} = \frac{3}{1} \Rightarrow 67.5 = 52.5 + 3x$ or $x = 5$.

**58.** Zinc in 20 kg. brass = 13 kg.

Zinc in 100 kg. brass = $\left(\frac{13}{20} \times 100\right)$ kg. = 65 kg.

**59.** In alloy $C$ (when one unit each of $A$ and $B$ is mixed),

gold = $\left(\frac{7}{9} + \frac{7}{18}\right) = \frac{21}{18}$ and copper = $\left(\frac{2}{9} + \frac{11}{18}\right) = \frac{15}{18}$.

$\therefore$ Ratio of gold and copper = $\frac{21}{18} : \frac{15}{18} = 7 : 5$.

**60.** By the rule of allegation, we have

Required ratio is 10 : 15 or 2 : 3

**61.** If $A$ gets Re.1, $B$ gets 90 paise.

Now, if $B$ gets Re.1, $C$ gets 110 paise.

If $B$ gets 90 paise, $C$ gets $\left(\frac{110}{100} \times 90\right)$ = 99 paise.

$\therefore A : B : C = 100 : 90 : 99$.

So, $B$'s share = Rs. $\left(86700 \times \frac{90}{289}\right)$ = Rs. 27000.

**62.** $A + B + C = 5625$ and $B = \frac{1}{4}(A + C)$ *i.e.* $A + C = 4B$.

$\therefore 4B + B = 5625$ or $B = 1125$.

Also, $A + C = 4B = 4 \times 1125 = 4500$.

Also, $A = \frac{1}{2}(B + C)$ or $B + C = 2A$ or $B = 2A - C$.

$\therefore 2A - C = 1125$.

Now, solving $A + C = 4500$ and $2A - C = 1125$, we get

$A = 1875$ and $C = 2625$.

$\therefore A - B = (1875 - 1125)$ = Rs. 750.

**63.** Let $\frac{13}{6} = \frac{23.4}{x}$. Then, $x = \left(\frac{23.4 \times 6}{13}\right)$ = 10.8 kg.

**64.** Let Ram, Gopal and Krishan have rupees $x$, $y$ and $z$ respectively.

Then, $\frac{x}{y} = \frac{7}{17}$ and $\frac{y}{z} = \frac{7}{17}$

$\therefore \frac{x}{z} = \left(\frac{x}{y} \times \frac{y}{z}\right) = \left(\frac{7}{17} \times \frac{7}{17}\right) = \frac{49}{289}$.

Thus, if Ram has Rs. 49, Krishan has Rs. 289.

If Ram has Rs. 490, Krishan has Rs. $\left(\frac{289}{49} \times 490\right)$ = Rs. 2890.

**65.** Age of youngest girl = $\left(60 \times \frac{3}{15}\right)$ = 12 years.

**66.** Let the present ages of father and son be $x$ and $y$ years respectively. Then,

$x - 5 = 5(y - 5)$ and $x + 2 = 3(y + 2)$

or $5y - x = 20$ and $x - 3y = 4$

Solving these equations, we get $x = 40$ and $y = 12$.

∴ Ratio of their present ages = 40 : 12 = 10 : 3.

**67.** Gayatri's present age = $\left(44 \times \frac{6}{11}\right)$ = 24 years.

Savitri's present age = (44 – 24) = 20 years.

Ratio of their ages after 8 years.

= (24 + 8) : (20 + 8) = 32 : 28 = 8 : 7.

**68.** Let the father's and son's present ages be $4x$ and $x$ years respectively. Then,

$4x \times x = 196$ or $x^2 = 49$ or $x = 7$

∴ Their present ages are 28 years and 7 years.

Ratio of their ages after 5 years = 33 : 12 or 11 : 4.

**69.** Let their ages be $2x$ and $3x$ years.

$3x - 2x = 6$ or $x = 6$.

∴ Sumit's age = 12 years, Prakash's age = 18 years.

So, the ratio of their ages after 6 years = 18 : 24 = 3 : 4.

**70.** Let their ages be $2x$ and $3x$ years.

Then, $\frac{2x + 12}{3x + 12} = \frac{11}{15} \Rightarrow 15(2x + 12) = 11(3x + 12)$.

∴ $30x + 180 = 33x + 132$ or $3x = 48$ or $x = 16$.

Amit's age = 48 years.

**71.** Let the speeds of the cars be $2x$, $3x$ & $4x$ km. per hour respectively and distance travelled by each be $y$ km.

∴ Ratio of times taken = $\frac{y}{2x} : \frac{y}{3x} : \frac{y}{4x} = \frac{1}{2} : \frac{1}{3} : \frac{1}{4}$

= 6 : 4 : 3.

**72.** Let the incomes of $A$ and $B$ be Rs. $3x$ and Rs. $2x$; and their expenditures be Rs. $5y$ and Rs. $3y$.

Then, savings are $(3x - 5y)$ and $(2x - 3y)$.

∴ $3x - 5y = 1000$ and $2x - 3y = 1000$.

Solving these equations, we get $x = 2000$

∴ $A$'s income = Rs. 6000.

73. Let the numbers be $3x$ and $5x$.

Then, $\frac{3x+10}{5x+10}=\frac{5}{7} \Rightarrow 7(3x+10)=5(5x+10) \Rightarrow x=5.$

So, the numbers are 15, 25.

74. Let the volume of each glass be $x$.

Then, milk $=\left(\frac{x}{3}+\frac{x}{4}\right)=\frac{7x}{12}$,

Water $=\left[\left(x-\frac{x}{3}\right)+\left(x-\frac{x}{4}\right)\right]=\left(\frac{2x}{3}+\frac{3x}{4}\right)=\frac{17x}{12}.$

$\therefore$ Milk : Water $=\frac{7x}{12}:\frac{17x}{12}=7:17.$

75. Ratio of amounts collected from 1st and 2nd class

$=(4\times 1:1\times 40)=(1:10).$

Amount collected from 1st class passengers

$=$ Rs. $\left(1100\times\frac{1}{11}\right)=$ Rs. 100.

76. Let the number of students in the class be $2x$, $3x$ and $5x$ respectively. Then,

$(2x+20):(3x+20):(5x+20)::4:5:7.$

$\frac{2x+20}{4}=\frac{3x+20}{5}=\frac{5x+20}{7}.$

Solving these equations, we get $x=10$.

$\therefore$ Total number of students in the class $=(2x+3x+5x)=10x=100.$

77. Let the salaries of Laxman and Gopal one year before be $x_1$, $y_1$ and now it be $x_2$, $y_2$ respectively. Then,

$\frac{x_1}{y_1}=\frac{3}{4}, \frac{x_1}{x_2}=\frac{4}{5}, \frac{y_1}{y_2}=\frac{2}{3}$ and $x_2+y_2=4160.$

Solving these equations, we get $x_2=1600$.

78. If the material costs Rs. 3, the cost of article = Rs. 8.

$\therefore$ Cost of article = Rs. $\left(67.50\times\frac{8}{3}\right)=$ Rs. 180.

79. Let $r_1$ and $r_2$ be the radii of the two spheres.

Then, $\frac{4\pi r_1^2}{4\pi r_2^2}=\frac{1}{4}$ or $\frac{r_1^2}{r_2^2}=\frac{1}{4}$ or $\frac{r_1}{r_2}=\frac{1}{2}$

Now, $\frac{v_1}{v_2}=\frac{\frac{4}{3}\pi r_1^3}{\frac{4}{3}\pi r_2^3}=\left(\frac{r_1}{r_2}\right)^3=\left(\frac{1}{2}\right)^3=\frac{1}{8}.$

80. Let the sides of the triangle be $\frac{x}{3}, \frac{x}{4}$ and $\frac{x}{5}$.

$\therefore \frac{x}{3}+\frac{x}{4}+\frac{x}{5}=94 \Rightarrow 20x+15x+12x=5640 \Rightarrow x=120.$

$\therefore$ Length of smallest side $=\frac{x}{5}=\frac{120}{5}=24$ cm.

**81.** Let the number of male and female workers be $5x$ and $3x$ respectively.

$5x-3x=40 \Rightarrow x=20.$

$\therefore$ Total number of workers $=8x=160.$

**82.** Mean proportion $=\sqrt{0.32 \times 0.02}=\sqrt{.0064}=\frac{8}{100}=.08.$

**83.** $\frac{a}{b}=\frac{2}{3}, \frac{b}{c}=\frac{4}{5}$ and $\frac{c}{d}=\frac{6}{7}.$

$\therefore \frac{a}{b}\times\frac{b}{c}\times\frac{c}{d}=\frac{2}{3}\times\frac{4}{5}\times\frac{6}{7}=\frac{16}{35}.$ Hence, $\frac{a}{d}=\frac{16}{35}.$

**84.** Let the salaries of $A, B, C$ be $2x$, $3x$ and $5x$ respectively.

Now, $5x-2x=1200 \Rightarrow x=400.$

$\therefore$ $B$'s salary $=3x=$ Rs. 1200.

**85.**

3.50          2.25
         2.75
0.50          0.75

$\therefore$ Required ratio is 0.50 : 0.75 or 50 : 75 or 2 : 3.

## ANSWERS (Exercise 8B)

| | | | | | |
|---|---|---|---|---|---|
| **1.** (*b*) | **2.** (*a*) | **3.** (*b*) | **4.** (*c*) | **5.** (a) | **6.** (*a*) |
| **7.** (*c*) | **8.** (*c*) | **9.** (*b*) | **10.** (*d*) | **11.** (*c*) | **12.** (*d*) |
| **13.** (*b*) | **14.** (*b*) | **15.** (*c*) | **16.** (*a*) | **17.** (*a*) | **18.** (*d*) |
| **19.** (*b*) | **20.** (*c*) | **21.** (*b*) | **22.** (*b*) | **23.** (*c*) | **24.** (*b*) |
| **25.** (*d*) | **26.** (*d*) | **27.** (*c*) | **28.** (*c*) | **29.** (*c*) | **30.** (*d*) |
| **31.** (*a*) | **32.** (*d*) | **33.** (*b*) | **34.** (*d*) | **35.** (*b*) | **36.** (*c*) |
| **37.** (*c*) | **38.** (*c*) | **39.** (*a*) | **40.** (*c*) | **41.** (*a*) | **42.** (*b*) |
| **43.** (*c*) | **44.** (*b*) | **45.** (*a*) | **46.** (*b*) | **47.** (*d*) | **48.** (*b*) |
| **49.** (*a*) | **50.** (a) | **51.** (*b*) | **52.** (*c*) | **53.** (*c*) | **54.** (*c*) |
| **55.** (*c*) | **56.** (*b*) | **57.** (*a*) | **58.** (*d*) | **59.** (*c*) | **60.** (*c*) |
| **61.** (*b*) | **62.** (*a*) | **63.** (*c*) | **64.** (*a*) | **65.** (*d*) | **66.** (*c*) |
| **67.** (*c*) | **68.** (*c*) | **69.** (*b*) | **70.** (*c*) | **71.** (*d*) | **72.** (*c*) |
| **73.** (*d*) | **74.** (*b*) | **75.** (*d*) | **76.** (*c*) | **77.** (*a*) | **78.** (*a*) |
| **79.** (*c*) | **80.** (*c*) | **81.** (*c*) | **82.** (*d*) | **83.** (*c*) | **84.** (*c*) |
| **85.** (*d*) | | | | | |

# 9

# PARTNERSHIP

**Introduction :** Sometimes two or more than two persons agree to run a business jointly. They are called partners and the deal is known as partnership.

(i) When capitals of all the partners are invested for the same time, the gains or losses in the business are divided among the partners in the ratio of their investments.

(ii) When capitals of partners are invested for different time, then equivalent capitals are obtained for 1 unit of time by multiplying the capital with the number of units. The gains or losses are now divided among partners in the ratio of these capitals.

**Remark.** A partner who manages the business is called a working partner while the one who simply invests money but does not look after the business is called a sleeping partner.

## SOLVED PROBLEMS :

**Ex. 1.** ***Three partners A, B, C agree to divide the profit or losses in the ratio 1.50 : 1.75 : 2.25. If, in a particular year, they earn a profit of Rs. 66,000, find the share of each.*** **(L.I.C. 1991)**

**Sol.** Ratio of profits = 1.50 : 1.75 : 2.25

= 150 : 175 : 225 = 6 : 7 : 9.

$\therefore$ $A$'s share = Rs. $\left(66000 \times \frac{6}{22}\right)$ = Rs. 18000 ;

$B$'s share = Rs. $\left(66000 \times \frac{7}{22}\right)$ = Rs. 21000 ;

*C's share = Rs. [66000 – (18000 + 21000)] = Rs. 27000.*

**Ex. 2.** ***Vinod, Ved and Arun started a business jointly by investing Rs. 380000, Rs. 400000 and Rs. 420000 respectively. Divide a net profit of Rs. 240000 among the partners.***

**Sol.** Ratio of investments = 380000 : 400000 : 420000

= 19 : 20 : 21.

$\therefore$ Ratio of profits = 19 : 20 : 21.

So, Vinod's share = Rs. $\left(240000 \times \frac{19}{60}\right)$ = Rs. 76000.

Ved's share = Rs. $\left(240000 \times \frac{20}{60}\right)$ = Rs. 80000.

$$\text{Arun's share} = \text{Rs.}\left(240000 \times \frac{21}{60}\right) = \text{Rs. } 84000.$$

**Ex. 3.** ***A, B, C enter into a partnership. A contributes Rs. 320000 for 4 months; B contributes Rs. 510000 for 3 months and C contributes Rs. 270000 for 5 months. If the total profit be Rs. 124800, how should they divide it among themselves ?***

**Sol.** Ratio of capitals of *A, B* and *C*

$= (320000 \times 4) : (510000 \times 3) : (270000 \times 5)$

$= 1280000 : 1530000 : 1350000$

$= 128 : 153 : 135.$

Sum of ratios $= (128 + 153 + 135) = 416.$

$$\therefore \text{A's share} = \text{Rs.}\left(124800 \times \frac{128}{416}\right) = \text{Rs. } 38400;$$

$$\text{B's share} = \text{Rs.}\left(124800 \times \frac{153}{416}\right) = \text{Rs. } 45900;$$

$$\text{C's share} = \text{Rs. }[124800 - (38400 + 45900)] = \text{Rs.} 40500.$$

**Ex. 4.** ***Vimla and Surjeet started a shop jointly by investing Rs. 9000 and Rs. 10500 respectively. After 4 months Jaya joined them by investing Rs. 12500 while Surjeet withdrew Rs. 2000. At the end of the year there was a profit of Rs. 4770. Find the share of each.*** **(L.I.C. 1991)**

**Sol.** Clearly Vimla invested Rs. 9000 for 12 months;

Surjeet invested Rs.10500 for 4 months & Rs. 8500 for 8 months;

Jaya invested Rs. 12500 for 8 months.

$\therefore$ Ratio of capitals of Vimla, Surjeet and Jaya

$= (9000 \times 12) : (10500 \times 4 + 8500 \times 8) : (12500 \times 8)$

$= 108000 : 110000 : 100000$

$= 108 : 110 : 100 = 54 : 55 : 50.$

Sum of ratios $= (54 + 55 + 50) = 159.$

$$\therefore \text{Vimla's share} = \text{Rs.}\left(4770 \times \frac{54}{159}\right) = \text{Rs. } 1620;$$

$$\text{Surjeet's share} = \text{Rs.}\left(4770 \times \frac{55}{159}\right) = \text{Rs. } 1650;$$

$$\text{Jaya's share} = \text{Rs.}[4770 - (1620 + 1650)] = \text{Rs. } 1500.$$

**Ex. 5.** ***A, B and C enter into a partnership, A putting in Rs. 2000 for the whole year, B putting Rs. 3000 at first and increasing it to Rs. 4000 at the end of 4 months, whilst C puts in at first Rs. 4000 but withdraws Rs. 1000 at the end of 9 months. How should they, at the end of a year, divide a profit of Rs. 8475 ?***

**Sol.** Calculating equivalent capital for 1 month for each :

*A's* capital = Rs. $(2000 \times 12)$ = Rs. 24000.

$B$'s capital = Rs. (3000 × 4 + 4000 × 8) = Rs. 44000.

$C$'s capital = Rs. (4000 × 9 + 3000 × 3) = Rs. 45000.

Ratio of capitals = 24000 : 44000 : 45000 = 24 : 44 : 45.

Sum of ratios = 24 + 44 + 45 = 113.

*A's share* = *Rs.* $\left(\frac{8475 \times 24}{113}\right)$ = *Rs. 1800.*

*B's share* = *Rs.* $\left(\frac{8475 \times 44}{113}\right)$ = *Rs. 3300.*

*C's share* = *Rs.* $\left(\frac{8475 \times 45}{113}\right)$ = *Rs. 3375.*

**Ex. 6.** ***A and B entered into partnership with capitals in the ratio of 4 and 5. After 3 months, A withdrew $\frac{1}{4}$ of his capital and B withdrew $\frac{1}{5}$ of his capital. The gain at the end of 10 months was Rs. 76000. Find their shares of profit.***

**Sol.** Suppose $A$ invests Rs. 4 and $B$ invests Rs. 5.

After 3 months, money withdrawn by $A$ = Rs. $\left(4 \times \frac{1}{4}\right)$ = Re. 1.

After 3 months, money withdrawn by $B$ = Rs. $\left(5 \times \frac{1}{5}\right)$ = Re. 1.

Thus, $A$ invested Rs. 4 for 3 months and Rs. 3 for 7 months; while $B$ invested Rs. 5 for 3 months and Rs .4 for 7 months.

∴ Ratio of their capitals

= (4 × 3 + 3 × 7) : (5 × 3 + 4 × 7) = 33 : 43.

*A's share* = *Rs.* $\left(76000 \times \frac{33}{76}\right)$ = *Rs. 33000 ;*

*B's share* = *Rs.* $\left(76000 \times \frac{43}{76}\right)$ = *Rs. 43000.*

**Ex. 7.** ***Three graziers A, B and C rent a piece of pasture for a month. A puts in 27 cattle for 19 days; B 21 for 17 days and C 24 for 23 days. If at the end of the month the rent amounts to Rs. 237, how much ought to be paid by each ?***

**Sol.** Calculating equivalent number of cattle for 1 day :—

$A$'s cattle = 27 × 19 = 513.

$B$'s cattle = 21 × 17 = 357.

$C$'s cattle = 24 × 23 = 552.

Ratio of cattle = 513 : 357 : 552 = 171 : 119 : 184.

Payment to be made by $A$ = Rs. $\frac{237 \times 171}{474}$ = Rs. 85.50.

Payment to be made by $B$ = Rs. $\dfrac{237 \times 119}{474}$ = Rs. 59.50.

*Payment to be made by C* = *Rs.* $\dfrac{237 \times 184}{474}$ = *Rs. 92.*

**Ex. 8.** ***A, B, C enter into a partnership and their capitals are in the proportion of*** $\frac{1}{2} : \frac{1}{3} : \frac{1}{4}$. ***B withdraws one half of his capital after 4 months. At the end of the year, a profit of Rs. 175000 is divided among them. Find the share of each.***

**Sol.** Initial ratio of investments = $\left(\frac{1}{2} : \frac{1}{3} : \frac{1}{4}\right) = (6 : 4 : 3)$.

Suppose that $A$, $B$, $C$ invest Rs. 6, Rs. 4 & Rs. 3 respectively.

Money withdrawn by $B$ after 4 months = $\frac{1}{2}$ of Rs. 4 = Rs. 2.

$\therefore$ Capital of $A$ is Rs. 6 for 12 months; capital of $B$ is Rs. 4 for 4 months & Rs. 2 for 8 months; capital of $C$ is Rs. 3 for 12 months.

$\therefore$ Ratio of capitals of $A$, $B$, $C$

$= (6 \times 12) : (4 \times 4 + 2 \times 8) : (3 \times 12)$

$= 72 : 32 : 36 = 18 : 8 : 9.$

Sum of ratios = 35.

$\therefore$ *A's share* = *Rs.* $\left(175000 \times \frac{18}{35}\right)$ = *Rs. 90000;*

*B's share* = *Rs.* $\left(175000 \times \frac{8}{35}\right)$ = *Rs. 40000;*

*C's share* = *Rs. [175000 – (90000 + 40000)]* = *Rs. 45000.*

**Ex. 9.** ***The ratio of investments of two partners is 11 : 12 and the ratio of their profits is 2 : 3. If A invested the money for 8 months, find for how much time B invested his money.***

**Sol.** Suppose $A$ invested Rs. 11 for 8 months and $B$ invested Rs. 12 for $x$ months.

Then, ratio of investments of $A$ and $B$

$= (11 \times 8) : (12 \times x) = 88 : 12x.$

$\therefore \dfrac{88}{12x} = \dfrac{2}{3}$ or $x = 11$.

*Hence, B invested the money for 11 months.*

**Ex. 10.** ***A, B and C enter into partnership in a business with capitals of Rs. 5000, Rs. 6000 and Rs. 4000 respectively. A gets 30% of the profit for managing the business and balance is divided in proportion to their capitals. At the end of the year, A gets Rs. 200 more than B and C together. Find the total profit and share of each.***

**Sol.** Let the total profit be Rs. $x$.

Amount obtained by $A$ for managing = Rs. $\left(\frac{30x}{100}\right)$ = Rs. $\left(\frac{3x}{10}\right)$.

Balance of profit = Rs. $\left(x-\frac{3x}{10}\right)$ = Rs. $\left(\frac{7x}{10}\right)$.

Ratio of capitals = 5000 : 6000 : 4000 or 5 : 6 : 4

$\therefore$ A's share = Rs. $\left[\left(\frac{7x}{10}\times\frac{5}{15}\right)+\frac{3x}{10}\right]$ = Rs. $\left(\frac{8x}{15}\right)$

$B$'s share = Rs. $\left(\frac{7x}{10}\times\frac{6}{15}\right)$ = Rs. $\left(\frac{7x}{25}\right)$

$C$'s share = Rs. $\left(\frac{7x}{10}\times\frac{4}{15}\right)$ = Rs. $\left(\frac{14x}{75}\right)$

$\therefore \frac{7x}{25}+\frac{14x}{75}+200=\frac{8x}{15}$ or $x = 3000$.

*Thus, the total profit is Rs. 3000.*

$\therefore$ *A's share* = *Rs.* $\left(\frac{8\times 3000}{15}\right)$ = *Rs. 1600,*

*B's share* = *Rs.* $\left(\frac{7\times 3000}{25}\right)$ = *Rs. 840,*

*C's share* = *Rs.* $\left(\frac{14\times 3000}{75}\right)$ = *Rs. 560.*

**Ex. 11.** ***Two partners invested Rs. 1250 and Rs. 850 respectively in a business. They distribute 60% of the profit equally and decide to distribute the remaining 40% as the interest on their capitals. If one partner received Rs. 30 more than the other, find the total profit.***

**Sol.** Let the total profit be Rs. $x$.

Then, 60% of the profit = Rs. $\left(\frac{60}{100}\times x\right)$ = Rs. $\left(\frac{3x}{5}\right)$

From this part of the profit each gets = Rs. $\left(\frac{3x}{10}\right)$

40% of total profit = Rs. $\left(\frac{40}{100}\times x\right)$ = Rs. $\left(\frac{2x}{5}\right)$

Now, this amount of Rs. $\left(\frac{2x}{5}\right)$ has been divided in the ratio of capitals namely, 1250 : 850 or 25 : 17 as interests.

$\therefore$ Interest on first capital = Rs. $\left(\frac{2x}{5}\times\frac{25}{42}\right)$ = Rs. $\left(\frac{5x}{21}\right)$

Interest on second capital = Rs. $\left(\frac{2x}{5}\times\frac{17}{42}\right)$ = Rs. $\left(\frac{17x}{105}\right)$

$$\text{Total money received by 1st investor} = \text{Rs.}\left(\frac{3x}{10}+\frac{5x}{21}\right)$$

$$= \text{Rs.}\left(\frac{113x}{210}\right)$$

$$\text{Total money received by 2nd} = \text{Rs.}\left(\frac{3x}{10}+\frac{17x}{105}\right) = \text{Rs.}\left(\frac{97x}{210}\right).$$

$$\therefore \frac{113x}{210}-\frac{97x}{210} = 30 \text{ or } x = 393.75.$$

*Hence, total profit = Rs. 393.75.*

**Ex. 12.** ***A and B enter into partnership. A supplies whole of the capital amounting to Rs. 45000 with the condition that the profits are to be equally divided and that B pays A interest on half the capital at 10% per annum, but receives Rs. 120 per month for carrying on the concern. Find their total yearly profit when B's income is one half of A's income.***

**Sol.** Let the yearly total profit be Rs. $x$.

Amount paid to $B$ as salary = Rs. (120 × 12) = Rs.1440.

$$\text{Share of each} = \text{Rs.}\left(\frac{x-1440}{2}\right).$$

$$\text{Interest paid by } B = \text{Rs.}\left(\frac{22500 \times 10}{100}\right) = \text{Rs. } 2250.$$

Total money received by $A$

$$= \text{Rs.}\left(\frac{x-1440}{2}+2250\right) = \text{Rs.}\left(\frac{x+3060}{2}\right)$$

Total money received by $B$.

$$= \text{Rs.}\left[\left(\frac{x-1440}{2}\right)+1440-2250\right] = \text{Rs.}\left(\frac{x-3060}{2}\right)$$

$$\frac{1}{2}\left(\frac{x+3060}{2}\right) = \left(\frac{x-3060}{2}\right) \text{ or } \frac{x+3060}{4} = \frac{x-3060}{2} \text{ or } x = 9180.$$

*Hence, the total profit = Rs. 9180.*

## EXERCISE 9 A (Subjective)

1. Harish and Hashmi started a business by investing Rs. 100000 and Rs. 150000 respectively. Find the share of each out of a profit of Rs. 36000.
2. $X$ and $Y$ invested in a business. They earned some profit and divided in the ratio 2 : 3. If $X$ invested Rs. 40, find the money invested by $Y$.
   **Hint.** *2 : 3 = 40 : y. Find y.* **(Railways 1991)**
3. $A$, $B$, $C$ are three partners in a business. If $A$'s capital is equal to twice $B$'s capital and $B$'s capital is 3 times $C$'s capital. Find the ratio of their capitals.
   **Hint.** *Let C's capital = Rs. x.*

*Then, B's capital = Rs. (3x) & A's capital = Rs. (6x).*

**4.** **Three partners A, B, C invest in a business. If 6 (A's capital) = 8 (B's capital) = 10 (C's capital), find their shares out of a profit of Rs. 94000.**

**Hint.** *6A = 8B = 10 C = x (say).*

*Then,* $A = \frac{x}{6}; B = \frac{x}{8}$ *and* $C = \frac{x}{10}$.

$\therefore A : B : C = \frac{x}{6} : \frac{x}{8} : \frac{x}{10} = 20 : 15 : 12.$

**5.** **A, B, C hire a meadow for Rs. 1467.30. A puts in 10 oxen for 20 days; B 30 oxen for 8 days and C 16 oxen for 9 days. Find the rent paid by each.**

**Hint.** *Ratio of rents to be paid by A, B, C*

$= (10 \times 20) : (30 \times 8) : (16 \times 9).$

**6.** **Jayant started a business, investing Rs. 6000. Six months later Madhu joined him, investing Rs. 4000. If they made a profit of Rs. 5200 at the end of the year, how much must be the share of Madhu ?**

**(Bank P.O. 1991)**

**Hint.** *Ratio of the capitals of Jayant & Madhu*

$= (6000 \times 12) : (4000 \times 6) = 3 : 1.$

**7.** **A and B started a business by investing Rs. 80000 and Rs. 100000 respectively. After 6 months, C joined them by investing Rs. 60000. At the end of 3 years, they had a profit of Rs. 96600. Find the share of each.**

**Hint.** *Ratio of capitals of A, B and C*

$= (80000 \times 3) : (100000 \times 3) : \left(60000 \times \frac{5}{2}\right) = 8 : 10 : 5.$

**8.** **Three partners A, B, C start a business. Twice the investment of A is equal to thrice the capital of B and the capital of B is four times the capital of C. Find the share of each out of a profit of Rs. 297000.**

**Hint.** *Let C's capital = Rs. x. Then, B's capital = Rs. 4x.*

*Now, 2 (A's capital) = (3 × 4x) or A's capital = Rs. 6x.*

$\therefore$ *Ratio of their capitals* = 6x : 4x : x = 6 : 4 : 1.

**9.** **A, B, C subscribe Rs. 470000 for a business. A subscribes Rs. 70000 more than B and B invests Rs. 50000 more than C. Out of a total profit of Rs. 94000, find the share of each.**

**Hint.** *Let C's capital = Rs. x. Then,*

*B's capital = Rs. (x + 50000) & A's capital = Rs. (x + 120000)*

$\therefore x + x + 50000 + x + 120000 = 470000 \Rightarrow x = 100000.$

$\therefore A : B : C = 220000 : 150000 : 100000 = 22 : 15 : 10.$

**10.** Deepak started a business by investing Rs. 27000. After some time Vikas joined him by investing Rs. 20250. At the end of one year, the profit was divided in the ratio 2 : 1. After how many months did Vikas join the business ?

**Hint.** *Suppose Vikas joined after x months.*

*Ratio of their capitals* = $(27000 \times 12) : [20250 \times (12 - x)]$.

$$\therefore \frac{27000 \times 12}{20250 \times (12 - x)} = \frac{2}{1}.$$

**11.** A starts a business by investing Rs.10000. After 3 months *B* also joins with an investment of Rs. 6000 and as soon as *B* joins, *A* withdraws Rs. 6000. After 3 months more, *C* joins and invests Rs. 5000. As soon as *C* joins, *A* withdraws Rs. 1000 and *B* withdraws Rs. 2000. At the end of the year, the total profit is Rs. 22000. Find the share of each in this profit.

**Hint.** *Ratio of capitals*

$= (10000 \times 3 + 4000 \times 3 + 3000 \times 6) : (6000 \times 3 + 4000 \times 6) : (5000 \times 6)$

$= 10 : 7 : 5.$

**12.** In a business *A* invests Rs. 600 more than *B*. The capital of *B* remained invested for $7\frac{1}{2}$ months, while the capital of *A* remained invested for 2 more months. If the total profit be Rs. 620 and *B* gets Rs. 140 less than what *A* gets, find the capital of each.

**Hint.** *Let B's capital be Rs. x, then A's capital = Rs. (x + 600).*

$\therefore$ *Ratio of capitals of A and B*

$$= \left[(x + 600) \times \frac{19}{2}\right] : \left[x \times \frac{15}{2}\right] = (19x + 11400) : (15x).$$

$$\therefore \frac{620 \times (19x + 11400)}{(34x + 11400)} - \frac{620 \times 15x}{(34x + 11400)} = 140; \text{ find } x.$$

**13.** *A, B, C* enter into a partnership. *A* contributes one-third of the whole capital while *B* contributes as much as *A* and *C* together contribute. If the profit at the end of the year is Rs. 84000, what would each receive ?

**Hint.** *Let total capital = Rs. x.*

*Then, A's capital* = Rs. $\left(\frac{1}{3}x\right)$.

*B's capital = (A + C)'s capital*

*or 2 (B's capital) = (A + B + C)'s capital = Rs. x.*

*or B's capital* = Rs. $\left(\frac{1}{2}x\right)$.

*C's capital* = Rs. $\left[x - \left(\frac{1}{3}x + \frac{1}{2}x\right)\right]$ = Rs. $\left(\frac{1}{6}x\right)$.

$$\therefore A : B : C = \frac{1}{3} : \frac{1}{2} : \frac{1}{6} = 2 : 3 : 1.$$

**14.** Two merchants *A* and *B* enter into partnership. *A* puts in Rs. 23250 and at the end of 4 months adds Rs. 3750 to the capital while *B* withdraws

Rs. 3000 at the end of 7 months. At the end of the year *A* and *B* receive equal shares of profits. Find how much *B* put in at first.

**Hint.** *Suppose B puts in Rs. x in the beginning.*

*Then,* $(23250 \times 4 + 27000 \times 8) = [7x + 5(x - 3000)]$. *Find x.*

## ANSWERS (EXERCISE 9 A)

1. Rs. 14400, Rs. 21600
2. Rs. 60
3. 6 : 3 : 1
4. Rs. 40000, Rs. 30000, Rs. 24000
5. Rs. 502.50, Rs. 603, Rs. 361.80
6. Rs. 1300
7. Rs. 33600, Rs. 42000, Rs. 21000
8. Rs. 162000, Rs. 10800 & Rs. 27000
9. Rs. 44000, Rs. 30000, Rs. 20000
10. 4 months
11. Rs. 10000, Rs. 7000, Rs. 5000
12. Rs. 3000, Rs. 2400
13. Rs. 24000, Rs. 36000, Rs. 12000
14. Rs. 27000

## EXERCISE 9 B

## OBJECTIVE TYPE QUESTIONS

**1.** *A*'s capital is equal to thrice *B*'s capital and *B*'s capital is 4 times *C*'s capital. The ratio of the capitals is :

(*a*) 1 : 3 : 12 (*b*) 12 : 4 : 1
(*c*) 3 : 1 : 4 (*d*) 1 : 3 : 4

**2.** 4 (*A*'s capital) = 6 (*B*'s capital) = 9 (*C*'s capital). Then the ratio of their capitals is :

(*a*) 4 : 6 : 9 (*b*) 12 : 15 : 20
(*c*) 9 : 6 : 4 (*d*) 6 : 8 : 10

**3.** *A, B, C* can do a work in 20, 25 and 30 days respectively. They undertook to finish the work together for Rs. 2220, then the share of *A* exceeds that of *B* by :

(*a*) Rs. 120 (*b*) Rs. 180
(*c*) Rs. 300 (*d*) Rs. 600 **(Central Excise, 1989)**

**4.** Three partners *A, B* and *C* invest Rs. 13000, Rs. 17000 and Rs. 5000 respectively in a business. They have a profit of Rs. 1750. *B*'s share of profit is :

(*a*) Rs. 650 (*b*) Rs. 850
(*c*) Rs. 250 (*d*) Rs. 750

**5.** Dilip, Ram and Amar started a shop by investing Rs. 2700, Rs. 8100 and Rs. 7200 respectively. At the end of one year, the profit was distributed. If Ram's share was Rs. 3600, their total profit was :

(*a*) Rs. 10800 (*b*) Rs. 11600
(*c*) Rs. 8000 (*d*) none of these

**(Bank Trainee Officer's Exam. 1988)**

**6.** *A* and *B* enter into partnership investing Rs. 12000 and Rs. 16000 respectively. After 8 months, *C* also joins the business with a capital of Rs. 15000. The share of *C* in a profit of Rs. 45600 after 2 years will be :

(*a*) Rs. 12000 (*b*) Rs. 14400
(*c*) Rs. 19200 (*d*) Rs. 21200

**(Central Excise 1988)**

**7.** *A* and *B* entered into a partnership investing Rs. 16000 and Rs. 12000 respectively. After 3 months, *A* withdrew Rs. 5000 while *B* invested Rs. 5000 more. After 3 more months, *C* joins the business with a capital of Rs. 21000. The share of *B* exceeds that of *C*, out of a total profit of Rs. 26,400 after one year, by :

(*a*) Rs. 1200 (*b*) Rs. 2400
(*c*) Rs. 3600 (*d*) Rs. 4800 **(Central Excise, 1989)**

**8.** Suresh invested Rs. 12000 in a shop and Dinesh joined him after 4 months by investing Rs. 7000. If the net profit after one year be Rs. 13300, Dinesh's share in the profit is :

(*a*) Rs. 9576 (*b*) Rs. 4900
(*c*) Rs. 8400 (*d*) none of these

**(S.B.I. P.O. Exam. 1987)**

**9.** *A, B, C* enter into a partnership with shares in the ratio $\frac{7}{2}:\frac{4}{3}:\frac{6}{5}$. After 4 months, *A* increases his share by 50%. If the total profit at the end of one year be Rs. 21600, then *B*'s share in the profit is :

(*a*) Rs. 2100 (*b*) Rs. 4000
(*c*) Rs. 3600 (*d*) Rs. 2400

**10.** *A, B* and *C* start a business. *A* invests 3 times as much as *B* invests and *B* invests two-third of what *C* invests. Then, the ratio of capitals of *A, B* and *C* is :

(*a*) 3 : 9 : 2 (*b*) 6 : 10 : 15
(*c*) 5 : 3 : 2 (*d*) 6 : 2 : 3

**11.** Rama and Pooja are partners in a business. Rama invests Rs. 5000 for 5 months and Pooja invests Rs. 6000 for 6 months. If the total profit is Rs. 610, then Pooja's share in the profit is :

(*a*) Rs. 250 (*b*) Rs. 360
(*c*) Rs. 520 (*d*) Rs. 630

**12.** Ashok invests Rs. 3000 for a year and Sunil joins him with Rs. 2000 after 4 months. After the year they receive a return of Rs. 2600. Sunil's share is :

(*a*) Rs. 800 (*b*) Rs. 1000
(*c*) Rs. 750 (*d*) Rs. 900 **(Railways, 1991)**

**13.** *A* starts a business with Rs. 30000 and 4 months later *B* joins. If at the end of the year, the profits are divided by *A* and *B* in the proportion of 9 : 4, *B*'s capital was :

(*a*) Rs. 20000 (*b*) Rs. 25000
(*c*) Rs. 20700 (*d*) Rs. 18000

**14.** *A, B* and *C* contract a work for Rs. 550. Together *A* and *B* are to do $\frac{7}{11}$ of the work. The share of *C* should be :

(*a*) Rs. 400 (*b*) Rs. 300
(*c*) Rs. 200 (*d*) Rs. $183\frac{1}{3}$

**(Clerical Grade, 1991)**

**15.** Jagmohan, Rooplal and Pandeyji rented a video cassette for one week at a rent of Rs. 350. If they use it for 6 hours, 10 hours and 12 hours respectively, the rent to be paid by Pandeyji is :

(*a*) Rs. 75 (*b*) Rs. 125
(*c*) Rs. 35 (*d*) Rs. 150

**(Bank P.O. Exam. 1988)**

**16.** *A* invests Rs. 3000 for one year in a business. How much *B* should invest in order that the profit after 1 year may be divided in the ratio 2 : 3 ?

(*a*) Rs. 2000 (*b*) Rs. 1800
(*c*) Rs. 3600 (*d*) Rs. 4500

**17.** *A* and *B* invest in a business in the ratio 3 : 2. 5% of total profit goes to charity. If *A*'s share is Rs. 855, total profit is :

(*a*) Rs. 1576 (*b*) Rs. 1537.50
(*c*) Rs. 1500 (*d*) Rs. 1425

**18.** Four milkmen rented a pasture. *A* grazed 18 cows for 4 months; *B* 25 cows for 2 months; *C* 28 cows for 5 months and *D* 21 cows for 3 months. If *A*'s share of rent is Rs. 360, the total rent of the field is :

(*a*) Rs. 1500 (*b*) Rs. 1600
(*c*) Rs. 1625 (*d*) Rs. 1650

**19.** Manoj got Rs. 6000 as his share out of the total profit of Rs. 9000 which he and Ramesh earned at the end of one year. If Manoj invested Rs. 20000 for 6 months, whereas Ramesh invested his amount for the whole year, the amount invested by Ramesh was :

(*a*) Rs. 60000 (*b*) Rs. 10000
(*c*) Rs. 40000 (*d*) Rs. 5000 **(P.O. Exam. 1991)**

**20.** *A, B* and *C* subscribe Rs. 4700 for a business. *A* subscribes Rs. 700 more than *B* and *B* Rs. 500 more than *C*. Out of a total profit of Rs. 940, *B* receives :

(*a*) Rs. 440 (*b*) Rs. 300
(*c*) Rs. 200 (*d*) Rs. 173.79

**21.** *A, B* and *C* enter into a partnership and their capitals are in the proportion of $\frac{1}{3}:\frac{1}{4}:\frac{1}{5}$. *A* withdraws half his capital at the end of 4 months. Out of a total annual profit of Rs 847, *A*'s share is :

(*a*) Rs. 252 (*b*) Rs. 280

(*c*) Rs. 315 (*d*) Rs. 412

**22.** *A, B* and *C* invest Rs. 2000, Rs. 3000 and Rs. 4000 in a business. After one year *A* removed his money. *B* and *C* continued the business for one more year. If the net profit after 2 years be Rs. 3200, then *A*'s share in the profit is :

(*a*) Rs. 1000 (*b*) Rs. 600

(*c*) Rs. 800 (*d*) Rs. 400 **(Asstt. Grade, 1987)**

**23.** Vijay began a business with Rs. 2100 and is joined afterwards by Asif Ali with Rs. 3600. After how many months did *B* join, if the profits at the end of the year are divided equally ?

(*a*) 3 months (*b*) 4 months

(*c*) 5 months (*d*) 6 months

**24.** *A* and *B* start a business with initial investments in the ratio 12 : 11 and their annual profits were in the ratio 4 : 1. If *A* invested the money for 11 months, *B* invested the money for

(*a*) 3 months (*b*) $3\frac{2}{3}$ months

(*c*) 4 months (*d*) 6 months

**25.** *A, B* and *C* enter into partnership. *A* invests some money at the beginning; *B* invests double the amount after 6 months and *C* invests thrice the amount after 8 months. If the annual profit be Rs. 18000, *C*'s share is :

(*a*) Rs. 7500 (*b*) Rs. 7200

(*c*) Rs. 6000 (*d*) Rs. 5750

**26.** *A* and *B* enter into partnership. *A* invests Rs.16000 for 8 months and *B* remains in the business for 4 months. Out of a total profit *B* claims (2/7) of the profit. *B* contributed :

(*a*) Rs. 11900 (*b*) Rs. 10500

(*c*) Rs. 12800 (*d*) Rs. 13600

**27.** *A* and *B* enter into partnership investing Rs. 12000 and Rs. 16000 respectively. After 8 months, *C* also joins the business with a capital of Rs. 15000. The share of *C* in a profit of Rs. 45600 after 2 years will be :

(*a*) Rs. 12000 (*b*) Rs. 14400

(*c*) Rs. 19200 (*d*) Rs. 21200

**(Central Excise & I. Tax, 1988)**

**28.** *A*'s share in a partnership is Rs. 7500 more than *B*'s but *A*'s capital is invested for 8 months while *B*'s for 12 months. If their annual shares of profit be equal, then *A*'s capital is :

(*a*) Rs. 25600 (*b*) Rs. 27500
(*c*) Rs. 22500 (*d*) Rs. 30000

**29.** **Kishan and Nandan started a joint firm. Kishan's investment was thrice the investment of Nandan and the period of his investment was two times the period of investment of Nandan. Nandan got Rs. 4000 as profit for his investment. Their total profit if the distribution of profit is directly proportional to the period and amount, is :**
(*a*) Rs. 24000 (*b*) Rs. 16000
(*c*) Rs. 28000 (*d*) Rs. 20000 **(Bank P.O. 1989)**

**30.** ***A, B* and *C* enter into partnership by making investments in the ratio 3 : 5 : 7. After a year, *C* invests another Rs. 337600 while *A* withdraws Rs. 45600. The ratio of investments then changes to 24 : 59 : 167. How much did *A* invest initially ?**
(*a*) Rs. 45600 (*b*) Rs. 96000
(*c*) Rs. 141600 (*d*) none of these
**(L.I.C. A. A.O., Exam. 1988)**

**31.** ***A* is a working and *B*, a sleeping partner in a business. *A* puts in Rs. 12000 and *B* Rs. 20000. *A* receives 10% of the profits for managing, the rest being divided in proportion to their capitals. Out of a total profit of Rs. 9600, the money received by *A* is :**
(*a*) Rs. 3240 (*b*) Rs. 4200
(*c*) Rs. 3600 (*d*) Rs. 4500

**32.** **In a partnership, *A* invests (1/6) of the capital for (1/6) of the time, *B* invests (1/3) of the capital for (1/3) of the time and *C*, the rest of the capital for the whole time. Out of a profit of Rs.4600, *B*'s share is :**
(*a*) Rs. 800 (*b*) Rs. 1000
(*c*) Rs. 650 (*d*) Rs. 960

## HINTS & SOLUTIONS (Exercise 9 B)

**1.** Let *C*'s capital = Rs. $x$. Then, *B*'s capital = Rs. $4x$.
$\therefore$ *A*'s capital = Rs. $(12x)$.
So, $A : B : C = 12x : 4x : x = 12 : 4 : 1$

**2.** Let $4A = 6B = 9C = x$. Then, $A = \frac{x}{4}; B = \frac{x}{6}$ & $C = \frac{x}{9}$.
$\therefore A : B : C = \frac{x}{4} : \frac{x}{6} : \frac{x}{9} = 9 : 6 : 4.$

**3.** Ratio of shares = Ratio of 1 day's work
$= \frac{1}{20} : \frac{1}{25} : \frac{1}{30} = 15 : 12 : 10.$
$\therefore$ *A*'s share = Rs. $\left(2220 \times \frac{15}{37}\right)$ = Rs. 900;

$B$'s share = Rs. $\left(2220 \times \frac{12}{37}\right)$ = Rs. 720.

$\therefore$ $A$'s share exceeds $B$'s share = Rs. 180.

**4.** Ratio of investments of $A, B, C$ = 13000 : 17000 : 5000 = 13 : 17 : 5.

$\therefore$ $B$'s share = Rs. $\left(1750 \times \frac{17}{35}\right)$ = Rs. 850.

**5.** Ratio of shares of *Dilip, Ram* and *Amar*

= 2700 : 8100 : 7200 = 3 : 9 : 8.

If *Ram's* share is Rs.9, total profit = Rs. 20

If *Ram's* share is Rs.3600, total profit = Rs. $\left(\frac{20}{9} \times 3600\right)$ = Rs. 8000.

**6.** Ratio of capitals of $A, B, C$

= (12000 × 24) : (16000 × 24) : (15000 × 16) = 6 : 8 : 5.

$C$'s share = Rs. $\left(45600 \times \frac{5}{19}\right)$ = Rs. 12000.

**7.** Ratio of capitals of $A, B$ and $C$

= (16000 × 3 + 11000 × 9) : (12000 × 3 + 17000 × 9) : (21000 × 6)

= 7 : 9 : 6.

$B$'s share = Rs. $\left(26400 \times \frac{9}{22}\right)$ = Rs. 10800;

C's share = Rs. $\left(26400 \times \frac{6}{22}\right)$ = Rs. 7200.

$\therefore$ $B$'s share exceeds $C$'s share by Rs.3600.

**8.** Ratio of investments = (12000 × 12) : (7000 × 8) = 18 : 7

*Dinesh's* share in the profit = Rs. $\left(13300 \times \frac{7}{25}\right)$ = Rs. 3724.

**9.** Given ratio = $\frac{7}{2} : \frac{4}{3} : \frac{6}{5}$ = 105 : 40 : 36.

Let them initially invest Rs 105, Rs.40 and Rs 36 respectively.

Ratio of investments

= [105 × 4 + (150% of 105) × 8] : (40 × 12) : (36 × 12)

= 1680 : 480 : 432 = 35 : 10 : 9.

$\therefore$ $B$'s share = Rs. $\left(21600 \times \frac{10}{54}\right)$ = Rs. 4000.

**10.** Suppose $C$ invests Rs. $x$. Then,

$B$ invests Rs. $\frac{2x}{3}$ and $A$ invests Rs. $2x$.

$\therefore$ Ratio of investments of $A, B, C$

$= 2x : \frac{2x}{3} : x$ or 6 : 2 : 3.

**11.** Ratio of capitals = (5000 × 5) : (6000 × 6) = 25 : 36.

$\therefore$ *Pooja's* share = Rs. $\left(610 \times \frac{25}{61}\right)$ = Rs. 250.

**12.** Ratio of capitals = (3000 × 12) : (2000 × 8) = 9 : 4.

$\therefore$ *Sunil's* share = Rs. $\left(\frac{4}{13} \times 2600\right)$ = Rs. 800.

**13.** Let *B*'s capital be Rs. $x$. Then,

$\frac{30000 \times 12}{x \times 8} = \frac{9}{4} \Rightarrow x = 20000.$

**14.** *C*'s share = Rs. $\left(550 \times \frac{4}{11}\right)$ = Rs. 200.

**15.** Ratio of rents = 6 : 10 : 12 = 3 : 5 : 6.

$\therefore$ *Pandey ji's* share of rent = Rs. $\left(350 \times \frac{6}{14}\right)$ = Rs. 150.

**16.** As the profits are always divided in the ratio of investments, we have

$\frac{A\text{'s capital}}{B\text{'s capital}} = \frac{2}{3}$ or $\frac{3000}{x} = \frac{2}{3}$ or $x = \frac{3000 \times 3}{2}$ = Rs. 4500.

**17.** Let the total profit be Rs.100.

After paying to charity, A's share = Rs. $\left(95 \times \frac{3}{5}\right)$ = Rs. 57.

$\therefore \frac{A\text{'s actual share}}{\text{Actual total profit}} = \frac{57}{100}$

i.e., total profit = Rs. $\left(\frac{855 \times 100}{57}\right)$ = Rs. 1500.

**18.** Ratio of rents

= (18 × 4) : (25 × 2) : (28 × 5) : (21 × 3) = 72 : 50 : 140 : 63.

If total rent is Rs. $x$, then, A's share = Rs. $\left(\frac{x \times 72}{325}\right)$

$\therefore \frac{72x}{325} = 360$ or $x = \frac{325 \times 360}{72} = 1625.$

**19.** Let amount invested by *Ramesh* (in rupees) be $x$.

$\therefore$ 20000 × 6 : 12 $x$ = 6000 : 3000.

i.e. $\frac{12x}{120000} = \frac{3000}{6000}$ or $x$ = Rs. 5000.

**20.** Let *C*'s capital be Rs. $x$. Then,

*B*'s capital = Rs. $(x + 500)$; and

*A*'s capital = Rs. $(x + 1200)$.

$\therefore x + x + 500 + x + 1200 = 4700$ or $x = 1000$.

So, the ratio of capitals

= 2200 : 1500 : 1000 or 22 : 15 : 10.

$\therefore$ *B*'s share = Rs. $\left(\frac{940 \times 15}{47}\right)$ = Rs. 300.

**21.** Ratio of capitals in the beginning $= \frac{1}{3} : \frac{1}{4} : \frac{1}{5}$ or $20 : 15 : 12$.

Ratio of investments for the whole year

$$= (20 \times 4 + 10 \times 8) : (15 \times 12) : (12 \times 12) = 40 : 45 : 36.$$

$\therefore$ *A*'s share = Rs. $\left(847 \times \frac{40}{21}\right)$ = Rs. 280.

**22.** Ratio of investments

$= (2000 \times 1) : (3000 \times 2) : (4000 \times 2) = 1 : 3 : 4.$

$\therefore$ *A*'s share = Rs. $\left(3200 \times \frac{1}{8}\right)$ = Rs. 400.

**23.** *Vijay's* capital = Rs. $(2100 \times 12)$ = Rs. 25200.

For equal profit, Asif's capital = Rs. 25200.

Number of months of investment $= \frac{25200}{3600} = 7.$

So, Asif joined the business after 5 months.

**24.** Suppose *B* invested the money for *x* months.

Then, ratio of investments $= (12 \times 11) : (11 \times x) = 12 : x.$

$\therefore \frac{12}{x} = \frac{4}{1}$ or $x = 3$ months.

**25.** Ratio of investments of *A*, *B* and *C*

$= (12 \times x) : (6 \times 2x) : (4 \times 3x) = 1 : 1 : 1.$

$\therefore$ *C*'s share = Rs. $\left(18000 \times \frac{1}{3}\right)$ = Rs. 6000.

**26.** *A*'s share $= \frac{5}{7}$ of total profit.

$\therefore$ Ratio of shares of profits of *A* and *B* $= \frac{5}{7} : \frac{2}{7} = 5 : 2.$

So, $\frac{1600 \times 8}{x \times 4} = \frac{5}{2}$ or $x = \frac{1600 \times 8 \times 2}{4 \times 5}$ = Rs. 12800

$\therefore$ *B* contributed Rs 12800.

**27.** Ratio of investments

$= (12000 \times 24) : (16000 \times 24) : (15000 \times 16) = 6 : 8 : 5.$

*C*'s share of profit = Rs $\left(46500 \times \frac{5}{19}\right)$ = Rs 12000 .

**28.** Let *A*'s capital be Rs *x*. Then,

$8x = (x - 7500) \times 12$ or $x = 22500.$

**29.** Let *Nandan's* investment be Rs *x* for *y* months.

Then, *Kishan's* investment is Rs $3x$ for $2y$ months.

$\therefore$ Ratio of their investments $= xy : 6xy = 1 : 6$.

*Nandan's* share = Rs 4000.

So, *Kishan's* share = Rs 24000.

$\therefore$ Total profit = Rs 28000.

**30.** Let the initial investments of *A, B, C* be Rs $3x$, Rs $5x$ and Rs $7x$ respectively. Then,

$(3x - 45600) : 5x : (7x + 337600) = 24 : 59 : 167$.

$$\therefore \frac{3x - 45600}{5x} = \frac{24}{59} \Rightarrow x = 47200.$$

$\therefore$ *A* invested initially Rs $(47200 \times 3)$ = Rs 141600.

**31.** For management *A* receives = Rs 960.

Balance = Rs $(9600 - 960)$ = Rs 8640

Ratio of their investments = $12000 : 20000 = 3 : 5$.

$$\therefore A\text{'s share} = \text{Rs}\left(8640 \times \frac{3}{8}\right) = \text{Rs } 3240.$$

So, *A* receives = Rs $(3240 + 960)$ = Rs 4200.

**32.** Suppose *A* invests Rs $\frac{x}{6}$ for $\frac{y}{6}$ months ; *B* invests Rs $\frac{x}{3}$ for $\frac{y}{3}$ months and

*C* invests $\left[x - \left(\frac{x}{6} + \frac{x}{3}\right)\right]$ for *y* months

$\therefore$ Ratio of their investments

$$= \left(\frac{x}{6} \times \frac{y}{6}\right) : \left(\frac{x}{3} \times \frac{y}{3}\right) : \left(\frac{x}{2} \times y\right) = \frac{1}{36} : \frac{1}{9} : \frac{1}{2} = 1 : 4 : 18.$$

$$\therefore B\text{'s share} = \text{Rs}\left(4600 \times \frac{4}{23}\right) = \text{Rs } 800.$$

## ANSWERS (EXERCISE 9 B)

| | | | | | |
|---|---|---|---|---|---|
| 1. (*b*) | 2. (*c*) | 3. (*b*) | 4. (*b*) | 5. (*c*) | 6. (*a*) |
| 7. (*c*) | 8. (*d*) | 9. (*b*) | 10. (*d*) | 11. (*a*) | 12. (*a*) |
| 13. (*a*) | 14. (*c*) | 15. (*d*) | 16. (*d*) | 17. (*c*) | 18. (*c*) |
| 19. (*d*) | 20. (*b*) | 21. (*b*) | 22. (*d*) | 23. (*c*) | 24. (*a*) |
| 25. (*c*) | 26. (*c*) | 27. (*a*) | 28. (*c*) | 29. (*c*) | 30. (*c*) |
| 31. (*b*) | 32. (*a*) | | | | |

# 10

# CHAIN RULE

*The method of finding the fourth proportional when the other three are given is called* **Simple proportion** *or* **Rule of Three.** *Repeated use of the rule of three is called* **Compound proportion.**

**Direct Proportion.** *Two quantities are said to be* **directly proportional** *if on the increase or decrease of the one, the other increases or decreases to the same extent.*

**Ex.** (i) Cost of articles is directly proportional to number of articles. i.e., more articles, more cost & less articles, less cost.

(ii) The work done is directly proportional to the number of men employed to do the work. i.e., more men, more work and less men, less work.

**Indirect Proportion.** *Two quantities are said to be* **indirectly proportional** *if on the increase of the one, the other decreases to the same extent and vice-versa.*

**Ex.** (i) Less number of days required to finish a work, more persons are to be employed.

(ii) The time taken to cover a distance is inversely proportional to the speed of the car. i.e., more speed, less is the time taken.

**Ex. 1.** ***If 15 dolls cost Rs. 35 what do 39 dolls cost ?***

**Sol.** Clearly more dolls, more cost. (Direct Proportion)

So, ratio of dolls is the same as ratio of costs.

Now, let the cost of 39 dolls be Rs. $x$.

Then ,15 : 39 :: 35 : $x$

$$\text{or } \frac{15}{39} = \frac{35}{x} \text{ or } x = \frac{35 \times 39}{15} = 91.$$

**Ex. 2.** ***If 36 men can do a certain piece of work in 25 days, in how many days will 15 men do it ?***

**Sol.** Clearly, less is the number of men employed, more will be the number of days taken to finish the work.

So, inverse ratio of men is equal to ratio of times taken.

Let the required number of days be $x$.

Then, 15 : 36 : : 25 : $x$

or $\frac{15}{36} = \frac{25}{x}$ or $x = \frac{36 \times 25}{15} = 60.$

$\therefore$ *Required number of days = 60.*

**Ex. 3.** ***If 20 men can build a wall 112 metres long in 6 days, what length of a similar wall can be built by 25 men in 3 days ?***

**Sol.** Since the length is to be found out, we compare each item with the length as shown below :—

More men, more length built (direct proportion)

Less days, less length built (direct proportion)

$$\therefore \left.\begin{array}{l}\text{Men } 20:25 \\ \text{Days } \ 6:3\end{array}\right\} :: 112 : x$$

or $20 \times 6 \times x = 25 \times 3 \times 112$ or $x = \frac{25 \times 3 \times 112}{20 \times 6} = 70.$

*Hence, required length of wall built = 70 metres.*

**Ex. 4.** ***If 8 men, working 9 hours a day can build a wall 18 metres long, 2 metres broad and 12 metres high in 10 days, how many men will be required to build a wall 32 metres long, 3 metres broad and 9 metres high, by working 6 hours a day, in 8 days ?***

**Sol.** Since the number of men is to be found out, we compare each item with the number of men, as shown below :—

More length, more men required (direct proportion)

More breadth, more men required (direct proportion)

More height, more men required (direct proportion)

Less daily working hrs., more men required (indirect proportion)

Less days to finish work, more men required (indirect proportion)

$$\left.\begin{array}{ll}\text{Length} & 18:32 \\ \text{Breadth} & 2:3 \\ \text{Height} & 12:9 \\ \text{Daily hrs.} & 6:9 \\ \text{Days} & 8:10\end{array}\right\} :: 8 : x$$

$\therefore 18 \times 2 \times 12 \times 6 \times 8 \times x = 32 \times 3 \times 9 \times 9 \times 10 \times 8$

or $x = \frac{32 \times 3 \times 9 \times 9 \times 10 \times 8}{18 \times 2 \times 12 \times 6 \times 8} = 30.$

$\therefore$ *Required number of men = 30.*

**Ex. 5.** ***A contract is to be completed in 56 days and 104 men were set to work, each working 8 hours a day. After 30 days $\frac{2}{5}$ of the work is completed. How many additional men may be employed, so that the work may be completed in time, each man now working 9 hours a day ?***

**Sol.** Remaining work $= \left(1 - \frac{2}{5}\right) = \frac{3}{5}$.

Remaining period = (56 – 30) = 26 days.

$$\left\{\begin{array}{ll}\text{more work, more men} & \text{(Direct)}\\ \text{more days, less men} & \text{(Indirect)}\\ \text{more hours per day, less men} & \text{(Indirect)}\end{array}\right.$$

$$\left.\begin{array}{ll}\therefore \text{Work} & \frac{2}{5}:\frac{3}{5}\\ \text{days} & 26:30\\ \text{Hours} & 9:8\end{array}\right\}::104:x$$

$$\text{or} \quad x = \frac{3 \times 30 \times 8 \times 104 \times 5}{5 \times 2 \times 26 \times 9} = 160.$$

*So, (160 – 104) i.e., 56 more men must be employed.*

**Ex. 6.** ***2 men and 7 boys together complete a certain work in 16 days, while 3 men and 8 boys together complete the same work in 12 days. Find, in how many days will 8 men and 8 boys together complete a work twice as big as the previous one.***

**Sol.** (2 men + 7 boys) complete the work in 16 days.

So, [(2 × 16) men + (7 × 16) boys] complete the work in 1 day.

i.e., (32 men + 112 boys) complete the work in 1 day.

Again, (3 men + 8 boys) complete the work in 12 days.

So, [(3 × 12) men + (8 × 12) boys] complete the work in 1 day.

i.e., (36 men + 96 boys) complete the work in 1 day.

∴ (32 men + 112 boys) = (36 men + 96 boys)

or 4 men = 16 boys or 1 man = 4 boys.

∴ (2 men + 7 boys) = 15 boys & 8 men + 8 boys = 40 boys.

Thus the problem becomes : 15 boys complete a work in 16 days, in how many days will 40 boys complete twice this work ?

More boys, less days. (indirect)

More work, more boys (direct)

$$\left.\begin{array}{ll}\text{Boys} & 40:15\\ \text{Work} & 1:2\end{array}\right\}::16:x$$

$$\therefore 40 \times 1 \times x = 15 \times 2 \times 16 \quad \text{or} \quad x = \frac{15 \times 2 \times 16}{40 \times 1} = 12.$$

*So, the required number of days = 12.*

**Ex. 7.** ***A garrison of 3300 men had provisions for 32 days, when given at the rate of 850 gms. per head. At the end of 7 days a reinforcement arrives and it was found that now the provisions will last 17 days more, when given at the rate of 825 gm. per head. What is the strength of the reinforcement ?***

**Sol.** The problem can be put in the form given below :—

3300 men taking 850 gm. per head have provisions for (32 – 7) or 25 days. How many men taking 825 gm. per head have provisions for 17 days ?

Let the number of men be $x$.

Now, Less ration per head, more men (indirect proportion)

Less days, more men (indirect proportion)

$$\therefore \left.\begin{matrix}\text{Ration} & 825 : 850 \\ \text{Days} & 17 : 25\end{matrix}\right\} :: 3300 : x$$

$$\therefore 825 \times 17 \times x = 850 \times 25 \times 3300 \quad \text{or} \quad x = \frac{850 \times 25 \times 3300}{825 \times 17} = 5000$$

*So, the strength of new reinforcement = (5000 – 3300) = 1700.*

**Ex. 8.** ***If 6 engines consume 15 metric tonnes of coal, when each is running 9 hours a day; how much coal will be required for 8 engines, each running 12 hours a day, it being given that 3 engines of the former type consume as much as 4 engines of latter type ?***

**Sol.** Since the quantity of coal is to be found out, we compare each item with the coal consumed.

Now, More engines, more coal (direct proportion)

More hours, more coal (direct proportion)

Let 3 engines (former type) consume = 1 unit

$\therefore$ 1 engine (former type) consumes $= \frac{1}{3}$ unit

Then, 4 engines (latter type) consume = 1 unit

$\therefore$ 1 engine (latter type) consumes $= \frac{1}{4}$ unit

Clearly, less rate of consumption, less coal (direct proportion)

Let the required quantity of coal consumed be $x$ tonnes.

$$\left.\begin{matrix}\text{Number of Engines} & 6 : 8 \\ \text{Hours of working} & 9 : 12 \\ \text{Rate of consumption} & \frac{1}{3} : \frac{1}{4}\end{matrix}\right\} :: 15 : x$$

$$\therefore 6 \times 9 \times \frac{1}{3} \times x = 8 \times 12 \times \frac{1}{4} \times 15 \quad \text{or} \quad x = \frac{8 \times 12 \times 1 \times 15 \times 3}{6 \times 9 \times 4} = 20$$

*Hence, the required quantity of coal consumed = 20 tonnes.*

## EXERCISE 10 A (Subjective)

1. If the cost of 19 metres of cloth is Rs.59.85, how much cloth can be purchased for Rs.151.20 ?

   **Hint.** *more money, more cloth. i.e., 59.85 : 151.20 :: 19 : x .*

2. If 30 labourers working 7 hours a day can finish a piece of work in 18 days, how many labourers working 6 hours a day can finish it in 30 days ?

**Hint.** *Less working hours, more labourers. (inverse proportion)*
*Less days, more labourers. (inverse proportion)*

$$\left.\begin{array}{ll}\text{Working hours per day} & 6:7\\ \text{Number of days} & 30:18\end{array}\right\} :: 30 : x.$$

**3.** **70 patients in a hospital consume 1350 litres of milk in 30 days. With the same rate, how many patients will consume 1710 litres in 28 days ?**
**Hint.** *More milk, more patients to consume it. (direct proportion)*
*Less days, more patients to consume it. (indirect proportion)*

$$\left.\begin{array}{ll}\text{Quantity of Milk} & 1350:1710\\ \text{Number of days} & 28:30\end{array}\right\} :: 70 : x$$

**4.** **If 17 men can dig a ditch 26 metres long in 18 days, working 8 hours a day; how many more men should be engaged to dig a similar ditch 39 metres long in 6 days, each man now working 9 hours a day ?**
**Hint.** *More length, more men required (direct proportion)*
*More hours per day, less men required (indirect proportion)*
*More days, less number of men (indirect proportion)*

$$\left.\begin{array}{ll}\therefore \text{Length} & 26:39\\ \text{Hours per day} & 9:8\\ \text{Number of days} & 6:18\end{array}\right\} :: 17 : x.$$

*Thus, x = 68, so reqd. more men = (68 – 17) = 51.*

**5.** **If 18 pumps can raise 2170 tonnes of water in 10 days, working 7 hours a day, in how many days will 16 pumps raise 1736 tonnes, working 9 hours a day ?**
**Hint.** *Less number of pumps, more number of days*
*Less quantity of water, less number of days*
*More working hours, less number of days*

$$\left.\begin{array}{ll}\therefore \text{Pumps} & 16:18\\ \text{Water} & 2170:1736\\ \text{Working hours} & 9:7\end{array}\right\} :: 10 : x.$$

**6.** **If 4 examiners can examine a certain number of answer books in 8 days by working 5 hours a day; for how many hours a day would 2 examiners have to work in order to examine twice the number of answer books in 20 days ?**
**Hint.** *Less examiners, more hours per day*
*More days, less hours per day*
*More answer books, more hours per day*

$$\left.\begin{array}{ll}\text{Examiners} & 2:4\\ \text{Days} & 20:8\\ \text{Ans. Books} & 1:2\end{array}\right\} :: 5 : x.$$

**7.**

7. If 9 men can do a piece of work in 20 days of $7\frac{1}{2}$ hours each, how long will it take for 12 men to do the same work, working 6 hours a day, assuming that 2 men in the latter case do as much as 3 men in the former ?

**Hint.** *Suppose 3 men of former type finish 1 unit of work, then 2 men of the latter type finish 1 unit of work. In other words, 1 man of former type finishes $\frac{1}{3}$ unit of work, while 1 man of latter type finishes $\frac{1}{2}$ unit of work.*

*Now, more men, less days taken to finish the work.*

*Less working hours, more days taken to finish the work.*

*Faster rate of working, less days taken to finish the work.*

$$\left.\begin{array}{ll} \therefore \textit{Men} & 12:9 \\ \textit{Working hours} & 6:\frac{15}{2} \\ \textit{Rate} & \frac{1}{2}:\frac{1}{3} \end{array}\right\} :: 20 : x$$

8. A contractor undertook to build a road in 200 days. He employed 140 men. After 60 days, he found that only $\frac{1}{4}$ of the road could be built. How many additional men should be employed to complete the work in time ?

**Hint.** *Let x men can build the rest of the road in (200 – 60) = 140 days.*

*Then, more days, less men required*

*more work, more men required*

$$\left.\begin{array}{ll} \therefore \textit{Days} & 140:60 \\ \textit{Work} & \frac{1}{4}:\frac{3}{4} \end{array}\right\} :: 140 : x.$$

*So, additional men = (x – 140).*

9. A contractor undertook to do a certain piece of work in 9 days. He employed certain number of labourers but 6 of them being absent from the very first day, the rest could finish the work in 15 days. Find the number of men originally employed.

**Hint.** $15 : 9 :: x : (x-6)$.

10. A contractor employed 15 men, each working 8 hours a day to do a certain piece of work in 19 days. At the end of 10 days, the work had to be suspended for 3 days owing to an accident, in which 4 men were disabled. How many more men must be engaged to complete the work in the specified time, all the men now working 9 hours a day ?

**Hint.** *15 men could finish the work in 9 days, working 8 hours a day.*

*Let x be the number of men who can finish the work in 6 days, working 9 hours a day.*

*Now, more hours a day, less men required*

*Less number of days, more men required*

$$\left.\begin{array}{ll}\therefore \textit{Hours per day} & 9:8 \\ \textit{Days} & 6:9\end{array}\right\} :: 15 : x, \textit{find } (x-11).$$

**11.** If 4 men or 6 boys can do a piece of work in 12 days working 7 hours a day; how many days will it take to complete a work twice as large with 10 men and 3 boys working together 8 hours a day ?

**Hint.** *4 men = 6 boys, so (10 men + 3 boys) = 18 boys.*

*Now, more boys, less days*

*more work, more days*

*more working hours, less days*

$$\left.\begin{array}{ll}\therefore \textit{Boys} & 18:6 \\ \textit{Work} & 1:2 \\ \textit{working hours} & 8:7\end{array}\right\} :: 12 : x.$$

**12.** 10 men and 6 boys can do a piece of work in 20 days, while 12 men and 4 boys can do this work in 18 days. What time will 8 men and 8 boys take to do the work ?

**13.** If 8 men and 12 boys can finish a piece of work in 12 days, in what time will 40 men and 45 boys finish another piece of work 3 times as great, supposing that 16 men can do as much work in 8 hours as 12 boys can do in 24 hours.

**Hint.** *16 men's 8 hours' work = 12 boy's 24 hour's work.*

(16 × 8) *men's 1 hours' work* = (12 × 24) *boy's 1 hour's work*

$$\therefore \textit{1 man} = \left(\frac{12 \times 24}{16 \times 8}\right) \textit{or } \frac{9}{4} \textit{boys}.$$

**14.** If the rent for grazing 40 cows for 20 days is Rs. 370, how many cows can graze for 30 days on Rs.111 ?

**Hint.** *Less rent, less cows*

*More days, less cows*

$$\left.\begin{array}{ll}\textit{Rent} & 370:111 \\ \textit{Days} & 30:20\end{array}\right\} :: 40 : x.$$

## ANSWERS (EXERCISE 10A)

| | | | | |
|---|---|---|---|---|
| **1.** 48 m. | **2.** 21 | **3.** 96 | **4.** 51 | **5.** 7 days |
| **6.** 8 hrs/day | **7.** $12\frac{1}{2}$ days | **8.** 40 men | **9.** 15 | **10.** 9 men |
| **11.** 7 days | **12.** $12\frac{1}{2}$ days | **13.** 6 days | **14.** 8 cows | |

# EXERCISE 10B (Objective Type Questions)

1. If 22.5 metres of a uniform iron rod weighs 85.5 kg., what will be the weight of 6 metres of the same rod ?

(*a*) 22.8 kg. (*b*) 25.6 kg.
(*c*) 28 kg. (*d*) none of these

2. On a scale of a map 0.8 cm. represents 8.8 km. If the distance between two points on the map is 80.5 cm., the distance between these points is approximately

(*a*) 9 km. (*b*) 70 km.
(*c*) 90 km. (*d*) 885 km.

**(S.B.I. P.O. Exam. 1987)**

3. If 40 persons consume 60 kg. of rice in 15 days, then in how many days will 30 persons consume 12 kg. of rice :

(*a*) $3\frac{3}{4}$ days (*b*) 4 days
(*c*) $6\frac{1}{4}$ days (*d*) 9 days

4. In a hospital there is a consumption of 1350 litres of milk for 70 patients for 30 days. How many patients will consume 1710 litres of milk in 28 days ?

(*a*) 59 (*b*) 85
(*c*) 95 (*d*) 105

5. If the rent for grazing 40 cows for 20 days is Rs. 370, how many cows can graze for 30 days on Rs.111 ?

(*a*) 6 (*b*) 8
(*c*) 5 (*d*) 12

6. If 18 binders bind 900 books in 10 days, how many binders will be required to bind 660 books in 12 days ?

(*a*) 55 (*b*) 14
(*c*) 13 (*d*) 11 **(Bank P.O. Exam 1988)**

7. If six men working 8 hours a day earn Rs. 840 per week, then 9 men working 6 hours a day will earn per week

(*a*) Rs. 840 (*b*) Rs. 945
(*c*) Rs. 1620 (*d*) Rs. 1680

**(Central Excise and I. Tax Exam. 1988)**

8. If 3 persons weave 168 shawls in 14 days, how many shawls will 8 persons weave in 5 days ?

(*a*) 90 (*b*) 105
(*c*) 126 (*d*) 160

9. If 20 men can build a wall 112 metres long in 6 days, what length of a similar wall can be built by 25 men in 3 days ?

(*a*) 140 metres (*b*) 44.8 metres
(*c*) 105 metres (*d*) 70 metres

**10.** If 300 men can do a piece of work in 16 days, how many men would do (1/5) of the work in 15 days ?
(*a*) 56 (*b*) 64
(*c*) 60 (*d*) 72

**11.** If 20 men working 7 hours a day can do a piece of work in 10 days, in how many days will 15 men working for 8 hours a day do the same piece of work ?
(*a*) $15\frac{5}{21}$ days (*b*) $11\frac{2}{3}$ days
(*c*) $6\frac{9}{16}$ days (*d*) $4\frac{1}{5}$ days

**12.** If 4 examiners can examine a certain number of answer books in 8 days by working 5 hours a day; for how many hours a day would 2 examiners have to work in order to examine twice the number of answer books in 20 days ?
(*a*) 6 hours (*b*) 8 hours
(*c*) 9 hours (*d*) $7\frac{1}{2}$ hours

**13.** If 18 pumps can raise 2170 tonnes of water in 10 days, working 7 hours a day, in how many days will 16 pumps raise 1736 tonnes, working 9 hours a day ?
(*a*) 9 days (*b*) 8 days
(*c*) 7 days (*d*) 6 days

**14.** 120 men had provisions for 200 days. After 5 days, 30 men died due to an epidemic. The remaining food will last for :
(*a*) 150 days (*b*) $146\frac{1}{4}$ days
(*c*) 245 days (*d*) 260 days

**15.** A garrison of 500 men had provisions for 24 days. However, a reinforcement of 300 men arrived. The food will now last for
(*a*) 18 days (*b*) $17\frac{1}{2}$ days
(*c*) 16 days (*d*) 15 days

**16.** A garrison had provisions for a certain number of days. After 10 days (1/5)th of the men desert and it is found that the provisions will now last just as long as before. How long was that :
(*a*) 35 days (*b*) 15 days
(*c*) 25 days (*d*) 50 days

**17.** 20 men complete one-third of a piece of work in 20 days. How many more men should be employed to finish the rest of the work in 25 more days ?

(*a*) 10 (*b*) 12

(*c*) 15 (*d*) 20

**18.** If 17 labourers can dig a ditch 26 metres long in 18 days, working 8 hours a day, how many more labourers should be engaged to dig a similar ditch 39 metres long in 6 days, each labourer working 9 hours a day ?

(*a*) 51 (*b*) 68

(*c*) 85 (*d*) 34

**19.** If $x$ men working $x$ hours per day can do $x$ units of a work in $x$ days, then $y$ men working $y$ hours per day would be able to complete in $y$ days :

(*a*) $\frac{x^2}{y^3}$ units of work (*b*) $\frac{x^3}{y^2}$ units of work

(*c*) $\frac{y^2}{x^3}$ units of work (*d*) $\frac{y^3}{x^2}$ units of work

**20.** If 5 men working 6 hours a day can reap a field in 20 days, in how many days will 15 men reap the field, working 8 hours a day ?

(*a*) 5 days (*b*) 6 days

(*c*) $7\frac{1}{2}$ days (*d*) 9 days

**21.** If 27 kg. of corn would feed 42 horses for 21 days, in how many days would 36 kg. of it feed 21 horses ?

(*a*) 28 days (*b*) 42 days

(*c*) 56 days (*d*) $31\frac{1}{2}$ days

**22.** If 12 boys can earn Rs. 240 in 5 days, how many boys can earn Rs. 420 in 21 days ?

(*a*) 15 boys (*b*) 5 boys

(*c*) 17 boys (*d*) none of these

**23.** A contractor undertook to do a certain piece of work in 9 days. He employed certain number of labourers but 6 of them being absent from the very first day, the rest could finish the work in 15 days. The number of men originally employed were :

(*a*) 12 (*b*) 15

(*c*) 18 (*d*) 24

**24.** A contractor undertakes to do a piece of work in 40 days. He engages 100 men at the beginning and 100 more after 35 days and completes the work in stipulated time. If he had not engaged the additional men, how many days behind schedule would it be finished ?

(*a*) 5 (*b*) 6

(*c*) 3 (*d*) 9

**25.** A contract is to be completed in 56 days and 104 men were set to work, each working 8 hours a day. After 30 days, $\frac{2}{5}$ of the work is completed. How many additional men may be employed, so that the work may be compeleted in time, each man now working 9 hours a day ?

(*a*) 60 (*b*) 56

(*c*) 70 (*d*) 42

**26.** 15 men take 21 days of 8 hours each to do a piece of work. How many days of 6 hours each would 21 women take, if 3 women do as much work as 2 men ?

(*a*) 20 (*b*) 25

(*c*) 18 (*d*) 30

**27.** If a certain number of workmen can do a piece of work in 25 days, in what time will another set of an equal number of men do a piece of work twice as great, supposing that 2 of the first set can do as much work in an hour as 3 of the second set can do in an hour ?

(*a*) 60 days (*b*) 75 days

(*c*) 90 days (*d*) 105 days

**28.** If 9 men working $7\frac{1}{2}$ hours a day can finish a work in 20 days; then how many days will be taken by 12 men, working 6 hours a day to finish the work; it being given that 3 men of latter type work as much as 2 men of the former type in the same time ?

(*a*) $12\frac{1}{2}$ (*b*) 13

(*c*) $9\frac{1}{2}$ (*d*) 11

**29.** If 5 engines consume 6 metric tonnes of coal when each is running 9 hours a day, how much coal will be needed for 8 engines, each running 10 hours a day, it being given that 3 engines of the former type consume as much as 4 engines of latter type ?

(*a*) 8 metric tonnes (*b*) $8\frac{8}{9}$ metric tonnes

(*c*) $3\frac{1}{8}$ metric tonnes (*d*) 6.48 metric tonnes

**30.** If 3 men or 6 boys can do a piece of work in 10 days, working 7 hours a day; how many days will it take to complete a work twice as large with 6 men and 2 boys working together for 8 hours a day ?

(*a*) $7\frac{1}{2}$ days (*b*) $8\frac{1}{2}$ days

(*c*) 9 days (*d*) 6 days

**31.** 2 men and 7 boys can do a piece of work in 14 days; 3 men and 8 boys can do the same in 11 days. 8 men and 6 boys can do 3 times the amount of this work in :

(*a*) 21 days (*b*) 18 days
(*c*) 24 days (*d*) 36 days

**32.** A contractor employed 30 men to do a piece of work in 38 days. After 25 days, he employed 5 men more and the work was finished one day earlier. How many days he would have been behind, if he had not employed additional men ?

(*a*) 1 day (*b*) $1\frac{1}{4}$ days
(*c*) $1\frac{3}{4}$ days (*d*) $1\frac{1}{2}$ days

**33.** A rope makes 140 rounds of the circumference of a cylinder whose radius of the base is 14 cms. How many times can it go round a cylinder with radius 20 cms. :

(*a*) 98 (*b*) 17
(*c*) 200 (*d*) none of these

**34.** If Raghu can walk a distance of 5 kms. in 20 minutes, how long he can go in 50 minutes :

(*a*) 10.5 km. (*b*) 12 km.
(*c*) 12.5 km. (*d*) 13.5 km.

**35.** If (4/5)th of a cistern is filled in 1 minute, how much more time will be required to fill the rest of it :

(*a*) 20 seconds (*b*) 15 seconds
(*c*) 12 seconds (*d*) 10 seconds

**36.** Ten pipes through which water flows at the same rate can fill a tank in 24 minutes. If two pipes go out of order, how long will the remaining pipes take to fill the tank ?

(*a*) 40 minutes (*b*) 45 minutes
(*c*) $19\frac{1}{5}$ minutes (*d*) 30 minutes

**37.** If 21 cows eat that much as 15 buffaloes, how many cows will eat that much as 35 buffaloes ?

(*a*) 49 (*b*) 56
(*c*) 45 (*d*) none of these

**38.** 16 men can reap a field in 30 days. In how many days will 20 men reap the field ?

(*a*) 25 days (*b*) 24 days

(c) $10\frac{2}{3}$ days (d) $37\frac{1}{2}$ days

**(Railway Board Exam 1989)**

## SOLUTIONS (Exercise 10B)

**1.** Less length, less weight.

$\therefore 22.5 : 6 :: 85.5 : x.$

So, $22.5 \times x = 6 \times 85.5$ or $x = \frac{6 \times 85.5}{22.5} = 22.8$ kg.

**2.** More distance on the map, more actual distance.

$\therefore 0.8 : 80.5 :: 8.8 : x$

So, $0.8 \times x = 80.5 \times 8.8$ or $x = \frac{80.5 \times 8.8}{0.8} = 885.5$ km.

$= 885$ km. (approx).

**3.** Less men, more days (indirect)

Less kg., less days (direct)

$$\left.\begin{matrix} \therefore \text{Men} & 30 : 40 \\ \text{kgs} & 60 : 12 \end{matrix}\right\} :: 15 : x$$

$$\therefore 30 \times 60 \times x = 40 \times 12 \times 15 \text{ or } x = \left(\frac{40 \times 12 \times 15}{30 \times 60}\right) = 4 \text{ days.}$$

**4.** More litres, more patients (direct)

Less days, more patients (indirect)

$$\left.\begin{matrix} \text{Litres} & 1350 : 1710 \\ \text{Days} & 28 : 30 \end{matrix}\right\} :: 70 : x$$

$$\therefore 1350 \times 28 \times x = 1710 \times 30 \times 70$$

$$\text{or } x = \left(\frac{1710 \times 30 \times 70}{1350 \times 28}\right) = 95 \text{ patients.}$$

**5.** More days, less cows (indirect)

Less rent, less cows (direct)

$$\left.\begin{matrix} \text{Days} & 30 : 20 \\ \text{Rent} & 370 : 111 \end{matrix}\right\} :: 18 : x \text{ or } x = \left(\frac{20 \times 111 \times 40}{30 \times 370}\right) = 8 \text{ cows.}$$

**6.** Less books, less number of binders (direct)

More days, less number of binders (indirect)

$$\left.\begin{matrix} \text{Books} & 900 : 660 \\ \text{Days} & 12 : 10 \end{matrix}\right\} :: 18 : x$$

$$\therefore x = \left(\frac{660 \times 10 \times 18}{900 \times 12}\right) = 11.$$

**7.** More men, more earning (direct)

Less hours, less earning (direct)

$$\left.\begin{matrix} \text{Men} & 6 : 9 \\ \text{Hours/day} & 8 : 6 \end{matrix}\right\} :: 840 : x$$

$\therefore x = \left(\frac{9 \times 6 \times 840}{6 \times 8}\right)$ = Rs. *945.*

**8.** More persons, more shawls (direct)

Less days, less shawls (direct)

$$\left.\begin{array}{ll}\text{Persons} & 3:8 \\ \text{Days} & 14:5\end{array}\right\} :: 168 : x$$

$\therefore x = \left(\frac{8 \times 5 \times 168}{3 \times 14}\right)$ = *160 shawls.*

**9.** More men, more length built (direct)

Less days, less length built (direct)

$$\left.\begin{array}{ll}\text{Men} & 20:25 \\ \text{Days} & 6:3\end{array}\right\} :: 112 : x$$

$\therefore x = \left(\frac{25 \times 3 \times 112}{20 \times 6}\right)$ = *70 metres.*

**10.** Less days, more men (indirect)

Less work, less men (direct)

$$\left.\begin{array}{ll}\text{Days} & 15:16 \\ \text{Work} & 1:\frac{1}{5}\end{array}\right\} :: 300 : x.$$

$\therefore x = \left(16 \times \frac{1}{5} \times 300 \times \frac{1}{15 \times 1}\right)$ = *64 men.*

**11.** Less men, more days (indirect)

More working hours, less days (indirect)

$$\left.\begin{array}{ll}\text{Men} & 15:20 \\ \text{Working hrs.} & 8:7\end{array}\right\} :: 10 : x$$

$\therefore x = \left(\frac{20 \times 7 \times 10}{15 \times 8}\right) = 11\frac{2}{3}$ *days.*

**12.** Less examiners, more hours per day (indirect)

More days, less hours per day (indirect)

More answer books, more hours per day (direct)

$$\left.\begin{array}{ll}\text{Examiners} & 2:4 \\ \text{Days} & 20:8 \\ \text{Ans. Books} & 1:2\end{array}\right\} :: 5 : x$$

$\therefore x = \left(\frac{4 \times 8 \times 2 \times 5}{2 \times 20 \times 1}\right)$ = *8 hours per day.*

**13.** Less pumps, more days (indirect)

Less water, less days (direct)

More working hrs., less days (indirect)

$$\left.\begin{array}{ll}\text{Pumps} & 16:18 \\ \text{Water} & 2170:1736 \\ \text{Working hours} & 9:7\end{array}\right\} :: 10 : x$$

$\therefore x = \left(\frac{18 \times 1736 \times 7 \times 10}{16 \times 2170 \times 9}\right) = 7$ *days.*

14. The remaining food is sufficient for 120 men for 195 days.
But, now remaining men = 90.
Less men, more days (indirect)
$\therefore 90 : 120 :: 195 : x$
or $x = \left(\frac{120 \times 195}{90}\right) = 260$ *days.*

15. More men, less number of days (indirect)
$800 : 500 :: 24 : x$
or $x = \left(\frac{500 \times 24}{800}\right) = 15$ *days.*

16. Let, initially there be $x$ men having provisions for $y$ days.
After 10 days, $x$ men had provisions for $(y - 10)$ days.
These provisions were for $\left(x - \frac{x}{5}\right)$ *i.e.*, $\frac{4x}{5}$ men for $y$ days.
$\therefore \; x(y - 10) = \frac{4x}{5} y$ or $xy - 50x = 0$
or $x(y - 50) = 0$ or $y - 50 = 0$ *i.e.* $y = 50$.

17. Work done $= \frac{1}{3}$, work to be done $= \frac{2}{3}$.
Now, more work, more men (direct)
More days, less men (indirect)

$$\left.\begin{array}{ll} \text{Work} & \frac{1}{3} : \frac{2}{3} \\ \text{Days} & 25 : 20 \end{array}\right\} :: 20 : x$$

$\therefore x = \left(\frac{2}{3} \times 20 \times 20 \times \frac{3}{25}\right) = 32$ men.
*So, 12 more men should be employed.*

18. More length, more labourers (direct)
More daily hours, less labourers (indirect)
Less days, more labourers (indirect)

$$\left.\begin{array}{ll} \text{Length} & 26 : 39 \\ \text{Daily hrs.} & 9 : 8 \\ \text{Days} & 6 : 18 \end{array}\right\} :: 17 : x$$

or $x = \left(\frac{39 \times 8 \times 18 \times 17}{26 \times 9 \times 6}\right) = 68$.
*So, more labourers to be engaged* $= (68 - 17) = 51$.

19. More men, more work (direct)
More working hrs., more work (direct)

More days, more work (direct)

$$\left.\begin{array}{ll}\text{Men} & x:y \\ \text{Working hrs.} & x:y \\ \text{Days} & x:y\end{array}\right\} :: x:z$$

$$\therefore z = \left(\frac{y \times y \times y \times x}{x \times x \times x \times x}\right) = \frac{y^3}{x^2} \textit{ units of work.}$$

**20.** More men, less days (indirect)
More working hrs., less days (indirect)

$$\left.\begin{array}{ll}\text{Men} & 15:5 \\ \text{Working hrs.} & 8:6\end{array}\right\} :: 20:x$$

$$\therefore x = \left(\frac{5 \times 6 \times 20}{15 \times 8}\right) = 5 \textit{ days.}$$

**21.** More corn, more days (direct)
Less horses, more days (indirect)

$$\left.\begin{array}{ll}\text{Corn} & 27:36 \\ \text{Horses} & 21:42\end{array}\right\} :: 21:x$$

$$\therefore x = \left(\frac{36 \times 42 \times 21}{27 \times 21}\right) = 56 \textit{ days.}$$

**22.** More money, more boys (direct)
More days, less boys (indirect)

$$\left.\begin{array}{ll}\text{Money} & 240:420 \\ \text{Days} & 21:5\end{array}\right\} :: 12:x$$

$$\therefore x = \left(\frac{420 \times 5 \times 12}{240 \times 21}\right) = 5 \textit{ boys.}$$

**23.** Let there be $x$ men at the beginning.
Now, less men would take more days.

$\therefore\ 15:9 :: x:(x-6)$

*or* $15 \times (x-6) = 9x$ *or* $x = 15$ *men.*

**24.** $[(100 \times 35) + (200 \times 5)]$ working for 1 day can finish the work.
Thus, 4500 men can finish it in 1 day.

So, 100 men can finish it in $\left(\frac{4500}{100}\right) = 45$ days,

*i.e. 5 days behind schedule.*

**25.** Remaining work $= \left(1 - \frac{2}{5}\right) = \frac{3}{5}$.

Remaining period $= (56 - 30) = 26$ days.

Now, the problem becomes : 104 men working 8 hrs a day can finish (2/5) work in 30 days, how many men working 9 hrs. a day can finish (3/5) work in 26 days ?

More work, more men (direct)

Less days, more men (indirect)

More hours, less men (indirect)

$$\left.\begin{array}{ll}\text{Work} & \frac{2}{5}:\frac{3}{5} \\ \text{Days} & 26:30 \\ \text{Hrs./day} & 9:8\end{array}\right\} :: 104 : x$$

$$\therefore x = \left(\frac{3}{5} \times 30 \times 8 \times 104 \times \frac{5}{2} \times \frac{1}{26} \times \frac{1}{9}\right) = 160.$$

*So, more men to be employed = (160 – 104) =56.*

**26.** 3 women = 2 men, so 21 women = 14 men.

Now, Less men, more days (indirect)

Less hours, more days (indirect)

$$\left.\begin{array}{ll}\text{Men} & 14:15 \\ \text{Working hrs.} & 6:8\end{array}\right\} :: 21 : x$$

$$\therefore x = \left(\frac{15 \times 8 \times 21}{14 \times 6}\right) = 30 \text{ days.}$$

**27.** Speed of doing work of first and second set of men is $\frac{1}{2}:\frac{1}{3}$.

Now, More work, more time (direct)

Less speed, more time (indirect)

$$\left.\begin{array}{ll}\text{Work} & 1:2 \\ \text{Speed} & \frac{1}{3}:\frac{1}{2}\end{array}\right\} :: 25 : x$$

$$\therefore x = \left(2 \times \frac{1}{2} \times 25 \times \frac{3}{1 \times 1}\right) = 75 \text{ days.}$$

**28.** More men, less days (indirect)

Less hours a day, more days (indirect)

More speed, less days (indirect)

$$\left.\begin{array}{ll}\text{Men} & 12:9 \\ \text{Hrs./day} & 6:7\frac{1}{2} \\ \text{Speed} & \frac{1}{2}:\frac{1}{3}\end{array}\right\} :: 20 : x$$

$$\therefore x = \left(9 \times \frac{15}{2} \times \frac{1}{3} \times 20 \times \frac{2}{12 \times 6 \times 1}\right) = 12\frac{1}{2} \text{ days.}$$

**29.** More engine, more coal (direct)

More hrs. a day, more coal (direct)

More rate of consumption, more coal (direct)

$$\left.\begin{array}{ll}\text{Engine} & 5:8\\ \text{Hrs./Day} & 9:10\\ \text{Rate} & \frac{1}{3}:\frac{1}{4}\end{array}\right\}::6:x$$

$$\therefore\ x=\left(8\times 10\times \frac{1}{4}\times 6\times \frac{1}{5}\times \frac{1}{9}\times 3\right)=8 \text{ metric tonnes.}$$

**30.** (6 men + 2 boys) = 14 boys.

Now, More work, more number of days (direct)

More boys, less number of days (indirect)

More hours a day, less number of days (indirect)

$$\left.\begin{array}{ll}\text{Work} & 1:2\\ \text{Boys} & 14:6\\ \text{Hrs./Day} & 8:7\end{array}\right\}::10:x$$

$$\therefore\ x=\frac{2\times 6\times 7\times 10}{1\times 14\times 8}=7\frac{1}{2}\text{ days.}$$

**31.** (2 × 14) men + (7 × 14) boys = (3 × 11) men + (8 × 11) boys

or 5 men = 10 boys or 1 man = 2 boys.

∴ 2 men + 7 boys = 11 boys.

and 8 men + 6 boys = 22 boys.

Now, More boys, less days (indirect)

More work, more days (direct)

$$\left.\begin{array}{ll}\text{Boys} & 22:11\\ \text{Work} & 1:3\end{array}\right\}::14:x$$

$$\therefore\ x=\frac{11\times 3\times 14}{22\times 1}=21\text{ days.}$$

**32.** After 25 days, 35 men complete the work in 12 days.

Now, 35 men can finish the remaining work in 12 days.

∴ 30 men can finish it in $\frac{12\times 35}{30}=14$ days.

*i.e. 1 day behind.*

**33.** More radius, less rounds (indirect)

20 : 14 : : 140 : $x$

or $x=\left(\frac{14\times 140}{20}\right)=98$ *times.*

**34.** More time, more distance covered (direct)

∴ 20 : 50 : : 5 : $x$

or $x=\left(\frac{50\times 5}{20}\right)=12.5$ *km.*

**35.** Remaining part = (1/5).

Less part to be filled, less time taken (direct)

$\therefore \frac{4}{5} : \frac{1}{5} = 1 : x$ or $x = \left(\frac{1}{5} \times 1 \times \frac{5}{4}\right) = \frac{1}{4}$ min.

$= 15$ *seconds.*

**36.** Less pipes, more time (indirect)

$8 : 10 :: 24 : x$

or $x = \left(\frac{10 \times 24}{8}\right) = 30$ minutes.

15 buffaloes = 21 cows

35 buffaloes $= \left(\frac{21}{15} \times 35\right) = 49$ cows.

**37.** More men; less days (indirect)

$20 : 16 :: 30 : x$

$\therefore x = \left(\frac{16 \times 30}{20}\right) = 24.$

## ANSWERS (Exercise 10 B)

| | | | | | |
|---|---|---|---|---|---|
| **1.** (*a*) | **2.** (*d*) | **3.** (*b*) | **4.** (*c*) | **5.** (*b*) | **6.** (*d*) |
| **7.** (*b*) | **8.** (*d*) | **9.** (*d*) | **10.** (*b*) | **11.** (*b*) | **12.** (*b*) |
| **13.** (*c*) | **14.** (*d*) | **15.** (*d*) | **16.** (*d*) | **17.** (*b*) | **18.** (*a*) |
| **19.** (*d*) | **20.** (*a*) | **21.** (*c*) | **22.** (*b*) | **23.** (*b*) | **24.** (*a*) |
| **25.** (*b*) | **26.** (*d*) | **27.** (*b*) | **28.** (*a*) | **29.** (*a*) | **30.** (*a*) |
| **31.** (*a*) | **32.** (*a*) | **33.** (*a*) | **34.** (*c*) | **35.** (*b*) | **36.** (*d*) |
| **37.** (*a*) | **38.** (*b*). | | | | |

# 11

# TIME & WORK

**General Rules :**

(i) *If A can finish a piece of work in n days, then A's 1 day's work* $= \frac{1}{n}$.

**Ex.** If $A$ can finish a piece of work in 18 days, then $A$'s 1 day's work $= \frac{1}{18}$.

(ii) If $A$ is twice as good a workman as $B$, then *Time taken by A to finish the work* $= \frac{1}{2} \times$ *(time taken by B).*

(iii) If the number of men to do a work be changed in the ratio 2 : 3, the time taken by them to finish the work is changed in the ratio 3 : 2.

**Solved Examples :**

**Ex. 1.** ***A can do a piece of work in 12 days and B alone can do it in 15 days. How much time both will take to finish the work ?***

**Sol.** $A$'s 1 day's work $= \frac{1}{12}$;

$B$'s 1 day's work $= \frac{1}{15}$.

$\therefore$ $(A + B)$'s 1 day's work $= \left(\frac{1}{12} + \frac{1}{15}\right) = \frac{9}{60} = \frac{3}{20}$.

$\therefore$ *Both together can finish the work in* $\frac{20}{3}$ *or* $6\frac{2}{3}$ *days.*

**Ex. 2.** ***A and B together can do a piece of work in 12 days, B alone can finish it in 30 days. In how many days can A alone finish the work ?***

**Sol.** $(A + B)$'s 1 day's work $= \frac{1}{12}$;

$B$'s 1 day's work $= \frac{1}{30}$.

$\therefore$ $A$'s 1 day's work $= \left(\frac{1}{12} - \frac{1}{30}\right) = \frac{1}{20}$.

Hence, $A$ alone can finish the work in 20 days .

**Ex. 3.** ***A and B can do a piece of work in 12 days; B and C in 15 days; C and A in 20 days. In how many days will they finish it together and separately ?***

**Sol.** $(A + B)$'s 1 day's work $= \frac{1}{12}$;

$(B + C)$'s 1 day's work $= \frac{1}{15}$;

$(C + A)$'s 1 day's work $= \frac{1}{20}$.

Adding, 2 $(A + B + C)$'s 1 day's work $= \left(\frac{1}{12} + \frac{1}{15} + \frac{1}{20}\right) = \frac{1}{5}$.

or $(A + B + C)$'s 1 day's work $= \frac{1}{10}$.

$\therefore$ *A, B, C together can finish the work in 10 days.*

Now, *C's 1 day's work*

$= [(A + B + C)\text{'s 1 day's work}] - [(A + B)\text{'s 1 day's work}]$

$= \left(\frac{1}{10} - \frac{1}{12}\right) = \frac{1}{60}$.

$\therefore$ *C alone can finish it in 60 days.*

Similarly, *B*'s 1 day's work $= \left(\frac{1}{10} - \frac{1}{20}\right) = \frac{1}{20}$.

$\therefore$ *B alone can finish it in 20 days.*

Also, *A*'s 1 day's work $= \left(\frac{1}{10} - \frac{1}{15}\right) = \frac{1}{30}$.

$\therefore$ *A alone can finish the work in 30 days.*

**Ex. 4.** ***A can do a piece of work in 25 days and B can finish it in 20 days. They work together for 5 days and then A goes away. In how many days will B finish the work ?***

**Sol.** $(A + B)$'s 5 day's work $= 5\left(\frac{1}{25} + \frac{1}{20}\right) = \frac{9}{20}$.

Remaining work $= \left(1 - \frac{9}{20}\right) = \frac{11}{20}$.

Now, $\frac{1}{20}$ work is finished by *B* in 1 day.

$\therefore$ $\frac{11}{20}$ *work will be finished by B in* $\left(\frac{20 \times 11}{20}\right)$ = *11 days.*

**Ex. 5.** ***A can do a piece of work in 12 days while B alone can do it in 15 days. With the help of C they can finish it in 5 days. If they are paid Rs.960 for the whole work, how should the money be divided among them ?***

**Sol.** *A*'s 1 day's work $= \frac{1}{12}$;

*B*'s 1 day's work $= \frac{1}{15}$;

$(A+B)$'s 1 day's work $= \left(\frac{1}{12}+\frac{1}{15}\right) = \frac{9}{60} = \frac{3}{20}$.

$(A+B+C)$'s 1 day's work $= \frac{1}{5}$.

$\therefore$ $C$'s 1 day's work $= \left(\frac{1}{5}-\frac{3}{20}\right) = \frac{1}{20}$.

So, the money must be divided among them in the ratio

$\frac{1}{12} : \frac{1}{15} : \frac{1}{20} = 5 : 4 : 3.$

$\therefore$ *A's share* = *Rs.* $\left(960 \times \frac{5}{12}\right)$ = *Rs. 400;*

*B's share* = *Rs.* $\left(960 \times \frac{4}{12}\right)$ = *Rs. 320;*

*C's share* = *Rs.* $[960-(400+320)]$ = *Rs. 240.*

**Ex. 6.** ***A can do a piece of work in 10 days; B in 15 days. They work together for 5 days. The rest of the work is finished by C in 2 days. If they get Rs.1920 for the whole work, how should they divide the money and what are their daily wages ?***

**Sol.** $A$'s 5 day's work $= \left(5 \times \frac{1}{10}\right) = \frac{1}{2}$;

$B$'s 5 day's work $= \left(5 \times \frac{1}{15}\right) = \frac{1}{3}$;

Remaining work $= \left[1-\left(\frac{1}{2}+\frac{1}{3}\right)\right] = \frac{1}{6}$.

$C$'s 2 day's work $= \frac{1}{6}$.

$\therefore$ Ratio of their shares $= \frac{1}{2} : \frac{1}{3} : \frac{1}{6} = 3 : 2 : 1.$

*So, A's share* = *Rs.* $\left(1920 \times \frac{3}{6}\right)$ = *Rs. 960;*

*B's share* = *Rs.* $\left(1920 \times \frac{2}{6}\right)$ = *Rs. 640;*

*C's share* = $[1920-(960+640)]$ = Rs. 320.

**Ex. 7.** ***A is thrice as good a workman as B and is therefore able to finish a piece of work in 60 days less than B. Find the time in which they can do it working together.***

**Sol.** Suppose $B$ finishes the work in $x$ days. Then, $A$ is able to finish it in $(x-60)$ days. Time taken by $A = \frac{1}{3}$ of the time taken by $B$.

$\therefore (x - 60) = \frac{x}{3}$ or $3x - 180 = x$ or $x = 90$.

So, time taken by $B$ = 90 days ;

Time taken by $A$ = 30 days.

$\therefore (A + B)$'s 1 day's work $= \left(\frac{1}{90} + \frac{1}{30}\right) = \frac{4}{90} = \frac{2}{45}$.

$\therefore$ *A & B together can finish the work in* $\frac{45}{2}$ *i.e.* $22\frac{1}{2}$ *days.*

**Ex. 8.** ***A can do a piece of work in 10 days, B in 12 days and C in 15 days. All begin together, but A leaves the work after 2 days and B leaves 3 days before the work is finished. How long did the work last ?***

**Sol.** $A$, $B$ and $C$ together work for 2 days. $C$ alone works for 3 days and the remaining work is done by $B$ & $C$ together.

Now, (A + B + C)'s 2 day's work $= 2 \times \left(\frac{1}{10} + \frac{1}{12} + \frac{1}{15}\right) = \frac{1}{2}$.

$C$'s 3 day's work $= \left(3 \times \frac{1}{15}\right) = \frac{1}{5}$

Remaining work $= 1 - \left(\frac{1}{2} + \frac{1}{5}\right) = \frac{3}{10}$

But, $(B + C)$'s 1 day's work $= \left(\frac{1}{12} + \frac{1}{15}\right) = \frac{27}{180} = \frac{3}{20}$

Now, $\frac{3}{20}$ work is done by $(B + C)$ in 1 day

$\therefore \frac{3}{10}$ work is done by $(B + C)$ in $\left(\frac{20}{3} \times \frac{3}{10}\right) = 2$ days.

*Hence, total time taken* = (2 + 3 + 2) = 7 *days.*

**Ex. 9.** ***A and B can do a piece of work in 45 and 40 days respectively. They began the work together but A leaves after some days and B finished the remaining work in 23 days. After how many days did A leave ?***

**Sol.** $B$'s 23 day's work $= \left(23 \times \frac{1}{40}\right) = \frac{23}{40}$

Remaining work $= \left(1 - \frac{23}{40}\right) = \frac{17}{40}$

Now, $(A + B)$'s 1 day's work $= \left(\frac{1}{45} + \frac{1}{40}\right) = \frac{17}{360}$

Thus, $\frac{17}{360}$ work is done by $(A + B)$ in 1 day.

$\therefore \frac{17}{40}$ work is done by $(A + B)$ in $\left(\frac{360}{17} \times \frac{17}{40}\right) = 9$ days.

*Hence, A left after 9 days. .*

**Ex. 10.** ***A can do a piece of work in 120 days and B can do it in 150 days. They work together for 20 days. Then, B leaves and A alone continues the work. 12 days after that C joins A and the work is completed in 48 days more. In how many days can C do it, if he works alone ?***

**Sol.** [(A + B)'s 20 day's work] + [A's 12 day's work]

$$= \left[20 \times \left(\frac{1}{120} + \frac{1}{150}\right) + \left(12 \times \frac{1}{120}\right)\right] = \frac{2}{5}.$$

$$\text{Remaining work} = \left(1 - \frac{2}{5}\right) = \frac{3}{5} = (A + C)\text{'s 48 day's work}$$

$$\therefore (A + C)\text{'s 1 day's work} = \left(\frac{3}{5} \times \frac{1}{48}\right) = \frac{1}{80}.$$

$$C\text{'s 1 day's work} = \left(\frac{1}{80} - \frac{1}{120}\right) = \frac{1}{240}.$$

*Hence C can finish the work in 240 days.*

**Ex. 11.** ***A and B can do a piece of work in 12 days. B and C together can do it in 15 days. If A is twice as good a workman as C, find in what time B alone can do it.***

**Sol.** A's 1 day's work = C's 2 day's work

∴ (A + B)'s 1 day's work = (B's 1 day's work) + (C's 2 day's work)

$$\text{or (B's 1 day's work) + (C's 2 day's work)} = \frac{1}{12}$$

$$\text{But (B's 1 days work) + (C's 1 day's work)} = \frac{1}{15}$$

$$\text{On subtraction, } C\text{'s 1 day's work} = \left(\frac{1}{12} - \frac{1}{15}\right) = \frac{1}{60}.$$

$$\therefore B\text{'s 1 day's work} = \left(\frac{1}{15} - \frac{1}{60}\right) = \frac{3}{60} = \frac{1}{20}.$$

*Hence, B alone can finish the work in 20 days.*

**Ex. 12.** ***A and B working alone can finish a piece of work in 8 and 7 days respectively. If they work at it alternately for a day, A beginning, in how many days will the work be finished ?***

**Sol.** Work done by A on first day $= \frac{1}{8}$;

Work done by B on second day $= \frac{1}{7}$.

$$\text{Work done in first 2 days} = \left(\frac{1}{8} + \frac{1}{7}\right) = \frac{15}{56}.$$

Work done in first 6 days i.e., (3 units of 2 days each)

$= \left(\frac{3 \times 15}{56}\right) = \frac{45}{56}.$

On 7th day, it is *A*'s turn, so work done in first 7 days

$$= \left(\frac{1}{8} + \frac{45}{56}\right) = \frac{13}{14}.$$

Remaining work $= \left(1 - \frac{13}{14}\right) = \frac{1}{14}$

Now on 8th day, it is *B*'s turn

(1/7) work is done by *B* in 1 day.

(1/14) work is done by *B* in (7/14) $= \frac{1}{2}$ day.

*Total time taken* $= \left(7 + \frac{1}{2}\right)$ *i.e.* $7\frac{1}{2}$ *days.*

**Ex. 13.** ***A certain number of men complete a piece of work in 60 days. If there were 8 men more, the work could be finished in 10 days less. How many men were originally there ?***

**Sol.** Let the original number of men be $x$.

Now $x$ men can finish the work in 60 days and $(x + 8)$ men can finish it in $(60 - 10)$ *i.e.* 50 days.

More-over, more men will finish the work in less days.

$\therefore (x + 8) : x :: 60 : 50$ or $\frac{x+8}{x} = \frac{60}{50}$ or $x = 40$

*Hence, the original number of men = 40.*

**Ex. 14.** ***If 4 men or 6 boys can finish a piece of work in 20 days, in how many days can 6 men and 11 boys finish it ?*** **(L.I.C. 1991)**

**Sol.** 4 men = 6 boys.

$\therefore$ 6 men $= \left(\frac{6}{4} \times 6\right) = 9$ boys.

$\therefore$ 6 men + 11 boys = 20 boys.

Now, 6 boys can finish the work in 20 days.

In how many days can 20 boys do it ?

Let it be done in $x$ days.

More boys, less days.

$\therefore 20 : 6 :: 20 : x$

*or* $20x = 6 \times 20$ *or* $x = \left(\frac{6 \times 20}{20}\right) = 6$ *days.*

**Ex. 15.** ***2 men and 3 women working 7 hours a day finish a work in 5 days; 4 men and 4 women working 3 hours a day do the same work in 7 days. Find the number of days in which the work is done by 7 men only, working 4 hours a day.***

**Sol.** [(2 × 7 × 5) men + (3 × 7 × 5) women] can do the work in 1 day of 1 hour.

[(4 × 3 × 7) men + (4 × 3 × 7) women] can do the work in 1 day of 1 hour.

∴ [(2 × 7 × 5) men + (3 × 7 × 5) women]

≡ [(4 × 3 × 7) men + (4 × 3 × 7) women]

or (70 men + 105 women) ≡ (84 men + 84 women).

or 1 woman ≡ $\frac{2}{3}$ men. So, (2 men + 3 women) ≡ 4 men.

Now, 4 men working 7 hours a day finish the work in 5 days. Let 7 men, working 4 hours a day, finish it in $x$ days.

Then more men, less days

and less working hours, more days

$$\left.\begin{array}{r} \therefore\ 7:4 \\ 4:7 \end{array}\right\} :: 5 : x$$

or $7 \times 4 \times x = 4 \times 7 \times 5$ or $x = 5$

∴ *Required number of days = 5.*

## EXERCISE 11A (Subjective)

1. *A* can do a piece of work in 7 hours while *B* alone can do it in 8 hours. Find the time taken by *A* and *B* together to finish the work.
2. *A* and *B* can do a piece of work in 6 days; *B* and *C* in 7 days; *C* and *A* in 8 days. Find the time taken by *A*, *B*, *C* working together to finish the work. Also, find the time taken by each alone to finish the work.
3. *A* can do a piece of work in 20 days, while *B* alone can do it in 25 days. *B* worked for 10 days and left. In how many days, *A* will now finish the remaining work ?
4. *A* can do a piece of work in 24 days, while *B* alone can do it in 16 days. With the help of *C* they finish the work in 8 days. Find, in how many days *C* alone can do the work.
5. *A* can do a piece of work in 25 days which *B* alone can do in 20 days. *A* started the work and was joined by *B* after 10 days. Find the time taken to finish the work.

**Hint.** *A's 10 day's work* = $\left(10 \times \frac{1}{25}\right) = \frac{2}{5}$.

*Remaining work* = $\left(1 - \frac{2}{5}\right) = \frac{3}{5}$.

*Now, (A + B)'s 1 day's work* = $\left(\frac{1}{25} + \frac{1}{20}\right) = \frac{9}{100}$.

*Thus,* $\frac{9}{100}$ *work is done by (A + B) in 1 day.*

$\therefore \frac{3}{5}$ *work is done by (A + B) in* $\left(\frac{100}{9} \times \frac{3}{5}\right) = \frac{20}{3}$ *days.*

$\therefore$ *Total time taken* $= \left(10 + 6\frac{2}{3}\right) = 16\frac{2}{3}$ *days.*

**6.** **A is twice as good a workman as B and is therefore able to finish a piece of work in 12 days less than B. If they work together, find the time in which they can finish the work.**

**Hint.** *Suppose B finishes the work in x days.*

*Then, A will finish it in (x – 12) days.*

*A's 1 day's work* $= \frac{1}{(x-12)}$, *B's 1 day's work* $= \frac{1}{x}$.

*Since A is twice as good a workman as B, so* $\frac{2}{x} = \frac{1}{x-12}$ or $x = 24$.

$\therefore$ *(A + B)'s 1 day's work* $= \left(\frac{1}{24} + \frac{1}{12}\right) = \frac{1}{8}$.

**7.** **A sum is sufficient to pay A's wages for 21 days or B's wages for 28 days. Find, for how many days the money is sufficient to pay the wages of both.**

**Hint.** *A's wages for 21 days = B's wages for 28 days = x*

*Then, A's wages for 1 day = Rs.* $\left(\frac{x}{21}\right)$;

*B's wages for 1 day = Rs.* $\left(\frac{x}{28}\right)$

*Ratio of their wages* $= \frac{1}{21} : \frac{1}{28}$

= *Ratio of their 1 day's work*

$\therefore$ *(A + B)'s 1 day's work* $= \left(\frac{1}{21} + \frac{1}{28}\right) = \frac{7}{84} = \frac{1}{12}$.

*So, they can finish the work in 12 days, i.e. money is sufficient to pay the wages of both for 12 days.*

**8.** **A and B can do a piece of work in 10 days; B and C in 15 days; C and A in 20 days. All of them work at it for 2 days, then A leaves. After 2 days of this, E also leaves. In how much time will now C complete the work ?**

**Hint.** *(A + B + C)'s 1 days work*

$$= \frac{1}{2}\left(\frac{1}{10} + \frac{1}{15} + \frac{1}{20}\right) = \frac{13}{120}.$$

*& C's 1 day's work* $= \left(\frac{13}{120} - \frac{1}{10}\right) = \frac{1}{120}$

*Work done by the C alone*

*Work done by the C alone*

$$= 1 - \left(2 \times \frac{13}{120} + 2 \times \frac{1}{15}\right) = \frac{13}{20}$$

*Now,* $\frac{1}{120}$ *work is done by C in 1 day*

$\therefore \frac{13}{20}$ *work is done by C in* $\left(\frac{120}{1} \times \frac{13}{20}\right) = 78$ *days.*

**9. If 2 men with 3 boys can earn Rs.252 in 7 days and 4 men with 7 boys can earn Rs.624 in 8 days, in what time 6 men with 8 boys earn Rs.1020 ?**

**Hint.** $\left[\left(\frac{2 \times 7}{252}\right) \text{men} + \left(\frac{3 \times 7}{252}\right) \text{boys}\right]$ *earn Re.1 in 1 day.*

*and* $\left[\left(\frac{4 \times 8}{624}\right) \text{men} + \left(\frac{7 \times 8}{624}\right) \text{boys}\right]$ *earn Re.1 in 1 day.*

*Thus,* $\left(\frac{1}{18} \text{men} + \frac{1}{12} \text{boys}\right) = \left(\frac{2}{39} \text{men} + \frac{7}{78} \text{boys}\right)$

*or* (156 men + 234 boys) = (144 men + 252 boys)

*or 12 men = 18 boys or 2 men = 3 boys.*

$\therefore$ *(2 men + 3 boys) = 6 boys & (6 men + 8 boys) = 17 boys.*

*Now* $\left.\begin{matrix} 252 : 1020 \\ 17 : 6 \end{matrix}\right\} :: 7 : x$

**10. 25 men are employed to do a piece of work which they could finish in 20 days, but the men drop off by 5 at the end of every 10 days. In what time will the work be completed ?**

**Hint.** *Work done by 25 men in 10 days* $= \left(10 \times \frac{1}{20}\right) = \frac{1}{2}$

*Remaining work* $= \frac{1}{2}$ *& remaining men* = 20

*Work done by 20 men in 10 days* $= \left(\frac{1}{25} \times \frac{1}{20} \times 20 \times 10\right) = \frac{2}{5}$

*Remaining work* $= \left(\frac{1}{2} - \frac{2}{5}\right) = \frac{1}{10}$ *& Remaining men* = 15

*15 men's 1 day's work* $= \frac{1}{25 \times 20} \times 15 = \frac{3}{100}$

*Now* $\frac{3}{100}$ *work is done by 15 men in 1 day*

$\therefore \frac{1}{10}$ *work is done by 15 men in* $\left(\frac{100}{3} \times \frac{1}{10}\right) = 3\frac{1}{3}$ *days.*

*Hence, the total time taken* $= \left(10 + 10 + 3\frac{1}{3}\right) = 23\frac{1}{3}$ *days.*

**11.** *A* can do half much again as *B* can do in the same time. *B* alone can do a piece of work in 18 days. Find, in how many days both together can finish the work.

**Hint.** *Ratio of working of A & B* $= \frac{3}{2} : 1$

*Ratio of time taken for same work* $= 1 : \frac{3}{2}$ or $2 : 3$

*Time taken by B alone to finish the work* = 18 *days*

*Time taken by A alone to finish the work* $= \left(18 \times \frac{2}{3}\right) = 12$ *days.*

$(A + B)$*'s 1 day's work* $= \left(\frac{1}{12} + \frac{1}{18}\right) = \frac{5}{36}$

$\therefore$ *Time taken by* $(A + B)$ *to finish the work* $= \frac{36}{5} = 7\frac{1}{5}$ *days.*

**12.** *A* is twice as good a workman as *B* and together they finish a piece of work in 14 days. In how many days can it be done by *A* alone ?

**13.** Three labourers *A, B, C* were given a contract of Rs.750 for doing a certain piece of work. All the three together can finish this work in 8 days. *A* and *C* together can do it in 12 days, while *A* and *B* together can do it in $13\frac{1}{3}$ days. How the money will be divided among them ?

**Hint.** *B's 1 day's work* = [$(A + B + C)$*'s 1 day's work*] – [$(A + C)$*'s 1 day's work*];

*C's 1 day's work* = [$(A + B + C)$*'s 1 day's work*] – [$(A + B)$*'s 1 day's work*]

& $(2A)$*'s 1 day's work* = [$(2A + B + C)$*'s 1 day's work*] – [B's 1 day's work + C' 1 day's work]

*Now, divide Rs.750 in the ratio of 1 day's work of A, B, C i.e.* $4 : 5 : 6$.

**14.** *A, B, C* together can finish a work in 12 days. *A* and *C* together can do twice as much as *B* does. Also, *A* and *B* can do thrice as much as *C* does. Find the time taken by each to finish the work.

**Hint.** *Suppose A, B & C separately can finish the work in x, y* and *z days respectively.*

*Then,* $\left(\frac{1}{x} + \frac{1}{y} + \frac{1}{z}\right) = \frac{1}{12}; \left(\frac{1}{x} + \frac{1}{z}\right) = \frac{2}{y}$ & $\left(\frac{1}{x} + \frac{1}{y}\right) = \frac{3}{z}$.

*Find x, y* and *z.*

## ANSWERS (Exercise 11A)

1. 3 hours 44 min.

2. together = $4\frac{44}{73}$ days, $C = 19\frac{13}{17}$ days; $A = 13\frac{11}{25}$ days, $B = 10\frac{26}{31}$ days.

3. 12 days
4. 48 days.
5. $16\frac{2}{3}$ days
6. 8 days
7. 12 days
8. 78 days
9. 10 days
10. $23\frac{1}{3}$ days
11. $7\frac{1}{5}$ days
12. 21 days
13. Rs.200, Rs.250, Rs.300
14. $28\frac{4}{5}$ days, 36 days, 48 days.

## EXERCISE 11B

## OBJECTIVE TYPE QUESTIONS

1. *A* can do a piece of work in 30 days while *B* can do it in 40 days. *A* and *B* working together can do it in

(*a*) 70 days (*b*) $42\frac{3}{4}$ days

(*c*) $27\frac{1}{7}$ days (*d*) $17\frac{1}{7}$ days **(Railways 1989)**

2. *A*, *B* and *C* can do a piece of work in 6, 12 and 24 days respectively. They altogether will complete the work in

(*a*) $3\frac{3}{7}$ days (*b*) $\frac{7}{24}$ days

(*c*) $4\frac{4}{5}$ days (*d*) $\frac{5}{24}$ days

**(Clerical Grade 1991)**

3. *A* can do (1/3) of a work in 5 days and *B* can do (2/5) of the work in 10 days. In how many days both *A* and *B* together can do the work ?

(*a*) $7\frac{3}{4}$ days (*b*) $9\frac{3}{8}$ days

(*c*) $8\frac{4}{5}$ days (*d*) 10 days **(Railways 1991)**

4. *A* and *B* can together do a piece of work in 15 days. *B* alone can do it in 20 days. *A* alone can do it in :

(*a*) 30 days (*b*) 40 days

(*c*) 45 days (*d*) 60 days **(Railways 1991)**

5. *A* and *B* finish a job in 12 days while *A*, *B* and *C* can finish it in 8 days. *C* alone will finish the job in :

(*a*) 20 days (*b*) 14 days

(*c*) 24 days (*d*) 16 days

**(Hotel Management 1991)**

6. *A* and *B* can do a piece of work in 18 days; *B* and *C* in 24 days; *A* and *C* in 36 days. In what time can they do it all working together :

(*a*) 16 days (*b*) 13 days
(*c*) 26 days (*d*) 12 days

7. The rates of working of *A* and *B* are in the ratio 3 : 4. The number of days taken by them to finish the work are in the ratio :

(*a*) 3 : 4 (*b*) 2 : 1
(*c*) 4 : 3 (*d*) none of these

8. *A* and *B* can do a piece of work in 12 days; *B* and *C* in 15 days; *C* and *A* in 20 days; *A* alone can do the work in :

(*a*) $15\frac{2}{3}$ days (*b*) 30 days
(*c*) 24 days (*d*) 40 days

9. Rahul can do a piece of work in 20 days while Santosh can do it in 12 days. Santosh worked at it for 9 days and left. Rahul can finish the remaining work in :

(*a*) 11 days (*b*) 7 days
(*c*) 5 days (*d*) 3 days

10. Mahesh and Umesh can complete a work in 10 and 15 days respectively. Umesh starts the work and after 5 days Mahesh also joins him. In all, the work would be completed in :

(*a*) 9 days (*b*) 7 days
(*c*) 11 days (*d*) none of these

**(Clerical Grade 1991)**

11. *A* can complete a job in 9 days, *B* in 10 days and *C* in 15 days. *B* and *C* start the work and are forced to leave after 2 days. The time taken to complete the remaining work is :

(*a*) 13 days (*b*) 10 days
(*c*) 9 days (*d*) 6 days

**(N.D.A. Exam 1987)**

12. *A* can do a piece of work in 80 days. He works at it for 10 days and then *B* alone finishes the work in 42 days. The two together could complete the work in :

(*a*) 24 days (*b*) 25 days
(*c*) 30 days (*d*) 35 days

**(Clerical Grade 1991)**

13. *A* and *B* can together finish a work in 30 days. They worked for it for 20 days and then *B* left. The remaining work was done by *A* alone in 20 more days. *A* alone can finish the work in :

(*a*) 54 days (*b*) 60 days
(*c*) 48 days (*d*) 50 days

**(Central Excise 1988)**

**14.** Ramesh can finish a job in 20 days. He worked for 10 days alone and completed the remaining job working with Dinesh, in 2 days. How many days would both Dinesh and Ramesh together take to complete the entire job ?

(*a*) 4 (*b*) 5
(*c*) 10 (*d*) 12 **(BSRB Exam 1991)**

**15.** *A* can do a piece of work in 14 days which *B* can do in 21 days. They begin together but 3 days before the completion of the work, *A* leaves off. The total number of days to complete the work is :

(*a*) $6\frac{3}{5}$ days (*b*) $8\frac{1}{2}$ days
(*c*) $10\frac{1}{5}$ days (*d*) $13\frac{1}{2}$ days

**16.** *A* can do a piece of work in 12 days. *B* is 60% more efficient than *A*. The number of days, it takes *B* to do the same piece of work, is :

(*a*) $7\frac{1}{2}$ (*b*) $6\frac{1}{4}$
(*c*) 8 (*d*) 6 **(C.B.I. Exam 1991)**

**17.** *A* is twice as good a workman as *B* and together they finish a piece of work in 14 days. *A* can finish the work in :

(*a*) 42 days (*b*) 11 days
(*c*) 28 days (*d*) 21 days

**18.** *A* is thrice as good a workman as *B* and takes 10 days less to do a piece of work than *B* takes. *B* can do the work in :

(*a*) 30 days (*b*) 20 days
(*c*) 15 days (*d*) 12 days

**19.** 12 men can complete a work within 9 days. After 3 days they started the work, 6 men joined them to replace 2 men. How many days will they take to complete the remaining work ?

(*a*) 2 (*b*) 3
(*c*) 4 (*d*) $4\frac{1}{2}$ **(BSRB Exam, 1991)**

**20.** *A* can complete a job in 9 days; *B* in 10 days and *C* in 15 days. *B* and *C* together start the work and are forced to leave after 2 days. The time taken by *A* to complete the remaining work is :

(*a*) $6\frac{1}{3}$ days (*b*) 6 days
(*c*) 10 days (*d*) 8 days

**21.** *A* and *B* can do a piece of work in 45 and 40 days respectively. They began to work together but *A* leaves after sometime and *B* finished the remaining work in 23 days. After how many days did *A* leave ?

(*a*) 6 days (*b*) 8 days
(*c*) 9 days (*d*) 12 days

**22.** **A can complete a work in 6 days and *B* in 5 days. They work together, finish the job and receive Rs.220 as wages. *B*'s share should be**
(*a*) Rs.120 (*b*) Rs.110
(*c*) Rs.100 (*d*) Rs.90 **(Clerical Grade 1991)**

**23.** ***A*, *B* and *C* together earn Rs.150 per day while *A* and *C* together earn Rs.94 and *B* and *C* together earn Rs.76. The daily earning of *C* is :**
(*a*) Rs.75 (*b*) Rs.56
(*c*) Rs.34 (*d*) Rs.20 **(Bank P.O. 1986)**

**24.** **A certain number of men complete a piece of work in 60 days. If there were 8 men more the work could be finished in 10 days less. How many men were originally there ?**
(*a*) 30 (*b*) 32
(*c*) 40 (*d*) 36

**25.** ***A* and *B* working separately can do a piece of work in 9 and 12 days respectively. If they work for a day alternately, *A* beginning, in how many days the work will be completed ?**
(*a*) $10\frac{1}{2}$ (*b*) $10\frac{1}{4}$
(*c*) $10\frac{2}{3}$ (*d*) $10\frac{1}{3}$

**26.** ***A* can do a piece of work in 5 hours, *B* in 9 hours and *C* in 15 hours. If *C* could work with them for 1 hour only, the time taken by *A* and *B* together to complete the work is :**
(*a*) 2 hours (*b*) 3 hours
(*c*) $3\frac{1}{2}$ hours (*d*) 4 hours **(Clerical Grade 1991)**

**27.** **Two men undertake to do a piece of work for Rs.200. One alone can do it in 6 days, the other in 8 days. With the help of a boy they finish it in 3 days. How much is the share of the boy ?**
(*a*) Rs.20 (*b*) Rs.25
(*c*) Rs.30 (*d*) Rs.40

**28.** ***A* does half as much work as *B* in three fourth of the time. If together they take 18 days to complete a work, how much time shall *B* take to do it ?**
(*a*) 40 days (*b*) 35 days
(*c*) 30 days (*d*) none of these
**(L.I.C. A.A.O. Exam 1988)**

**29.** **A sum of money is sufficient to pay *A*'s wages for 21 days or *B*'s wages for 28 days. The money is sufficient to pay the wages of both for :**

(*a*) $12\frac{1}{4}$ days (*b*) 12 days

(*c*) 14 days (*d*) none of these

**30.** If 5 men or 9 women can do a piece of work in 19 days; 3 men and 6 women will do the same work in :

(*a*) 10 days (*b*) 12 days

(*c*) 13 days (*d*) 15 days

**31.** 10 men can finish a piece of work in 10 days whereas it takes 12 women to finish it in 10 days. If 15 men and 6 women undertake to complete the work how many days will they take to complete it.

(*a*) 11 (*b*) 5

(*c*) 4 (*d*) 2 **(Bank P.O. 1991)**

**32.** 2 men and 3 women can finish a piece of work in 10 days, while 4 men can do it in 10 days. 3 men and 3 women will finish it in

(*a*) 6 days (*b*) 10 days

(*c*) 8 days (*d*) 14 days

**33.** 8 children and 12 men complete a certain piece of work in 9 days. Each child takes twice the time by a man to finish the work. In how many days will 12 men finish the same work ?

(*a*) 8 (*b*) 9

(*c*) 12 (*d*) 15 **(Bank P.O. 1988)**

**34.** 8 men can dig a pit in 20 days. If a man works half as much again as a boy, then 4 men and 9 boys can dig it in :

(*a*) 10 days (*b*) 12 days

(*c*) 15 days (*d*) 16 days

**35.** 3 men and 4 boys do a piece of work in 8 days, while 4 men and 4 boys finish it in 6 days. 2 men and 4 boys will finish it in

(*a*) 9 days (*b*) 10 days

(*c*) 12 days (*d*) 14 days

**36.** If 1 man or 2 women or 3 boys can do a piece of work in 44 days, 1 man, 1 woman and 1 boy will do it in :

(*a*) 21 days (*b*) 24 days

(*c*) 26 days (*d*) 33 days

**37.** 4 men and 6 women finish a job in 8 days, while 3 men and 7 women finish it in 10 days. 10 women working alone will finish it in :

(*a*) 24 days (*b*) 32 days

(*c*) 36 days (*d*) 40 days

## HINTS & SOLUTIONS (Exercise 11 B)

**1.** $(A + B)$'s 1 day's work $= \left(\frac{1}{30} + \frac{1}{40}\right) = \frac{7}{120}$.

$\therefore$ Both together will finish the work in $\frac{120}{7}$ days i.e. $17\frac{1}{7}$ days.

2. $(A+B+C)$'s 1 day's work $= \left(\frac{1}{6}+\frac{1}{12}+\frac{1}{24}\right) = \frac{7}{24}$.

$\therefore$ They altogether will complete the work in $\frac{24}{7}$ i.e. $3\frac{3}{7}$ days.

3. $A$'s one day's work $= \frac{1}{15}$.

$B$'s one day's work $= \frac{1}{25}$.

$(A+B)$'s 1 day's work $= \left(\frac{1}{15}+\frac{1}{25}\right) = \frac{8}{75}$.

$\therefore$ Both together will finish the work in $\frac{75}{8}$ i.e. $9\frac{3}{8}$ days.

4. $A$'s 1 day's work
$= [(A+B)$'s 1 day's work $- B$'s 1 day's work$]$
$= \left[\frac{1}{15}-\frac{1}{20}\right] = \frac{1}{60}$.
$\therefore$ $A$ alone can do the work in 60 days.

5. $C$'s 1 day's work
$[(A+B+C)$'s 1 day's work $- (A+B)$'s 1 day's work$]$
$= \frac{1}{8}-\frac{1}{12} = \frac{1}{24}$.
$\therefore$ $C$ alone will finish the work in 24 days.

6. $[(A+B)+(B+C)+(A+C)]$'s 1 day's work
$= \left(\frac{1}{18}+\frac{1}{24}+\frac{1}{36}\right) = \frac{1}{8}$.
or 2 $(A+B+C)$'s 1 day's work = (1/8)
or $(A+B+C)$'s 1 day's work = (1/16).
So, they all together will finish the work in 16 days.

7. Clearly, number of days taken are in a reverse ratio i.e. 4 : 3.

8. Adding, 2 $(A+B+C)$'s 1 day's work
$$= \left(\frac{1}{12}+\frac{1}{15}+\frac{1}{20}\right) = \frac{1}{5}.$$
$\therefore$ $(A+B+C)$'s 1 day's work = (1/10).

So, $A$'s 1 day's work$= \left(\frac{1}{10}-\frac{1}{15}\right) = \frac{1}{30}$.

$\therefore$ $A$ alone can finish the work in 30 days.

9. Santosh's 9 day's work $= \frac{9}{12} = \frac{3}{4}$.

Remaining work $= \left(1 - \frac{3}{4}\right) = \frac{1}{4}$.

Now, (1/20) work is done by Rahul in 1 day.

So, (1/4) work will be done by Rahul in 5 days.

**10.** Umesh's 5 day's work $= \frac{5}{15} = \frac{1}{3}$.

Remaining work $= \left(1 - \frac{1}{3}\right) = \frac{2}{3}$.

Now, $\left(\frac{1}{10} + \frac{1}{15}\right)$ work is done by $A$ & $B$ in 1 day.

$\therefore \frac{2}{3}$ work will be done by $A$ &$B$ in $\left(6 \times \frac{2}{3}\right) = 4$ days .

So, the work would be completed in 9 days.

**11.** $(B + C)$'s 2 day's work $= 2\left(\frac{1}{10} + \frac{1}{15}\right) = \frac{1}{3}$.

Remaining work $= \left(1 - \frac{1}{3}\right) = \frac{2}{3}$.

Now, (1/9) work is done by $A$ in 1 day.

$\therefore$ (2/3) work will be done by $A$ in $\left(9 \times \frac{2}{3}\right) = 6$ days.

**12.** $A$'s 10 day's work $= \frac{10}{80} = \frac{1}{8}$.

$\left(1 - \frac{1}{8}\right)$ work is done by $B$ in 42 days.

Therefore, $B$ alone can finish the work in $\left(\frac{42 \times 8}{7}\right) = 48$ days.

$\therefore$ $B$'s 1 day's work $= \frac{1}{48}$.

$(A + B)$'s 1 day's work $= \left(\frac{1}{80} + \frac{1}{48}\right) = \frac{1}{30}$.

$\therefore$ $A$ and $B$ together can finish the work in 30 days.

**13.** $(A + B)$'s 20 day's work $= \frac{20}{30} = \frac{2}{3}$.

Remaining work $= \left(1 - \frac{2}{3}\right) = \frac{1}{3}$.

Now, (1/3) work is done by $A$ alone in 20 days.

The whole work will be done by $A$ in $(20 \times 3) = 60$ days.

**14.** Ramesh alone finished $\frac{1}{2}$ of the work in 10 days

Remaining $\frac{1}{2}$ of the job was finished by Ramesh and Dinesh together in 2 days.

Therefore, they both together can finish the complete job in 4 days.

**15.** $B$'s 3 day's work $= \frac{3}{21} = \frac{1}{7}$.

Remaining work $= \left(1 - \frac{1}{7}\right) = \frac{6}{7}$.

Now, $\left(\frac{1}{14} + \frac{1}{21}\right)$ *i.e.* $\frac{5}{42}$ work is done by $A$ & $B$ in 1 days.

$\therefore$ $\frac{6}{7}$ work is done by $(A + B)$ in $\left(\frac{42}{5} \times \frac{6}{7}\right) = \frac{36}{5}$ days.

$\therefore$ Total time taken $= \left(3 + 7\frac{1}{5}\right) = 10\frac{1}{5}$. days.

**16.** $A$'s 1 day's work $= \frac{1}{12}$.

$B$'s 1 day's work $= \frac{1}{12} + 60\%$ of $\frac{1}{12} = \frac{2}{15}$.

$\therefore$ $B$ can do the work in $\frac{ddd15}{2}$ *i.e.* $7\frac{1}{2}$ days

**17.** ($A$'s 1 day's work) : ($B$'s 1 day's work) $= 2 : 1$.

$(A + B)$'s 1 day's work $= \frac{1}{14}$.

$\therefore$ $A$'s 1 day's work $= \left(\frac{1}{14} \times \frac{2}{3}\right) = \frac{1}{21}$.

[Divide (1/14) in ratio 2 : 1]

So, $A$ can finish the work in 21 days.

**18.** Ratio of time taken by $A$ and $B = 1 : 3$

If the difference of time is 2 days, $B$ takes 3 days.

If the difference of time is 10 days, $B$ takes $\left(\frac{3}{2} \times 10\right) = 15$ days.

**19.** 12 men can complete $\frac{1}{3}$ of the work in 3 days and the remaining $\frac{2}{3}$ of the work in 6 days.

1 man can complete $\frac{2}{3}$ of the wcrk in $(12 \times 6) = 72$ days.

$\therefore$ $12 - 2 + 6 = 16$ men can complete $\frac{2}{3}$ of the work in $\frac{72}{16} = 4\frac{1}{2}$ days.

**20.** $(B + C)$'s 2 day's work $= \left(\frac{2}{10} + \frac{2}{15}\right) = \frac{1}{3}$

Remaining work $= \left(1 - \frac{1}{3}\right) = \frac{2}{3}$.

Now, (1/9) of the work is done by $A$ in 1 day.

$\therefore \frac{2}{3}$ of the work will be done by $A$ in $\left(9 \times \frac{2}{3}\right)$ i.e. 6 days.

**21.** $B$'s 23 day's work $= \frac{23}{40}$.

Remaining work $= \left(1 - \frac{23}{40}\right) = \frac{17}{40}$.

Now, $(A + B)$'s 1 day's work $= \left(\frac{1}{45} + \frac{1}{40}\right) = \frac{17}{360}$.

$\frac{17}{360}$ work is done by $A$ and $B$ in 1 day.

$\therefore \frac{17}{40}$ work is done by $A$ and $B$ in $\frac{360 \times 17}{17 \times 40} = 9$ days.

So, $A$ left after 9 days.

**22.** Ratio of time taken by $A$ & $B = 6 : 5$.

Ratio of work done in same time $= 5 : 6$.

So, the money is to be divided among $A$ and $B$ in the ratio $5 : 6$.

$\therefore B$'s share = Rs. $\left(220 \times \frac{6}{11}\right)$ = Rs. 120.

**23.** $B$'s daily earning = Rs. (150 – 94) = Rs. 56.

$A$'s daily earning = Rs. (150 – 76) = Rs. 74.

$C$'s daily earning = Rs. [150 – (56 + 74)] = Rs.20.

**24.** Let the original number of men be $x$

Now, $x$ men finish the work in 60 days and $(x + 8)$ men can finish it in 50 days.

$\therefore$ More men, less days

So, $(x + 8) : x :: 60 : 50$

or $\frac{x+8}{x} = \frac{60}{50}$ or $x = 40$ men.

**25.** $(A + B)$'s 2 day's work $= \left(\frac{1}{9} + \frac{1}{12}\right) = \frac{7}{36}$.

Evidently, work done by $A$ and $B$ during 5 pairs of days

$= \left(5 \times \frac{7}{36}\right) = \frac{35}{36}$.

Remaining work $= \left(1 - \frac{35}{36}\right) = \frac{1}{36}$.

Now, on 11th day it is $A$'s turn .

Now, $\frac{1}{9}$ work is done by $A$ in 1 day.

$\therefore \frac{1}{36}$ work will be done by $A$ in $9 \times \frac{1}{36} = \frac{1}{4}$ day.

So, total time taken $= 10\frac{1}{4}$ days.

Remark : In such questions we take the pair of days in largest integer smaller than time taken to finish it together.

**26.** $\left(\frac{1}{5} + \frac{1}{9} + \frac{1}{15}\right)$ *i.e.* $\frac{17}{45}$ work is finished in 1 hour.

$\therefore$ Remaining work $= 1 - \frac{17}{45} = \frac{28}{45}$.

$(A + B)$'s 1 hour's work $= \frac{1}{5} + \frac{1}{9} = \frac{14}{45}$.

$\frac{14}{45}$ work is done by $A$ & $B$ in 1 hour .

$\frac{28}{45}$ work will be done by $A$ and $B$ in $\left(\frac{45}{14} \times \frac{28}{45}\right) = 2$ hours.

**27.** 1 boy's 1 day's work $= \frac{1}{3} - \left(\frac{1}{6} + \frac{1}{8}\right) = \frac{1}{24}$.

Ratio of 1 day's work of 1st man, 2nd man and the boy

$$= \frac{1}{6} : \frac{1}{8} : \frac{1}{24} = 4 : 3 : 1$$

$\therefore$ Boy's share = Rs. $\left(200 \times \frac{1}{8}\right)$ = Rs. 25.

**28.** Suppose $B$ takes $x$ days to do the work.

$\therefore A$ takes $\left(2 \times \frac{3}{4}x\right)$ *i.e.* $\frac{3x}{2}$ days to do it.

Now, $(A + B)$'s 1 day's work $= \frac{1}{18}$.

$\therefore \frac{1}{x} + \frac{2}{3x} = \frac{1}{18}$ or $x = 30$.

**29.** $A$'s 1 day's wages = (1/21) of whole money.

$B$'s 1 day's wages = (1/28) of whole money.

$(A + B)$'s 1 day's wages $= \left(\frac{1}{21} + \frac{1}{28}\right) = \frac{1}{12}$ of whole money.

So, the money is sufficient to pay the wages of both for 12 days.

**30.** 5 men = 9 women or 1 man $= \frac{9}{5}$ women.

$\therefore$ 3 men + 6 women = $\left(3 \times \frac{9}{5} + 6\right)$ *i.e.* $\frac{57}{5}$ women.

Now, 9 women can do the work in 19 days.

$\therefore \frac{57}{5}$ women can do it in $\left(\frac{19 \times 9 \times 5}{57}\right)$ = 15 days.

**31.** 10 men = 12 women.

5 men = 6 women.

$\therefore$ 15 men and 6 women = 20 men.

10 men can finish the work in 10 days.

$\therefore$ 20 men can finish the work in 5 days.

**32.** Clearly, 4 men = 2 men + 3 women

or 2 men = 3 women

$\therefore$ 1 woman = $\frac{2}{3}$ man.

So, 3 men + 3 women = 3 men + $\left(3 \times \frac{2}{3}\right)$ men = 5 men.

$\therefore$ 5 men can do it in $\left(\frac{10 \times 4}{5}\right)$ = 8 days.

**33.** 2 children = 1 man.

$\therefore$ (8 children + 12 men) = 16 men.

Now, less men, more days (indirect)

12 : 16 : : 9 : $x$

$\therefore x = \left(\frac{16 \times 9}{12}\right)$ = 12 days.

**34.** 1 man = $\frac{3}{2}$ boys, so (4 men + 9 boys) = 15 boys.

Also, 8 men = $\frac{3}{2} \times 8$ *i.e.* 12 boys.

Now, 12 boys can dig the pit in 20 days.

$\therefore$ 15 boys can dig it in $\left(\frac{20 \times 12}{15}\right)$ = 16 days.

**35.** (3 men + 4 boys)'s 1 day's work = (1/8)

(4 men + 4 boys)'s 1 day's work = (1/6)

Subtracting, 1 man's 1 day's work = $\left(\frac{1}{6} - \frac{1}{8}\right) = \frac{1}{24}$.

(2 men + 4 boys)'s 1 day's work = $\left(\frac{1}{8} - \frac{1}{24}\right) = \frac{1}{12}$.

Thus, 2 men and 4 boys will finish it in 12 days.

**36.** 1 man = 2 women = 3 boys

$\therefore$ 1 woman $= \frac{1}{2}$ man and 1 boy $= \frac{1}{3}$ man

So, (1 man + 1 woman + 1 boy) $= \left(1 + \frac{1}{2} + \frac{1}{3}\right) = \frac{11}{6}$ men.

Now, 1 man do the work in 44 days

$\therefore \frac{11}{6}$ men can do it in $\left(44 \times \frac{6}{11}\right) = 24$ days.

37. Let 1 man's and 1 woman's one day's work be $x$ and $y$ respectively. Then, $4x + 6y = (1/8)$ and $3x + 7y = (1/10)$.
Solving these equations, we get $y = (1/400)$.
Thus, 1 woman can finish it in 400 days.
$\therefore$ 10 women can finish it in (400 ÷ 10) = 40 days.

## ANSWERS (EXERCISE 11B)

| | | | | | |
|---|---|---|---|---|---|
| **1.** (*d*) | **2.** (*a*) | **3.** (*b*) | **4.** (*d*) | **5.** (*c*) | **6.** (*a*) |
| **7.** (*c*) | **8.** (*b*) | **9.** (*c*) | **10.** (*a*) | **11.** (*d*) | **12.** (*c*) |
| **13.** (*b*) | **14.** (*a*) | **15.** (*c*) | **16.** (*a*) | **17.** (*d*) | **18.** (*c*) |
| **19.** (*d*) | **20.** (*b*) | **21.** (*c*) | **22.** (*a*) | **23.** (*d*) | **24.** (*c*) |
| **25.** (*b*) | **26.** (*a*) | **27.** (*b*) | **28.** (*c*) | **29.** (*b*) | **30.** (*d*) |
| **31.** (*b*) | **32.** (*c*) | **33.** (*c*) | **34.** (*d*) | **35.** (*c*) | **36.** (*b*) |
| **37.** (*d*). | | | | | |

# 12

# PIPES & CISTERNS

A pipe connected with a cistern is called an *inlet,* if it fills the cistern.

A pipe connected with a cistern is called an *outlet,* if it empties the cistern.

*Filled part of a tank during a certain period*

= *(Sum of work done by inlets) – (Sum of work done by outlets).*

## Solved Problems

**Ex. 1.** ***Two pipes can fill a cistern in 9 hours and 12 hours respectively. In how much time will they fill the cistern when opened together ?***

**Sol.** Work done by 1st pipe in 1 hour $= \frac{1}{9}$;

Work done by 2nd pipe in 1 hour $= \frac{1}{12}$.

Work done by both in 1 hour $= \left(\frac{1}{9} + \frac{1}{12}\right) = \frac{7}{36}$.

$\therefore$ *Both the pipes together will fill the cistern in* $\left(\frac{36}{7}\right)$ *hours i.e.*

$5\frac{1}{7}$ *hours.*

**Ex. 2.** ***There is a leak in the bottom of a cistern. When the cistern is in thorough repair, it would be filled in $3\frac{1}{2}$ hours. It now takes half an hour longer. If the cistern is full, how long would the leak take to empty the cistern ?***

**Sol.** Work done by inlet (without leak) in 1 hr. $= \frac{2}{7}$.

Time taken to fill the cistern (with leak) = 4 hours.

Work done by inlet (with leak) in 1 hour $= \frac{1}{4}$.

Work done by the leak in 1 hour $= \left(\frac{2}{7} - \frac{1}{4}\right) = \frac{1}{28}$.

$\therefore$ *Time taken by the leak to empty the cistern = 28 hours.*

**Ex. 3.** ***Two pipes can fill a cistern in 14 and 16 hours respectively. The pipes are opened simultaneously and it is found that due to leakage in the bottom 32 minutes extra are taken for the cistern to be filled up. If the cistern is full, in what time will the leak empty it ?***

**Sol.** Work done by both the pipes in 1 hour $= \frac{1}{14} + \frac{1}{16} = \frac{15}{112}$.

Time taken by the pipes to fill the cistern (without leakage)

$$= \frac{112}{15} \text{ hrs.} = 7 \text{ hrs. } 28 \text{ mts.}$$

Time taken by the 2 pipes and the leak to fill it

$$= (7 \text{ hrs. } 28 \text{ mts.}) + (32 \text{ mts.}) = 8 \text{ hours.}$$

$\therefore$ Work done by 2 pipes & leak in 1 hour $= \frac{1}{8}$.

Work done by leak in 1 hour $= \left(\frac{15}{112} - \frac{1}{8}\right) = \frac{1}{112}$.

$\therefore$ *Time taken by the leak to empty the cistern* = 112 *hours.*

**Ex. 4.** ***Two pipes A and B can fill a cistern in 20 minutes and 25 minutes respectively. Both are opened together, but at the end of 5 minutes, B is turned off. How much longer will the cistern take to fill ?***

**Sol.** Cistern filled by *A* & *B* in 5 minutes

$$= 5\left(\frac{1}{20} + \frac{1}{25}\right) = \frac{9}{20}.$$

Part unfilled $= \left(1 - \frac{9}{20}\right) = \frac{11}{20}$.

Now $\frac{1}{20}$ part is filled by *A* in 1 minute

$\therefore \frac{11}{20}$ part will be filled by *A* in $\frac{20 \times 11}{20} = 11$ minutes.

*Total time taken to fill the cistern* = 5 + 11 = 16 *minutes.*

**Ex. 5.** ***A pipe can fill a cistern in 12 minutes and another pipe in 15 minutes, but a third pipe can empty it in 6 minutes. The first two pipes are kept open for 5 minutes in the beginning and then the third pipe is also opened. In what time is the cistern emptied ?***

**Sol.** Cistern filled in 5 minutes $= 5\left(\frac{1}{12} + \frac{1}{15}\right) = \frac{3}{4}$.

Net work done by 3 pipes in 1 minute

$= \left(\frac{1}{12} + \frac{1}{15}\right) - \frac{1}{6} = -\frac{1}{60}$ *i.e.,* $\frac{1}{60}$ part is emptied in 1 minute

$\therefore$ *$\frac{3}{4}$ part is emptied in* $\frac{60 \times 3}{4}$ = 45 *minutes.*

**Ex. 6.** ***Two pipes A and B can fill a tank in 15 hours & 20 hours respectively while a third pipe C can empty the full tank in 25 hours. All the three pipes are opened in the beginning. After 10 hours C is closed. Find, in how much time will the tank be full ?***

**Sol.** Tank filled in 10 hours

$= 10\left(\frac{1}{15}+\frac{1}{20}-\frac{1}{25}\right) = \frac{23}{30}$.

Remaining part $= \left(1-\frac{23}{30}\right) = \frac{7}{30}$.

Work done by $(A+B)$ in 1 hour $= \left(\frac{1}{15}+\frac{1}{20}\right) = \frac{7}{60}$.

Now, $\frac{7}{60}$ part is filled by $(A+B)$ in 1 hour

$\therefore \frac{7}{30}$ part will be filled by $(A+B)$ in $\left(\frac{60}{7}\times\frac{7}{30}\right)$ hrs.

$= 2$ hours.

$\therefore$ *Total time in which the tank is full* $= (10+2) = 12$ *hours.*

**Ex. 7.** ***A cistern can be filled separately by two pipes A and B in 36 minutes and 45 minutes respectively. A tap C at the bottom A and B can empty the full cistern in 30 minutes. If the tap C is opened 7 minutes after the pipes A and B are opened, find when the cistern becomes full.***

**Sol.** Net filling by $(A+B)$ in 7 minutes $= 7\times\left(\frac{1}{36}+\frac{1}{45}\right) = \frac{7}{20}$.

Unfilled part $= \left(1-\frac{7}{20}\right) = \frac{13}{20}$.

Net filling in 1 mt. by $(A+B+C) = \left(\frac{1}{36}+\frac{1}{45}-\frac{1}{30}\right) = \frac{1}{60}$

Now, net filling of $\frac{1}{60}$ part is done in 1 minute.

$\therefore$ Net filling of $\frac{13}{20}$ part is done in $\left(\frac{60\times 13}{20}\right) = 39$ minutes.

*Hence, the total time taken to fill it* $= (7+39) = 46$ *minutes.*

**Ex. 8.** ***Two pipes A and B can separately fill a cistern in $7\frac{1}{2}$ and 5 minutes respectively and a waste pipe C can carry off 14 litres per minute. If all the pipes are opened when the cistern is full, it is emptied in 1 hour. How many litres does it hold ?***

**Sol.** Cistern filled by $(A+B)$ in 1 minute $= \left(\frac{2}{15}+\frac{1}{5}\right) = \frac{1}{3}$.

Net, emptying work done by $(A+B+C)$ in 1 min. $= \frac{1}{60}$.

Work done by $C$ in 1 minute $= \left(\frac{1}{60}+\frac{1}{3}\right) = \frac{7}{20}$.

$\therefore$ $C$ alone can empty the cistern in $\left(\frac{20}{7}\right)$ min.

Water thrown out by $C$ in 1 min. = 14 litres

Water thrown out by $C$ in $\left(\frac{20}{7}\right)$ min. = $\left(14 \times \frac{20}{7}\right)$ = 40 litres.

$\therefore$ *Capacity of the cistern = 40 litres.*

**Ex. 9.** ***A cistern can be filled by one of two pipes in 30 minutes and by the other in 36 minutes. Both pipes are opened together for a certain time but being particularly clogged, only $\frac{5}{6}$ of the full quantity of water flows through the former and only $\frac{9}{10}$ through the latter. The obstructions, however, being suddenly removed, the cistern is filled in $15\frac{1}{2}$ minutes from that moment. How long was it before the full flow of water began ?***

**Sol.** Net filling in last $15\frac{1}{2}$ minutes = $\frac{31}{2}\left(\frac{1}{30}+\frac{1}{36}\right) = \frac{341}{360}$.

Now, suppose they remained clogged for $x$ minutes.

Net filling in these $x$ minutes

$= \left(\frac{x}{30}\times\frac{5}{6}+\frac{x}{36}\times\frac{9}{10}\right) = \frac{19x}{360}$.

Remaining part = $\left(1-\frac{19x}{360}\right) = \left(\frac{360-19x}{360}\right)$.

$\therefore \frac{360-19x}{360} = \frac{341}{360}$ or $x = 1$.

*Hence, the pipes remained clogged for 1 minute.*

**Ex.10.** ***Three pipes A, B and C are attached to a cistern, A can fill it in 20 minutes and B in 30 minutes. C is a waste pipe meant for emptying it. After opening both the pipes A and B, a man leaves the cistern and returns when the cistern should have been just full. Finding however that the waste pipe had been left open, he closes it and the cistern now gets filled in 3 minutes. In how much time would the waste pipe C, if opened alone, empty the full cistern ?***

**Sol.** $(A + B)$'s one minute's filling work = $\left(\frac{1}{20}+\frac{1}{30}\right) = \frac{1}{12}$.

So, $A$ and $B$ together can fill the cistern in 12 minutes.

Suppose $C$ can empty the full cistern in $x$ minutes.

Then $C$'s one minute's emptying work = $\frac{1}{x}$.

$(A + B + C)$'s one minute's resultant filling work = $\left(\frac{1}{12}-\frac{1}{x}\right)$.

$(A + B + C)$'s resultant filling work in 12 mts.

$$= 12\left(\frac{1}{12} - \frac{1}{x}\right) = \left(\frac{x-12}{x}\right).$$

Also $(A + B)$'s filling work in 3 mts. $= 3 \times \left(\frac{1}{12}\right) = \frac{1}{4}$.

But $[(A + B + C)$'s resultant filling in 12 mts.$] + [(A + B)$'s filling in 3 mts.$] = 1$.

$$\therefore \frac{x-12}{x} + \frac{1}{4} = 1 \text{ or } 4x - 48 + x = 4x \text{ or } x = 48.$$

*Hence, C alone can empty the cistern in 48 minutes.*

## EXERCISE 12A (Subjective)

1. Two inlets $A$ and $B$ can fill a water tank in 36 hours and 48 hours respectively. In how much time can both together fill the tank ?
2. A cistern which could be filled in 9 hours takes 1 hour more to be filled owing to a leak in its bottom. If the cistern is full, in what time will leak empty it ?

   **Hint.** *Work done by leak in 1 hour* $= \left(\frac{1}{9} - \frac{1}{10}\right) = \frac{1}{90}$.
3. A cistern can be filled by two pipes $A$ and $B$ in 12 minutes and 16 minutes respectively and it can be emptied by a third pipe $C$ in 8 minutes. If all the taps be turned on at the same time, in how much time will the cistern be full ?

   **Hint.** *Net filling in 1 minute* $= \left(\frac{1}{12} + \frac{1}{16} - \frac{1}{8}\right) = \frac{1}{48}$.
4. A cistern has a leak which would empty it in 6 hours. $A$ tap is turned on, which admits 4 litres a minute into the cistern and is now emptied in 8 hours. What is the capacity of the cistern ?

   **Hint.** *Work done by inlet & leak in 1 hour* $= \frac{1}{8}$.

   *Work done by the leak in 1 hour* $= \frac{1}{6}$.

   *Net work done by inlet in 1 hour* $= \left(\frac{1}{6} - \frac{1}{8}\right) = \frac{1}{24}$.

   *But, net volume of water sent by inlet in 1 hour to the cistern*

   $= (4 \times 60)$ *litres* $= 240$ *litres.*

   $\therefore \frac{1}{24}$ *part of the cistern contains* $= 240$ *litres.*

   *Full cistern contains* $(240 \times 24)$ *litres* $= 5760$ *litres.*
5. Two pipes $A$ and $B$ can separately fill a cistern in 60 minutes and 75 minutes respectively. There is a waste pipe $C$ in the bottom of the cistern.

If all the three pipes are simultaneously opened, then the cistern is full in 50 minutes. In how much time the waste pipe can empty the cistern, when it is full ?

**Hint.** *Work done by waste pipe in 1 min.* $= \left(\frac{1}{60} + \frac{1}{75} - \frac{1}{50}\right)$.

**6.** Three pipes $A, B$ and $C$ can fill a cistern in 10 hours; 12 hours and 15 hours respectively. First $A$ was opened. After 1 hour, $B$ was opened and after 2 hours from the start of $A$, $C$ was also opened. Find the time in which the cistern is just full.

**Hint.** *[(A's 1 hour work) + (A + B)'s 1 hour work]*

$$= \frac{1}{10} + \left(\frac{1}{10} + \frac{1}{12}\right) = \frac{17}{60}.$$

*Remaining part* $= \left(1 - \frac{17}{60}\right) = \frac{43}{60}$.

*Now,* $(A + B + C)$*'s 1 hour work* $= \left(\frac{1}{10} + \frac{1}{12} + \frac{1}{15}\right) = \frac{1}{4}$.

$\frac{1}{4}$ *part is filled by 3 pipes in 1 hour.*

$\frac{43}{60}$ *part will be filled by them in* $\left(4 \times \frac{43}{60}\right)$ *hrs. = 2 hours 52 min.*

$\therefore$ *Total time taken to fill the cistern = 4 hours 52 min.*

**7.** Two pipes $A$ and $B$ can fill a cistern in 24 minutes and 32 minutes respectively. Both the pipes being opened, find when the second pipe must be turned off so that the cistern may be just full in 18 minutes.

**Hint.** *Let it be turned off after x minutes.*

*Work done by* $(A + B)$ *in x min.* $= x\left(\frac{1}{24} + \frac{1}{32}\right) = \frac{7x}{96}$.

*Work done by A in* $(18 - x)$ *min.* $= \frac{(18 - x)}{24}$.

$\therefore \frac{7x}{96} + \frac{(18 - x)}{24} = 1$, find $x$.

**8.** In what time would a cistern be filled by three pipes whose diameters are 1 cm., $1\frac{1}{3}$ cm., 2 cm., running together, when the largest alone will fill it in 61 minutes, the amount of water flowing in by each pipe being proportional to the square of its diameter ?

**Hint.** *Ratio of the amount of water flown in*

$$= \left(1 : \frac{16}{9} : 4\right) \text{ or } \left(\frac{1}{4} : \frac{4}{9} : 1\right).$$

*Work done by third pipe in 1 min.* $= \frac{1}{61}$.

*Work done by first pipe in 1 min.* $= \left(\frac{1}{4} \times \frac{1}{61}\right)$.

*Work done by second pipe in 1 min.* $= \left(\frac{4}{9} \times \frac{1}{61}\right)$.

$\therefore$ *Total work done by three pipes in 1 min.*

$$= \left(\frac{1}{61} + \frac{1}{244} + \frac{4}{549}\right) = \frac{1}{36}.$$

**9.** **A bath can be filled by the cold water pipe in 10 minutes and by the hot water pipe in 15 minutes. A person leaves the bath room after turning on both pipes simultaneously and returns at the moment when the bath should be full. Finding, however, that the waste pipe has been open, he now closes it. In 4 minutes, more, the bath is full. In what time would the waste pipe empty it ?**

**Hint.** *Work done by both in 1 min.* $= \left(\frac{1}{10} + \frac{1}{15}\right) = \frac{1}{6}$.

*Work done by both in 4 min.* $= \left(4 \times \frac{1}{6}\right) = \frac{2}{3}$.

$\therefore$ *Work done by waste pipes in 6 min.* $= \frac{2}{3}$.

*Work done by waste pipe in 1 min.* $= \left(\frac{2}{3} \times \frac{1}{6}\right) = \frac{1}{9}$.

$\therefore$ *Waste pipe can empty the bath in 9 min.*

**10.** **$A, B, C$ are pipes attached to a cistern. $A$ and $B$ can fill it in 20 and 30 minutes respectively, while $C$ can empty it in 15 minutes. If $A, B, C$ be kept open successively for 1 minute each, how soon will the cistern be filled ?**

**Hint.** *Work done in 3 minutes* $= \left(\frac{1}{20} + \frac{1}{30} - \frac{1}{15}\right) = \frac{1}{60}$.

*Clearly,* $\frac{55}{60}$ *part of cistern is filled in* $3 \times 55$ *or 165 min.*

*Remaining part* $= \left(1 - \frac{55}{60}\right) = \frac{5}{60} = \frac{1}{12}$.

*Now,* $\frac{1}{20}$ *part is filled by A in 1 min.*

*and* $\left(\frac{1}{12} - \frac{1}{20}\right)$ *i.e.* $\frac{1}{30}$ *part is filled by B in 1 min.*

## ANSWERS (Exercise 12A)

**1.** $20\frac{4}{7}$ **hours** **2.** **90 hours** **3.** **48 hours**

**4.** **5760 litres** **5.** **100 min.** **6.** **4 hrs. 52 min.**

**7.** 8 min. **8.** 36 min. **9.** 9 min.
10.167 min.

## EXERCISE 12B (Objective Type Questions)

1. A pipe can fill a tank in $x$ hours and another can empty it in $y$ hours. They can together fill it in $(y > x)$ :
(*a*) $(x - y)$ hours (*b*) $(y - x)$ hours
(*c*) $\frac{xy}{(x-y)}$ hours (*d*) $\frac{xy}{(y-x)}$ hours

2. A tap can fill a cistern in 8 hours and another can empty it in 16 hours. If both the taps are opened simultaneously, the time (in hours) to fill the tank is :
(*a*) 8 (*b*) 10
(*c*) 16 (*d*) 24 (Clerical Grade. 1991)

3. Two pipes can fill a tank in 10 hours and 15 hours respectively. Both together can fill it in :
(*a*) $12\frac{1}{2}$ hours (*b*) 5 hours
(*c*) 6 hours (*d*) none of these

4. A cistern can be filled by pipes *A* and *B* in 20 hours and 30 hours respectively. When full, the tank can be emptied by pipe *C* in 40 hours. If all the taps be turned on at the same time, the cistern will be full in :
(*a*) 10 hours (*b*) 90 hours
(*c*) 16 hours (*d*) $17\frac{1}{7}$ hours

5. A cistern can be filled by a pipe in 15 hours. But due to a leak in the bottom the cistern is just full in 20 hours. When the cistern is full, the leak can empty it in :
(*a*) 60 hours (*b*) 40 hours
(*c*) 45 hours (*d*) 30 hours

6. A cistern is normally filled in 8 hrs but takes 2 hrs longer to fill because of a leak in its bottom. If the cistern is full, the leak will empty it in :
(*a*) 16 hrs. (*b*) 40 hrs.
(*c*) 25 hrs. (*d*) 20 hrs. (Railways 1991)

7. Two taps can separately fill a cistern in 10 minutes and 15 minutes respectively and when the waste pipe is opened they can together fill it in 18 minutes. The waste pipe can empty the full cistern in :
(*a*) 6 minutes (*b*) 9 minutes
(*c*) 13 minutes (*d*) 23 minutes

8. Pipes $A$ and $B$ can fill a tank in 20 minutes and 25 minutes respectively. If both the taps are opened and after 5 minutes, pipe $B$ is turned off, the tank will be completely filled in :
(*a*) $17\frac{1}{2}$ min. (*b*) 12 min.
(*c*) 11 min. (*d*) 6 min.

9. A tap can fill a tank in 16 minutes and another can empty it in 8 minutes. If the tank is already half full and both the tanks are opened together, the tank will be :
(*a*) filled in 12 min. (*b*) emptied in 12 min.
(*c*) filled in 8 min. (*d*) emptied in 8 min.

10. A cistern can be filled by two pipes $A$ and $B$ in 24 minutes and 32 minutes respectively. If both the pipes are opened together, then after how much time $B$ should be closed so that the tank is full in 18 minutes ?
(*a*) 6 minutes (*b*) 8 minutes
(*c*) 10 minutes (*d*) 12 minutes

11. Tap $A$ and tap $B$ can fill a tank in 12 minutes and 15 minutes respectively and tap $C$ can empty it in 6 minutes. Taps $A$ and $B$ are opened first and after 5 minutes, tap $C$ is also opened. The tank is emptied in :
(*a*) 30 min. (*b*) 33 min.
(*c*) $37\frac{1}{2}$ min. (*d*) 45 min.

12. The cistern is filled in 12 hours when two pipes are made to function simultaneously. If one pipe fills the cistern 10 hours faster than the other, the second pipe fills the cistern in :
(*a*) 35 hours (*b*) 30 hours
(*c*) 28 hours (*d*) 25 hours

13. Two pipes $X$ and $Y$ can fill a tank in 15 hours and 20 hours respectively while a third pipe $Z$ can empty the full tank in 25 hours. All the three pipes are opened. After 10 hours, $Z$ is closed. The tank will be full in :
(*a*) 12 hours (*b*) $13\frac{1}{2}$ hours
(*c*) 16 hours (*d*) 18 hours

14. A cistern has a leak which would empty it in 8 hours. A tap is turned on, which admits 6 litres a minute into the cistern and it is now emptied in 12 hours. How many litres does the cistern hold ?
(*a*) 8640 litres (*b*) 5760 litres
(*c*) 3890 litres (*d*) 6480 litres

15. Two pipes $P$ and $Q$ can fill a cistern in 12 minutes and 16 minutes respectively. Both are opened together, but at the end of 4 minutes $Q$ is turned off. The cistern will be full in :

(*a*) 9 minutes (*b*) 14 minutes
(*c*) 8 minutes (*d*) 24 minutes

## HINTS & SOLUTIONS (EXERCISE 12B)

1. Net filling in 1 hour $= \left(\frac{1}{x} - \frac{1}{y}\right) = \left(\frac{y-x}{xy}\right)$.

$\therefore$ Time taken to fill the tank $= \left(\frac{xy}{y-x}\right)$ hours.

2. Net filling in 1 hour $= \left(\frac{1}{8} - \frac{1}{16}\right) = \frac{1}{16}$.

$\therefore$ Time taken to fill the cistern = 16 hours.

3. Work done by both pipes in 1 hour $= \left(\frac{1}{10} + \frac{1}{15}\right) = \frac{1}{6}$.

So, both the pipes will fill it in 6 hours.

4. Net filling in 1 hour $= \left(\frac{1}{20} + \frac{1}{30} - \frac{1}{40}\right) = \frac{7}{120}$

$\therefore$ Time taken to fill the cistern $= \frac{120}{7}$ hours $= 17\frac{1}{7}$ hours.

5. Work done by the leak in 1 hour $= \left(\frac{1}{15} - \frac{1}{20}\right) = \frac{1}{60}$

So, the leak can empty the cistern in 60 hours.

6. Work done by the leak in 1 hr. $= \left(\frac{1}{8} - \frac{1}{10}\right) = \frac{1}{40}$

So, the leak can empty the cistern in 40 hours.

7. Work done by waste pipe in 1 minute

$= \left(\frac{1}{10} + \frac{1}{15} - \frac{1}{18}\right) = \frac{1}{6}$.

$\therefore$ Waste pipe can empty the full cistern in 6 minutes.

8. Part filled in 5 min. by *A* and *B*

$= 5\left(\frac{1}{20} + \frac{1}{25}\right) = \frac{9}{20}$

Remaining part $= \left(1 - \frac{9}{20}\right) = \frac{11}{20}$.

Now, $\frac{1}{20}$ part is filled by *A* in 1 minute.

$\therefore \frac{11}{20}$ part will be filled by *A* in $\left(\frac{20 \times 11}{20}\right)$ min.= 11 min.

9. Clearly, the rate of waste pipe is more, so the tank will be emptied, if both the pipes are opened.

Net work done by both the pipes in 1 minute .

$= \left(\frac{1}{16} - \frac{1}{8}\right) = -\frac{1}{16}$. (*– ve sign means emptying*).

So, both the pipes together will empty the full tank in 16 minutes. Hence, they will empty half full tank in 8 minutes.

**10.** Let $B$ be closed after $x$ min.

Part filled by both in $x$ min. $= x\left(\frac{1}{24} + \frac{1}{32}\right) = \frac{7x}{96}$

Part filled by $A$ in $(18 - x)$ min. $= \frac{18 - x}{24}$

$\therefore \frac{7x}{96} + \frac{18 - x}{24} = 1$ or $x = 8$ min.

**11.** Tank filled in 5 minutes $= 5\left(\frac{1}{12} + \frac{1}{15}\right) = \frac{3}{4}$

Net work done by three pipes in 1 min.

$$= \left(\frac{1}{12} + \frac{1}{15} - \frac{1}{6}\right) = -\frac{1}{60}$$

Thus, $\frac{1}{60}$ part is emptied in 1 min.

$\therefore \frac{3}{4}$ part will be emptied in $\left(60 \times \frac{3}{4}\right) = 45$ minutes.

**12.** Let the cistern be filled by first pipe in $x$ hours.
Then, it will be filled by second pipe in $(x + 10)$ hrs

$\therefore \frac{1}{x} + \frac{1}{x + 10} = \frac{1}{12}$ or $12(x + 10 - x) = x(x + 10)$

or $x^2 - 14x - 120 = 0$ or $(x - 20)(x + 6) = 0$

or $x = 20$.

So, the second pipe takes 30 hours to fill the cistern.

**13.** Tank filled in 10 hours $= 10\left(\frac{1}{15} + \frac{1}{20} - \frac{1}{25}\right) = \frac{23}{30}$

Remaining part $= \left(1 - \frac{23}{30}\right) = \frac{7}{30}$.

Net work done by $X$ and $Y$ in 1 hour $= \left(\frac{1}{15} + \frac{1}{20}\right) = \frac{7}{60}$.

Now, $\frac{7}{60}$ is filled in 1 hr.

$\therefore \frac{7}{30}$ part will be filled in $\left(\frac{60}{7} \times \frac{7}{30}\right) = 2$ hours.

Hence, the tank will be full in 12 hours.

**14.** Work done by inlet in 1 hour $= \left(\frac{1}{8} - \frac{1}{12}\right) = \frac{1}{24}$.

Work done by inlet in 1 min. $= \left(\frac{1}{24 \times 60}\right) = \frac{1}{1440}$

$\therefore$ Volume of $\frac{1}{1440}$ part = 6 litres.

Hence, volume of the tank = (1440 × 6) = 8640 litres.

**15.** Part filled in 4 min. by both the pipes

$$= 4\left(\frac{1}{12} + \frac{1}{16}\right) = \frac{7}{12}.$$

Remaining part $= \left(1 - \frac{7}{12}\right) = \frac{5}{12}.$

Now, $\frac{1}{12}$ is filled by first pipe in 1 minute.

$\therefore \frac{5}{12}$ will be filled by it in $\left(\frac{12 \times 5}{12}\right)$ min. = 5 min.

$\therefore$ The cistern will be full in 9 minutes.

## ANSWERS (EXERCISE 12B)

| | | | | | |
|---|---|---|---|---|---|
| 1. (*d*) | 2. (*c*) | 3. (*c*) | 4. (*d*) | 5. (*a*) | 6. (*b*) |
| 7. (*a*) | 8. (*c*) | 9. (*d*) | 10. (*b*) | 11. (*d*) | 12. (*b*) |
| 13. (*a*) | 14. (*a*) | 15. (*a*) | | | |

# 13

# TIME & DISTANCE

**FORMULAE :**

(i) Speed = $\frac{Distance}{Time}$.

(ii) Distance = (*Speed* × *Time*).

(iii) Time = $\frac{Distance}{Speed}$.

(iv) To convert a speed in km. per hour to metres per second, we multiply it by $\frac{5}{18}$.

**Ex.** 54 km. per hour = $\left(54 \times \frac{5}{18}\right)$ m/sec. = 15 m/sec.

(v) To convert a speed in metres per second to km. per hour, we multiply it by $\frac{18}{5}$.

**Ex.** 35 metres per second = $\left(35 \times \frac{18}{5}\right)$ km/hr.= 126 km/hr.

(vi) If the speed of a body is changed in the ratio 2 : 3, then the ratio of the time taken changes in the ratio 3 : 2.

**Solved Examples :**

**Ex. 1.** ***Express a speed of :***

(i) 18 km per hour in metres per second;

(ii) 3 metres per second in km. per hour.

**Sol.** (i) 18 km/hour = $\left(18 \times \frac{5}{18}\right)$ m/sec. = 5 m/sec.

(ii) 3 m/sec. = $\left(3 \times \frac{18}{5}\right)$ km/hr. = 10.8 km/hr.

**Ex. 2.** ***A and B are two cities. A man travels on cycle from A to B at a speed of 15 km/hr. and returns back at the rate of 10 km/hr. Find his average speed for the whole journey.***

**Sol.** Let the distance between *A* and *B* be *x* km.

Then, time taken to cover *x* km. from *A* to *B* = $\frac{x}{15}$ hrs.

Time taken to cover *x* km. from *B* to *A* = $\frac{x}{10}$ hrs.

$\therefore$ Time taken to cover $2x$ km. $= \left(\frac{x}{15} + \frac{x}{10}\right)$ hrs. $= \left(\frac{x}{6}\right)$ hrs.

$\therefore$ Average speed $= \frac{2x}{(x/6)}$ km/hr. = 12 km/hr.

**Remark.** *If a certain distance is covered at x km/hr. and then one returns back to the starting point at y km/hr., then*

*average speed* $= \left(\frac{2xy}{x+y}\right)$ *km/hr.*

**Ex. 3.** ***A man performs*** $\frac{2}{15}$ ***of his total journey by rail,*** $\frac{9}{20}$ ***by a tonga and the remaining 10 km. on foot. Find his total journey.***

**Sol.** Let total journey be $x$ km.

Then, $\frac{2x}{15} + \frac{9x}{20} + 10 = x \Rightarrow x = 24.$

$\therefore$ *Total journey = 24 km.*

**Ex. 4.** ***Two men, starting from the same place walk at the rate of 4 km/hr and 4.5 km/hr respectively. How many km will they be apart at the end of*** $3\frac{1}{2}$ ***hrs., if (i) they walk in opposite directions***

***and (ii) they walk in the same direction.***

**Sol.** (i) If they walk in opposite directions, they will be (4 + 4.5) km. i.e., 8.5 km. apart in 1 hour.

$\therefore$ In $3\frac{1}{2}$ hrs. they will be $8.5 \times \frac{7}{2} = 29.75$ km apart.

(ii) If they walk in the same direction, then they are (4.5 – 4) = 0.5 km. apart in 1 hour.

*So, in* $3\frac{1}{2}$ *hours, they will be* $0.5 \times \frac{7}{2} = 1.75$ *km apart.*

**Ex. 5.** ***Two men starting from the same place walk at the rate of 4 km/hr and 4.5 km/hr respectively. What time will they take to be 17 km. apart if (i) they walk in opposite directions and (ii) they walk in the same direction.***

**Sol.** (i) If they walk in the opposite directions, then they are 4.5 + 4 = 8.5 km. apart in 1 hour. Thus,

They are 8.5 km. apart in 1 hr.

*So, they are 17 km. apart in* $\frac{1}{8.5} \times 17 = 2$ *hours.*

(ii) If they walk in the same direction, then they are 4.5 – 4 = 0.5 km. apart in 1 hour. Thus,

They are 0.5 km. apart in 1 hour.

$\therefore$ *They are 17 km. apart in* $\frac{1}{0.5} \times 17 = 34$ *hours.*

**Ex. 6.** ***I start from my house for the college at a certain fixed time. If I walk at the rate of 5 km. an hour, I am late by 7 minutes. However, if I walk at the rate of 6 km. an hour, I reach the college 5 minutes earlier than the scheduled time. Find the distance of the college from my house.***

**Sol.** Let the distance be 1 km.

Time taken to cover 1 km. at 5 km/hr $= \frac{1}{5}$ hr.

Time taken to cover 1 km. at 6 km/hr $= \frac{1}{6}$ hr.

Difference of times in covering 1 km. $= \left(\frac{1}{5} - \frac{1}{6}\right) = \frac{1}{30}$ hr.

Actual difference of time in covering whole distance

$= (7 + 5)$ mts. $= 12$ minutes $= \frac{1}{5}$ hr.

If the difference of times is (1/30) hr., distance = 1 km.

If the difference of times is $\frac{1}{5}$ hr distance $= \left(\frac{30 \times 1}{5}\right) = 6$ km.

$\therefore$ *Distance between college and house = 6 km.*

**Ex. 7.** ***Distance between two towns A and B is 110 km. A motor cycle rider starts from A towards B at 8 a.m. at a speed of 20 km. per hour. Another motor cycle rider starts from B towards A at 9 a.m. at a speed of 25 km. per hour. Find, when will they cross each other.***

**Sol.** Suppose they meet after $x$ hours from 8 a.m.

Distance travelled by first rider in $x$ hrs. $= 20x$ km.

Distance travelled by second rider in $(x - 1)$ hours $= 25(x - 1)$ km.

$\therefore 20x + 25(x - 1) = 110$ or $x = 3$ hours.

*So, they meet each other at 11 a.m.*

**Ex. 8.** ***Walking (6/7) of his usual speed, a man is 25 minutes too late. Find his usual time.***

**Sol.** New speed $= \frac{6}{7}$ of usual speed.

$\therefore$ New time taken $= \frac{7}{6}$ of usual time.

$\therefore \frac{7}{6}$ of usual time = (usual time) + 25 min.

or $\frac{1}{6}$ of usual time = 25 min.

$\therefore$ *Usual time* $= (25 \times 6)$ *min.* $= 2$ *hrs. 30 min.*

**Ex. 9.** ***A thief is spotted by a policeman from a distance of 200 metres. When the policeman starts the chase, the thief also starts running. Assuming the speed of the thief to be 10 km. per hour and that of policeman 12 km. per hour, how far the thief will have run before he is overtaken ?***

**Sol.** Suppose the thief has run $x$ km. before he is overtaken.

Time taken by the thief to cover $x$ km. = $(x/10)$ hours.

Time taken by the policeman to cover $\left(x+\frac{1}{5}\right)$ km.

$$= \left(\frac{5x+1}{5 \times 12}\right) \text{hrs.} = \frac{(5x+1)}{60} \text{hrs.}$$

$$\therefore \frac{x}{10} = \frac{5x+1}{60} \text{ or } 60x = 10(5x+1) \text{ or } x = 1.$$

*Thus, the thief has run 1 km. before he is overtaken.*

**Ex. 10.** ***A man covers a certain distance on bicycle. Had he moved 3 km. per hour faster, he would have taken 40 minutes less. If he had moved 2 km. per hour slower, he would have taken 40 minutes more. Find the distance.***

**Sol.** Let distance = $x$ km & usual rate = $y$ km/hr

Time taken to cover $x$ km at $y$ km/hr = $\left(\frac{x}{y}\right)$ hrs.

Time taken to cover $x$ km at $(y+3)$ km/hr = $\left(\frac{x}{y+3}\right)$ hrs.

$$\therefore \frac{x}{y} - \frac{x}{y+3} = \frac{2}{3} \text{ or } 2y(y+3) = 9x \qquad \text{... (i)}$$

Time taken to cover $x$ km. at $(y-2)$ km./hr. = $\left(\frac{x}{y-2}\right)$ hrs.

$$\therefore \frac{x}{y-2} - \frac{x}{y} = \frac{2}{3} \text{ or } y(y-2) = 3x \qquad \text{... (ii)}$$

Dividing (i) by (ii), we get $\frac{2(y+3)}{(y-2)} = 3$ or $y = 12$.

Putting $y = 12$ in (i), we get $x = 40$.

$\therefore$ *Required distance* = 40 *km.*

**Ex. 11.** ***A man takes 6 hours 30 minutes in walking to a certain place and riding back. He would have gained 2 hours 10 minutes by riding both ways. How long would he take to walk both ways ?***

**Sol.** Suppose the distance is $x$ kms.

Time taken to cover $2x$ km. by riding

$$= \left(\frac{13}{2} - \frac{13}{6}\right) \text{hrs.} = \frac{13}{3} \text{hrs.}$$

Time taken to cover $x$ km by riding

$$= \left(\frac{13}{3 \times 2}\right) \text{hrs.} = \frac{13}{6} \text{hrs.}$$

Time taken to cover $x$ km by walking & $x$ km by riding $= \frac{13}{2}$ hours.

Time taken to cover $x$ km by walking

$$= \left(\frac{13}{2} - \frac{13}{6}\right) \text{hrs.} = \frac{13}{3} \text{hrs.}$$

*Time taken to cover 2x km. by walking*

$$= \left(\frac{13}{3} \times 2\right) \textit{hours} = \textit{8 hrs. 40 min.}$$

**Ex. 12.** ***A person has to reach a place 40 km. away. He walks at the rate of 4 km/hour for the first 16 km and then he travels in a tonga for the rest of the journey. However, if he had travelled by the tonga for the first 16 km. and the remaining distance on foot at 4 km/hr, he would have taken an hour longer to complete the journey. Find the speed of tonga.***

**Sol.** Let the speed of tonga be $x$ km./hr.

Time taken to cover 16 km on foot $= \left(\frac{1}{4} \times 16\right)$ hrs. = 4 hrs.

$\therefore$ Total time taken to cover whole journey $= \left(4 + \frac{24}{x}\right)$ hrs.

Again, time taken to cover 16 km by tonga $= \left(\frac{16}{x}\right)$ hrs.

And, time taken to cover 24 km on foot $= \left(\frac{1}{4} \times 24\right)$ hrs. = 6 hrs.

$\therefore$ Total time taken to cover whole journey $= \left(6 + \frac{16}{x}\right)$ hrs.

Thus, $\left(6 + \frac{16}{x}\right) - \left(4 + \frac{24}{x}\right) = 1$ or $\left(\frac{24}{x} - \frac{16}{x}\right) = 1$ or $x = 8$.

*Hence, the speed of tonga = 8 km./hour.*

**Ex. 13.** ***A cyclist starts at 8 a.m. for a town 11.5 km away. At 8.45 a.m. there is a breakdown in his bicycle and he spends 30 minutes in trying to get it repaired, but realizing that repair will take much longer time, he starts walking to reach his destination. If he cycles at 10 km an hour and walks at 4 km an hour, at what time will he reach his destination ?***

**Sol.** Distance covered on cycle in 45 mts. $= \left(\frac{10}{60} \times 45\right)$ km. = 7.5 km.

Remaining distance = (11.5 – 7.5) km = 4 km.

Time taken to cover 4 km by walking = 1 hour.

Time spent on waiting = 30 minutes.

Total time taken to reach the destination = (45 + 60 + 30) mts. hrs.

= 2 hrs. 15 mts.

*So, the cyclist reaches his destination at 10.15 a.m.*

**Ex. 14.** ***A man travels 360 km in 4 hours, partly by air and partly by train. If he had travelled all the way by air, he would have saved $\frac{4}{5}$ of the time he was in train and would have arrived his destination 2 hours early. Find the distance he travelled by air and the time, he spent in train.***

**Sol.** Clearly, $\frac{4}{5} \times$ [total time in train] = 2 hours.

$\therefore$ Total time in train = $\left(2 \times \frac{5}{4}\right)$ hrs. = $\frac{5}{2}$ hours.

Consequently, time spent in travelling by air = $\left(4 - \frac{5}{2}\right)$ hrs. = $\frac{3}{2}$ hrs.

By given hypothesis, if 360 km is covered by air, then time taken is 2 hours.

$\therefore$ In 2 hours, distance covered by air = 360 km.

In $\frac{3}{2}$ hours, distance covered by air = $\left(\frac{360}{2} \times \frac{3}{2}\right)$ km = 270 km.

$\therefore$ *Distance covered by the train* = (360 – 270) = *90 km.*

**Ex. 15.** ***A monkey climbing up a greased pole ascends 12 metres and slips down 5 metres in alternate minutes. If the pole is 63 metres high, how long will it take him to reach the top?***

**Sol.** In 1 minute the monkey ascends 12 metres but then he takes 1 minute to slip down 5 metres. Thus, at the end of 2 minutes the net ascending of the monkey is (12 - 5) = 7 metres. Thus, to have a net ascending of 7 metres, the process of ascending & then slipping happens once. So, to cover 63 metres, the above process is repeated $\frac{63}{7}$ or 9 times. It is clear that in 9 such happenings, the monkey will slip 8 times, because on 9th time, he will ascend at the top.

Thus, in climbing 8 times & slipping 8 times, he covers (8 × 7) or 56 metres.

Time taken to cover 56 metres = $\left(\frac{56 \times 2}{7}\right)$ = 16 minutes .

Remaining distance = (63 – 56) = 7 metres.

Time taken to ascend 7 metres = $\frac{7}{12}$ min.

$\therefore$ Total time taken = $16\frac{7}{12}$ min.

**Ex. 16.** ***A hare, pursued by a grey hound is 50 of her own leaps before him. While the hare takes 4 leaps, the grey hound takes 3 leaps, In one leap, the hare goes 1.75 metres and the grey hound 2.75 metres. In how many leaps, will the grey hound overtake the hare ?***

**Sol.** In the same time grey hound & hare make 3 leaps and 4 leaps respectively.

Distance covered by grey hound in 3 leaps = $(3 \times 2.75)$ m. = 8.25 m.

Distance covered by hare in 4 leaps = $(4 \times 1.75)$ m. = 7 metres.

Distance gained by grey hound in 3 leaps = $(8.25 - 7)$ m. = 1.25 m.

Distance covered by hare in 50 leaps = $50 \times 1.75$ m. = 87.5 m.

Now 1.25 m is gained by hound in 3 leaps.

$\therefore$ *87.5 m is gained by hound in* $\left(\frac{3}{1.25} \times 87.5\right) = 210$ *leaps* .

## EXERCISE 13A (Subjective)

**1.** Convert :

(i) 63 km per hour into metres per second;

(ii) 15 metres per second into km per hour.

**2.** A man starts from Meerut to Delhi on scooter at a uniform speed of 35 km per hour and returns back to the starting point at a uniform speed of 28 km per hour. Find his average speed during the whole journey.

**Hint.** *Average speed* = $\left(\frac{2 \times 35 \times 28}{35 + 28}\right)$ *km/hr.*

**3.** If Hariya cycles at 16 km per hour, he covers a distance in 3 hours 45 min. Find the distance.

**Hint.** *Distance = Time × Speed.*

**4.** Puneet starts from his home to school on cycle at 10 km/hr and reaches 4 minutes earlier than the scheduled time. If he cycles at 8 km/hr, he is late by 5 minutes. What is the distance between his residence and the school ?

**Hint.** *Let the distance be x km. Then,* $\frac{x}{8} - \frac{x}{10} = \frac{9}{60}$. *Find x.*

**5.** A car travels a distance of 840 km. at a uniform speed. If the speed of the car is 10 km/hr more, then it takes 2 hours less to cover the same distance. Find the original speed of the car.

**Hint.** *Let the original speed be x km/hr*

*Then* $\frac{840}{x} - \frac{840}{(x + 10)} = 2$, *find x.*

**6.** Walking at 4 km an hour, a clerk reaches his office 5 minutes late. If he walks at 5 km per hour, he will be $2\frac{1}{2}$ minutes too early. Find the distance of his office from his residence.

**7.** A man travels 35 km. partly at 4 km/hr and partly at 5 km/hr. If he covers former distance at 5 km/hr and later distance at 4 km/hr, then he could cover 2 km. more in the same time. Find the times taken to cover these distances.

**Hint.** *Suppose the man covers the first distance in x hours and the second distance in y hours. Then,*

$$4x + 5y = 35 \ \& \ 5x + 4y = 37. \text{ Find } x \ \& \ y.$$

**8.** Two men start together to walk a certain distance, one at 3.75 km. an hour and another at 3 km an hour. The former arrives half an hour before the later. Find the distance.

**Hint.** *Let the distance be x km. Then,* $\left(\frac{x}{3} - \frac{x}{3.75}\right) = \frac{1}{2}$, *find x.*

**9.** A man cycles from *A* to *B*, a distance of 21 km. in 1 hour 40 minutes. The road from *A* is level for 13 km. and then it is uphill to *B*. The man's average speed on level is 15 km/hr. Find his average uphill speed per hour.

**Hint.** *Let the uphill speed be x km/hr.*

Then, $\left(\frac{13}{15} + \frac{8}{x}\right) = \frac{5}{3}$, *find x.*

**10.** Vikas can cover a certain distance in 42 minutes. He covers two third of it at 4 km. an hour and the remaining at 5 km. an hour. Find the total distance.

**Hint.** *Let the total distance be x km.*

*Then,* $\left(\frac{2x}{3 \times 4} + \frac{x}{3 \times 5}\right) = \frac{42}{60}$, *find x.*

**11.** A thief steals a car at 1.30 p.m. and drives it at 40 km. an hour. The theft is discovered at 2 p.m. and the owner sets off in another car at 50 km. an hour. When will he overtake the thief ?

**Hint.** *Suppose that the thief was overtaken x hrs. after 1.30 p.m. Then, distance covered by thief in x hrs.* = (40 x) *km.*

*and, distance covered by owner in* $\left(x - \frac{1}{2}\right)$ *hrs.* = (50x – 25) *km.*

∴ 40 *41x* = (50x – 25), *find x.*

**12.** An insect crawls at the rate of 3 metres/min. up a vertical pole of height 15 metres. Every minute it slides down 1 metre. How long will it take to reach the top ?

13. A man takes 120 steps a minute. If his stride is 75 cm. long, find his speed in km. per hour.

## ANSWERS (Exercise 13A)

1. (i) 17.5 m/sec. (ii) 54 km/hr.
2. $31\frac{1}{9}$ km/hr.
3. 60 km.
4. 6 km.
5. 60 km/hr.
6. 2.5 km.
7. 8 hours
8. 7.5 km.
9. 10 km/hr.
10. 3 km.
11. 4 p.m.
12. 7 min.
13. 5.4 km/hr.

## EXERCISE 13B(Objective Type Questions)

1. A man crosses a street 600 m. long in 5 minutes. His speed in kilometres per hour is :

(*a*) 7.2 (*b*) $\frac{5}{36}$

(*c*) 3.6 (*d*) 10 **(Clerical Grade, 1991)**

2. A speed of 6 m/sec. is the same as :

(*a*) 15.4 km/hr. (*b*) 16.5 km/hr.

(*c*) 21.6 km/hr. (*d*) 22.5 km/hr.

3. If a man covers $10\frac{1}{5}$ km in 3 hours, the distance covered by him in 5 hours is :

(*a*) 18 km (*b*) 15 km

(*c*) 16 km (*d*) 17 km **(Astt. Grade 1987)**

4. A train travels 92.4 km/hr. How many metres will it travel in 10 minutes ?

(*a*) 15400 (*b*) 1540

(*c*) 154 (*d*) 15.40 **(Bank P.O. 1991)**

5. Ajit takes 4 hours to cover a distance of 9 km. The time taken by him to travel 24 km. is :

(*a*) 10 hours 50 min. (*b*) 10 hours 40 min.

(*c*) 10 hours 30 min. (*d*) 9 hours 40 min.

6. A man goes to a place at the rate of 4 km/hr. He comes back on a bicycle at 16 km/hr. His average speed for the entire journey is :

(*a*) 5 km/hr. (*b*) 6.4 km/hr.

(*c*) 8.5 km/hr. (*d*) 10 km/hr.

7. A man covers a certain distance in 2 hours 45 min, when he walks at the rate of 4 km/hr. How much time will he take to cover the same distance if he runs at a speed of 6.5 km/hr. ?

(*a*) 40 min. (*b*) 41 min. 15 sec.

(*c*) 45 min. (*d*) 100 min.

**8.** Suresh travelled 1200 km. by air which formed (2/5) of his trip. The part of his trip which was one third of the whole trip, he travelled by car. The rest of the journey was performed by train. The distance travelled by train was :

(*a*) 1600 km. (*b*) 800 km.

(*c*) 480 km. (*d*) 1800 km.

**(Hotel Management, 1991)**

**9.** Two men start together to walk to a certain destination, one at 3.75 km an hour and another at 3 km. an hour. The former arrives half an hour before the latter. The distance is :

(*a*) 9.5 km. (*b*) 8 km.

(*c*) 7.5 km. (*d*) 6 km.

**10.** Sharad can cover a certain distance in 42 minutes. He covers two third of it at 4 km. an hour and the remaining at 5 km. an hour. The total distance is :

(*a*) 2.5 km. (*b*) 4.6 km.

(*c*) 4 km. (*d*) 3 km.

**11.** Two cyclists start from the same place in the opposite directions. One goes towards north at 18 km/hr. and the other goes towards south at the speed of 20 km/hr. What time will they take to be 95 km. apart ?

(*a*) $4\frac{1}{2}$ hrs. (*b*) 4 hrs. 45 min.

(*c*) 5 hrs. 16 min. (*d*) 2 hrs. 30 min.

**12.** A boy goes to school with the speed of 3 km. an hour and returns with a speed of 2 km/hr. If he takes 5 hours in all, the distance in kms. between the village and the school is :

(*a*) 6 (*b*) 7

(*c*) 8 (*d*) 9 **(N.D.A. 1990)**

**13.** A man performs $\frac{2}{15}$ of his total journey by rail, $\frac{9}{20}$ by tonga and the remaining 10 km. on foot. His total journey is :

(*a*) 15.6 km. (*b*) 12.8 km.

(*c*) 16.4 km. (*d*) 24 km.

**14.** *A* travels for 14 hours 40 min. He covers one half of the journey by train at the rate of 60 km/hr. and the rest half by road at the rate of 50 km/hr. The distance travelled by *A* is :

(*a*) 1000 km. (*b*) 960 km.

(*c*) 800 km. (*d*) 720 km.

**15.** A train goes from a station *A* to another station *B* at a speed of 64 km/hr. but returns to *A* at a slower speed. If its average speed for the trip is 56 km/hr., the return speed of the train is nearly :

(*a*) 48 km/hr. (*b*) 50 km/hr.
(*c*) 52 km/hr. (*d*) 47.4 km/hr.

**(Hotel Management, 1991)**

**16.** A car covers four successive three km. stretches at speeds of 10 km/ph, 20 km/hr., 30 km/hr. and 60 km/hr. respectively. Its average speed over this distance is :
(*a*) 10 km/hr. (*b*) 20 km/hr.
(*c*) 30 km/hr. (*d*) 25 km/hr.

**17.** A car completes a certain journey in 8 hours. It covers half the distance at 40 km/hr. and the rest at 60 km/hr. The length of the journey in kilometres is :
(*a*) 350 (*b*) 384
(*c*) 400 (*d*) 420 **(Clerical Grade, 1991)**

**18.** A train leaves Meerut at 6 a.m. and reaches Delhi at 10 a.m. Another train leaves Delhi at 8 a.m. and reaches Meerut at 11.30 a.m. At what time do the two trains cross one another ?
(*a*) 9.26 a.m. (*b*) 9 a.m.
(*c*) 8.30 a.m. (*d*) 8.56 a.m.

**19.** *X* and *Y* are two stations 500 km. apart. A train starts from *X* and moves towards *Y* at the rate of 20 km/hr. Another train starts from *Y* at the same time and moves towards *X* at the rate of 30 km/hr. How far from *X* will they cross each other ?
(*a*) 200 km. (*b*) 300 km.
(*c*) 120 km. (*d*) 40 km.

**20.** A thief steals a car at 1.30 p.m. and drives it at 40 km. an hour. The theft is discovered at 2 p.m. and the owner sets off in another car at 50 km. an hour. He will overtake the thief at :
(*a*) 4 p.m. (*b*) 3.30 p.m.
(*c*) 6 p.m. (*d*) 4.30 p.m.

**21.** A man walking at the rate of 3 km/hr. crosses a square field diagonally in 2 min. The area of the field is :
(*a*) 25 ares (*b*) 30 ares
(*c*) 50 ares (*d*) 60 ares

**22.** By walking at $\frac{3}{4}$ of his usual speed, a man reaches office 20 minutes later than usual. His usual time is :
(*a*) 30 minutes (*b*) 60 minutes
(*c*) 75 minutes (*d*) $1\frac{1}{2}$ hours **(Railways, 1991)**

**23.** Ram travels at the rate of 3 km/hr and he reaches 15 minutes late. If he travels at the rate of 4 km/hr, he reaches 15 minutes earlier. The distance Ram has to travel is :

(*a*) 1 km. (*b*) 6 km.
(*c*) 7 km. (*d*) 12 km. **(C.D.S. Exam 1989)**

**24.** If a train runs at 40 km/hr., it reaches its destination late by 11 min. but if it runs at 50 km/hr., it is late by 5 min. only. The correct time for the train to complete its journey is :

(*a*) 13 min. (*b*) 15 min.
(*c*) 21 min. (*d*) 19 min.

**25.** Excluding stoppages, the speed of a bus is 54 km/hr. and including stoppages, it is 45 km/hr. For how many minutes does the bus stop per hour ?

(*a*) 9 (*b*) 10
(*c*) 12 (*d*) 20

**26.** The ratio between the rates of walking of *A* and *B* is 3 : 4. If the time taken by *B* to cover a certain distance is 24 minutes, time taken by *A* to cover that much distance is :

(*a*) 18 min. (*b*) 32 min.
(*c*) $10\frac{6}{7}$ min. (*d*) $13\frac{5}{7}$ min.

**27.** A certain distance is covered at a certain speed. If half this distance is covered in double the time, the ratio of the two speeds is :

(*a*) 4 : 1 (*b*) 1 : 4
(*c*) 2 : 1 (*d*) 1 : 2 **(Bank P.O. 1986)**

**28.** *A* is twice as fast as *B* and *B* is thrice as fast as *C* is. The journey covered by *C* in 42 minutes, will be covered by *A* in :

(*a*) 14 min. (*b*) 28 min.
(*c*) 63 min. (*d*) 7 min.

**29.** The ratio between the rates of travelling of *A* and *B* is 2 : 3 and therefore *A* takes 10 minutes more than the time taken by *B* to reach a destination. If *A* had walked at double the speed, he would have covered the distance in :

(*a*) 30 min. (*b*) 25 min.
(*c*) 15 min. (*d*) 20 min.

**30.** A bus travels a distance of 840 km at a uniform speed. If the speed of the car is 10 km/hr. more, then it takes 2 hours less to cover the same distance. The original speed of the car was :

(*a*) 45 km/hr. (*b*) 50 km/hr.
(*c*) 60 km/hr. (*d*) 75 km/hr.

**31.** If a boy takes as much time in running 10 metres as a car takes in covering 25 metres; the distance covered by the boy during the time the car covers 1 km., is :

(*a*) 400 metres (*b*) 40 metres

(*c*) 4 metres (*d*) 2.5 metres

**(Astt. Grade Exam. 1987)**

**32.** Suresh started cycling along the boundaries of a square field from corner point *A*. After half an hour, he reached the corner point *C*, diagonally opposite to *A*. If his speed was 8 km. per hour, the area of the field in square km. is :

(*a*) 64 (*b*) 8

(*c*) 4 (*d*) cannot be determined

**(Bank Trainee Officer's Exam. 1988)**

**33.** On a tour a man travels at the rate of 64 km. an hour for the first 160 km., then travels the next 160 km. at the rate of 80 km. an hour. The average speed in km. per hour for the first 320 km. of the tour, is :

(*a*) 35.55 (*b*) 71.11

(*c*) 36 (*d*) 72 **(S.B.I. P.O. Exam. 1988)**

**34.** A man travels 35 kms. partly at 4 km/hr. and partly at 5 km/hr. If he covers former distance at 5 km/hr. and later distance at 4 km/hr., he could cover 2 km. more in the same time. The time taken to cover the whole distance, at original rate is :

(*a*) 9 hours (*b*) 7 hours

(*c*) 4.5 hours (*d*) 8 hours

**35.** Laxman has to cover a distance of 6 km. in 45 minutes. If he covers one half of the distance in $\frac{2}{3}$rd time, what should be his speed in km/hr. to cover the remaining distance in the remaining time ?

(*a*) 12 (*b*) 16

(*c*) 3 (*d*) 8 **(Bank P.O. Exam. 1991)**

**36.** A train covers a distance in 50 minutes if it runs at a speed of 48 km. per hour on an average. The speed at which the train must run to reduce the time of journey to 40 minutes, will be :

(*a*) 50 km/hour (*b*) 55 km/hour

(*c*) 60 km/hour (*d*) 70 km/hour

**(Central Excise & I. Tax. 1988)**

**37.** A man takes 6 hours to walk to a certain place and riding back. However, he could have gained $2\frac{1}{2}$ hours, if he had covered both ways by riding. How long would he have taken to walk both ways :

(*a*) $8\frac{1}{2}$ hours (*b*) $3\frac{1}{2}$ hours

(*c*) 11 hours (*d*) $5\frac{1}{2}$ hours

**38.** Two trains start at the same time from Aligarh and Delhi and proceed towards each other at the rate of 16 km. and 21 km. per hour respectively. When they meet, it is found that one train has travelled 60 km. more than the other. The distance between two stations is :

(*a*) 445 km. (*b*) 444 km.

(*c*) 440 km. (*d*) 450 km.(Bank P.O. Exam. 1988)

## HINTS & SOLUTIONS (Exercise 13B)

**1.** Required speed = $\left(\frac{600 \times 60}{1000 \times 5}\right)$ km/hr. = 7.2 km/hr.

**2.** 6 metres/sec. = $\left(6 \times \frac{18}{5}\right)$ km/hr. = 21.6 km./hr.

**3.** Distance covered in 5 hours = $\left(\frac{51}{5 \times 3} \times 5\right)$ km = 17 km.

**4.** Speed of train = $\left(92.4 \times \frac{5}{18}\right)$ m/sec. = $\frac{77}{3}$ m/sec.

Distance covered in (10 × 60) sec.

$= \left(\frac{77}{3} \times 10 \times 60\right)$ metres = 15400 metres.

**5.** More distance, more time taken.

$\therefore$ 9 : 24 :: 4 : $x$ or $x = \left(\frac{24 \times 4}{9}\right)$ = 10 hrs. 40 min.

**6.** **Formula :** *Average speed* = $\left(\frac{2xy}{x+y}\right)$ *km/hr.*

$\therefore$ Average speed $= \frac{(2 \times 4 \times 16)}{(4 + 16)}$ km/hr. = $\frac{32}{5}$ km/hr.

= 6.4 km/hr.

**7.** Distance = (speed × time) = $\left(4 \times \frac{11}{4}\right)$ km = 11 km.

$\therefore$ Time taken = $\frac{11}{16.5}$ hrs. = 40 minutes.

**8.** Let the total distance be $x$ km. Then,

$\frac{2}{5}x = 1200$ or $x = 3000$ km.

Distance covered by car = $\left(3000 \times \frac{1}{3}\right)$ km. = 1000 km.

$\therefore$ Distance travelled by train = [3000 – (1000 + 1200)] km. = 800 km.

**9.** Let the distance be $x$ km. Then,

$$\left(\frac{x}{3}-\frac{x}{3.75}\right)=\frac{1}{2} \text{ or } \frac{0.75\times x}{11.25}=\frac{1}{2} \text{ or } x=7.5 \text{ km.}$$

**10.** Let the distance be $x$ km. Then,

$$\left(\frac{2x}{3}\times\frac{1}{4}\right)+\left(\frac{x}{3}\times\frac{1}{5}\right)=\frac{42}{60} \text{ or } \frac{7x}{30}=\frac{7}{10} \text{ or } x=3 \text{ km.}$$

**11.** They are 38 km. apart in 1 hour.

They will be 95 km. apart in $\left(\frac{1}{38}\times 95\right)$ hours

$= 2$ hrs. 30 min.

**12.** Let the distance be $d$ km. Then,

$$\frac{2\times3\times2}{3+2}=\frac{2d}{5}\left[\because \text{Average speed}=\frac{2xy}{x+y}\right]$$

$\therefore 10d=60$ or $d=6$ km.

**13.** Let the total journey be $x$ km. Then,

$$\frac{2x}{15}+\frac{9x}{20}+10=x \text{ or } 25x=600 \text{ or } x=24 \text{ km.}$$

**14.** Average speed $=\left(\frac{(2\times60\times50)}{(60+50)}\right)$ km/hr. $=\frac{600}{11}$ km/hr.

$\therefore$ Distance = (speed × time)

$=\left(\frac{600}{11}\times\frac{44}{3}\right)$ km. $= 800$ km.

**15.** Let the return speed of the train be $y$ km/hr.

Then, $\frac{(2\times64\times y)}{(64+y)}=56\left[\because \text{Average speed}=\left(\frac{2xy}{x+y}\right)\text{ km/hr.}\right]$

$\Rightarrow 2\times64\times y=56(64+y)\Rightarrow y=49.8=50$ (nearly)

**16.** Total time taken $=\left(\frac{3}{10}+\frac{3}{20}+\frac{3}{30}+\frac{3}{60}\right)$ hrs. $=\frac{3}{5}$ hrs.

Average speed $=\left\{\frac{12}{(3/5)}\right\}$ km./hr. $=\left(\frac{12\times5}{3}\right)$ km/hr.

$= 20$ km/hr.

**17.** Let the distance travelled during each half be $x$ km. Then,

$$\frac{x}{40}+\frac{x}{60}=8 \text{ or } x=192.$$

$\therefore$ Length of journey $=(2\times192)=384$ km.

**18.** Suppose the total distance is $y$ km. and they meet $x$ hrs. after the start of the train from Meerut.

Average speeds of these trains are $(y/4)$ km/hr. and $(2y/7)$ km/hr. respectively.

$$\therefore \frac{xy}{4}+\frac{2y(x-2)}{7}=y \text{ or } \frac{x}{4}+\frac{2x-4}{7}=1.$$

This gives, $15x = 44$ or $x = 2$ hours 56 min.

So, the trains cross each other at 8.56 A.M.

**19.** Suppose they meet at a distance of $x$ km. from $X$. Then,

$$\frac{x}{20} = \left(\frac{500 - x}{30}\right) \text{ or } x = 200 \text{ km.}$$

So, they meet at a distance of 200 km. from $X$.

**20.** Distance covered by thief in half an hour = 20 km.

Now, 20 km. is compensated by the owner at a relative speed of 10 km/hr. in 2 hours.

So, he overtakes the thief at 4 p.m.

**21.** $3 \text{ km/hr.} = \left(3 \times \frac{5}{18}\right) \text{ metres/sec.} = \frac{5}{6} \text{ metres/sec.}$

Distance travelled in 2 min. $= \left(\frac{5}{6} \times 120\right) = 100$ metres .

$\therefore$ Diagonal of the field = 100 metres.

$\therefore$ Area of the field $= \left(\frac{1}{2} \times 100 \times 100\right) m^2 = 5000\ m^2 = 50$ ares.

**22.** With a speed of (3/4)th of usual speed, the time taken is (4/3) of the usual time.

$$\therefore \left[\left(\frac{4}{3} \text{ usual time}\right) - (\text{usual time})\right] = 20 \text{ min.}$$

$$\text{or} \quad \left(\frac{1}{3} \text{ of usual time}\right) = 20 \text{ min.}$$

or usual time = 60 min.

**23.** Let the distance be $x$ km. Then,

$$\left(\frac{x}{3} - \frac{x}{4}\right) = \frac{30}{60} \text{ or } \frac{x}{12} = \frac{1}{2} \text{ or } x = 6 \text{ km.}$$

**24.** Suppose the correct time to complete the journey is $x$ minutes. Then, the distance covered in $(x + 5)$ min. at 50 km/hr. is equal to the distance covered in $(x + 11)$ min. at 40 km/hr.

$$\therefore \frac{50}{60}(x + 5) = \frac{40}{60}(x + 11) \text{ or } x = 19 \text{ min.}$$

**25.** Due to stoppages, it covers 9 km. less.

Now, time taken to cover 9 km. $= \left(\frac{9}{54} \times 60\right)$ min. = 10 min.

**26.** The ratio of times taken by $A$ and $B$ is 4 : 3.

Now, $4 : 3 :: x : 24$ or $x = \left(\frac{4 \times 24}{3}\right)$ min. = 32 min.

**27.** Let $x$ km. be covered in $y$ hours. Then,

1st speed = $(x/y)$ km. per hour.

2nd speed $= \left(\dfrac{\frac{1}{2}x}{2y}\right) = \left(\dfrac{x}{4y}\right)$ km. per hour.

Ratio of speeds $= \dfrac{x}{y} : \dfrac{x}{4y} = 1 : \dfrac{1}{4} = 4 : 1.$

**28.** Time taken by $B = \left(\dfrac{1}{3} \times 42\right)$ min. = 14 min.

$\therefore$ Time taken by $A = \left(\dfrac{1}{2} \times 14\right)$ min. = 7 min.

**29.** The ratio of times taken by $A$ and $B$ is 3 : 2.
If difference is 1 min., $A$ takes 3 minutes.
If difference is 10 min., $A$ takes 30 minutes.
$\therefore$ Time taken by $A$ with double speed = 15 min.

**30.** Let the original speed be $x$ km./hr. Then,

$\left[\dfrac{840}{x} - \dfrac{840}{(x+10)}\right] = 2$ or $x^2 + 10x - 4200 = 0$

or $(x + 70)(x - 60) = 0$ or $x = 60$ km/hr.

**31.** $25 : 10 :: 1000 : x$

or $x = \dfrac{10 \times 1000}{25} = 400$ metres.

**32.** Distance covered in half an hour = 4 km.
2 × (side of the square) = 4 km.
or side of the square = 2 km.
$\therefore$ Area of the square field = 4 sq. km.

**33.** **Formula :** *Average speed* $= \dfrac{2xy}{(x+y)}$ *km/hr.*

$\therefore$ Average speed $= \left(\dfrac{2 \times 64 \times 80}{64 + 80}\right)$ km./hr. = 71.11 km/hr.

**34.** Suppose the man covers first distance in $x$ hours and the second distance in $y$ hours.
Then, $4x + 5y = 35$ and $5x + 4y = 37$.
Solving these equations, we get $x = 5$ and $y = 3$.
So, total time taken = 8 hours.

**35.** Time left $= \left(\dfrac{1}{3} \text{ of } \dfrac{45}{60} \text{ hours}\right) = \dfrac{1}{4}$ hour.
Distance left = 3 km.
$\therefore$ Required speed = (3 × 4) km/hr. = 12 km/hr.

**36.** Distance = (speed × time) $= \left(48 \times \dfrac{50}{60}\right) = 40$ km.

Speed $= \dfrac{\text{Distance}}{\text{Time}} = \left(\dfrac{40 \times 60}{40}\right)$ km/hr. = 60 km/hr.

**37.** One way walking time + one way riding time = 6 hrs.
2 ways walking time + 2 ways riding time = 12 hrs.
2 ways walking time = (12 hrs.) – (2 ways riding time)
$$= \left(12 - 3\frac{1}{2}\right) \text{ hours} = 8\frac{1}{2} \text{ hrs.}$$

**38.** Suppose they meet after $x$ hours. Then,
$21x - 16x = 60$ or $x = 12$.
Now, distance between the stations
$= (16 \times 12 + 21 \times 12)$ km. = 444 km.

## ANSWERS (Exercise 13B)

| | | | | | |
|---|---|---|---|---|---|
| **1.** (*a*) | **2.** (*c*) | **3.** (*d*) | **4.** (*a*) | **5.** (*b*) | **6.** (*b*) |
| **7.** (*a*) | **8.** (*b*) | **9.** (*c*) | **10.** (*d*) | **11.** (*d*) | **12.** (*a*) |
| **13.** (*d*) | **14.** (*c*) | **15.** (*b*) | **16.** (*b*) | **17.** (*b*) | **18.** (*d*) |
| **19.** (*a*) | **20.** (*a*) | **21.** (*c*) | **22.** (*b*) | **23.** (*b*) | **24.** (*d*) |
| **25.** (*b*) | **26.** (*b*) | **27.** (*a*) | **28.** (*b*) | **29.** (*c*) | **30.** (*c*) |
| **31.** (*a*) | **32.** (*c*) | **33.** (*b*) | **34.** (*d*) | **35.** (*a*) | **36.** (*c*) |
| **37.** (*a*) | **38.** (*b*) | | | | |

# 14

# TRAINS

In solving problems on trains, the following points should be kept in mind.

(i) *The time taken by a train x metres long, in passing* a signal post or a pole or a standing man is the same as the time taken by the train to cover a distance of $x$ metres.

(ii) The time taken by a train $x$ metres long, in passing a bridge or a tunnel or a train at rest or a platform of length $y$ metres is the same as the time taken by the train to cover a distance of $(x + y)$ metres.

(iii) If two trains or two bodies are moving in the same direction with a speed of $s$ km/hr. and $t$ km/hr respectively such that $s > t$, then : their relative speed = $(s - t)$ km/hour.
If faster train has length $x$ km. and slower train has length $y$ km., then :
time taken by faster train to cross the slower train (moving in the same direction) = $\left(\frac{x + y}{s - t}\right)$ hrs.

(iv) If two trains or two bodies are moving in opposite directions with a speed of $s$ km/hr and $t$ km/hr respectively, then
relative speed = $(s + t)$ km/hr.
If $x$ km & $y$ km be the lengths of the trains, then time taken by the trains to cross each other (moving in opposite directions)
$= \left(\frac{x + y}{s + t}\right)$ hours.

(v) If two trains start at the same time from two points $A$ and $B$ towards each other and after crossing, they take $a$ and $b$ hours in reaching $B$ and $A$ respectively. Then,
($A$'s speed) : ($B$'s speed) = $(\sqrt{b} : \sqrt{a})$.

**Ex. 1.** ***A train 135 m. long is running with a speed of 54 km. per hour. In what time will it pass a telegraph post ?***

**Sol.** Speed of the train = 54 km/hr.

$$= \left(54 \times \frac{5}{18}\right) \text{ m/sec.} = 15 \text{ m/sec.}$$

Time taken by the train to pass a telegraph post

= Time taken by it to cover 135 m at 15 m/sec.

$= \left(\frac{135}{15}\right)$ sec = 9 sec.

**Ex. 2.** ***A train 540 metres long is running with a speed of 72 km per hour. In what time will it pass a tunnel 160 metres long ?***

**Sol.** Speed of the train = 72 km/hr.

$= \left(72 \times \frac{5}{18}\right)$ m/sec = 20 m/sec.

Sum of the lengths of the train and tunnel = (540 + 160) m = 700 m.

∴ Time taken to pass the tunnel

= Time taken to cover 700 m at 20 m/sec.

$= \left(\frac{700}{20}\right)$ *sec. = 35 sec.*

**Ex. 3.** ***A man is standing on a railway bridge which is 50 metres long. He finds that a train crosses the bridge in $4\frac{1}{2}$ seconds but himself in 2 seconds. Find the length of the train and its speed.***

**Sol.** Let the length of the train be $x$ metres.

Then, the time taken by the train to cover $(x + 50)$ metres is $4\frac{1}{2}$ seconds.

$\therefore$ Speed of the train $= \frac{(x+50)}{\left(\frac{9}{2}\right)}$ m/sec $= \left(\frac{2x+100}{9}\right)$ m/sec ... (i)

Again, the time taken by the train to cover $x$ metres is 2 seconds.

$\therefore$ Speed of the train $= \left(\frac{x}{2}\right)$ m/sec. ... (ii)

Thus, from (i) and (ii), we get :

$\frac{2x+100}{9} = \frac{x}{2}$ or $5x = 200$ or $x = 40$ .

∴ *Length of the train = 40 metres .*

*Speed of the train* $= \left(\frac{40}{2}\right)$ *m/sec* $= \left(\frac{40}{2} \times \frac{18}{5}\right)$ *km/hr = 72 km/hr.*

**Ex. 4.** ***A train 100 metres long is running with a speed of 70 km per hour. In what time will it pass a man who is running at 10 km. per hour in the same direction in which the train is going ?***

**Sol.** Speed of train relative to man

$= (70 - 10)$ km/hr $= \left(60 \times \frac{5}{18}\right)$ m/sec $= \frac{50}{3}$ m/sec.

Length of the train = 100 m.

Time taken by the train to cross the man

$$= \text{Time taken by the train to cover 100 m at } \frac{50}{3} \text{ m/sec.}$$

$$= \left(100 \times \frac{3}{50}\right) \text{ sec} = 6 \text{ sec.}$$

**Ex. 5.** ***A train 110 metres long is running with a speed of 60 km per hour. In what time will it pass a man who is running at 6 km per hour in the direction opposite to that in which the train is going ?***

**Sol.** Speed of the train relative to man

$$= (60 + 6) \text{ km/hr} = \left(66 \times \frac{5}{18}\right) \text{ m/sec} = \frac{55}{3} \text{ m/sec}$$

Length of the train = 110 m.

Time taken by the train to cross the man

$$= \text{Time taken to cover 110 m at } \frac{55}{3} \text{ m/sec.}$$

$$= \left(110 \times \frac{3}{55}\right) \text{ sec.} = 6 \text{ sec.}$$

**Ex. 6.** ***Two trains 132 metres and 108 metres in length are running towards each other on parallel lines, one at the rate of 32 km per hour and another at 40 km per hour. In what time will they be clear of each other from the moment they meet ?***

**Sol.** Relative speed = (32 + 40) km/hr = 72 km/hr

$$= \left(72 \times \frac{5}{18}\right) \text{ m/sec} = 20 \text{ m/sec.}$$

Sum of lengths of the trains = (132 + 108) m = 240 m

Time taken by the trains in passing each other

$$= \left(\frac{240}{20}\right) \text{ sec} = 12 \text{ sec.}$$

**Ex. 7.** ***A train running at 25 km per hour takes 18 seconds to pass a platform. Next, it takes 12 seconds to pass a man walking at the rate of 5 km/hr in the same direction. Find the length of the train and that of the platform.***

**Sol.** Let the length of train be $x$ metres and the length of platform be $y$ metres.

$$\text{Speed of the train} = \left(25 \times \frac{5}{18}\right) \text{ m/sec} = \frac{125}{18} \text{ m/sec.}$$

$$\text{Time taken by the train to pass the platform} = \left[(x + y) \times \frac{18}{125}\right] \text{ sec.}$$

$$\therefore (x+y)\times\frac{18}{125}=18 \text{ or } x+y=125 \qquad \text{... (i)}$$

Speed of the train relative to man

$$=(25+5)\text{ km/hr}=\left(30\times\frac{5}{18}\right)\text{ m/sec}=\frac{25}{3}\text{ m/sec.}$$

Time taken by the train to pass the man

$$=\left(x\times\frac{3}{25}\right)\text{ sec}=\frac{3x}{25}\text{ sec.}$$

$$\therefore \frac{3x}{25}=12\Rightarrow x=\left(\frac{25\times 12}{3}\right)=100\text{ metres.}$$

Putting $x = 100$ in (i), we get, $y = 25$ metres.

$\therefore$ *Length of train = 100 m and the length of the platform = 25 m.*

**Ex. 8.** ***A train 100 metres long takes $7\frac{1}{5}$ seconds to cross a man walking at the rate of 5 km/hr. in a direction opposite to that of the train. Find the speed of the train.***

**Sol.** Let the speed of the train be $x$ km/hr.

Speed of train relative to man = $(x + 5)$ km/hr.

Now, time taken to cover 100 metres i.e., (1/10) km.

$$=\left[\frac{1}{(x+5)}\times\frac{1}{10}\times 60\times 60\right]\text{ sec.}=\left(\frac{360}{x+5}\right)\text{ sec.}$$

$$\therefore \frac{360}{x+5}=\frac{36}{5} \text{ or } x=45.$$

*Hence, the speed of the train = 45 km/hr.*

**Ex. 9.** ***A man sitting in the train which is travelling at the rate of 50 km per hour observes that it takes 9 seconds for a goods train travelling in the opposite direction to pass him. If the goods train is 187.5 metres long, find its speed.***

**Sol.** In 9 seconds, the goods train covers 187.5 metres.

$$\therefore \text{In 1 hour, the goods train covers }\left(\frac{187.5}{9}\times\frac{60\times 60}{1000}\right)\text{ km}=75\text{ km.}$$

So, the speed of goods train relative to man = 75 km/hour.

$\therefore$ *Speed of goods train = (75 – 50) km/hr. = 25 km/hour.*

**Ex. 10.** ***Two trains 200 metres and 175 metres long are running on parallel lines. They take $7\frac{1}{2}$ seconds when running in opposite directions and $37\frac{1}{2}$ seconds when running in the same direction, to pass each other. Find their speeds in km per hour.***

**Sol.** Let the speeds of the trains be $x$ metres/sec. and $y$ metres/sec.

respectively.

Sum of the lengths of the trains = (200 + 175) m. = 375 m.

When trains are moving in opposite directions :

Relative speed = $(x + y)$ m/sec.

In this case, the time taken by the trains to cross each other

$$= \frac{375}{(x+y)} \text{ sec.}$$

$$\therefore \quad \frac{375}{x+y} = \frac{15}{2} \quad \text{or} \quad x + y = 50 \qquad \text{... (i)}$$

When trains are moving in the same direction :

Relative speed = $(x - y)$ m./sec.

In this case, the time taken by the trains to cross each other

$$= \frac{375}{(x-y)} \text{ sec.}$$

$$\frac{375}{(x-y)} = \frac{75}{2} \quad \text{or} \quad x - y = 10 \qquad \text{.. (ii)}$$

Solving (i) and (ii), we get $x = 30, y = 20$.

$\therefore$ Speeds of the trains are $\left(30 \times \frac{18}{5}\right)$ km/hr. & $\left(20 \times \frac{18}{5}\right)$ km/hr.

*which is 108 km/hr. and 72 km/hr.*

**Ex. 11.** ***The driver of a car driving at the speed of 38 km. per hour locates a bus ahead of him. After 20 seconds, the bus is 60 metres behind. Find the speed of the bus.***

**Sol.** Let the speed of the bus be $x$ km/hr.

Speed of car relative to bus = $(38 - x)$ km/hr.

Time taken to cover (40 + 60) or 100 metres or (1/10) km

$$= \left\{\frac{1}{10} \times \frac{1}{(38-x)}\right\} \text{hrs.} = \frac{360}{(38-x)} \text{ seconds.}$$

$$\therefore \frac{360}{(38-x)} = 20 \text{ or } 38 - x = 18 \text{ or } x = 20.$$

*Hence, the speed of the bus is 20 km/hour.*

**Ex. 12.** ***A train 75 metres long overtook a person who was walking at the rate of 6 km. an hour and passed him in $7\frac{1}{2}$ seconds. Subsequently, it overtook a second person and passed him in $6\frac{3}{4}$ seconds. At what rate was the second person travelling ?***

**Sol.** Let the speed of the train be $x$ metres/sec.

$$\text{Speed of 1st man} = \left(6 \times \frac{5}{18}\right) \text{m/sec.} = \frac{5}{3} \text{ m/sec.}$$

Speed of train relative to 1st man = $\left(x - \frac{5}{3}\right)$ m/sec.

$\therefore$ Time taken by the train to cross the 1st man

$$= \frac{75}{\left(x - \frac{5}{3}\right)} \text{ sec.} = \left(\frac{225}{3x-5}\right) \text{ sec.}$$

$$\therefore \frac{225}{3x-5} = \frac{15}{2} \text{ or } x = \left(\frac{35}{3}\right) \text{ m/sec.}$$

Let the speed of 2nd man be $y$ metres/sec.

Relative speed = $\left(\frac{35}{3} - y\right)$ m/sec. = $\left(\frac{35-3y}{3}\right)$ m/sec.

$\therefore$ Time taken by the train to cross the 2nd man

$$= \frac{75}{\left(\frac{35-3y}{3}\right)} \text{ sec.} = \left(\frac{225}{35-3y}\right) \text{ sec.}$$

$$\therefore \frac{225}{35-3y} = \frac{27}{4} \text{ or } y = \frac{5}{9}.$$

$\therefore$ Speed of second man = $\left(\frac{5}{9}\right)$ m/sec.

$$= \left(\frac{5}{9} \times \frac{18}{5}\right) \text{ km/hr.} = 2 \text{ km/hr.}$$

**Ex. 13.** ***Two trains A and B start from stations X and Y and proceed towards Y and X respectively. After passing each other, they take 4 hours 48 minutes and 3 hours 20 minutes to reach Y and X respectively. At what rate the train B is moving, if the train A is running at 45 km. per hour ?***

**Sol.** $\frac{A\text{'s rate}}{B\text{'s rate}} = \sqrt{\left(\frac{\text{Time taken by } B \text{ to reach } X \text{ after crossing}}{\text{Time taken by } A \text{ to reach } Y \text{ after crossing}}\right)}$

or $\frac{45}{B\text{'s rate}} = \sqrt{\left(\frac{10}{3}\right) + \left(\frac{24}{5}\right)} = \sqrt{\left(\frac{10}{3} \times \frac{5}{24}\right)} = \frac{5}{6}.$

$\therefore$ *B's rate* = $\left(\frac{45 \times 6}{5}\right)$ km/hr. = 54 km/hr.

## EXERCISE 14A (Subjective)

1. A train 550 metres long is running with a speed of 55 km. per hour. In what time will it pass a signal post ?
2. A train 200 metres long is running with a speed of 72 km. per hour. In what time will it pass a platform 160 metres long ?

3. A train 160 metres long passes a standing man in 18 seconds. What is the speed of the train ?
**Hint.** *The train covers 160 m in 18 seconds.*

4. A train 240 metres long passes a bridge 120 metres long in 24 seconds. Find the speed with which the train is moving.
**Hint.** *speed* = $\left(\frac{240 + 120}{24}\right)$ *m/sec.* = *54 km/hr.*

5. A railway train travelling at a uniform speed, clears a platform 200 metres long in 10 seconds and passes a telegraph post in 6 seconds. Find the length of the train and its speed.
**Hint.** *Let the length of the train be x metres and its speed be y metres per second. Then,*
$\left(\frac{x}{y} = 6 \Rightarrow x = 6y\right)$ *and* $\left(\frac{x + 200}{y} = 10 \Rightarrow 10y - x = 200\right)$.
*Solving* $x = 6y$ *and* $10y - x = 200$, *we get* $x = 300$, $y = 50$.

6. A train 150 metres long passes a telegraph post in 12 seconds. Find, in what time, it will pass a bridge 250 metres long.
**Hint.** *Speed of the train* = $\left(\frac{150}{12}\right)$ *m/sec.*
*Time taken to cross the bridge* = $\left[(150 + 250) \times \frac{12}{150}\right]$ *sec.*

7. A train 550 metres long is running with a speed of 55 km. per hour. In what time will it pass a man who is walking at 10 km. per hour in the same direction in which the train is going ?
**Hint.** *Relative speed* = *(55 – 10) km./hr.* = $\left(\frac{25}{2}\right)$ *m/sec.*
∴ *Time taken by the train to pass the man*
$= \left(550 \times \frac{2}{25}\right)$ *sec.* = *44 sec.*

8. A train 135 metres long is running with a speed of 49 km. per hour. In what time will it pass a man who is walking at 5 km. per hour in the direction opposite to that of the train ?
**Hint.** *Relative speed* = (49 + 5) *km/hr.* = 15 *m/sec.*
∴ *Time taken by the train to pass the man* = $\left(\frac{135}{15}\right)$ *sec.* = *9 sec.*

9. A train 100 metres long meets a man going in opposite direction at 5 km. per hour and passes him in $7\frac{1}{5}$ seconds, At what rate is the train going ?
**Hint.** *Let the speed of the train be x metres/sec.*

*Speed of man* = $\left(5 \times \frac{5}{18}\right)$ *m/sec.* = $\left(\frac{25}{18}\right)$ *m/sec.*

*Relative speed* = $\left(x + \frac{25}{18}\right)$ *m/sec.* = $\left(\frac{18x + 25}{18}\right)$ *m/sec.*

$\therefore \frac{100}{\left(\frac{18x+25}{18}\right)} = \frac{36}{5} \Rightarrow x = \left(\frac{225}{18}\right)$ *m/sec.* = $\left(\frac{225}{18} \times \frac{18}{5}\right)$ *km/hr.*

**10.** **Two trains are moving in the same direction at the speed of 50 km/hr. and 30 km/hr. respectively. The faster train crosses a man in the slower train in 18 seconds. Find the length of the faster train.**

**Hint.** *Find the distance covered in 18 seconds at the rate of (50 – 30) km/hr. i.e., 20 km/hr.*

**11.** **Two trains 127 metres and 113 metres in length respectively are running in opposite directions, one at the rate of 46 km. per hour and another at the rate of 26 km. per hour. In what time will they be clear of each other from the moment they meet ?**

**Hint.** *Relative speed* = (46 + 26) *km/hr.* = 20 *m/sec.*

*Time taken by the trains in passing each other*

= *Time taken to cover (127 + 113) m at 20 m/sec.*

**12.** **Two trains of length 110 metres and 90 metres are running on parallel lines in the same direction with a speed of 35 km per hour and 40 km. per hour respectively. In what time will they pass each other ?**

**Hint.** *Relative speed* = (40 – 35) *km/hr.* = $\left(\frac{25}{18}\right)$ *m/sec.*

*Time taken by the trains in passing each other*

= *Time taken to cover (110 + 90) m at* $\left(\frac{25}{18}\right)$ *m/sec.*

**13.** **If a train overtakes two persons who are walking in the same direction in which the train is going at the rate of 2 km/hr. and 4 km/hr. and passes them completely in 9 and 10 seconds. Find the length of train and its speed in km. per hour.**

**Hint.** *Let the length of train be x metres and its speed be y metres per sec.*

*Speed of 1st man* = 2 *km/hr.* = $\frac{5}{9}$ *m/sec.*

*Relative speed* = $\left(y - \frac{5}{9}\right)$ *m/sec.* = $\left(\frac{9y - 5}{9}\right)$ *m/sec.*

$\therefore \frac{x}{\left(\frac{9y-5}{9}\right)} = 9 \Rightarrow 9y - 5 = x \Rightarrow 9y - x = 5$ ... (i)

$$\therefore \frac{x}{\left(\frac{9y-5}{9}\right)} = 9 \Rightarrow 9y - 5 = x \Rightarrow 9y - x = 5 \quad \text{... (i)}$$

*Speed of 2nd man* = 4 *km/hr.* = $\left(\frac{10}{9}\right)$ *m/sec.*

*Relative speed* = $\left(y - \frac{10}{9}\right)$ *m/sec.* = $\left(\frac{9y-10}{9}\right)$ *m/sec.*

$$\therefore \frac{x}{\left(\frac{9y-10}{9}\right)} = 10 \Rightarrow 90y - 9x = 100 \quad \text{... (ii)}$$

*Solving (i) and (ii), we get* $x = 50,\ y = \frac{55}{9}$.

**14.** **Two trains, one travelling at 60 km/hr. and the other at 45 km/hr. leave two stations 70 km apart at the same time and travel towards each other. When & where will they meet ?**

**Hint.** *Suppose they meet x hrs. after the start.*

*Distance covered by 1st train in x hrs.* = (60x) *km.*

*Distance covered by 2nd train in x hrs.* = (45x) *km.*

$\therefore$ 60x + 45x = 70, *find x.*

**15.** **Two trains each 80 metres long, pass each other on parallel lines. If they are going in the same direction, the faster one takes one minute to pass the other completely. If they are going in opposite directions, they completely pass each other in 3 seconds. Find the rate of each train in metres per second.**

**16.** **A train leaves a station and travels at 32 km. an hour. After 45 minutes, a fast train leaves the station and travels in the same direction. At what speed must it travel so as to overtake the slower train in 80 minutes ?**

## ANSWERS (Exercise 14A)

| | | |
|---|---|---|
| **1.** 36 sec. | **2.** 18 sec. | **3.** 32 km/hr. |
| **4.** 54 km/hr. | **5.** 300 m., 180 km/hr. | **6.** 32 seconds |
| **7.** 44 sec. | **8.** 9 sec. | **9.** 45 km/hr. |
| **10.** 100 metres | **11.** 12 sec. | **12.** 144 sec. |

**13.** 50 metres, 22 km/hr.

**14.** 40 metres after the start and 40 km. away from the station from which the faster train started

**15.** 28 m/sec., $25\frac{1}{3}$ m/sec. **16.** 50 km/hr.

## EXERCISE 14B (Objective Type Questions)

**1.** A speed of 54 km/hr. is the same as :

(*a*) 14 m/sec. (*b*) 15 m/sec.
(*c*) 21 m/sec. (*d*) 27 m/sec.

2. A speed of $33\frac{1}{3}$ metres/second is the same as :
(*a*) 100 km./hr. (*b*) 80 km./hr.
(*c*) 120 km./hr. (*d*) 75 km./hr. **(Railways, 1991)**

3. If a man running at the rate of 15 km/hr. crosses a bridge in 5 minutes, the length of the bridge (in metres) is :
(*a*) $1333\frac{1}{3}$ (*b*) 1000
(*c*) 7500 (*d*) 1250 **(Clerical Grade, 1991)**

4. A train 150 metres long, running with a speed of 30 km/hr., will pass a standing man in :
(*a*) 18 seconds (*b*) 45 seconds
(*c*) 12 seconds (*d*) 5 seconds

5. A train 280 metres long is moving at a speed of 60 km/hr. The time taken by the train to cross a platform 220 metres long is :
(*a*) 20 seconds (*b*) 25 seconds
(*c*) 30 seconds (*d*) 35 seconds
**(R.R.B. Exam, 1989)**

6. A train 120 m long, crosses a pole in 10 seconds. The speed of the train is :
(*a*) 40 km/hr. (*b*) 43.2 km/hr.
(*c*) 45 km/hr. (*d*) none of these
**(Clerical Grade, 1989)**

7. A train 50 metres long passes a platform 100 metres long in 10 seconds. The speed of the train in metres/second is :
(*a*) 150 (*b*) 50
(*c*) 10 (*d*) 15 **(Railways, 1989)**

8. A train running at the rate of 36 km/hr. passes a standing man in 8 seconds. The length of the train is :
(*a*) 45 metres (*b*) 288 metres
(*c*) 80 metres (*d*) 48 metres

9. A train 700 metres long, is running at the speed of 72 km/hr. If it crosses a tunnel in 1 minute, then the length of the tunnel (in metres) is :
(*a*) 700 (*b*) 600
(*c*) 550 (*d*) 500 **(N.D.A. Exam, 1990)**

10. A train speeds past a pole in 15 seconds and speeds past a platform of 100 metres in 30 seconds. Its length (in metres) is :
(*a*) 200 (*b*) 100
(*c*) 50 (*d*) Data inadequate
**(Bank P.O. 1989)**

11. A train 120 metres long travels at 60 km/hr. A man is running at 6 km/hr. in the same direction in which the train is going. The train will cross the man in :

(a) 6 seconds (b) 7 seconds

(c) $6\frac{2}{3}$ seconds (d) 8 seconds

12. A train 270 metres long is moving at a speed of 25 km/hr. It will cross a man coming from the opposite direction at a speed of 2 km./hr. in :

(a) 36 seconds (b) 32 seconds

(c) 28 seconds (d) 24 seconds **(Railways, 1989)**

13. A train 150 metres long crosses a man walking at a speed of 6 km/hr. in the opposite direction in 6 seconds. The speed of the train is :

(a) 96 km/hr. (b) 84 km/hr.

(c) 106 km/hr. (d) 66 km/hr. **(BSRB Exam., 1990)**

14. A train 120 metres long passes a standing man in 5 seconds. How long will it take to cross a bridge 180 metres long ?

(a) $3\frac{1}{3}$ seconds (b) $12\frac{1}{2}$ seconds

(c) $7\frac{1}{2}$ seconds (d) 6 seconds

15. A train travelling at 36 km/hr. takes 10 seconds to pass a telegraph pole. How long would it take to cross a platform 55 metres long ?

(a) 12 seconds (b) $5\frac{1}{2}$ seconds

(c) $16\frac{1}{2}$ seconds (d) $15\frac{1}{2}$ seconds

16. A train 100 metres long completely crosses a bridge 300 metres long in 30 seconds. How many seconds will it take to pass a telegraph pole ?

(a) 10 seconds (b) $7\frac{1}{2}$ seconds

(c) 9 seconds (d) 3 seconds

17. A train crosses a platform in 60 seconds at a speed of 45 km/hr. How much time will it take to cross an electric pole if the length of the platform is 100 metres ?

(a) 8 seconds (b) 1 minute

(c) 52 seconds (d) none of these **(Bank P.O. 1990)**

18. Two trains 90 metres and 120 metres long are running in opposite directions, one at the rate of 48 km/hr. and another one at the rate of 60 km/hr. From the moment they meet, they will cross each other in :

(a) 6 seconds (b) 7 seconds

(c) 8 seconds (d) 9 seconds

**19.** **A train of length 150 metres, takes 10 seconds to pass over another train 100 metres long coming from the opposite direction. If the speed of the first train be 30 km/hr., the speed of the second train is :**

(*a*) 54 km/hr. (*b*) 60 km/hr.

(*c*) 72 km/hr. (*d*) 36 km/hr. **(Railways, 1991)**

**20.** **A train 300 metres long is running at a speed of 90 km/hr. How many seconds will it take to cross a 200 m long train running in the opposite direction at a speed of 60 km/hr. ?**

(*a*) 60 (*b*) $7\frac{1}{5}$

(*c*) 12 (*d*) 20 **(Bank P.O. 1989)**

**21.** **Two trains are running on parallel lines in the same direction at a speed of 50 km. and 30 km per hour respectively. The faster train crosses a man in slower train in 18 seconds. The length of the faster train is :**

(*a*) 98 metres (*b*) 170 metres

(*c*) 100 metres (*d*) 85 metres

**22.** **Two trains of equal length are running on parallel lines in the same direction at the rate of 46 km/hr. and 36 km./hr. The faster train passes the slower train in 36 seconds. The length of each train is :**

(*a*) 50 metres (*b*) 80 metres

(*c*) 72 metres (*d*) 82 metres

**23.** **Two trains *X* and *Y* start from stations *A* and *B* towards *B* and *A* respectively. After passing each other, they take 4 hours 48 minutes and 3 hours 20 minutes to reach *B* and *A* respectively. If train *X* is moving at 45 km/hr., the speed of train *Y* is :**

(*a*) 60 km/hr. (*b*) 54 km/hr.

(*c*) 64.8 km/hr. (*d*) 37.5 km/hr.

**24.** **A train 100 metres in length passes a mile stone in 10 seconds and another train of the same length travelling in opposite direction in 8 seconds. The speed of the second train is :**

(*a*) 36 km/hr. (*b*) 48 km/hr.

(*c*) 54 km/hr. (*d*) 60 km/hr.

**25.** **Two trains travelling in the same direction at 40 km/hr. and 22 km/hr. completely pass one another in 1 minute. If the length of the first train is 125 metres, the length of the second train is :**

(*a*) 125 metres (*b*) 150 metres

(*c*) 175 metres (*d*) 200 metres

**26.** **If a train overtakes two persons who are walking in the same direction in which the train is going, at the rate of 2 km/hr. and 4 km/hr. and passes them completely in 9 and 10 seconds respectively, the length of the train is :**

(*a*) 45 metres (*b*) 50 metres
(*c*) 54 metres (*d*) 72 metres

**27.** A train travelling at 36 km/hr. completely crosses another train having half its length and travelling in the opposite direction at 54 km/hr. in 12 seconds. If it also passes a railway platform in $1\frac{1}{2}$ minutes, the length of the platform is :
(*a*) 750 metres (*b*) 700 metres
(*c*) 620 metres (*d*) 560 metres

**28.** Two stations *A* and *B* are 110 kms apart on a straight line. One train starts from *A* at 7 A.M. and travels towards *B* at 20 km/hr. speed. Another train starts from *B* at 8 A.M. and travels towards *A* at 25 km/hr. speed. At what time will they meet ?
(*a*) 9 A.M. (*b*) 10 A.M.
(*c*) 11 A.M. (*d*) none of these **(Railways, 1989)**

**29.** Two trains are running in opposite directions with the same speed. If the length of each train is 135 metres and they cross each other in 18 seconds, the speed of each train is :
(*a*) 104 km/hr. (*b*) 27 km/hr.
(*c*) 54 km/hr. (*d*) none of these **(Railways, 1991)**

**30.** A train travels a certain distance without stoppages with an average speed of 90 km/hr.; but with stoppages at an average speed of 80 km/hr. The number of minutes per hour that the train stops, is :
(*a*) $13\frac{1}{3}$ (*b*) 12
(*c*) 8 (*d*) $6\frac{2}{3}$ **(Clerical Grade, 1991)**

## HINTS & SOLUTIONS (Exercise 14B)

**1.** 54 km/hr. $= \left(54 \times \frac{5}{18}\right)$ metres/sec. = 15 m/sec.

**2.** $33\frac{1}{3}$ metres/sec. $= \left(\frac{100}{3} \times \frac{18}{5}\right)$ km/hr. = 120 km/hr.

**3.** Rate of running $= \left(15 \times \frac{5}{18}\right)$ m/sec. $= \frac{25}{6}$ m/sec.

Length of bridge = (speed × time)

$= \left(\frac{25}{6} \times 5 \times 60\right)$ metres = 1250 m.

**4.** Speed of train $= \left(30 \times \frac{5}{18}\right)$ m/sec. $= \frac{25}{3}$ m/sec.

Requisite time $= \left(\frac{150 \times 3}{25}\right)$ sec. = 18 sec.

**5.** Speed of train $= \left(60 \times \frac{5}{18}\right)$ m/sec. $= \frac{50}{3}$ m/sec.

Time taken by the train to cover (220 + 280) metres

$= \left(500 \div \frac{50}{3}\right)$ sec. $= \left(\frac{500 \times 3}{50}\right)$ sec. = 30 sec.

**6.** Distance covered in one second = 12 m.

$\therefore$ Speed = 12 m/sec. $= \left(12 \times \frac{18}{5}\right)$ km/hr = 43.2 km/hr.

**7.** The train covers 150 metres in 10 seconds. So, its speed is 15 metres/sec.

**8.** Speed of train $= \left(36 \times \frac{5}{18}\right)$ metres/sec. = 10 m/sec.

Length of train = (speed × time) = 80 metres.

**9.** Let the length of tunnel be $x$ metres.

Speed of train $= \left(72 \times \frac{5}{18}\right)$ m/sec. = 20 m/sec.

Time taken by the train to cover $(700 + x)$ m $= \left(\frac{700 + x}{20}\right)$ sec.

$\therefore \frac{700 + x}{20} = 60 \Rightarrow x = 500.$

**10.** Let the length of train be $x$ metres. Then,

$\frac{x}{15} = \frac{x + 100}{30} \Rightarrow 30x = 15x + 1500 \Rightarrow x = 100.$

**11.** Speed of train relative to man

$= (60 - 6)$ km/hr. $= \left(54 \times \frac{5}{18}\right)$ m/sec. = 15 m/sec.

Time taken by the train to cross the man $= \left(\frac{120}{15}\right)$ sec. = 8 seconds.

**12.** Speed of train relative to man = 27 km/hr.

$= \left(27 \times \frac{5}{18}\right)$ metres/sec. $= \left(\frac{15}{2}\right)$ metres/sec.

Requisite time $= \left(270 \times \frac{2}{15}\right)$ sec. = 36 sec.

**13.** In 6 seconds, the man walking at the rate of 6 km. per hour, covers 10 metres.

So, the train has to move actually (150 – 10) i.e. 140 metres in 6 seconds to cross the man.

Hence, speed of the train $= \frac{140}{6}$ m/sec.

$$= \left(\frac{140}{3} \times \frac{18}{5}\right) \text{ km/hr.} = 84 \text{ km/hr.}$$

**14.** Speed of train $= \left(\frac{120}{5}\right)$ m/sec. = 24 m/sec.

Time taken to cross the bridge $= \left(\frac{300}{24}\right)$ sec. $= 12\frac{1}{2}$ sec.

**15.** Speed of train = 36 km/hr. $= \left(36 \times \frac{5}{18}\right)$ m/sec = 10 m/sec.

Length of train = (10 × 10) metres = 100 metres .

Sum of lengths of train & platform = 155 metres .

∴ Time taken to cross the platform $= \left(\frac{155}{10}\right)$ sec. $= 15\frac{1}{2}$ seconds.

**16.** Sum of lengths of train and bridge = 400 metres

Speed of train $= \left(\frac{400}{30}\right)$ m/sec. $= \left(\frac{40}{3}\right)$ m/sec.

Time taken to pass a telegraph pole

$$= \left(100 \times \frac{3}{40}\right) \text{ sec.} = 7\frac{1}{2} \text{ seconds.}$$

**17.** Distance covered by the train in crossing the platform

$= \left(\frac{45 \times 60}{3600}\right)$ km. $= \frac{3}{4}$ km. = 750 m.

∴ Length of train = (750 – 100) m. = 650 metres

∴ Time taken to cross the pole $= \left(650 \div \frac{750}{60}\right)$ sec.

$$= \left(650 \times \frac{60}{750}\right) \text{ sec.} = 52 \text{ sec.}$$

**18.** Relative speed of the trains = (48 + 60) km./hr.

$$= \left(108 \times \frac{5}{18}\right) \text{ m/sec.} = 30 \text{ m/sec.}$$

Requisite time $= \left(\frac{90 + 120}{30}\right)$ sec. = 7 sec.

**19.** Let the speed of second train be $x$ km/hr.

Then, relative speed = $(30 + x)$ km/hr.

∴ Time taken to cover (150 + 100) metres at $(30 + x)$ km/hr.

= 10 seconds

$$\therefore \frac{250}{(30 + x) \times \frac{5}{18}} = 10 \Rightarrow \frac{250 \times 18}{150 + 5x} = 10 \Rightarrow x = 60 \text{ km/hr.}$$

**20.** Relative speed = (90 – 60) km/hr. = 30 km/hr.

$$= \left(30 \times \frac{5}{18}\right) \text{ m/sec.} = \frac{25}{3} \text{ m/sec.}$$

Time taken to cover (300 + 200) m at $\frac{25}{3}$ m/sec. = $\left(\frac{500 \times 3}{25}\right)$ sec. = 60 sec.

**21.** The man is moving with the slower train at the rate of 30 km/hr. in the direction of faster train.

Relative speed of faster train w.r.t. man

= (50 – 30) km/hr. = $\left(20 \times \frac{5}{18}\right)$ m/sec. = $\frac{50}{9}$ m/sec.

Time taken to cross the man

= Time taken to cover the length of faster train at relative speed.

$\therefore l \times \frac{9}{50} = 18$ or $l = \left(\frac{50 \times 18}{9}\right) = 100$ metres.

**22.** Speed of faster train relative to slower one

= (46 – 36) km/hr. = $\left(10 \times \frac{5}{18}\right)$ m/sec. = $\frac{25}{9}$ m/sec.

Distance covered in 36 seconds = $\left(\frac{25}{9} \times 36\right)$ metres = 100 metres.

$\therefore$ 2 (length of each train) = 100 metres .

or length of each train = 50 metres.

**23.** $\frac{X\text{'s rate}}{Y\text{'s rate}} = \sqrt{\left(\frac{\text{Time taken by } Y \text{ to reach } A}{\text{Time taken by } X \text{ to reach } B}\right)}$

or $\frac{45}{Y\text{'s rate}} = \sqrt{\left(\frac{10}{3} \times \frac{5}{24}\right)} = \frac{5}{6}$.

$\therefore$ $Y$'s rate = $\left(\frac{45 \times 6}{5}\right)$ km/hr. = 54 km/hr.

**24.** Speed of first train = $\left(\frac{100}{10}\right)$ m/sec. = 10 m/sec.

Let the speed of second train be $x$ m/sec. Then,

$\frac{200}{x + 10} = 8$ or $x = 15$ m/sec. = $\left(15 \times \frac{18}{5}\right)$ km/hr. = 54 km/hr.

**25.** Relative speed of faster train = (40 – 22) km/hr.

= $\left(18 \times \frac{5}{18}\right)$ m/sec. = 5 m/sec.

Now, $\frac{125 + l}{5} = 60$ or $l = (300 - 125) = 175$ metres.

**26.** Let the speed of the train be $x$ km/hr. Then,

$\frac{(x-2) \times 1000}{60 \times 60} \times 9 = \frac{(x-4) \times 1000}{60 \times 60} \times 10$

or $\frac{5x-10}{2}=\frac{25x-100}{9}$ or $x = 22$ km/hr.

$\therefore$ Length of train $= \frac{(5x-10)}{2}$ metres = 50 metres.

**27.** Relative speed = (36 + 54) km/hr. = 90 km/hr.

$= \left(90 \times \frac{5}{18}\right)$ m/sec. = 25 m/sec.

Let $l$ be the length of slower train.

Now, $\frac{\left(l+\frac{l}{2}\right)}{25} = 12$ or $l = 200$ m.

Also, speed of slower train $= \left(36 \times \frac{5}{18}\right)$ m/sec. = 10 m/sec.

Now, if $x$ be the length of platform, then

$\frac{x+200}{10} = 90$ or $x = 700$ metres.

**28.** Suppose they meet $x$ hours after 7 A.M.

Distance travelled by train from $A$ in $x$ hours = $(20x)$ km.

Distance travelled by train from $B$ in $(x-1)$ hours = $25(x-1)$ km.

$\therefore 20x + 25(x-1) = 110$ or $x = 3$ hours. So, the trains meet at 10 A.M.

**29.** Let the speed of each train be $(x)$ m/sec.

Then, $\frac{135+135}{x+x} = 18$.

$\therefore x = \frac{15}{2}$ m/sec. $= \left(\frac{15}{2} \times \frac{18}{5}\right)$ km/hr. = 27 km/hr.

**30.** Let the distance be 720 km. (L.C.M. of 90 and 80)

Time taken to cover it (without stoppages) = 8 hrs.

Time taken to cover it (with stoppages) = 9 hrs.

So, during 9 hours, it stops for 1 hour

During 1 hour, it stops for $\left(\frac{1}{9} \times 60\right)$ min. $= 6\frac{2}{3}$ min.

## ANSWERS (EXERCISE 14B)

| | | | | | |
|---|---|---|---|---|---|
| **1.** (*b*) | **2.** (*c*) | **3.** (*d*) | **4.** (*a*) | **5.** (*c*) | **6.** (*b*) |
| **7.** (*d*) | **8.** (*c*) | **9.** (*d*) | **10.** (*b*) | **11.** (*d*) | **12.** (*a*) |
| **13.** (*b*) | **14.** (*b*) | **15.** (*d*) | **16.** (*b*) | **17.** (*c*) | **18.** (*b*) |
| **19.** (*b*) | **20.** (*a*) | **21.** (*c*) | **22.** (*a*) | **23.** (*b*) | **24.** (*c*) |
| **25.** (*c*) | **26.** (*b*) | **27.** (*b*) | **28.** (*b*) | **29.** (*b*) | **30.** (*d*) |

# 15

# BOATS & STREAMS

**Formulae :**

1. If the speed of a boat in still water be $x$ km/hr. and the speed of the stream be $y$ km/hr., then :

   (i) Speed of boat along the stream (i.e. downstream)
   $= (x + y)$ km/hour.

   (ii) Speed of boat against the stream (i.e. upstream)
   $= (x - y)$ km/hour.

2. If a man rows in still water at $x$ km/hour and the rate of current (or stream) is $y$ km/hour, then

   (i) man's rate with the current $= (x + y)$ km/hr.

   (ii) man's rate against the current $= (x - y)$ km/hr.

3. (i) Rate in still water

   $= \frac{1}{2}$ [(rate with the current) + (rate against the current)]

   (ii) Rate of current

   $= \frac{1}{2}$ [(rate with the current) – (rate against the current)]

**Solved Examples :**

**Ex. 1.** ***A man can row downstream at the rate of 14 km/hr. and upstream at 5 km/hr. Find man's rate in still water and the rate of current.***

**Sol.** *Rate in still water* $= \frac{1}{2}(14 + 5)$ *km/hr.* $= 9.5$ *km/hr.*

*Rate of current* $= \frac{1}{2}(14 - 5)$ *km/hr.* $= 4.5$ *km/hr.*

**Ex. 2.** ***A man rows upstream 16 km. and downstream 27 km. taking 5 hours each time. What is the velocity of current ?***

**Sol.** Man's rate upstream $= (16/5)$ km/hr.

Man's rate downstream $= (27/5)$ km/hr.

*Velocity of current* $= \frac{1}{2}\left(\frac{27}{5} - \frac{16}{5}\right)$ *km/hr.* $= 1.1$ *km/hr.*

**Ex. 3.** ***A man can row 4.5 km/hr. in still water and he finds that it takes him twice as long to row up as to row down the river. Find the rate of stream.***

**Sol.** Clearly, man's rate with the current is twice man's rate against the current. Let man's rate against current be $x$ km/hr. Then, man's rate with the current = $2x$ km/hr.

$\therefore \frac{1}{2}(2x + x) = 4.5$ or $x = 3$.

So, man's rate against current = 3 km/hr.

Man's rate with the current = 6 km/hr.

$\therefore$ *Rate of current* = $\frac{1}{2}(6 - 3)$ *km/hr.* = *1.5 km/hr.*

**Ex. 4.** ***A man can row 5 km/hr. in still water. If the river is running at 1.5 km/hr., it takes him 1 hour to row to a place and back. How far is the place ?***

**Sol.** Man's rate downstream = (5 + 1.5) km/hr. = 6.5 km/hr.

Man's rate upstream = (5 – 1.5) km/hr. = 3.5 km/hr.

Let the distance be $x$ km.

Total time taken to row $x$ km & back = $\left(\frac{x}{6.5} + \frac{x}{3.5}\right)$ hrs.

$\therefore \frac{x}{6.5} + \frac{x}{3.5} = 1$ or $\frac{2x}{13} + \frac{2x}{7} = 1$ or $x = 2.275$ km.

*Hence, the distance is 2 km. 275 metres.*

**Ex. 5.** ***The current of a stream runs at the rate of 2 km per hour. A motor boat goes 10 km. upstream and back again to the starting point in 55 minutes. Find the speed of the motor boat in still water.***

**Sol.** Let the speed of boat in still water be $x$ km/hr.

Speed downstream = $(x + 2)$ km/hr.

speed upstream = $(x - 2)$ km/hr.

Time taken to row 10 kms & back

$$= \left(\frac{10}{(x+2)} + \frac{10}{(x-2)}\right) \text{ hrs.}$$

$\therefore \frac{10}{x+2} + \frac{10}{x-2} = \frac{55}{60}$ or $11x^2 - 240x - 44 = 0$

or $11x^2 - 242x + 2x - 44 = 0$ or $(x - 22)(11x + 2) = 0$

$\therefore x = 22$, since – ve value of $x$ is not admissible.

$\therefore$ Speed of boat in still water = 22 km/hr.

**Ex. 6.** ***A man can row 30 km upstream and 44 km downstream in 10 hours, while he can row 40 km upstream and 55 km downstream in 13 hours. Find the rate of the current and the speed of the man in still water.***

**Sol.** Let the rates upstream and downstream be $x$ km/hr. and $y$ km/hr. respectively. Then,

$$\frac{30}{x} + \frac{44}{y} = 10 \qquad \text{... (i)}$$

$$\frac{40}{x} + \frac{55}{y} = 13 \qquad \text{.. (ii)}$$

Solving (i) and (ii) for $\frac{1}{x}$ and $\frac{1}{y}$, we get :

$\frac{1}{x} = \frac{1}{5}$ and $\frac{1}{y} = \frac{1}{11}$.

$\therefore x = 5$ km/hr. and $y = 11$ km/hr.

*So, rate of current* $= \frac{1}{2}(11 - 5)$ *km/hr.* $= 3$ *km/hr.*

*Rate in still water* $= \frac{1}{2}(11 + 5)$ *km/hr.* $= 8$ *km/hr.*

## EXERCISE 15A

1. A man can row downstream at the rate of 16 km. per hour and upstream at 11 km. per hour. Find man's rate in still water and the rate of currem.

2. A boat moves downstream at the rate of one km. in $7\frac{1}{2}$ minutes and upstream at the rate of 5 km. an hour. What is the velocity of current ?

   **Hint.** *Rate downstream* $= \left(\frac{2}{15} \times 60\right)$ *km/hr.* $= 8$ *km/hr.*

   *Rate upstream* $= 5$ *km/hr.*

   *Velocity of current* $= \frac{1}{2}(8 - 5)$ *km/hr.*

3. A person rows a kilometre down the stream in 10 minutes and uptream in 30 minutes. Find the velocity of the stream.

4. A man can row three quarters of a km. against the stream in 11 minutes 15 seconds and return in 7 minutes 30 seconds. Find the speed of the man in still water and also the speed of the stream.

5. The current of a stream runs at the rate of 4 km. an hour. A boat goes 6 km. and back to its starting point in 2 hours. Find the speed of the boat in still water.

   **Hint.** *Let the speed of the boat in still water be x km/hr.*

   *Then,* $\frac{6}{(x + 4)} + \frac{6}{(x - 4)} = 2$, *find x with positive value.*

6. A man can row 8 km. per hour in still water. If the river is running at 2 km. an hour, it takes him 48 minutes to row to a place and back, how far is the place ?

7. A man rows to a place 48 km. distant and back in 14 hours. He finds that he can row 4 km. with the stream in the same time as 3 km. against the stream. Find the rate of the stream.

**Hint.** *Suppose that the man takes x hours to cover 4 km. down stream and x hours to cover 3 km. upstream.*

*Then,* $\frac{48x}{4}+\frac{48x}{3}=14$ *or* $x=\frac{1}{2}$.

$\therefore$ *Rate upstream = 6 km/hr. & rate downstream = 8 km/hr.*

8. A boat's man goes 48 km. downstream in 8 hours and returns back in 12 hours. Find the speed of the boat in still water and the rate of the stream.

9. A man can row at the rate of 3.5 km. per hour in still water. If the time taken to row a certain distance upstream is $2\frac{1}{2}$ times as much as to row the same distance downstream, find the speed of the current.

10. Fill in the blanks :

(i) A boat moves with a speed of 11 km. per hour along the stream and 7 km. per hour against the stream. The rate of the stream is (......) km/hr.

(ii) The speed of a boat in still water is 12 km. per hour. Going downstream it moves at the rate of 19 km. per hour. The speed of the boat against the stream is (......) km/hr.

(iii) A man can row 15 km. downstream in 3 hours and 5 km. upstream in $2\frac{1}{2}$ hours. His speed in still water is (......) km/hr.

(iv) A man rows upstream 11 km. and downstream 26 km. taking 5 hours each time. The velocity of the current is (......) km/hr.

(v) A man can row 5 km. per hour in still water. If the river is running at 1 km. an hour, it takes him 75 minutes to row to a place and back. The place is at a distance of (......) km. from the starting point.

## ANSWERS (EXERCISE 15A)

**1.** 13.5 km/hr., 2.5 km/hr. **2.** 1.5 km/hr. **3.** 2 km/hr.
**4.** 5 km/hr., 1 km/hr. **5.** 8 km/hr. **6.** 3 km
**7.** 1 km/hr. **8.** 5 km/hr, 1 km/hr. **9.** 1.5 km/hr.
**10.** (i) 2 km/hr. (ii) 5 km/hr. (iii) 3.5 km/hr.
(iv) 1.5 km/hr. (v) 3 km

## EXERCISE 15B (Objective Type Questions)

1. A man can row with the stream at 10 km/hr. and against the stream at 5 km/hr. Man's rate in still water is :

(*a*) 5 km/hr. (*b*) 2.5 km/hr.

(c) 7.5 km/hr. (d) 15 km/hr.

2. The speed of a boat in still water is 2 km/hr. If its speed upstream be 1 km/hr., then speed of the stream is :
(a) 2 km/hr. (b) 3 km/hr.
(c) 1 km/hr. (d) none of these
(Astt. Grade. 1987)

3. The speed of a boat in still water is 10 km./hr. If its speed downstream be 13 km/hr., then speed of the stream is :
(a) 1.5 km/hr. (b) 3 km/hr.
(c) 11.5 km/hr. (d) 5.75 km/hr.

4. A boat moves downstream at the rate of 1 km in 6 minutes and upstream at the rate of 1 km in 10 minutes. The speed of the current is :
(a) 2 km/hr. (b) 1 km/hr.
(c) 1.5 km/hr. (d) 2.5 km/hr.

5. The speed of a boat in still water is 15 km/hr. and the rate of current is 3 km/hr. The distance travelled downstream in 12 minutes is :
(a) 3.6 km (b) 2.4 km
(c) 1.2 km (d) 1.8 km

6. A man rows upstream 11 km. and downstream 26 km taking 5 hours each time. The velocity of the current is :
(a) 7.5 km/hr. (b) 2.5 km/hr.
(c) 1.5 km/hr. (d) 1 km/hr.

7. A boat goes 14 km. upstream in 56 minutes. The speed of stream is 2 km/hr. The speed of boat in still water is :
(a) 6 km/hr. (b) 15 km/hr.
(c) 14 km/hr. (d) 17 km/hr.

8. The speed of a boat downstream is 15 km/hr. and the speed of the stream is 1.5 km/hr. The speed of the boat upstream is :
(a) 13.5 km/hr. (b) 16.5 km/hr.
(c) 12 km/hr. (d) 8.25 km/hr.

9. A boat goes 40 km upstream in 8 hours and a distance of 36 km downstream in 6 hours. The speed of the boat in standing water is :
(a) 6.5 km/hr. (b) 6 km/hr.
(c) 5.5 km/hr. (d) 5 km/hr.

10. If a man rows at the rate of 5 km/hr. in still water and his rate against the current is 3.5 km/hr., then the man's rate along the current is :
(a) 8.5 km/hr. (b) 6.5 km/hr.
(c) 6 km/hr. (d) 4.25 km/hr.

11. If a man's rate with the current is 12 km/hr. and the rate of current is 1.5 km/hr., then the man's rate against the current is :
(a) 9 km/hr. (b) 6.75 km/hr.

(*c*) 5.25 km/hr. (*d*) 7.5 km/hr.

**12.** Speed of a boat in standing water is 7 km/hr. and the speed of the stream is 1.5 km/hr. A distance of 7.7 km, going upstream is covered in

(*a*) 1 hr. 15 minutes (*b*) 1 hr. 12 minutes

(*c*) 1 hr. 24 minutes (*d*) 2 hr. 6 minutes

**13.** A man can row 5 km/hr. in still water. If the river is running at 1 km/hr., it takes him 1 hour to row to a place and back. How far is the place ?

(*a*) 2.5 km (*b*) 2.4 km

(*c*) 3 km (*d*) 3.6 km

**14.** A boat travels upstream from *B* to *A* and downstream from *A* to *B* in 3 hours. If the speed of boat in still water is 9 km./hr. and the speed of current is 3 km/hr., the distance between *A* and *B* (in km.) is :

(*a*) 4 (*b*) 6

(*c*) 8 (*d*) 12 **(BSRB Exam. 1990)**

## HINTS & SOLUTIONS (Exercise 15B)

**1.** Man's rate in still water

$= \frac{1}{2}$ [man's rate with current + man's rate against current]

$$= \frac{1}{2}(10 + 5) \text{ km./hr.} = 7.5 \text{ km/hr.}$$

**2.** Speed of stream = (speed in still water) – (speed upstream)

$= (2 - 1)$ km/hr. = 1 km/hr.

**3.** Speed of stream = (speed downstream) – (speed in still water)

$= (13 - 10)$ km/hr. = 3 km/hr.

**4.** Speed downstream $= \frac{1}{6}$ km/minute = 10 km/hr.

Speed upstream $= \frac{1}{10}$ km/minute = 6 km/hr.

Speed of the current $= \frac{1}{2}$ [(speed downstream) – (speed upstream)]

$$= \frac{1}{2}[(10 - 6)] \text{ km/hr.} = 2 \text{ km/hr.}$$

**5.** Speed downstream = 18 km/hr.

∴ Distance travelled in 12 minutes $= \left(\frac{18}{20} \times 12\right)$ km = 3.6 km.

**6.** Speed upstream $= \frac{11}{5}$ km/hr.

Speed downstream $= \frac{26}{5}$ km/hr.

**Velocity of current** $= \frac{1}{2}$ **[(speed downstream) – (speed upstream)]**

$= \frac{1}{2}\left(\frac{26}{5} - \frac{11}{5}\right)$ **km/hr. = 1.5 km/hr.**

**7.** **Speed upstream** $= \left(\frac{14}{56} \times 60\right)$ **km/hr. = 15 km/hr.**

**Speed in still water = (speed upstream) + (speed of stream)**

**= (15 + 2) km/hr. = 17 km/hr.**

**8.** **Speed in still water = (speed downstream) – (speed of stream)**

**= (15 – 1.5) km/hr. = 13.5 km/hr.**

**∴ Speed upstream = (13.5 – 1.5) km/hr. = 12 km/hr.**

**9.** **Speed upstream = (40 ÷ 8) km/hr. = 5 km/hr.**

**Speed downstream = (36 ÷ 6) km/hr. = 6 km/hr.**

**Speed of boat in standing water** $= \frac{1}{2}(5 + 6)$ **km/hr. = 5.5 km/hr.**

**10.** **Let the man's rate along the current be** $x$ **km/hr.**

**Then,** $5 = \frac{1}{2}(x + 3.5)$ **or** $x + 3.5 = 10$ **or** $x = 6.5$ **km/hr.**

**11.** **Let the man's rate against the current be** $x$ **km/hr.**

$\therefore 1.5 = \frac{1}{2}(12 - x)$ **or** $12 - x = 3$ **or** $x = 9$ **km/hr.**

**12.** **Speed upstream = (7 – 1.5) km/hr. = 5.5 km/hr.**

**Time taken to cover 7.7 km.** $= \frac{7.7}{5.5}$ **hr. = 1 hr. 24 minutes.**

**13.** **Man's rate downstream = 6 km/hr.**

**Man's rate upstream = 4 km/hr.**

**Let the distance be** $x$ **km.**

**Then,** $\frac{x}{6} + \frac{x}{4} = 1$ **or** $x = 2.4$ **km.**

**14.** **Speed downstream = (9 + 3) km/hr. = 12 km/hr.**

**Speed upstream = (9 – 3) km/hr. = 6 km/hr.**

**Suppose distance between** $A$ **and** $B = x$ **km.**

$\therefore \frac{x}{6} + \frac{x}{12} = 3$ **or** $x = 12.$

## ANSWERS (EXERCISE 15B)

| | | | | | |
|---|---|---|---|---|---|
| **1.** (*c*) | **2.** (*c*) | **3.** (*b*) | **4.** (*a*) | **5.** (*a*) | **6.** (*c*) |
| **7.** (*d*) | **8.** (*c*) | **9.** (*c*) | **10.** (*b*) | **11.** (*a*) | **12.** (*c*) |
| **13.** (*b*) | **14.** (*d*) | | | | |

# 16

# PROFIT & LOSS

The Cost Price (abbreviated as C.P.) of an article is the price or the money for which the article has been purchased.

The selling price (abbreviated as S.P.) of an article is the price or the money for which the article has been sold.

If S.P. is greater than the C.P., then the seller is said to have a profit or gain. On the other hand, if the C.P. is greater than the S.P. then the seller has a loss.

**Remarks.**

(i) Loss & Gain are always reckoned on cost price.

(ii) Gain or loss on Rs. 100 is called gain or loss percent respectively.

**Formulae.**

(i) Gain = (S.P.) – (C.P.); Loss = (C.P.) – (S.P.);

(ii) $\text{Gain\%} = \dfrac{\text{Gain} \times 100}{\text{C.P.}}$; $\text{Loss\%} = \dfrac{\text{Loss} \times 100}{\text{C.P.}}$.

(iii) When the cost price and gain percent are given :–

$$\text{S.P.} = \left(\frac{100 + \text{Gain\%}}{100}\right) \times (\text{C.P.})$$

(iv) When the cost price & loss percent are given :–

$$\text{S.P.} = \left(\frac{100 - \text{Loss\%}}{100}\right) \times (\text{C.P.})$$

(v) When the selling price & gain percent are given :–

$$\text{C.P.} = \frac{100}{(100 + \text{Gain\%})} \times (\text{S.P.})$$

(vi) When the selling price & loss percent are given :–

$$\text{C.P.} = \frac{100}{(100 - \text{Loss\%})} \times (\text{S.P.})$$

**Solved Problems.**

**Ex. 1.**

(i) Given : C.P. = Rs. 275 & S.P. = Rs. 286. Find gain %.

(ii) Given : C.P. = Rs. 490 & S.P. = 465.50. Find loss %.

(iii) Given : C.P. = Rs. 470 & Gain = 10%. Find S.P.

(iv) Given : S.P. = Rs. 585 & Loss = 10%. Find C.P.

**Sol.** (i) Gain = (S.P.) – (C.P.) = Rs. (286 – 275) = Rs. 11.

$$\text{Gain\%} = \left(\frac{11}{275} \times 100\right)\% = 4\%.$$

(ii) Loss = (C.P.) – (S.P.) = Rs. (490 – 465.50) = Rs. 24.50.

$$\therefore \text{Loss\%} = \left(\frac{24.50}{490} \times 100\right)\% = 5\%.$$

$$\text{(iii) S.P.} = \frac{(100 + \text{Gain\%})}{100} \times (\text{C.P.})$$

$$= \text{Rs.} \left(\frac{110}{100} \times 470\right) = \text{Rs. } 517.$$

$$\text{(iv) C.P.} = \frac{100}{(100 - \text{Loss\%})} \times (\text{S.P.})$$

$$= \text{Rs.} \left(\frac{100}{90} \times 585\right) = \text{Rs. } 650.$$

**Ex. 2.** ***By selling a watch for Rs. 144 a man loses 10%. At what price should he sell it in order to gain 15% ?***

**Sol.** S.P. = Rs. 144, Loss = 10%.

$$\therefore \text{C.P.} = \text{Rs.} \left[\frac{100}{(100 - 10)} \times 144\right] = \text{Rs.} \left(\frac{100}{90} \times 144\right) = \text{Rs. } 160.$$

Gain Required = 15%.

$$\therefore S.P. = Rs. \left(\frac{115}{100} \times 160\right) = Rs.\ 184.$$

**Ex. 3.** ***A man sold two horses for Rs.924 each. On one he gains 12% and on another he loses 12%. How much does he gain or lose in the whole transaction ?***

**Sol.** S.P. of one horse = Rs. 924

Gain = 12%

∴ C.P. of this horse

$$= \text{Rs.} \left(\frac{100}{112} \times 924\right) = \text{Rs. } 825$$

S.P. of another horse = Rs. 924.

Loss = 12%

∴ C.P. of this horse

$$= \text{Rs.} \left(\frac{100}{88} \times 924\right) = \text{Rs. } 1050$$

Total C.P. of both the horses = Rs. (825 + 1050) = Rs. 1875.

Total S.P. of both the horses = Rs. (2 × 924) = Rs. 1848.

$$\therefore Loss\% = \left(\frac{27}{1875} \times 100\right)\% = 1.44\%.$$

**Ex. 4.** ***If the selling price of 10 articles is the same as the cost price of 11 articles, then find gain percent.***

**Sol.** Let the C.P. of 1 article be Re. 1.

Then, C.P. of 10 articles = Rs. 10.

C.P. of 11 articles = Rs. 11.

$\therefore$ S.P. of 10 articles = C.P. of 11 articles = Rs. 11.

Thus, C.P. = Rs. 10, S.P. = Rs. 11 and so, gain = Re. 1.

$\therefore$ *Gain%* $= \left(\frac{1}{10} \times 100\right)\% = 10\%.$

**Ex. 5.**

(i) A vendor sells 10 oranges for a rupee gaining thereby 40%. How many oranges did he buy for a rupee ?

(ii) A vendor bought oranges at 6 for a rupee. How many for a rupee must he sell to gain 20% ?

(iii) A vendor bought a number of oranges at 6 for a rupee and sold them at 4 for a rupee. Find his gain percent. If his total gain is Rs. 26, find the number of oranges, he bought.

**Sol.** (i) S.P. of 10 oranges = Re. 1, Gain = 40%.

$\therefore$ C.P. of 10 oranges = Rs. $\left(\frac{100}{140} \times 1\right)$ = Rs. $\frac{5}{7}$.

Now, Rs. $\frac{5}{7}$ is C.P. of 10 oranges.

$\therefore$ Re. 1 is C.P. of $\frac{10 \times 7}{5}$ or 14 oranges.

***So, he bought 14 oranges for a rupee.***

(ii) C.P. of 6 oranges = Re. 1, Gain = 20%

$\therefore$ S.P. of 6 oranges = Rs. $\left(\frac{120}{100} \times 1\right)$ = Rs. $\frac{6}{5}$.

Now, Rs. $\frac{6}{5}$ is S.P. of 6 oranges.

$\therefore$ Re. 1 is S.P. of $\left(\frac{6 \times 5}{6}\right)$ or 5 oranges.

***So, he must sell them at 5 for a rupee.***

(iii) Suppose he bought 12 (L.C.M. of 6 & 4) oranges.

C.P. of 12 oranges = Rs. $\left(\frac{1}{6} \times 12\right)$ = Rs. 2.

S.P. of 12 oranges = Rs. $\left(\frac{1}{4} \times 12\right)$ = Rs. 3.

Gain% $= \left(\frac{1}{2} \times 100\right)\% = 50\%.$

Now, Gain on 12 oranges = Rs. (3 – 2) = Re. 1.

$\therefore$ If the gain is Re. 1, he bought = 12 oranges.

*If the gain is Rs. 26, he bought* = (12 × 26) *or 312 oranges.*

**Ex. 6.** ***A man sells an article at a profit of 20%. If he had bought it at 20% less and sold it for Rs. 5 less, he would have gained 25%. Find the cost price of the article.***

**Sol.** Let the C.P. be Rs. 100.

Then, gain = 20%. So, S.P. = Rs. 120.

New C.P. = 20% less = Rs. 80.

If gain is 25%, then S.P. = Rs. $\left(\frac{125}{100} \times 80\right)$ = Rs. 100.

Difference in S.P. = Rs. (120 – 100) = Rs. 20.

∴ if difference in S.P. is Rs. 20, then C.P. = Rs. 100.

*and if difference in S.P. is Rs. 5, then C.P.*

$$= Rs. \left(\frac{100}{20} \times 5\right) = Rs.\ 25.$$

**Ex. 7.** ***A manufacturer sells goods to an agent at a profit of 20%. The agent's whole sale price to a shopkeeper is at a profit of 10% and the shopkeeper retails his goods at a profit of*** $12\frac{1}{2}$ ***%. Find the manufacturing cost of goods bought from the shop for Rs. 14.85.***

**Sol.** Let the manufacturing cost be Rs. 100.

Then, agent's C.P. = Rs. 120.

Shopkeeper's C.P. = Agent's S.P.

$$= \text{Rs.} \left(\frac{110}{100} \times 120\right) = \text{Rs. } 132.$$

Shopkeeper's S.P. = Rs. $\left(\frac{225}{2 \times 100} \times 132\right)$ = Rs. $\frac{297}{2}$.

If retail price is Rs. $\frac{297}{2}$, manufacturing cost = Rs. 100.

*If retail price is Rs. 14.85, manufacturing cost*

$$= Rs. \left(\frac{100 \times 2}{297} \times 14.85\right) = Rs.\ 10.$$

**Ex. 8.** ***A tradesman marks his goods at such a price that after allowing a discount of*** $12\frac{1}{2}$ ***% for cash, he makes a profit of 20%. What is the marked price of an article which costs him Rs. 210 ?***

**Sol.** Let the marked price be Rs. 100.

S.P. = (Marked price) – (Discount)

$$= \text{Rs.} \left(100 - \frac{25}{2}\right) = \text{Rs.} \left(\frac{175}{2}\right).$$

Profit made = 20%.

$\therefore$ C.P. = Rs. $\left(\frac{100}{120} \times \frac{175}{2}\right)$ = Rs. $\left(\frac{875}{12}\right)$

If C.P. is Rs. $\left(\frac{875}{12}\right)$, marked price = Rs. 100.

If C.P. is Rs. 210, marked price

$= Rs. \left(\frac{100 \times 12}{875} \times 210\right) = Rs. 288.$

**Ex. 9.** ***A radio dealer marks a radio with a price which is 20% more than the cost price and allows a discount of 10% on cash. Find his gain percent.***

**Sol.** Let C.P. be Rs. 100.

Then, marked price = Rs. 120.

Discount allowed = Rs. $\left(\frac{10}{100} \times 120\right)$ = Rs. 12.

S.P. of radio = Rs. (120 – 12) = Rs. 108.

*Gain% = (108 – 100) = 8%.*

**Ex. 10.** ***A reduction of 20% in the price of sugar enables a purchaser to obtain 8 kg. more for Rs. 160. What is the reduced price per kg. ? What was the price per kg. before reduction ?***

**Sol.** Let the original rate be Rs. $x$ per kg.

Quantity bought for Rs. 160 = $(160/x)$ kg.

Reduced rate per kg. = $\left(\frac{80}{100} x\right)$ = Rs. $\frac{4x}{5}$

Quantity bought for Rs. 160 = $\left(\frac{160 \times 5}{4x}\right)$ kg. = $\left(\frac{200}{x}\right)$ kg.

$\therefore \left(\frac{200}{x} - \frac{160}{x}\right) = 8$ or $x = 5$.

So, original rate = Rs. 5 per kg.

*Reduced rate* = *Rs.* $\left(\frac{4 \times 5}{5}\right)$ = *Rs. 4 per kg.*

**Ex. 11.** ***A dishonest dealer professes to sell his goods at cost price, but he uses a weight of 960 gm. for a kg. Find his gain percent.***

**Sol.** Let the cost price of 1 gm. be Re. 1.

Then, C.P. of 960 gm. = Rs. 960

S.P. of 960 gm. = C.P. of 1 kg. = Rs. 1000

$\therefore$ Profit on Rs. 960 = Rs. (1000 – 960) = Rs. 40

*Hence, gain%* = $\left(\frac{40}{960} \times 100\right)\% = 4\frac{1}{6}\%$

**Ex. 12.** ***A tradesman bought 500 metres of electric wire at 75 paise per metre. He sold 60% of it at a profit of 8%. At what gain percent should he sell the remainder so as to gain 12% on the whole transaction ?***

**Sol.** C.P. of wire = Rs. $\left(\frac{500 \times 75}{100}\right)$ = Rs. 375

Now, 60% of 500 metres = $\left(\frac{60}{100} \times 500\right)$ = 300 metres.

C.P. of 300 metres of wire = Rs. $\left(\frac{300 \times 75}{100}\right)$ = Rs. 225

Profit made on this wire = 8%

$\therefore$ S.P. of 300 metres of wire = Rs. $\left(\frac{108}{100} \times 225\right)$ = Rs. 243

But, gain desired on whole transaction = 12%

$\therefore$ S.P. of whole must be = Rs. $\left(\frac{112}{100} \times 375\right)$ = Rs. 420.

So, the S.P. of remaining 200 metres = Rs. (420 – 243) = Rs. 177

Also, C.P. of 200 metres = Rs. $\left(\frac{200 \times 75}{100}\right)$ = Rs. 150

$\therefore$ *Gain% on this wire* = $\left(\frac{27}{150} \times 100\right)$ = *18%.*

**Ex. 13.** ***A merchant buys 200 kg. of rice at Rs. 7.25 per kg., 400 kg. of rice at Rs. 5.75 per kg. He mixes them and sells one third of the mixture at Rs. 6 per kg. At what rate should he sell the remaining mixture so that he may earn a profit of 20% on the whole outlay ?***

**Sol.** C.P. of 200 kg. of rice = Rs. (7.25 × 200) = Rs. 1450

C.P. of 400 kg. of rice = Rs. (5.75 × 400) = Rs. 2300

Total C.P. of 600 kg. of mix. = Rs. (1450 + 2300)

= Rs. 3750

Profit required on whole outlay= 20%

$\therefore$ S.P. must be = Rs. $\left(\frac{120}{100} \times 3750\right)$ = Rs. 4500

Now, one third of mixture = $\left(\frac{1}{3} \times 600\right)$ = 200 kg.

S.P. of this 200 kg. of mix. = Rs. (6 × 200) = Rs. 1200.

S.P. of remaining 400 kg. = Rs. (4500 – 1200) = Rs. 3300

$\therefore$ Rate of selling the remaining = Rs. $\left(\frac{3300}{400}\right)$

= Rs. 8.25 per kg.

**Ex. 14.** ***A merchant bought 200 eggs, out of which 38 were broken. He sold the remaining eggs at the rate of Rs. 4.80 per dozen and thus gained 8%. Find his total investment.***

**Sol.** Number of saleable eggs = (200 – 38) = 162.

S.P. of 162 eggs = Rs. $\left(\frac{4.80}{12} \times 162\right)$ = Rs. 64.80

Profit made = 8%

$\therefore$ C.P. = Rs. $\left(\frac{100}{108} \times 64.80\right)$ = Rs. 60

*Hence, the total investment = Rs. 60.*

**Ex. 15.** ***A man bought two cows for Rs. 1040. He sold one at a loss of 15% and other at a profit of 36% and then he found that each cow was sold for the same price. Find the cost price of each cow.***

**Sol.** Let the C.P. of one cow be Rs. $x$.

Then, C.P. of another cow = Rs. $(1040 - x)$.

Now, C.P. of Ist cow = Rs. $x$

Loss = 15%

$\therefore$ S.P. = Rs. $\left(\frac{85}{100} \times x\right)$

= Rs. $\left(\frac{17x}{20}\right)$

C.P. of IInd cow

= Rs. $(1040 - x)$

Profit = 36%

$\therefore$ S.P. = Rs. $\left[\frac{136}{100} \times (1040 - x)\right]$

Since the cows were sold for the same price, so

$\frac{17x}{20} = \left[\frac{136}{100} \times (1040 - x)\right]$ or $x = 640$.

$\therefore$ C.P. of one cow = Rs. 640

*C.P. of another cow = Rs. (1040 – 640) = Rs. 400.*

**Ex. 16.** ***A cycle dealer buys 30 bicycles, of which 8 are first grade and the rest are second grade, for Rs. 4725. Find at what price must he sell the first grade bicycles, so that if he sells the second grade bicycles at three quarter of this price, he makes a profit of 40% on his outlay.***

**Sol.** C.P. of all the bicycles = Rs. 4725 & Profit made = 40%

$\therefore$ S.P. of all the bicycles = $\left(\frac{140}{100} \times 4725\right)$

= Rs. 6615

Let the selling rate of first grade bicycles be Rs. $x$.

Then, the selling rate of second grade bicycles = Rs. $\left(\frac{3x}{4}\right)$

$\therefore$ S.P. of 8 first grade bicycles = Rs. 8 $x$

S.P. of 22 second grade bicycles = Rs. $\left(\frac{3x}{4}\times 22\right)$ = Rs. $\left(\frac{33x}{2}\right)$

$\therefore \left(8x+\frac{33x}{2}\right) = 6615$ or $x = 270$

*Hence, the rate of first grade bicycles = Rs. 270 per bicycle.*

**Ex. 17.** ***A shopkeeper uses a false weighing balance. While purchasing, he deceives the seller by 20% and while selling he deceives the customer by 20%. Find his gain percent.***

**Sol.** Clearly, he invests Rs 100 and purchases goods worth of Rs. 120.

Again, goods worth of Rs. 100 are sold by him in Rs. 120.

$\therefore$ Goods worth of Rs. 120 will be sold by him in

Rs. $\left(\frac{120}{100}\times 120\right)$ = Rs. 144.

Thus, investment = Rs. 100 & output = Rs. 144.

$\therefore$ *Gain = 44%.*

**Ex. 18.** ***If an article is sold at a gain of 6% instead of being sold at a loss of 6%, one gets Rs. 6 more. Find the cost price of the article.***

**Sol.** Let the cost price of the article be Rs. 100.

S.P. when sold at 6% loss = Rs. 94

S.P. when sold at 6% gain = Rs. 104

Difference of two selling prices = Rs. (104 – 94) = Rs. 10.

Now, if the difference of two S.P.'s is Rs. 10, C.P. = Rs. 100

If the difference of two S.P.'s is Rs. 6, C.P. = $\left(\frac{100}{10}\times 6\right)$ = Rs. 60.

## EXERCISE 16A (Subjective)

1. Find the unknown term :
   (i) Cost price = Rs. 48, Gain = 25%, Selling price = ?
   (ii) Selling price = Rs. 99, Loss = 10%, Cost price = ?
2. If cost price of 10 articles is equal to the selling price of 9 articles, then find gain percent.
3. By selling a watch for Rs. 1140, a man loses 5%. For how much should he sell it to gain 5% ?

**Hint.** *C.P. of watch = Rs.* $\left(\frac{100}{95}\times 1140\right)$ *= Rs. 1200*.

$\therefore$ *S.P. for a gain of 5% = Rs.* $\left(\frac{105}{100}\times 1200\right)$ *= Rs. 1260.*

4. A bicycle is sold at a gain of 16%. If it were sold for Rs. 20 more, there would have been a gain of 20%. Find the cost price of the bicycle.
**Hint.** *Let C.P. be Rs. 100.*
*Then, S.P. for a gain of 16% = Rs. 116 .*
*and S.P. for a gain of 20% = Rs. 120 :*
*Now if the difference in S.P.'s is Rs. 4, C.P. = Rs. 100 .*
*And, if the difference in S.P.'s is Rs. 20, C.P.*

$$= Rs. \left(\frac{100}{4} \times 20\right) = Rs. 500.$$

5. If an article is sold at a certain price one gains 10%.
What will be the gain percent, if sold for double the price ?
**Hint.** *Let the C.P. be Rs. 100*
*Then, S.P. for a gain of 10% = Rs. 110 .*
*S.P. for double the price = Rs. 220 and so gain = 120%.*

6. If I sell a horse for Rs. 900 and a goat for Rs. 200, I gain 10% on the original cost of both, but, if the horse is sold for Rs. 750 and the goat for its original cost, I lose 10%. Find the original cost of both.
**Hint.** *Total S.P. of both = Rs. 1100, profit made = 10% .*

$$\therefore \text{ Total C.P. of both} = Rs. \left(\frac{100}{110} \times 1100\right) = Rs. 1000 .$$

*Loss in second case = 10% .*

$$\text{Total S.P. in this case} = Rs. \left(\frac{90}{100} \times 1000\right) = Rs. 900 .$$

$\therefore$ *C.P. of goat = Rs. (900 – 750) = Rs. 150.*

7. How much percent above the cost price should a shopkeeper mark his goods so that after allowing a discount of 10%, he gains 26% ?
**Hint.** *If C.P. is Rs. 100, S.P. must be Rs. 126.*
*Now, if Rs. 90 is the S.P., then marked price = Rs. 100*
*and if Rs. 126 is the S.P., then marked price*

$$= Rs. \left(\frac{100}{90} \times 126\right) = Rs. 140.$$

8. A fruit seller lost 4% by selling oranges at the rate of 21 for a rupee. How many for a rupee should he sell to gain 12% ?

9. By selling 200 oranges, a man gains the selling price of 40 oranges. Find his gain percent.
**Hint.** *S.P. of 160 oranges = C.P. of 200 oranges.*

10. The catalogue price of a radio is Rs. 720. If it is sold at a discount of $16\frac{2}{3}\%$ of the catalogue price, the gain is 25%. If it is sold for Rs. 160 below the catalogue price, find the gain or loss percent.

**11.** A merchant purchased 80 metres of silk at Rs. 5.50 per metre and sold three fourth of it at a gain of 6%. At what gain percent should he sell the remainder to gain 10% on the whole ?

**12.** A man purchased a horse and a carriage. If he sells the horse at 10% loss and carriage at 20% profit, he neither gains nor loses. If he sells the horse at 5% gain and carriage at 5% loss, he loses Rs. 10 in all. Find the cost price of the horse and the carriage.

**Hint.** *Let the cost price of the horse and the carriage be Rs. x and Rs. y respectively. Then,*

$$(x+y)=\left(\frac{90x}{100}+\frac{120y}{100}\right) \text{ and } \left(\frac{105x}{100}+\frac{95y}{100}+10\right)=(x+y).$$

*Find x and y.*

**13.** A man buys eggs at 5 for a rupee and an equal number at 4 for a rupee. If he sells all the eggs at a uniform rate of 28 paise per egg, find the number of eggs, he must sell to earn a gain of Rs. 11 ?

**Hint.** *Suppose he buys $5 \times 4$ or 20 eggs. Find the gain by selling these 20 eggs. This comes out to be Rs. 1.10.*

*Now to gain Rs. 1.10, he sells 20 eggs.*

$\therefore$ *To gain Rs. 11, he would sell* $\left(\frac{20}{1.10}\times 11\right)$ *eggs.*

**14.** By selling an article for Rs. 9, a person loses as much percent as is its cost price. Find the cost price of the article.

**Hint.** *Let cost price be Rs. x. Then* $\frac{(x-9)\,100}{x}=x$ *Find x.*

**15.** A man sold a horse at a loss of 7%. Had he been able to sell it at a gain of 9%, it would have fetched Rs. 128 more. What was the cost price ?

**16.** A merchant professes to lose 4% on coffee, but he uses a weight equal to 840 gms., instead of 1 kg. Find his real loss or gain %.

**Hint.** *Let C.P. of 1 kg. be Rs. 100.*

*Now if C.P. is Rs. 84, then S.P. = Rs. 96.*

**17.** A merchant buys goods at a uniform price and sells 80% of them at a profit of 15% and the remainder at a loss of 5%. His total profit is Rs. 9350. Find, what did he pay for the goods.

**18.** A merchant imported 5 type writers for Rs. 23000 and paid 10% as excise duty. He marks a price in his catalogue and after deducting a discount of 8%, he gains 12%. What is the marked price of each type-writer ?

**Hint.** *C.P. of each type-writer = Rs. (4600 + 460) = Rs. 5060 .*

*S.P. for a gain of 12% = Rs. 5667.20*

$\therefore$ *Marked price* $= Rs. \left(\frac{100}{92} \times 5667.20\right) = Rs. 6160.$

**19.** Ramu purchased 150 quintals of wheat. One-fourth of the total quantity he sold at a loss of 10%. At what gain percent should he sell the remaining wheat to gain 10% on the whole transaction ?

**Hint.** *Let C.P. of 150 quintals be Rs. 150.*

*Then, C.P. of* $\frac{1}{4}$ *of the total quantity* $= Rs. \left(\frac{150}{4}\right) = Rs. 37.50.$

*S.P. of this quantity at 10% loss* $= Rs. \left(\frac{90}{100} \times 37.50\right) = Rs. 33.75$

*Required S.P. of the whole* $= Rs. \left(\frac{110}{100} \times 150\right) = Rs. 165$ .

$\therefore$ *S.P. of* $\frac{3}{4}$ *of the total quantity* $= Rs. (165 - 33.75) = Rs. (131.25)$ .

*C.P. of this quantity* $= Rs. \left(\frac{3}{4} \times 150\right) = Rs. 112.50$

$\therefore$ *Gain%* $= \left(\frac{18.75}{112.50} \times 100\right) = 16\frac{2}{3}\%.$

## ANSWERS (Exercise 16A)

| | | |
|---|---|---|
| **1.** (i) Rs. 60 (ii) Rs. 110 | **2.** $11\frac{1}{9}\%$ | **3.** Rs. 1260 |
| **4.** Rs. 500 | **5.** 120% | **6.** Rs. 850, Rs. 150 |
| **7.** Rs. 140 | **8.** 18 | **9.** 25% |
| **10.** $16\frac{2}{3}\%$ | **11.** 22% | **12.** Rs. 400, Rs. 200 |
| **13.** 200 | **14.** Rs.10 or Rs. 90 | **15.** Rs. 800 |
| **16.** $14\frac{2}{7}\%$ | **17.** Rs. 85000 | **18.** Rs. 6160 |
| **19.** $16\frac{2}{3}\%$ | | |

## EXERCISE 16B (Objective Type Questions)

**1.** By selling an article for Rs. 100, one gains Rs. 10. Then, the gain percent is :

(*a*) 10% (*b*) 9%

(*c*) $11\frac{1}{9}\%$ (*d*) none of these

**2.** By selling an article for Rs. 100, one loses Rs. 10. Then, the loss percent is :

(*a*) $9\frac{1}{11}\%$ (*b*) 10%

(*c*) $11\frac{1}{9}\%$ (*d*) none of these

**3.** A man sold his horse for Rs. 1980 and gained 10%. The horse was bought for :

(*a*) Rs. 1800 (*b*) Rs. 1782

(*c*) Rs. 2178 (*d*) none of these

**4.** A loss of 5% was suffered by selling a plot for Rs. 4,085. The cost price of the plot was :

(*a*) Rs. 4350 (*b*) Rs. 4259.25

(*c*) Rs. 4200 (*d*) Rs. 4300 **(RRB Exam. 1991)**

**5.** On selling an article for Rs. 240, a trader loses 4%. In order to gain 10%, he must sell that article for :

(*a*) Rs. 264.00 (*b*) Rs. 273.20

(*c*) Rs. 275.00 (*d*) Rs. 280.00

**(N.D.A. Exam. 1990)**

**6.** A man purchased a watch for Rs. 400 and sold it at a gain of 20% of the selling price. The selling price of the watch is :

(*a*) Rs. 300 (*b*) Rs. 320

(*c*) Rs. 440 (*d*) Rs. 500

**(Clerks' Grade Exam. 1990)**

**7.** By selling a table for Rs. 30 instead of Rs. 40, 5% more is lost. The cost of the table is :

(*a*) Rs. 250 (*b*) Rs. 210

(*c*) Rs. 200 (*d*) Rs. 225 **(RRB Exam. 1991)**

**8.** If 5% more is gained by selling an article for Rs. 350 than by selling it for Rs. 340, the cost of the article is :

(*a*) Rs. 50 (*b*) Rs. 160

(*c*) Rs. 200 (*d*) Rs. 225

**(Clerks' Grade Exam. 1991)**

**9.** The selling price of 12 articles is equal to the cost price of 15 articles. The gain percent is :

(*a*) 25% (*b*) 20%

(*c*) 80% (*d*) $6\frac{2}{3}\%$

**10.** If the cost price of 15 tables be equal to the selling price of 20 tables, the loss percent is :

(*a*) 25% (*b*) 37.5%

(*c*) 35% (*d*) 20%

**11.** By selling 36 bananas, a vendor loses the selling price of 4 bananas. His loss percent is :

(a) $11\frac{1}{9}\%$ (b) 10%

(c) $12\frac{1}{2}\%$ (d) none of these

**12.** By selling 100 bananas, a fruit seller gains the selling price of 20 bananas. His gain percent is :

(a) 10% (b) 15%

(c) 20% (d) 25%

**13.** By selling a book for Rs. 10, the publisher loses (1/11) of what it costs him. His C.P. is :

(a) Rs. 12 (b) Rs. 11

(c) Rs. 10 (d) Rs. 9

**14.** Profit after selling a commodity for Rs. 425 is same as loss after selling it for Rs. 355. The cost of the commodity is :

(a) Rs. 385 (b) Rs. 390

(c) Rs. 395 (d) Rs. 400 **(Bank P.O. 1989)**

**15.** The cost price of an article, which on being sold at a gain of 12% yields Rs. 6 more than when it is sold at a loss of 12%, is :

(a) Rs. 30 (b) Rs. 25

(c) Rs. 20 (d) Rs. 24 **(C.B.I. Exam. 1990)**

**16.** The C.P. of an article which is sold at a loss of 25% for Rs. 150, is :

(a) Rs. 125 (b) Rs. 175

(c) Rs. 200 (d) Rs. 225 **(RRB Exam. 1991)**

**17.** The cost of 2 almirahs and a radio is Rs. 7000, while 2 radios and one almirah together cost Rs. 4250. The cost of an almirah is :

(a) Rs. 3000 (b) Rs. 3160

(c) Rs. 3240 (d) none of these

**18.** If by selling an article for Rs. 144, a man loses $\frac{1}{7}$ of his outlay. By selling it for Rs. 168, his gain or loss percent is :

(a) 20% loss (b) 20% gain

(c) $4\frac{1}{6}\%$ gain (d) none of these

**19.** A shopkeeper mixes two varieties of tea, one costing Rs. 35 per kg. and another at Rs. 45 per kg. in the ratio 3 : 2. If he sells the mixed variety at Rs. 41.60 per kg., his gain or loss percent is :

(a) 4% gain (b) 4% loss

(c) $6\frac{2}{3}\%$ gain (d) $6\frac{2}{3}\%$ loss

**20.** Alok bought 25 kg. of rice at the rate of Rs. 6.00 per kg. and 35 kg. of rice at the rate of Rs. 7.00 per kg. He mixed the two and sold the mixture at the rate of Rs. 6.75 per kg. What was his gain/loss in this transaction?

(*a*) Rs. 16.00 gain (*b*) Rs. 16.00 loss
(*c*) Rs. 20.00 gain (*d*) none of these

**(P.O. Exam. 1990)**

**21.** A dishonest dealer professes to sell his goods at cost price but he uses a false weight and thus gains $6\frac{18}{47}\%$. For a kg., he uses a weight of :

(*a*) 953 gm. (*b*) 940 gm.
(*c*) 960 gm. (*d*) 947 gm.

**22.** A dealer professing to sell at cost price, uses a 900 gm. weight for a kilogram. His gain percent is :

(*a*) 10 (*b*) 11
(*c*) 9 (*d*) $11\frac{2}{9}$

**(RRB Exam. 1991)**

**23.** *A* sold a watch at a gain of 5% to *B* and *B* sold it to *C* at a gain of 4%. If *C* paid Rs. 91 for it, then *A* paid :

(*a*) Rs. 82.81 (*b*) Rs. 83
(*c*) Rs. 83.33 (*d*) none of these

**24.** When the price of pressure cooker was increased by 15%, its sale fell down by 15%. The effect on the money receipt was :

(*a*) no effect (*b*) 15% decrease
(*c*) 7.5% increase (*d*) 2.25% decrease

**(S.B.I. P.O. Exam. 1987)**

**25.** A man sells two houses at the rate of Rs. 1.995 lakh each. On one he gains 5% and on the other he loses 5%. His gain or loss percent in the whole transaction is :

(*a*) 0.25% loss (*b*) 0.25% gain
(*c*) 2.5% loss (*d*) neither loss nor gain

**(Central Excise & I. Tax 1988)**

**26.** The price of sugar having gone down by 10%, a consumer can buy 5 kg. more sugar for Rs. 270. The difference between the original and the reduced price per kg. is :

(*a*) 60 paise (*b*) 75 paise
(*c*) 53 paise (*d*) 62 paise

**27.** An article is sold at a certain price. By selling it at $\frac{2}{3}$ of that price, one loses 10%. The gain percent at original price is :

(*a*) $33\frac{1}{3}\%$ (*b*) 35%
(*c*) 40% (*d*) 20%

**28.** A man sells 320 mangoes at the cost price of 400 mangoes. His gain percent is :

(*a*) 10% (*b*) 25%
(*c*) 15% (*d*) 20%

**(Asstt. Grade Exam. 1987)**

**29.** A fruit seller buys lemons at 2 for a rupee and sells them at 5 for three rupees. His gain per cent is :

(*a*) 10% (*b*) 15%
(*c*) 20% (*d*) none of these

**30.** By selling 45 oranges for Rs. 40, a man loses 20%. How many should he sell for Rs. 24 so as to gain 20% ?

(*a*) 16 (*b*) 18
(*c*) 20 (*d*) 22

**31.** By selling 12 oranges for one rupee a man loses 20%. How many for a rupee should he sell to get a gain of 20% ?

(*a*) 5 (*b*) 8
(*c*) 10 (*d*) 15 **(C.D.S. Exam. 1989)**

**32.** A boy buys oranges at Rs. 2 for 3 oranges and sells them at a rupee each. To make a profit of Rs. 10, he must sell :

(*a*) 10 oranges (*b*) 20 oranges
(*c*) 30 oranges (*d*) 40 oranges

**(C.D.S. Exam. 1991)**

**33.** Toffees are bought at the rate of 8 for a rupee. To gain 60%, they must be sold at :

(*a*) 6 for a rupee (*b*) 5 for a rupee
(*c*) 9 for 2 rupees (*d*) 24 for Rs. 5

**34.** A man buys 10 articles for Rs. 8 and sells them at the rate of Rs. 1.25 per article. His gain percent is :

(*a*) 50% (*b*) $56\frac{1}{4}\%$
(*c*) $19\frac{1}{2}\%$ (*d*) 20% **(C.D.S. Exam. 1991)**

**35.** A man sells an article at a gain of 10%. If he had bought it at 10% less and sold it for Rs. 132 less, he would have gained 10%. The C.P. of the article is :

(*a*) Rs. 1200 (*b*) Rs. 1320
(*c*) Rs. 1188 (*d*) none of these

**36.** By selling sugar at Rs. 5.58 per kg. a man loses 7%. To gain 7% it must be sold at the rate of :

(*a*) Rs. 5.62 per kg. (*b*) Rs. 6.42 per kg.
(*c*) Rs. 7.32 per kg. (*d*) Rs. 6.62 per kg.

**37.** A person bought an article and sold it at a loss of 10%. If he had bought it for 20% less and sold it for Rs. 55 more, he would have had a profit of 40%. The C.P. of the article is :

(*a*) Rs. 200 (*b*) Rs. 225
(*c*) Rs. 250 (*d*) none of these

**(Central Excise & I. Tax. 1988)**

**38.** A dealer sells a transistor at a gain of 16%. If he had sold it for Rs. 20 more, he would have gained 20%. The C.P. of the transistor is :

(*a*) Rs. 600 (*b*) Rs. 500
(*c*) Rs. 400 (*d*) Rs. 350

**39.** If the cost of 8 pens and 6 pencils is Rs. 9.30 and the cost of 5 pens and 4 pencils is Rs. 6, the cost of one pen & one pencil is :

(*a*) Rs. 4.05 (*b*) Rs. 2.25
(*c*) Rs. 1.45 (*d*) Rs. 1.35

**40.** If a commodity is sold at a gain of 6% instead of at a loss of 6% then the seller gets Rs. 6 more. The C.P. of the commodity is :

(*a*) Rs. 50 (*b*) Rs. 94
(*c*) Rs. 100 (*d*) Rs. 106

**41.** A tradesman by means of a false balance defrauds to the extent of 8% in buying goods and also defrauds to 8% in selling. His gain percent is :

(*a*) 16% (*b*) 15.48%
(*c*) 16.64% (*d*) none of these

**42.** A radio dealer sold a radio at a loss of 2.5%. Had he sold it for Rs. 100 more, he would have gained $7\frac{1}{2}\%$. In order to gain $12\frac{1}{2}\%$, he should sell it for :

(*a*) Rs. 850 (*b*) Rs. 925
(*c*) Rs. 1080 (*d*) Rs. 1125

**43.** A man sells two horses for Rs. 4000 each, neither losing nor gaining in the deal. If he sold one horse at a gain of 25%, the other horse is sold at a loss of :

(*a*) 25% (*b*) $18\frac{2}{9}\%$
(*c*) $16\frac{2}{3}\%$ (*d*) none of these

**44.** The percent profit made when an article is sold for Rs. 78 is twice as when it is sold for Rs. 69. The C.P. of the article is :

(*a*) Rs. 49 (*b*) Rs. 51
(*c*) Rs. 57 (*d*) Rs. 60

**45.** A man sold an article for Rs. 75 and lost something. Had he sold it for Rs. 96, his gain would have been double the former loss. The C.P. of the article is :

(*a*) Rs. 81 (*b*) Rs. 82
(*c*) Rs. 83 (*d*) Rs. 85.50

**46.** A grocer sells rice at a profit of 10% and uses weights which are 20% less than the market weight. The total gain earned by him will be :

(*a*) 30% (*b*) 35%
(*c*) 37.5% (*d*) none of these

**47.** Due to an increase of 30% in the price of eggs, 3 eggs less are available for Rs. 7.80. The present rate of eggs per dozen is :

(*a*) Rs. 8.64 (*b*) Rs. 8.88
(*c*) Rs. 9.36 (*d*) none of these

**48.** By selling an article for Rs. 144, a man gained such that the percentage gain equals the C.P. The C.P. of the article is :

(*a*) Rs. 60 (*b*) Rs. 64
(*c*) Rs. 72 (*d*) Rs. 80

**49.** A man gains 10% by selling an article for a certain price. If he sells it at double the price, the profit made is :

(*a*) 20% (*b*) 60%
(*c*) 100% (*d*) 120%

**50.** A man purchased sugar worth of Rs. 400. He sold (3/4)th at a loss of 10% and the remainder at a gain of 10%. On the whole, he gets :

(*a*) a loss of 5% (*b*) a gain of $5\frac{1}{2}\%$
(*c*) a loss of $5\frac{1}{19}\%$ (*d*) a loss of $5\frac{5}{19}\%$

**51.** Subhash purchased a tape recorder at $\frac{9}{10}$th of its selling price and sold it at 8% more than its selling price. His gain is :

(*a*) 8% (*b*) 10%
(*c*) 18% (*d*) 20% **(S.B.I. P.O. Exam. 1987)**

**52.** 6% more is gained by selling a radio for Rs. 475 than by selling for Rs. 451. The C. P. of the radio is :

(*a*) Rs. 400 (*b*) Rs. 434
(*c*) Rs. 446.50 (*d*) none of these

**53.** A merchant professes to lose 6% on a certain tea, but he uses a weight equal to 900 gm. instead of one kg. His real gain is :

(*a*) $5\frac{1}{9}\%$ (*b*) 6%

(*c*) 4% (*d*) $4\frac{4}{9}\%$ **(Clerks' Exam. 1991)**

**54.** At what price must Kantilal sell a mixture of 80 kg. sugar at Rs. 6.75 per kg. with 120 kg. at Rs. 8 per kg. to gain 20% ?

(*a*) Rs. 7.50 per kg. (*b*) Rs. 9 per kg.
(*c*) Rs. 8.20 per kg. (*d*) Rs. 8.85 per kg.

**(S.B.I. P.O. Exam. 1987)**

**55.** A man sells a car to his friend at 10% loss. If the friend sells it for Rs. 54000 and gains 20%, the original C.P. of the car was :

(*a*) Rs. 25000 (*b*) Rs. 37500
(*c*) Rs. 50000 (*d*) Rs. 60000 **(S.S.C. Exam. 1987)**

**56.** Bhajan Singh purchased 120 reams of paper at Rs. 80 per ream. He spent Rs. 280 on transportation, paid octroi at the rate of 40 paise per ream and paid Rs. 72 to the coolie. If he wants to have a gain of 8%, the selling price per ream must be :

(*a*) Rs. 86 (*b*) Rs. 89
(*c*) Rs. 90 (*d*) Rs. 87.48

**(Bank P.O. Exam. 1988)**

**57.** A shopkeeper earns a gain of $22\frac{1}{2}\%$ of his C.P. If his monthly sale for a particular item be Rs. 3920, his gain is :

(*a*) Rs. 700 (*b*) Rs. 720
(*c*) Rs. 880 (*d*) Rs. 882

**58.** The loss incurred on selling an article for Rs. 270 is as much as the profit made after selling it at 10% profit. The C.P. of the article is :

(*a*) Rs. 90 (*b*) Rs. 110
(*c*) Rs. 363 (*d*) Rs. 300

**(Bank P.O. Exam. 1989)**

**59.** If I purchased 11 books for Rs. 10 and sold all the books at the rate of 10 books for Rs. 11, the profit percent is :

(*a*) 10% (*b*) 11%
(*c*) 21% (*d*) 100% **(R.R.B. Exam. 1989)**

**60.** An item costing Rs. 200 is being sold at 10% loss. If the price is further reduced by 5%, the selling price will be :

(*a*) Rs. 179 (*b*) Rs. 175
(*c*) Rs. 171 (*d*) Rs. 170 **(N.D.A. Exam. 1987)**

**61.** A trader lists his articles 20% above C.P. and allows a discount of 10% on cash payment. His gain percent is :

(*a*) 10% (*b*) 6%
(*c*) 8% (*d*) 5% **(R.R.B. Exam. 1991)**

**62.** The marked price is 10% higher than C.P. A discount of 10% is given on marked price. In this kind of sale the seller

(a) bears no loss, no gain (b) gains
(c) loses (d) loses 1%

**(R.R.B. Exam. 1991)**

**63.** An umbrella marked at Rs. 80 is sold for Rs. 68. The rate of discount is :
(a) 12% (b) 15%
(c) $17\frac{11}{17}\%$ (d) 20% **(R.R.B. Exam. 1989)**

**64.** A discount series of 10%, 20% and 40% is equal to a single discount of :
(a) 50% (b) 56.80%
(c) 70% (d) 70.28%

**(Central Excise & I. Tax. 1989)**

**65.** Rashmi purchased a cooker at $\frac{9}{10}$th of its marked price and sold it for 8% more than its marked price. Her profit percent is :
(a) 16% (b) 18%
(c) 20% (d) none of these

**66.** A retailer purchases a sewing machine at a discount of 15% and sells it for Rs. 1955. In the bargain he makes a profit of 15%. How much is the discount which he got from the whole sale ?
(a) Rs. 270 (b) Rs. 290
(c) Rs. 300 (d) none of these

**(L.I.C. A.A.O. Exam. 1988)**

**67.** While selling a watch a shopkeeper gives a discount of 5%. If he gives a discount of 7%, he earns Rs. 15 less as profit. What is the marked price of the watch ?
(a) Rs. 697.50 (b) Rs. 712.50
(c) Rs. 787.50 (d) none of these

**(L.I.C. A.A.O. Exam. 1988)**

**68.** The marked price of a radio is Rs. 480. The shopkeeper allows a discount of 10% and gains 8%. If no discount is allowed, his gain percent would be :
(a) 18% (b) 18.5%
(c) 20% (d) 20.5%

**69.** A shopkeeper earns a profit of 12% after selling a book at 10% discount on the printed price. The ratio of the cost price and printed price of the book is :
(a) 45 : 56 (b) 50 : 61
(c) 99 : 125 (d) none of these

**(G.I.C. Exam. 1988)**

**70.** A fan is listed at Rs. 150, with a discount of 20%. What additional discount must be offered to the customer to bring the net price to Rs. 108 ?

(*a*) 8% (*b*) 10%
(*c*) 15% (*d*) none of these

**71.** A trader marks his goods in such a way that after allowing a discount of 10%, he gains 15%. If an article costs him Rs. 72, then its marked price is :

(*a*) Rs. 74.52 (*b*) Rs. 84.62
(*c*) Rs. 90 (*d*) Rs. 92

**72.** Tarun bought a T.V. with 20% discount on the labelled price. Had he bought it with 25% discount, he would have saved Rs. 500. At what price did he buy the T.V. ?

(*a*) Rs. 5000 (*b*) Rs. 10000
(*c*) Rs. 12000 (*d*) none of these

**(P.O. Exam. 1990)**

**73.** If a commission of 10% is given on the marked price of a book, the publisher gains 20%. If the commission is increased to 15%, the gain is :

(*a*) $16\frac{2}{3}$ % (*b*) $13\frac{1}{3}$ %
(*c*) $15\frac{1}{6}$ % (*d*) none of these

**74.** A shopkeeper marks his goods in such a way that after allowing a discount of 10%, he gains 26%. How much percent above the cost price is the marked price ?

(*a*) 36% (*b*) 38%
(*c*) 40% (*d*) 28%

**75.** The list price of an electric iron is Rs. 75. A customer is given two successive discounts. He pays Rs. 63.45 for it. If the rate of first discount is 10%, the second discount is :

(*a*) 4% (*b*) 5%
(*c*) 6% (*d*) 6.5%

**76.** A tradesman's'prices are 20% above C.P. He allows his customers some discount on his bill and makes a profit of 8%. The rate of discount is :

(*a*) 10% (*b*) 12%
(*c*) 14% (*d*) 16%

**77.** A reduction of 20% in the price of mangoes enables a person to purchase 12 more for Rs. 15. The price of 16 manoges before reduction was :

(*a*) Rs. 5 (*b*) Rs. 6
(*c*) Rs. 7 (*d*) Rs. 9 **(R.R.B. Exam. 1989)**

**78.** There would be 10% loss if rice is sold at Rs. 5.40 per kg. At what price per kg. should it be sold to earn a profit of 20% ?

(*a*) Rs. 7.20 (*b*) Rs. 7.02

(*c*) Rs. 6.48 (*d*) Rs. 6 **(S.B.I. P.O. Exam. 1988)**

## HINTS & SOLUTIONS

**1.** S.P. = Rs. 100; C.P. = Rs. (100 – 10) = Rs. 90.

$$\therefore \text{ Gain\%} = \left(\frac{10}{90} \times 100\right)\% = 11\frac{1}{9}\ \%.$$

**2.** C.P. = (S.P. + loss) = Rs. 110.

$$\therefore \text{ Loss \%} = \left(\frac{10}{110} \times 100\right)\% = 9\frac{1}{11}\ \%.$$

**3.** $\text{C.P.} = \text{Rs.}\left(\frac{100}{110} \times 1980\right) = \text{Rs. } 1800.$

**4.** $\text{C.P.} = \text{Rs.}\left(\frac{100}{95} \times 4085\right) = \text{Rs. } 4300.$

**5.** Let it be sold for Rs. $x$. Then,

96 : 240 :: 110 : $x$.

$$\text{or } x = \left(\frac{240 \times 110}{96}\right) = \text{Rs. } 275.$$

**6.** Let S.P. = Rs. $x$

$$\text{Then, gain} = \text{Rs.}\left(\frac{20}{100} \times x\right) = \text{Rs. } \frac{x}{5}$$

$$\text{C.P.} = \text{Rs.}\left(x - \frac{x}{5}\right) = \text{Rs. } 400$$

$$\Rightarrow \frac{4}{5}x = 400 \Rightarrow x = \text{Rs. } 500.$$

**7.** Let C.P. be Rs. $x$

Then if S.P. = Rs. 30, loss = Rs. $(x - 30)$

if S.P. = Rs. 40, loss = Rs. $(x - 40)$

$$\therefore (x - 30) - (x - 40) = \left(\frac{5}{100} \times x\right)$$

$$\Rightarrow 10 = \frac{5}{100} \times x \Rightarrow x = \text{Rs. } 200.$$

**8.** Let C.P. be Rs. $x$

Then, if S.P. = Rs. 350, gain = Rs. $(350 - x)$.

if S.P. = Rs. 340, gain = Rs. $(340 - x)$

$$\therefore (350 - x) - (340 - x) = \frac{5}{100}x$$

$$\Rightarrow 10 = \frac{5}{100}x \Rightarrow x = \text{Rs. } 200$$

**9.** Let C.P. of each article be Re. 1. Then,
C.P. = Rs. 12 and S.P. = Rs. 15.

$$Gain\ \% = \left(\frac{3}{12} \times 100\right)\% = 25\%.$$

**10.** Let C.P. of each table be Re. 1.
C.P of 20 tables = Rs. 20.
S.P. of 20 tables = Rs. 15.

$$Loss\ \% = \left(\frac{5}{20} \times 100\right)\% = 25\%.$$

**11.** Loss = (C.P. of 36 bananas) – (S.P. of 36 bananas)
∴ S.P. of 4 bananas
= (C.P. of 36 bananas) – (S.P. of 36 bananas)
or S.P. of 40 bananas = C.P. of 36 bananas.
Now, let C.P. of each banana be Re. 1. Then,
C.P. of 40 bananas = Rs. 40.
S.P. of 40 bananas = Rs. 36.

$$Loss\ \% = \left(\frac{4}{40} \times 100\right)\% = 10\%.$$

**12.** Gain = (S.P. of 100 bananas) – (C.P. of 100 bananas)
or S.P. of 20 bananas
= (S.P. of 100 bananas) – (C.P. of 100 bananas)
or S.P. of 80 bananas = C.P. of 100 bananas.
Now, let C.P. of each banana be Re. 1. Then,
C.P. of 80 bananas = Rs. 80.
S.P. of 80 bananas = Rs. 100.

$$Gain\ \% = \left(\frac{20}{80} \times 100\right)\% = 25\%.$$

**13.** Let C.P. be Rs. $x$. Then, loss $= \frac{x}{11}$.

$$\therefore \text{S.P.} = \left(x - \frac{x}{11}\right) = \frac{10x}{11}.$$

$$So, \frac{10x}{11} = 10 \Rightarrow x = 11.$$

**14.** Let C.P. = Rs. $x$. Then,
$425 - x = x - 355 \Rightarrow 2x = 780 \Rightarrow x = 390.$

**15.** Let the C.P. be Rs. $x$
Then, S.P. when gain is 12% $= \left(\frac{12x}{100} + x\right) = \frac{112x}{100}$.

$$\therefore \frac{112x}{100} - \frac{88x}{100} = 6.$$

$\Rightarrow \frac{24x}{100} = 6 \Rightarrow x = \frac{600}{24} = 25$.

**16.** C.P. = Rs. $\left(\frac{100}{75} \times 150\right)$ = *Rs.* 200.

**17.** Let the cost of each almirah be Rs. $x$ and that of each radio be Rs. $y$. Then, $2x + y = 7000$ and $x + 2y = 4520$.
*Solving these equations, we get $x$ = Rs. 3160.*

**18.** Let C.P. be Rs. $x$. Then, loss = Rs. $(x/7)$.
S.P. = Rs. $\left(x - \frac{x}{7}\right)$ = Rs. $\left(\frac{6x}{7}\right)$.
$\therefore \frac{6x}{7} = 144$ or $x = \left(\frac{144 \times 7}{6}\right)$ = Rs. 168.
Now, C.P. = Rs. 168 and New S.P. = Rs. 168.
$\therefore$ *Gain or loss % = 0%.*

**19.** C.P. of 5 kg. mix. = Rs. $(35 \times 3 + 45 \times 2)$ = Rs. 195
S.P. of 5 kg. mix. = Rs. $(41.60 \times 5)$ = Rs. 208.
*Gain %* = $\left(\frac{13}{195} \times 100\right)\% = 6\frac{2}{3}\%$.

**20.** C.P. of total 60 kg. of rice = Rs. $(6 \times 25 + 7 \times 35)$ = Rs. 395.
S.P. of total 60 kg. of rice = Rs. $(6.75 \times 60)$ = Rs. 405.
*Gain = Rs. (405 – 395) = Rs. 10.*

**21.** Suppose he uses a weight of $x$ gm. for 1 kg.
Then, $\frac{x}{1000 - x} \times 100 = \frac{300}{47}$ or $\frac{x}{1000 - x} = \frac{3}{47}$.
or $47x = 3000 - 3x$ or $50x = 3000$ or $x = 60$.
*So, he uses a weight of (1000 – 60) = 940 gm. for 1 kg.*

**22.** Let the C.P. of 1 gm. be Re. 1.
Then, C.P. of 900 gm. = Rs. 900.
S.P. of 900 gm. = C.P. of 1000 gm. = Rs. 1000.
$\therefore$ Profit = Rs. (1000 – 900) = Rs. 100.
*Hence, gain %* = $\left(\frac{100}{900} \times 100\right)\% = 11\frac{1}{9}\%$.

**23.** C.P. of $C$ = (104% of C.P. of $B$)
= 104% of (105% of C.P. of $A$)
or Rs. 91 = $\left(\frac{104}{100} \times \frac{105}{100}\right)$ of C.P. of $A$.
$\therefore$ *C.P. of A* = *Rs.* $\left(\frac{91 \times 100 \times 100}{104 \times 105}\right)$ = *Rs. 83.33.*

**24.** Let the original cost of each cooker be Re. 1 and let the number sold originally be 100.

Total sale proceed = Rs. $(100 \times 1)$ = Rs. 100.

New rate = (115% of Re. 1) = Rs. 1.15.

Number sold now = 85.

$\therefore$ Sale proceed now = Rs. $(1.15 \times 85)$ = Rs. 97.75.

*So, there is a decrease of 2.25% in the money receipt.*

**25.** In such problems, there is always a loss.

Formula : $\text{Loss } \% = \frac{(\text{common loss and gain }\%)^2}{10}$

$$= \left(\frac{5}{10}\right)^2 = \frac{1}{4} = 0.25\% .$$

**26.** Let the original rate be Rs. $x$ per kg.

Quantity purchased for Rs. 270 = $\frac{270}{x}$ kg.

New rate = (90% *of* $x$) = Rs. $\frac{9x}{10}$ per kg.

Quantity purchased for Rs. 270 = $\left(\frac{270 \times 10}{9x}\right)$ kg. = $\frac{300}{x}$ kg.

$\therefore \frac{300}{x} - \frac{270}{x} = 5$ or $5x = 30$ or $x = 6$.

Difference between the original and the reduced rate

$= \left(x - \frac{9x}{10}\right) = \frac{x}{10} = Rs. \frac{6}{10}$ *per kg. = 60 paise per kg.*

**27.** Let C.P. be Rs. 100.

Then, S.P. at 10% Loss = Rs. 90

$\therefore$ (2/3) of actual S.P. = Rs. 90.

*or Actual S.P. = Rs.* $\left(\frac{90 \times 3}{2}\right)$ *= Rs. 135.So, gain % = 35%.*

**28.** Let C.P. of each mango be Re. 1. Then,

C.P. of 400 mangoes = Rs. 400.

$\therefore$ C.P. of 320 mangoes = Rs. 320.

And, S.P. of 320 mangoes = Rs. 400.

$\therefore$ *Gain %* $= \left(\frac{80}{320} \times 100\right)\% = 25\%.$

**29.** Suppose that he buys 10 lemons. Then,

C.P. = Rs. 5; S.P. = Rs. 6.

$\therefore$ *Gain %* $= \left(\frac{1}{5} \times 100\right)\% = 20\%.$

**30.** S.P. of 1 orange = Rs. $\frac{40}{45}$ = Rs. $\frac{8}{9}$.

80% of C.P. $= \frac{8}{9}$ or C.P. = Rs. $\left(\frac{8}{9} \times \frac{100}{80}\right)$ = Rs. $\frac{10}{9}$.

Now, S.P. = 120% of C.P. = Rs. $\left(\frac{120}{100} \times \frac{10}{9}\right)$ = Rs. $\frac{4}{3}$.

For Rs. $\frac{4}{3}$, he sells 1 orange.

*For Rs. 24, he would sell* $\left(\frac{3}{4} \times 24\right)$ *= 18 oranges.*

**31.** S.P. = Re. 1, Loss = 20% ⇒ C.P. = Rs. $\left(\frac{100}{80} \times 1\right)$ = Rs. $\frac{5}{4}$.

Now, C.P. = Rs. $\frac{5}{4}$, gain = 20% ⇒ S.P. = Rs. $\left(\frac{120}{100} \times \frac{5}{4}\right)$ = Rs. $\frac{3}{2}$.

For Rs. $\frac{3}{2}$, he must sell 12 oranges.

*For Re. 1, he must sell* $\left(12 \times \frac{2}{3}\right)$ *= 8 oranges.*

**32.** Suppose he sells $x$ oranges

Then, C.P. of $x$ oranges = Rs. $\frac{2}{3}x$.

S.P. of $x$ oranges = Rs. $x$

Profit on $x$ oranges = Rs. $\left(x - \frac{2}{3}x\right)$ = Rs. $\frac{x}{3}$

$\therefore \frac{x}{3} = 10 \Rightarrow x = 30.$

**33.** C.P. of 1 toffee = Re. (1/8)

S.P. = $\left(160\% \text{ of } \frac{1}{8}\right)$ = Rs. $\left(\frac{160}{100} \times \frac{1}{8}\right)$ = Rs. $\frac{1}{5}$.

*So, they must be sold at five for a rupee.*

**34.** C.P. of 10 articles = Rs. 8.

S.P. of 10 articles = Rs. (1.25 × 10) = Rs. 12.50

Profit = Rs. (12.50 – 8) = Rs. 4.50

*Gain %* $= \left(\frac{4.50}{8} \times 100\right) = 56\frac{1}{4}\%.$

**35.** Let C.P. be Rs. $x$. Then,

S.P. = (110% of $x$) = Rs. $\frac{11x}{10}$.

New C.P. = $\left(\frac{90}{100} \times x\right)$ = Rs. $\frac{9x}{10}$.

New S.P. = $\left(110\% \text{ of } \frac{9x}{10}\right)$ = Rs. $\frac{99x}{100}$.

$\therefore \frac{11x}{10} - \frac{99x}{100} = 132$ or $x = 1200$.

$\therefore$ *C.P. = Rs. 1200.*

**36.** 93% of C.P. = Rs. 5.58.

$\therefore$ C.P. = Rs. $\left(\frac{100}{93} \times 5.58\right)$ = Rs. 6.

*New S.P. = (107% of Rs. 6) = Rs. 6.42 per kg.*

**37.** Let C.P. = Rs. $x$

Then, S.P. = 90% of $x$ = Rs. $\left(\frac{9x}{10}\right)$.

New C.P. = 80% of $x$ = Rs. $\left(\frac{4x}{5}\right)$

Now, gain = 40%

$\therefore$ New S.P. = $\left(140\% \text{ of } \frac{4x}{5}\right)$ = Rs. $\frac{28x}{25}$.

*Thus,* $\frac{28x}{25} - \frac{9x}{10} = 55$ *or* $x = 250$.

**38.** (20% of C.P.) – (16% of C.P.) = Rs. 20.

or 4% of C.P. = 20 or $\left(\frac{4}{100} \times \text{C.P.}\right) = 20$.

*or C.P. = Rs.* $\left(\frac{100 \times 20}{4}\right)$ *= Rs. 500.*

**39.** Let the cost of a pen and a pencil be $x$ paise and $y$ paise respectively.

Then, $8x + 6y = 930$ and $5x + 4y = 600$

Solving the two equations, we get

$x = 60$ and $y = 75$

$\therefore$ *Cost of 1 pen and 1 pencil = Rs. 1.35.*

**40.** (6% of C.P. + 6% of C.P.) = Rs. 6.

or 12% of C.P. = Rs. 6 or $\frac{12}{100} \times \text{C.P.} = 6$

*or C.P. = Rs.* $\left(\frac{100 \times 6}{12}\right)$ *= Rs. 50.*

**41.** Rule : In such questions we adopt the rule :

$$\text{Gain \%} = \frac{(100 + \text{common gain \%})^2}{100} - 100$$

$$= \left\{\frac{(108)^2}{100} - 100\right\} \% = 16.64\%.$$

**42.** Clearly, (2.5% of C.P. + 7.5% of C.P.) = 100

or 10% of C.P. = 100 or $\frac{1}{10}$ of C.P. = 100.

$\therefore$ C.P. = Rs. 1000.

Now, gain = $12\frac{1}{2}$ % and therefore,

$S.P. = 112\frac{1}{2}\%\ of\ C.P. = Rs.\left(\frac{225}{200}\times 1000\right) = Rs.\ 1125.$

**43.** C.P. of 2 horses = Rs. 8000.

Now, S.P. of one horse = Rs. 4000, gain = 25%

$\therefore$ C.P. of this horse = Rs. $\left(\frac{100}{125}\times 4000\right)$ = Rs. 3200.

Now, C.P. of another horse = Rs. (8000 – 3200)

= Rs. 4800.

S.P. of this horse = Rs. 4000.

$\therefore Loss\ \% = \left(\frac{800}{4800}\times 100\right)\% = 16\frac{2}{3}\ \%$

**44.** Let the C.P. be Rs. $x$.

Then, $\frac{2\,(69-x)}{100} = \frac{78-x}{100}$

or $138 - 2x = 78 - x$ or $x = 60$. C.P. = Rs. 60.

**45.** Let the loss be Rs. $x$.

Then, 75 = C.P. – $x$ and 96 = C.P. + $2x$

On subtracting, we get $3x = 21$ or $x = 7$.

$\therefore$ C.P. = 75 + $x$ = Rs. 82.

**46.** Let us consider a packet of rice marked 1 kg.

Then, its actual weight = (80% of 1 kg.) = 0.8 kg.

Let C.P. of 1 kg. be Rs. $x$.

Then, C.P. of 0.8 kg. = Rs. $0.8\,x$.

Now, S.P. = 110% of C.P. of 1 kg.

$= \left(\frac{110}{100}\times x\right)$ = Rs. $1.1\,x$.

$Gain\ \% = \left(\frac{0.3x}{0.8x}\times 100\right)\% = 37.5\%.$

**47.** Let the original rate be $x$ paise per egg.

Number of eggs bought for Rs. 7.80 = $(780/x)$.

New rate = (130% of $x$)paise per egg = $\frac{13x}{10}$ paise per egg.

Number of eggs bought for Rs. 7.80 = $\frac{780\times 10}{13x} = \frac{600}{x}$.

$\therefore \frac{780}{x} - \frac{600}{x} = 3$ or $x = 180$ or $3\,x = 60$.

So, present rate $= \left(\frac{13 \times 60}{10}\right)$ paise per egg

$= 78$ *paise per egg* $=$ *Rs. 9.36 per dozen*

**48.** Let C.P. be Rs. $x$. Then,

$x + x\%$ of $x = 144$ or $x + \frac{x^2}{100} = 144$.

or $x^2 + 100x - 14400 = 0$. or $(x + 180)(x - 80) = 0$.

$\therefore x = 80$ (neglecting value $x = -180$) So, C.P. = Rs. 80.

**49.** 1st C.P. = (110% of C.P.) $= \left(\frac{11}{10}\text{ of C.P.}\right)$

2nd S.P. $= 2\left(\frac{11}{10}\text{ of C.P.}\right) = \frac{11}{5}$ of C.P.

$= \left(\frac{11}{5} \times 100\right)\%$ of C.P. $= 220\%$ of C.P.

$\therefore$ *New gain* $= 120\%$.

**50.** S.P. $= \left[90\% \text{ of } \left(\frac{3}{4}\text{ of Rs. } 400\right)\right] + \left[110\% \text{ of } \left(\frac{1}{4}\text{ of Rs. } 400\right)\right]$

$=$ Rs. $\left(\frac{90}{100} \times 30 + \frac{110}{100} \times 100\right) =$ Rs. 380.

*Loss* % $= \left(\frac{20}{400} \times 100\right)\% = 5\%$.

**51.** Let the S.P. be Rs. $x$.

Then, C.P. paid by Subhash = Rs. $\frac{9x}{10}$.

S.P. received by Subhash = (108% of Rs. $x$) = Rs. $\frac{27x}{25}$.

$\therefore$ Gain = Rs. $\left(\frac{27x}{25} - \frac{9x}{10}\right) =$ Rs. $\left(\frac{9x}{50}\right)$.

*Hence, gain* % $= \left(\frac{9x}{50} \times \frac{10}{9x} \times 100\right)\% = 20\%$.

**52.** Difference between two selling prices = Rs. 24.

$\therefore$ 6% of C.P. = Rs. 24.

*Hence, C.P. = Rs.* $\left(\frac{24 \times 100}{6}\right) =$ *Rs. 400.*

**53.** Let C.P. = Rs. 100. Then, S.P. = Rs. 94.

Thus, S.P. of 900 gm. = Rs. 94

S.P. of 1000 gm. = Rs. $\left(\frac{94}{900} \times 1000\right) =$ Rs. $\frac{940}{9}$.

*Gain* $= \left(\frac{940}{9} - 100\right) = \frac{40}{9} = 4\frac{4}{9}\%$.

**54.** Total C.P. of 200 kg. sugar

$$= \text{Rs. } (80 \times 6.75 + 120 \times 8) = \text{Rs. } 1500$$

C.P. of 1 kg. = Rs. $\left(\frac{1500}{200}\right)$ = Rs. 7.50.

Gain required = 20%.

$\therefore$ S.P. of 1 kg. = (120% of Rs. 7.50)

$$= Rs. \left(\frac{120}{100} \times 7.50\right) = Rs.\ 9 \text{ per kg.}$$

**55.** S.P. = Rs. 54000 and Gain earned = 20%.

C.P. = Rs. $\left(\frac{100}{120} \times 54000\right)$ = Rs. 45000.

Now, S.P. = Rs. 45000 and loss = 10%.

$\therefore$ *C.P. = Rs.* $\left(\frac{100}{90} \times 45000\right)$ *= Rs. 50000.*

**56.** Total C.P. of 120 reams

= Rs. $(120 \times 80 + 280 + 72 + 120 \times 40)$ = Rs. 10000.

C.P. of 1 ream = $(10000 \div 120)$ = Rs. $\left(\frac{250}{3}\right)$.

$\therefore$ *S.P. of 1 ream* = $\left[108\% \text{ of Rs. } \frac{250}{3}\right]$ *= Rs. 90.*

**57.** If S.P. is Rs. $\frac{245}{2}$, C.P. = Rs. 100.

If S.P. is Rs. 3920, C.P. = Rs. $\left(\frac{100 \times 2}{245} \times 3920\right)$ = Rs. 3200.

$\therefore$ *Gain* = $\left(\frac{45}{2} \times \frac{1}{100} \times 3200\right)$ *= Rs. 720.*

**58.** Let C.P. be Rs. $x$. Then,

$x - 270 = 10\%$ *of* $x = \frac{x}{10}$ *or* $x = 300$.

**59.** C.P. = Rs. 10

S.P. = Rs. $\left(\frac{11}{10} \times 11\right)$ = Rs. $\frac{121}{10}$

Gain = Rs. $\left(\frac{121}{10} - 10\right)$ = Rs. $\frac{21}{10}$.

$\therefore$ *Gain = Rs.* $\left(\frac{21}{10 \times 10} \times 100\right)$ *= 21%.*

**60.** S.P. = 90% of Rs. 200 = Rs. 180.

*Further, S.P. = (95% of Rs. 180) = Rs. 171.*

**61.** Let C.P. = Rs. 100. Then, M.P. = Rs. 120.

S.P. = 90% of Rs. 120 = Rs. 108.

*Gain % = 8%.*

**62.** Let C.P. = Rs. 100. Then, M.P. = Rs. 110.

S.P. = 90% of Rs. 110 = 99.

*Loss = 1%.*

**63.** Discount = $\left(\frac{12}{80} \times 100\right)\% = 15\%$.

**64.** Let original price = Rs. 100.

Price after 1st discount = Rs. 90.

Price after 2nd discount = Rs. $\left(\frac{80}{100} \times 90\right)$ = Rs. 72.

Price after 3rd discount = Rs. $\left(\frac{60}{100} \times 72\right)$ = Rs. 43.20

$\therefore$ *Single discount = (100 – 43.20) = 56.8%.*

**65.** Let the marked price be Rs. $x$. Then,

C.P. = $\frac{9}{10}x = 0.9x$ and S.P. = (108% of $x$) = $1.08x$.

$\therefore$ *Gain %* = $\left(\frac{.18}{.9} \times 100\right) = 20\%$.

**66.** Let the marked price be Rs. $x$.

Discount availed by the retailer = 15% of Rs. $x$.

$\therefore$ C.P. of the machine by the retailer

= ($x$ – 15% of $x$) = Rs. $\frac{17x}{20}$.

So, 15% of $\frac{17x}{20} = 1955 - \frac{17x}{20}$.

$\therefore \frac{51x}{400} + \frac{17x}{20} = 1955$ or $x = 2000$.

*Discount received by retailer = (15% of Rs. 2000) = Rs. 300.*

**67.** Let the marked price be Rs. $x$.

Then, (7% of $x$) – 15 = (5% of $x$)

*or* $\frac{7x}{100} - \frac{5x}{100} = 15$ *or* $x = 750$.

**68.** S.P. = (90% of Rs. 480) = Rs. $\left(\frac{90}{100} \times 480\right)$ = Rs. 432.

Gain earned on it = 8%.

$\therefore$ C.P. = Rs. $\left(\frac{100}{108} \times 432\right) = 400$.

If no discount is allowed, S.P. = Rs. 480.

*Gain %* = $\left(\frac{80}{400} \times 100\right)\% = 20\%$.

**69.** Let the printed price of the book be Rs. 100.
After a discount of 10%, S.P. = Rs. 90.
Profit earned = 12%.

$\therefore$ C.P. of the book = Rs. $\left(\frac{100}{112} \times 90\right)$ = Rs. $\frac{1125}{14}$.

*Hence, (C.P.) : (Printed price) =* $\frac{1125}{14} : 100$ *or* $45 : 56$.

**70.** Price after 20% discount = (80% of Rs. 150) = Rs. 120.
Further discount needed = Rs. (120 – 108) = Rs. 12.
Let $x$% of 120 = 12.

*Then,* $x = \left(\frac{12}{120} \times 100\right) = 10\%$.

**71.** 90% of marked price = 115% of Rs. 72.

$\therefore \left(\frac{90}{100} \times x\right) = \left(\frac{115}{100} \times 72\right)$ *or* $x = \left(\frac{115 \times 72}{100} \times \frac{100}{90}\right)$ = *Rs.* 92.

**72.** Let S.P. of T.V. (by trader) = Rs. 100.
If S.P. is Rs. 80, M.P. = Rs. 100.

If S.P. is Rs. 100, M.P. = Rs. $\left(\frac{100}{80} \times 100\right)$ = Rs. 125.

Now, if discount is 25%, then S.P. = (75% of Rs. 125) = Rs. $\frac{375}{4}$.

Diff. between two S.P. = Rs. $\left(100 - \frac{375}{4}\right)$ = Rs. $\frac{25}{4}$.

If diff. is Rs. $\frac{25}{4}$, S.P. = Rs. 100

*If diff. is Rs. 500, S.P. = Rs.* $\left(100 \times \frac{4}{25} \times 500\right)$ = *Rs. 8000.*

**73.** Let marked price be Rs. 100.
S.P. with 10% discount = Rs. 90.
$\therefore$ 120% of C.P. = Rs. 90.

or C.P. = Rs. $\left(\frac{90 \times 100}{120}\right)$ = Rs. 75.

S.P. with 15% commission = Rs. 85.

$\therefore$ *Gain in this case* = $\left(\frac{10}{75} \times 100\right)\% = 13\frac{1}{3}\%$.

**74.** Let C.P. be Rs. 100. Then, S.P. = Rs. 126.
Now, 90% of marked price = Rs. 126.

$\therefore$ Marked price = Rs. $\left(\frac{126 \times 100}{90}\right)$ = Rs. 140.

*So, he must mark his goods 40% above C.P.*

**75.** Let second discount be $x\%$. Then,

$(100-x)\ \%$ of [90% of Rs. 75] = 63.45

or $\left(\frac{100-x}{100}\times\frac{90}{100}\times 75\right)=63.45$

or $27\ (100-x)=2538$ or $100-x=\frac{2538}{27}=94.$

$\therefore\ x=6\%.$

**76.** Let the C.P. be Rs. 100.

Then, marked price = Rs. 120 and S.P. = Rs. 108.

$\therefore$ *Discount* $=\left(\frac{12}{120}\times 100\right)\%=10\%.$

**77.** Suppose the price of 1 mango be $x$ paise.

Number of mangoes for Rs. 15 $=\frac{1500}{x}$.

New price of one mango = (80% of $x$) $=\frac{4x}{5}$ paise.

$\therefore$ Number of mangoes for Rs. 15 $=\left(\frac{1500\times 5}{4x}\right)$.

$\therefore\ \frac{7500}{4x}-\frac{1500}{x}=12$ or $x=31.25$

$\therefore$ Cost of 16 mangoes before reduction

$$= Rs.\left(\frac{31.25\times 16}{100}\right)= Rs.\ 5.$$

**78.** Let C.P. per kg. be Rs. $x$. Then,

$x-10\%$ of $x=5.40$ or $x=6.$

$\therefore$ *S.P. = Rs. [6 + 20% of 6] = Rs. 7.20.*

## ANSWERS (Exercise 16B)

| | | | | | |
|---|---|---|---|---|---|
| 1. (c) | 2. (a) | 3. (a) | 4. (d) | 5. (c) | 6. (d) |
| 7. (c) | 8. (c) | 9. (a) | 10. (a) | 11. (b) | 12. (c) |
| 13. (b) | 14. (b) | 15. (b) | 16. (c) | 17. (b) | 18. (d) |
| 19. (c) | 20. (d) | 21. (b) | 22. (d) | 23. (c) | 24. (d) |
| 25. (a) | 26. (a) | 27. (b) | 28. (b) | 29. (c) | 30. (b) |
| 31. (b) | 32. (c) | 33. (b) | 34. (b) | 35. (a) | 36. (b) |
| 37. (c) | 38. (b) | 39. (d) | 40. (a) | 41. (c) | 42. (d) |
| 43. (c) | 44. (d) | 45. (b) | 46. (c) | 47. (c) | 48. (d) |
| 49. (d) | 50. (a) | 51. (d) | 52. (a) | 53. (d) | 54. (b) |
| 55. (c) | 56. (c) | 57. (b) | 58. (d) | 59. (c) | 60. (c) |
| 61. (c) | 62. (d) | 63. (b) | 64. (b) | 65. (c) | 66. (c) |
| 67. (d) | 68. (c) | 69. (a) | 70. (b) | 71. (d) | 72. (d) |
| 73. (b) | 74. (c) | 75. (c) | 76. (a) | 77. (a) | 78. (a) |

17

# ALLIGATION OR MIXTURE

**Alligation** *is the rule that enables us to find the proportion in which the two or more ingradients at the given price must be mixed to produce a mixture at a given price.*

*Cost price of unit quantity of the mixture is called the* **Mean Price.**

**Rule of Alligation.** If two ingradients are mixed in a ratio, then

$$\frac{\text{(Quantity of cheaper)}}{\text{(Quantity of dearer)}} = \frac{\text{(C.P. of dearer)} - \text{(Mean Price)}}{\text{(Mean Price)} - \text{(C.P. of cheaper)}}$$

We represent it as under :

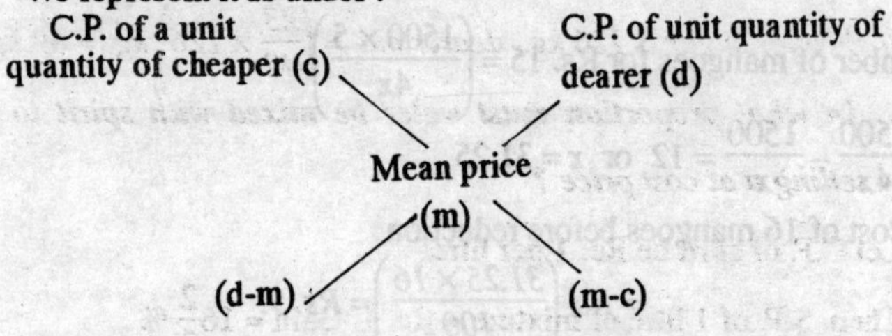


(Cheaper quantity) : (dearer quantity) = (d – m) : (m – c).

## Solved Problems.

**Ex. 1.** ***In what proportion must rice at Rs. 3.10 per kg. be mixed with rice at Rs. 3.60 per kg., so that the mixture be worth Rs. 3.25 a kg. ?***

**Sol.**

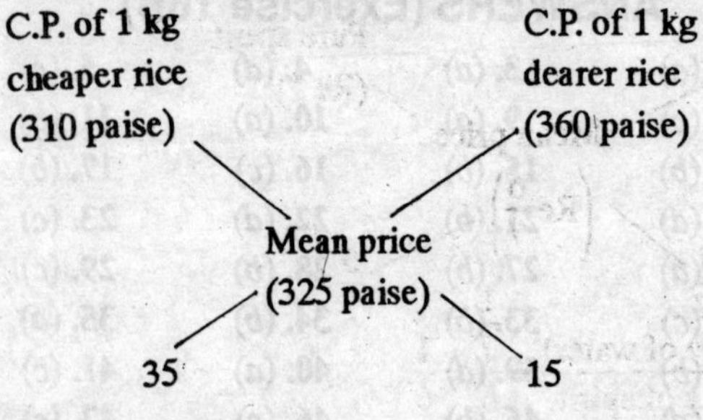


By the alligation rule :

$$\frac{\text{(Quantity of cheaper rice)}}{\text{(Quantity of dearer rice)}} = \frac{35}{15} = \frac{7}{3}.$$

∴ *They must be mixed in the ratio 7 : 3.*

**Ex. 2.** ***How many kilograms of sugar costing Rs. 6.10 per kg. must be mixed with 126 kg. of sugar costing Rs. 2.85 per kg. so that 20% may be gained by selling the mixture at Rs. 4.80 per kg. ?***

**Sol.** S.P. of 1 kg. of mixture = Rs. 4.80, Gain = 20%.

$\therefore$ C.P. of kg. of mixture = Rs. $\left(\frac{100}{120} \times 4.80\right)$ = Rs. 4.

C.P. of 1 kg of Cheaper sugar (285 paise) — C.P. of 1 kg of dearer sugar (610 paise)

Mean price (400 paise)

210 — 115

$$\therefore \frac{\text{(Quantity of cheaper sugar)}}{\text{(Quantity of dearer sugar)}} = \frac{210}{115} = \frac{42}{23}.$$

If cheaper sugar is 42 kg., dearer one = 23 kg.

*If cheaper sugar is 126 kg., dearer one =* $\left(\frac{23}{42} \times 126\right)$ *kg. = 69 kg.*

**Ex. 3.** ***In what proportion must water be mixed with spirit to gain*** $16\frac{2}{3}$ ***% by selling it at cost price ?***

**Sol.** Let C.P. of sprit be Re. 1 per litre.

Then, S.P. of 1 litre of mixture = Re. 1, Gain = $16\frac{2}{3}$ %.

C.P. of 1 litre of mixture = Rs. $\left(\frac{100 \times 3 \times 1}{350}\right)$ = Rs. $\left(\frac{6}{7}\right)$.

C.P. of 1 kg water (Re 0) — C.P. of 1 kg Pure spirit (Re 1)

Mean price $\left(\text{Re } \frac{6}{7}\right)$

$\frac{1}{7}$ — $\frac{6}{7}$

$$\frac{\text{(Quantity of water)}}{\text{Quantity of spirit}} = \frac{\frac{1}{7}}{\frac{6}{7}} = \frac{1}{6}.$$

*or Ratio of water and spirit = 1 : 6.*

**Ex. 4.** ***In what ratio must a person mix three kinds of wheat costing him Rs. 1.20, Rs. 1.44 and Rs. 1.74 per kg., so that the mixture may be worth Rs. 1.41 per kg. ?***

**Sol. Step I.** Mix wheats of first and third kind to get a mixture worth Rs 1.41 per kg. ?

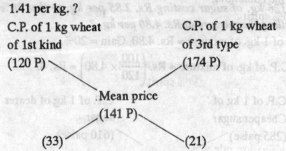

C.P. of 1 kg wheat of 1st kind (120 P) — C.P. of 1 kg wheat of 3rd type (174 P)

Mean price (141 P)

(33) — (21)

By alligation rule :

$$\frac{\text{(Quantity of 1st kind of wheat)}}{\text{(Quantity of 3rd kind of wheat)}} = \frac{33}{21} = \frac{11}{7}$$

i.e., they must be mixed in the ratio 11 : 7.

**Step II.** Mix wheats of 1st and 2nd kind to obtain a mixture worth of Rs. 1.41 per kg.

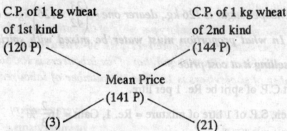

C.P. of 1 kg wheat of 1st kind (120 P) — C.P. of 1 kg wheat of 2nd kind (144 P)

Mean Price (141 P)

(3) — (21)

$\therefore$ By alligation rule :

$$\frac{\text{(Quantity of 1st kind of wheat)}}{\text{(Quantity of 2nd kind of wheat)}} = \frac{3}{21} = \frac{1}{7}.$$

i.e., they must be mixed in the ratio 1 : 7.

Thus, $\frac{\text{(Quantity of 2nd kind of wheat)}}{\text{(Quantity of 3rd kind of wheat)}}$

$$= \frac{\text{(Quantity of 2nd kind of wheat)}}{\text{(Quantity of 1st kind of wheat)}} \times \frac{\text{(Quantity of 1st kind of wheat)}}{\text{(Quantity of 3rd kind of wheat)}}$$

$$= \left(\frac{7}{1} \times \frac{11}{7}\right) = \left(\frac{11}{1}\right).$$

$\therefore$ *Quantities of wheat of (1st kind : 2nd kind : 3rd kind)*

$$= \left(1 : 7 : \frac{7}{11}\right) = (11 : 77 : 7).$$

**Ex. 5.** ***A butler stole wine from a butt of sherry which contained 40% of spirit and he replaced, what he had stolen by wine containing only 16% spirit. The butt was then of 24% strength only. How much of the butt did he steal ?***

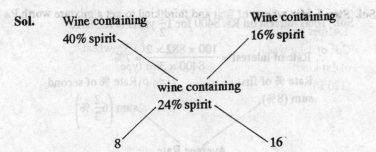

**Sol.** Wine containing 40% spirit — Wine containing 16% spirit

wine containing 24% spirit

8 — 16

∴ By alligation rule :

$$\frac{\text{(Wine with 40\% Spirit)}}{\text{(Wine with 16\% Spirit)}} = \frac{8}{16} = \frac{1}{2}.$$

*i.e.*, they must be mixed in the ratio (1 : 2).

*Thus $\frac{1}{3}$ of the butt of sherry was left and hence the butler drew out $\frac{2}{3}$ of the butt.*

**Ex. 6.** ***The average weekly salary per head of the entire staff of a factory consisting of supervisors and the loboure rs is Rs. 60- The average salary per head of the supervisors is Rs. 400 and that of the labourers is Rs. 56. Given that the number of supervisors is 12, find the number of labourers in the factory.***

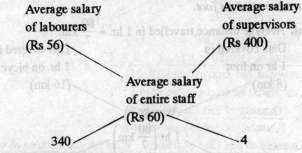

**Sol.** Average salary of labourers (Rs 56) — Average salary of supervisors (Rs 400)

Average salary of entire staff (Rs 60)

340 — 4

By alligation rule :

$$\frac{\text{(Number of labourers)}}{\text{(Number of supervisors)}} = \frac{340}{4} = \frac{85}{1}$$

Thus, if the number of supervisors is 1, number of labourers = 85

∴ *If the number of supervisors is 12, number of labourers = 85 × 12 = 1020.*

**Ex. 7.** ***A man possessing Rs. 8400 lent a part of it at 8% simple interest and the remaining at $6\frac{2}{3}$ % simple interest. His total income after $1\frac{1}{2}$ years was Rs. 882. Find the sum lent at different rates.***

**Sol.** Total interest on Rs. 8400 for $1\frac{1}{2}$ years is Rs. 882.

$\therefore$ Rate of interest $= \frac{100 \times 882 \times 2}{8400 \times 3} = 7\%$

Rate % of first sum (8%) — Rate % of second sum $\left(6\frac{2}{3}\%\right)$

Average Rate (7%)

$\frac{1}{3}$ — 1

Now

By alligation rule :

$$\frac{\text{Money Given at 8\% S.I.}}{\text{Money Given at } 6\frac{2}{3}\text{ \% S.I.}} = \frac{1}{3} : 1 = 1 : 3$$

$\therefore$ Money lent at 8% = Rs. $\left(8400 \times \frac{1}{4}\right)$ = Rs. 2100

*Money lent at* $6\frac{2}{3}$ % *= Rs.* $\left(8400 \times \frac{3}{4}\right)$ *= Rs. 6300*

**Ex. 8.** ***A man travelled a distance of 80 km. in 7 hours partly on foot at the rate of 8 km. per hour and partly on bicycle at 16 km. per hour. Find the distance travelled on foot.***

**Sol.** Average distance travelled in 1 hr. $= \frac{80}{7}$ km.

Dist. covered in 1 hr on foot (8 km) — Dist covered in 1 hr. on bicycle (16 km)

Average in 1 hr $\left(\frac{80}{7}\text{ km}\right)$

$\frac{32}{7}$ — $\frac{24}{7}$

By alligation rule :

$$\frac{\text{Time taken on foot}}{\text{Time taken on bicycle}} = \frac{32}{24} = 4 : 3.$$

Thus out of 7 hours in all, he took 4 hours to travel on foot.

*Distance covered on foot in 4 hours* = (4 × 8) *km* = *32 km.*

**Ex. 9.** ***A sum of Rs. 41 was divided among 50 boys and girls. Each boy gets 90 paise and a girl 65 paise. Find the number of boys and girls.***

**Sol.** Average money received by each = Rs. $\frac{41}{50}$ = 82 P.

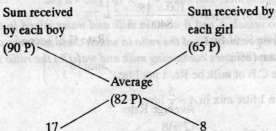

Sum received by each boy (90 P) — Sum received by each girl (65 P)

Average (82 P)

17 — 8

By alligation rule :

*Ratio of boys and girls = 17 : 8.*

**Ex. 10.** ***A lump of two metals weighing 18 gms. is worth Rs. 87 but if their weights be interchanged, it would be worth Rs. 78.60. If the price of one metal be Rs. 6.70 per gm., find the weight of the other metal in the mixture.***

**Sol.** If one lump is mixed with another lump with the quantities of metals interchanged then the mixture of the two lumps would contain 18 gm. of first metal and 18 gm. of second metal and the price of the mixture would be Rs. (87 + 78.60) or Rs. 165.60.

$\therefore$ cost of (18 gm. of 1st metal + 18 gm. of 2nd metal)

= Rs. 165.60

So, cost of (1 gm. of 1st metal + 1 gm. of 2nd metal)

= Rs. $\frac{165.60}{18}$ = Rs. 9.20.

(cost of 1 gm. of 1st metal) + (cost of 1 gm. of 2nd metal)

= Rs. 9.20

Cost of 1 gm. of 2nd metal = Rs. (9.20 – 6.70) = Rs. 2.50

Now, mean price of lump = Rs. $\left(\frac{87}{18}\right)$ per gm. = Rs. $\left(\frac{29}{6}\right)$.

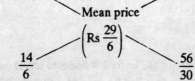

C.P. of 1 gm of 1st metal (Rs 6.70) — C.P. of 1 gm of 2nd metal (Rs 2.50)

Mean price $\left(\text{Rs } \frac{29}{6}\right)$

$\frac{14}{6}$ — $\frac{56}{30}$

$\therefore$ By alligation rule :

$$\frac{\text{Quantity of 1st metal}}{\text{Quantity of 2nd metal}} = \frac{14}{6} : \frac{56}{30} = 5 : 4.$$

In 9 gm. of mix., 2nd metal = 4 gm.

*In 18 gm. of mix., 2nd metal* = $\left(\frac{4}{9} \times 18\right)$ *gm. = 8 gm.*

**Ex. 11.** ***Two vessels A and B contain milk and water mixed in the ratio 5 : 2 and 8 : 5 respectively. Find the ratio in which these mixtures are to be mixed to get a new mixture containing milk and water in the ratio 9 : 4.***

**Sol.** Let the C.P. of milk be Re. 1 per litre.

Milk in 1 litre mix in $A = \frac{5}{7}$ litre.

Milk in 1 litre mix. in $B = \frac{8}{13}$ litre.

Milk in 1 litre mix. of this mix. $= \frac{9}{13}$ litre.

C.P. of 1 litre mix. in $A$ = Rs. $\frac{5}{7}$

C.P. of 1 litre mix. in $B$ = Rs. $\frac{8}{13}$.

Mean price = Rs. $\left(\frac{9}{13}\right)$

C.P. of 1 lt. mixture in A — C.P. of 1 lt. mixture in B

$\left(\text{Rs } \frac{5}{7}\right)$ — $\left(\text{Rs } \frac{8}{13}\right)$

Mean price

$\left(\text{Rs } \frac{9}{13}\right)$

$\frac{1}{13}$ — $\frac{2}{91}$

$\therefore$ (Mix. in $A$) : (Mix in $B$) $= \frac{1}{13} : \frac{2}{91} = 7 : 2.$

**Ex. 12** ***A container contains 80 kg. of milk. From this container, 8 kg. of milk was taken out and replaced by water. This process was further repeated two times. How much milk is now contained by the container ?***

**Remarks.** Amount of liquid left after $n$ operations, when the container originally contains $x$ units of liquid, from which $y$ units is taken out each time is $\left[x\left(1-\frac{y}{x}\right)^n\right]$ *units.*

**Sol.** Amount of milk left

$$80\left[\left(1-\frac{8}{80}\right)^3\right] \text{ kg.} = 58.34 \text{ kg.}$$

## EXERCISE 17A (Subjective)

1. In what proportion must wheat at Rs. 1.60 per kg. be mixed with wheat at Rs. 1.45 per kg., so that the mixture be worth Rs. 1.54 per kg. ?
2. Milk and water are mixed in a vessel *A* in the ratio 4 : 1 and in a vessel *B* in the ratio 5 : 2. Find, what quantities should be taken from the vessels to have a new mixture consisting of 7 kg. of milk and 2 kg. of water.
3. A merchant buys spirit at a certain rate per gallon and after mixing it with water, sells it again at the same rate. Find, how much water is there in one gallon of the mixture, if the merchant makes a profit of 25%.

   **Hint.** *Let the C.P. of spirit be Re. 1 per gallon.*

   *Then, S.P. of the mixture = Re. 1 per gallon.*

   *Gain = 25%, So, C.P. of mix*

   $$= \text{Rs.} \left(\frac{100}{125} \times 1\right) = Rs. \left(\frac{4}{5}\right).$$

   ∴ *By alligation rule :–*

   ∴ *Ratio of spirit and water* $= \left(\frac{4}{5} : \frac{1}{5}\right) = (4 : 1).$

4. A merchant has 100 kg. of sugar, part of which he sells at 7% profit and the rest at 17% profit. He gains 10% on the whole. How much is sold at 17% ?

   **Hint.**

   *Ratio of (Sugar sold at 7% gain and Sugar sold at 17% gain = 7 : 3*

   *out of a total of 10 kg., sugar sold at 17% gain = 3 kg.*

   *out of a total of 100 kg., sugar sold at 17% gain*

   $$= \left(\frac{3}{10} \times 100\right) kg. = 30\ kg.$$

5. A mixture of 40 litres of milk and water contains 10% of water. How much water must be added to make the water 20% in the new mixture ?

   **Hint.** *Quantity of water in given mix* $= \left\{\frac{10}{100} \times 40\right\} = 4$ *litres.*

   *Quantity of milk in given mix.. = (40 – 4) = 36 litres.*

   *Let x litres of water be added to it.*

   *Then, milk =36 litres, water = (4 + x) litres*

   *and Total mix. = (40 + x ) litres.*

   $$\therefore \frac{(4+x)}{(40+x)} \times 100 = 20, \text{ find } x.$$

6. A merchant borrowed Rs. 2500 from two money lenders. For one loan he paid 8% per annum and for the other 6% per annum. The total interest paid for 1 year was Rs. 180. How much did he borrow at each rate ?

**7.** How many kgs. of rice worth Rs. 3.60 per kg. must be mixed with 8 kg. of rice worth Rs. 4.20 per kg., so that by selling the mixture at Rs. 4.40 per kg., there may be a gain of 10% ?

**Hint.** *C.P. of 1 kg. of mix.* $= Rs. \left(\frac{100}{110} \times 4.40\right) = Rs. 4.$

**8.** Two bottles *A* and *B* contain diluted sulphuric acid. In the bottle *A*, the amount of water is double the amount of acid while in bottle *B*, the amount of acid is 3 times that of water. How much mixture should be taken from each bottle in order to prepare 5 litres of diluted sulphuric acid containing equal amount of acid and water ?

**Hint.** *Quantity of water in bottle A* $= \frac{2}{3}$ *of the whole.*

*Quantity of water in bottle B* $= \frac{1}{4}$ *of the whole.*

*Quantity of water in resultant mixture* $= \frac{1}{2}$ *of the whole.*

$\therefore$ *Ratio of mixtures from A and B* $= \frac{1}{4} : \frac{1}{6} = 3 : 2.$

**9.** A tea merchant buys two kinds of tea, the price of the first kind being twice that of the second. He sells the mixture at Rs. 17.50 per kg., thereby making a profit of 25%. If the ratio of first and second kinds of tea in the mixture be 2 : 3, find the cost price of each kind of tea.

**Hint.** *In 1 kg. of mixture :–*

*tea of 1st kind* $= \frac{2}{5}$ *kg. and tea of 2nd kind* $= \frac{3}{5}$ *kg.*

*Let the price of first kind of tea be Rs. x per kg.*

*Then, price of second kind* $= Rs. \left(\frac{x}{2}\right)$ *per kg.*

$\therefore$ *Price of 1 kg. mix.* $= Rs. \left(\frac{2x}{5} + \frac{3x}{10}\right) = Rs. \left(\frac{7x}{10}\right)$

*But, C.P. of mix. per kg.* $= Rs. \left(\frac{17.50 \times 100}{125}\right) = Rs. 14.$

$\therefore \frac{7x}{10} = 4$, *find x.*

**10.** A cup of milk contains 3 parts pure milk and one part of water. How much of the mixture must be withdrawn and substituted by water in order that the resulting mixture may be half milk and half water.

**Hint.** *Let the total quantity of mixture be 1 litre. Then, it contains* $\frac{3}{4}$ *litre of milk and* $\frac{1}{4}$ *litre of water. Let x kg. of mixture be taken out and substituted by water. Then,*

*Milk in x litres* $= \left(\frac{3}{4}x\right)$ *litres and water in x litres* $= \left(\frac{x}{4}\right)$ *litres.*

$\therefore$ *Milk in remaining mix.* $= \left(\frac{3}{4} - \frac{3x}{4}\right) = \left(\frac{(3-3x)}{4}\right)$ *litres.*

*And, water in remaining mix.* $= \left(\frac{1}{4} - \frac{x}{4} + x\right) = \frac{(3x+1)}{4}$ *litres.*

$\therefore \frac{3-3x}{4} = \frac{3x+1}{4}$ *or* $3 - 3x = 3x + 1$ *or* $x = \frac{1}{3}$

**11.** Four gallons are drawn from a cask full of wine. It is then filled with water. Four gallons of mixture are again drawn and the cask is again filled with water. The quantity of wine now left in the cask is to that of wine in it is 36 : 13. How much does the cask hold ?

**Hint.** *Let the Cask hold x litres.*

$\therefore$ *Wine left after 2 operations* $= \left(1 - \frac{4}{x}\right)^2$ *gallons.*

$\therefore \left(1 - \frac{4}{x}\right)^2 = \frac{36}{49} = \left(\frac{6}{7}\right)^2$ *or* $1 - \frac{4}{x} = \frac{6}{7}$*, find x.*

**12.** In what ratio should water be mixed with a liquid costing Rs. 12 per litre, so as to make a profit of 25% by selling the diluted liquid at Rs. 13.75 per litre ?

**13.** A sum of Rs. 18.45 is made up of 90 coins which are either 10 paise coins or 25 paise coins. Find the number of each type of coins.

**Hint.** *Average value of coins* $= \frac{1845}{90} = \frac{41}{2}$ *paise.*

*So, ratio of coins*$= \frac{9}{2} : \frac{21}{2} = 3 : 7.$

**14.** A dishonest milkman professes to sell his milk at cost price, but he mixes it with water and thereby gains 25%. Find the percentage of water in the mixture.

**Hint.** *Let the C.P. of 1 litre of milk be Re. 1.*

*Then, S.P. of 1 litre of mixture = Re. 1.*

$\therefore$ *C.P. of 1 litre of mixture* = *Rs.* $\left(\frac{100}{125} \times 1\right)$ = *Rs.* $\left(\frac{4}{5}\right)$

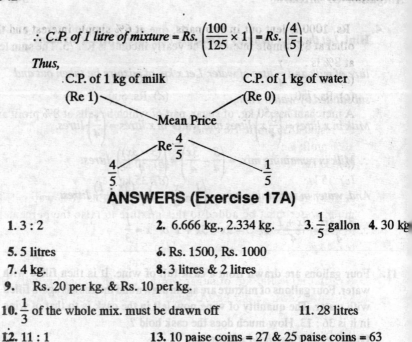

*Thus,*

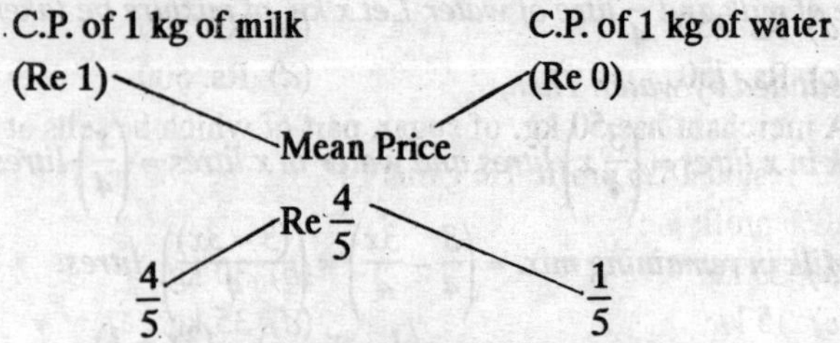


## ANSWERS (Exercise 17A)

**1.** 3 : 2 **2.** 6.666 kg., 2.334 kg. **3.** $\frac{1}{5}$ gallon **4.** 30 kg

**5.** 5 litres **6.** Rs. 1500, Rs. 1000

**7.** 4 kg. **8.** 3 litres & 2 litres

**9.** Rs. 20 per kg. & Rs. 10 per kg.

**10.** $\frac{1}{3}$ of the whole mix. must be drawn off **11.** 28 litres

**12.** 11 : 1 **13.** 10 paise coins = 27 & 25 paise coins = 63

**14.** 20% .

## EXERCISE 17B (Objective Type Questions)

**1.** In what proportion must wheat at Rs. 1.60 per kg. be mixed with wheat at Rs. 1.45 per kg., so that the mixture be worth Rs. 1.54 per kg. ?

(*a*) 2 : 3 (*b*) 3 : 2

(*c*) 3 : 4 (*d*) 4 : 3

**2.** 15 litres of a mixture contains 20% alcohol and the rest water. If 3 litres of water be mixed in it, the percentage of alcohol in the new mixture will be :

(*a*) 17 (*b*) $16\frac{2}{3}$

(*c*) $18\frac{1}{2}$ (*d*) 15 **(Clerical Grade 1991)**

**3.** A grocer buys two kind of rice at Rs. 1.80 and Rs. 1.20 per kg. respectively. In what proportion should these be mixed, so that by selling the mixture at Rs. 1.75 per kg., 25% may be gained ?

(*a*) 2 : 1 (*b*) 3 : 2

(*c*) 3 : 4 (*d*) 1 : 2

**Hint.** Mean Price = Rs. 1.40.

4. Rs. 1000 is lent out in two parts, one at 6% simple interest and the other at 8% simple interest. The yearly income is Rs. 75. The sum lent at 8% is :

(*a*) Rs. 250 (*b*) Rs. 500
(*c*) Rs. 750 (*d*) Rs. 600

5. A merchant has 50 kg. of sugar, part of which he sells at 8% profit and the rest at 18% profit. He gains 14% on the whole. The quantity sold at 18% profit is :

(*a*) 20 kg. (*b*) 30 kg.
(*c*) 15 kg. (*d*) 35 kg.

6. A mixture of 20 kg. of spirit and water contains 10% water. How much water must be added to this mixture to raise the percentage of water to 25% ?

(*a*) 4 kg. (*b*) 5 kg.
(*c*) 8 kg. (*d*) 30 kg.

7. Kantilal mixes 80 kg. of sugar worth of Rs. 6.75 per kg. with 120 kg. worth of Rs. 8 per kg. At what rate shall he sell the mixture to gain 20% ?

(*a*) Rs. 7.50 (*b*) Rs. 9
(*c*) Rs. 8.20 (*d*) Rs. 8.85

**(S.B.I. P.O. Exam. 1987)**

8. A jar full of whisky contains 40% of alcohol. A part of this whisky is replaced by another containing 19% alcohol and now the percentage of alcohol was found to be 26. The quantity of whisky replaced is :

(*a*) $\frac{2}{5}$ (*b*) $\frac{1}{3}$
(*c*) $\frac{2}{3}$ (*d*) $\frac{3}{5}$ **(Hotel Management, 1991)**

9. Two vessels *A* and *B* contain milk and water mixed in the ratio 5 : 3 and 2 : 3. When these mixtures are mixed to form a new mixture containing half milk and half water, they must be taken in the ratio :

(*a*) 2 : 5 (*b*) 3 : 5
(*c*) 4 : 5 (*d*) 7 : 3

10. The ratio of milk and water in 66 kg. of adulterated milk is 5 : 1. Water is added to it to make the ratio 5 : 3. Tho quantity of water added is :

(*a*) 22 kg. (*b*) 24.750 kg.
(*c*) 16.500 kg. (*d*) 20 kg.

11. Some amount out of Rs. 7000 was lent at 6% p.a. and the remaining at 4% p.a. If the total simple interest from both the fractions in 5 years was Rs. 1600, the sum lent at 6% p.a. was :

(*a*) Rs. 2000 (*b*) Rs. 5000

(*c*) Rs. 3500 (*d*) none of these

**(Bank P.O. 1988)**

**12.** 729 ml. of a mixture contains milk and water in the ratio 7 : 2. How much more water is to be added to get a new mixture containing milk and water in ratio 7 : 3 ?

(*a*) 600 ml. (*b*) 710 ml.

(*c*) 520 ml. (*d*) none of these **(Railways, 1991)**

**13.** A dishonest milkman professes to sell his milk at C.P. but he mixes it with water and thereby gains 25%. The percentage of water in the mixture is :

(*a*) 25% (*b*) 20%

(*c*) 4% (*d*) none of these

**14.** A sum of Rs. 41 was divided among 50 boys and girls. Each boy gets 90 paise and a girl 65 paise. The number of boys is :

(*a*) 16 (*b*) 34

(*c*) 14 (*d*) 36

**15.** A can contains a mixture of two liquids *A* and *B* in proportion 7 : 5. When 9 litres of mixture are drawn off and the can is filled with *B*, the proportion of *A* and *B* becomes 7 : 9. How many litres of liquid *A* was contained by the can initially ?

(*a*) 25 (*b*) 10

(*c*) 20 (*d*) 21 **(Railways, 1991)**

**16.** In a mixture of 60 litres, the ratio of milk and water is 2 : 1. If the ratio of the milk and water is to be 1 : 2, then the amount of water to be further added is :

(*a*) 20 litres (*b*) 30 litres

(*c*) 40 litres (*d*) 60 litres **(N.D.A. Exam. 1990)**

## HINTS & SOLUTIONS (Exercise 17B)

**1.**

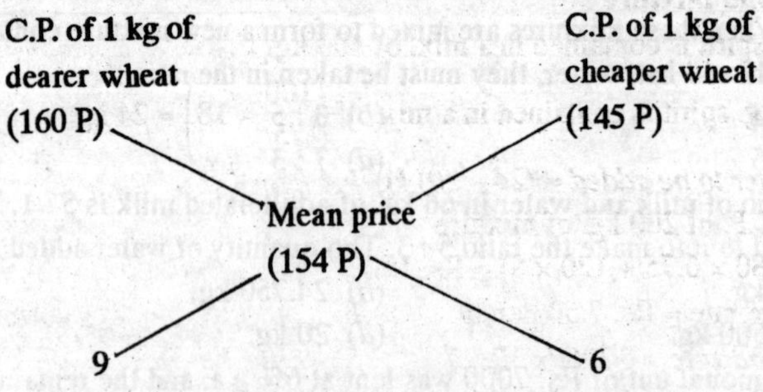


∴ *(Dearer wheat) : (cheaper wheat) = 9 : 6 = 3 : 2.*

**2.** Initially, the mixture contains 3 litres of alcohol and 12 litres of water. Afterwards, the mixture contains 3 litres of alcohol and 15 litres of water.

$\therefore$ *Percentage of alcohol* $= \left(\frac{3}{18} \times 100\right)\% = 16\frac{2}{3}\%$.

**3.** S.P. of 1 kg. mixture = Rs. 1.75, Gain = 25%.

$\therefore$ Mean price = Rs. $\left(\frac{1.75 \times 100}{125}\right)$ = Rs. 1.40.

$\therefore$ *(Dearer rice) : (cheaper rice) = 20 : 40 = 1 : 2.*

**4.** Total interest = Rs. 75.

Average rate $= \left(\frac{100 \times 75}{1000 \times 1}\right)\% = 7\frac{1}{2}\%$

$\therefore$ (Sum at 6%) : (Sum at 8%) $= \frac{1}{2} : \frac{3}{2} = 1 : 3$.

*Hence, sum at 8% = Rs.* $\left(1000 \times \frac{3}{4}\right)$ *= Rs. 750.*

**5.**

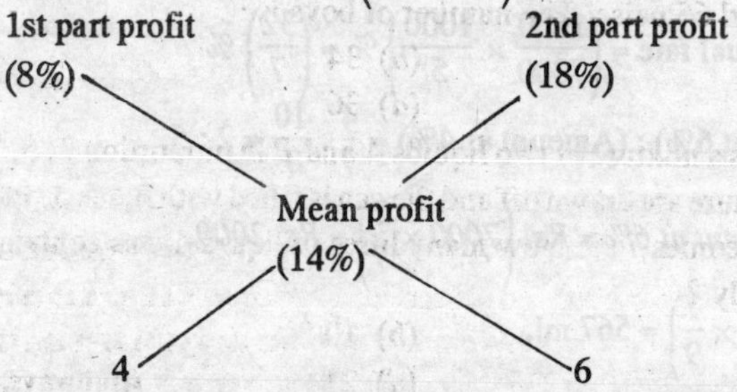


Ratio of 1st and 2nd part = 4 : 6 = 2 : 3.

$\therefore$ *Quantity sold at 18%* $= \left(50 \times \frac{3}{5}\right)$ *kg. = 30 kg.*

**6.** **In first mixture :**

water $= \left(\frac{10}{100} \times 20\right)$ kg. and spirit = 18 kg.

**In second mixture :**

75 kg. spirit is contained in a mix. of 100 kg.

$\therefore$ 18 kg. spirit is contained in a mix. of $\left(\frac{100}{75} \times 18\right)$ = 24 kg.

*So, water to be added = (24 – 20) kg. = 4 kg.*

**7.** Total C.P. of 200 kg. of mixture

= Rs. (80 × 6.75 + 120 × 8) = Rs. 1500.

Average rate = Rs. 7.50 per kg.

*Required rate = 120% of Rs. 7.50 = Rs. 9 per kg.*

**8.** Using the method of alligation.

Required ratio = 7 : 14 = 1 : 2

$\therefore$ *Required quantity* $= \frac{2}{3}$.

**9.** Milk in $A = \frac{5}{8}$ of whole, Milk in $B = \frac{2}{5}$ of whole, Milk in mixture of $A$ and $B = \frac{1}{2}$.

$\because$ *By alligation rule, (Mix. in A) : (Mix in B)* $= \frac{1}{10} : \frac{1}{8} = 4 : 5$.

**10.** **In first mixture :**

Milk $= \left(\frac{66 \times 5}{6}\right) = 55$ kg. and water = 11 kg.

**In second mixture :**

If milk is 55 kg., then water $= \left(\frac{3}{5} \times 55\right) = 33$ kg.

$\therefore$ water to be added = 22 kg.

**11.** Average annual rate $= \left(\frac{1600}{7000} \times \frac{1000}{5}\right)\% = \left(\frac{32}{7}\right)\%$

$\therefore$ (Amount at 6%) : (Amount at 4%) $= \frac{4}{7} : \frac{10}{7} = 2 : 5$.

*Hence, sum lent at 6% = Rs.* $\left(7000 \times \frac{2}{7}\right)$ *= Rs. 2000.*

**12.** Milk $= \left(729 \times \frac{7}{9}\right) = 567$ ml.

Water = (729 – 567) = 162 ml.

Now, $\frac{567}{162 + x} = \frac{7}{3} \Rightarrow x = 81$.

**13.** Let C.P. of 1 litre of milk be Re. 1.

Then, S.P. of 1 litre of mixture = Re.1.

Gain = 25%.

$\therefore$ C.P. of 1 litre of mixture = Re. $\left(\frac{100}{125} \times 1\right) =$ Re. $\frac{4}{5}$

$\therefore$ Ratio of milk and water $= \frac{4}{5} : \frac{1}{5} = 4 : 1$.

*Hence, percentage of water in the mixture* $= \left(\frac{100 \times 1}{5}\right)\% = 20\%$.

**14.** Average money received by each = Rs. (41/50) = 82 paise.

Ratio of boys and girls = 17 : 8.

$\therefore$ *Number of boys* $= \left(50 \times \frac{17}{25}\right) = 34$.

**15.** Let the can initially contain $7x$ litres and $5x$ litres of mixtures $A$ and $B$ respectively. Thus, out of $12x$ litres of total mixture, 9 litres were taken out.

Quantity of $A$ in mix. left

$$= \left(7x - \frac{9}{12x} \times 7x\right) = \left(\frac{28x - 21}{4}\right) \text{ litres.}$$

Quantity of B in mix. left

$$= \left(5x - \frac{9}{12x} \times 5x\right) = \left(\frac{20x - 15}{4}\right) \text{ litres.}$$

$$\therefore \left(\frac{28x - 21}{4} : \frac{20x - 15}{4} + 9\right) :: (7 : 9) \text{ or } x = 3.$$

**16.** Ratio of milk and water in mixture of 60 litres = 2 : 1.

∴ Quantity of milk = 40 litres.

Quantity of water = 20 litres.

If ratio of milk and water is to be 1 : 2, then in 40 litres of milk, water should be 80 litres.

∴ *Quantity of water to be added = 60 litres.*

## ANSWERS (Exercise 17B)

| | | | | | |
|---|---|---|---|---|---|
| **1.** (*b*) | **2.** (*b*) | **3.** (*d*) | **4.** (*c*) | **5.** (*b*) | **6.** (*a*) |
| **7.** (*b*) | **8.** (*c*) | **9.** (*c*) | **10.** (*a*) | **11.** (*a*) | **12.** (*d*) |
| **13.** (*b*) | **14.** (*b*) | **15.** (*d*) | **16.** (*d*). | | |

# 18

# SIMPLE INTEREST

*If a person borrows some money from some one for a certain period, then the borrower has to pay some extra money, called* **Interest on the money borrowed for that period.** *The money borrowed is called* **the Principal** *and the total sum of principal and the interest is called* **the Amount.**

*If the interest on a certain sum borrowed for a certain period is reckoned uniformly, then it is called* **Simple Interest.**

**Formulae.** The simple interest (S.I.) on a principal $P$ at $R\%$ per annum for $T$ years is given by :

$$S.I. = \frac{P \times R \times T}{100}$$

$$\therefore P = \frac{100 \times S.I.}{R \times T}, R = \frac{100 \times S.I.}{P \times T} \text{ and } T = \frac{100 \times S.I.}{P \times R}.$$

## Solved Problems.

**Ex. 1.** (i) Find simple interest on Rs. 625 at $6\frac{1}{2}$ % per annum for $2\frac{1}{2}$ years.

(ii) Find the amount of Rs. 585 at $4\frac{3}{4}$ % p.a. after 6 months.

(iii) Find S.I. on Rs. 700 at 9% per annum for the period from 5th Feb., 1984 to 18th April, 1984.

(iv) If S.I. = Rs. 108, amount = Rs. 540, rate = 5% per annum, find time.

(v) Amount = Rs. 1110, Time = 100 days, Rate = 5% p.a., Simple Interest = ?

**Sol.** (i) S.I. = Rs. $\left(\frac{625 \times 13 \times 5}{100 \times 2 \times 2}\right)$ = Rs. 101.56.

(ii) S.I. = Rs. $\left(\frac{585 \times 19 \times 1}{100 \times 4 \times 2}\right)$ = Rs. 13.89.

$\therefore$ Amount = Rs. (585 + 13.89) = Rs. 598.89.

(iii) Feb. March April

Time = 24 + 31 + 18 = 73 days = $\frac{1}{5}$ year.

$\therefore$ S.I. = Rs. $\left(\frac{700 \times 9 \times 1}{100 \times 5}\right)$ = Rs. 12.60.

(iv) Principal = (Amount) – (S.I.) = Rs. (540 – 108) = Rs. 432.

$\therefore$ Time $= \left(\frac{100 \times 108}{432 \times 5}\right)$ years $\doteq$ 5 years.

(v) Let the principal be Rs. 100

Then, S.I. = Rs. $\left(100 \times \frac{100}{365} \times 5 \times \frac{1}{100}\right)$ = Rs. $\left(\frac{100}{73}\right)$

$\therefore$ Amount = Rs. $\left(100 + \frac{100}{73}\right)$ = Rs. $\left(\frac{7400}{73}\right)$

Now, if amount is Rs. $\left(\frac{7400}{73}\right)$, then sum = Rs. 100

$\therefore$ If amount is Rs. 1110, then sum

= Rs. $\left(\frac{100 \times 73 \times 1110}{7400}\right)$ = Rs. 1095.

*Hence, simple interest = Rs. (1110 – 1095) = Rs.15.*

**Ex. 2.** ***A sum when reckoned simple interest at*** $3\frac{1}{2}$***% per annum amounts to Rs. 364.80 after 4 years. Find the sum.***

**Sol.** Let the sum be Rs. 100. Then,

S.I. on Rs. 100 = Rs. $\left(\frac{100 \times 7 \times 4}{100 \times 2}\right)$ = Rs. 14.

$\therefore$ Amount = Rs. (100 + 14) = Rs. 114.

If amount is Rs. 114, sum = Rs. 100.

*If amount is Rs. 364.80, sum = Rs.* $\left(\frac{100}{114} \times 364.80\right)$ *= Rs. 320.*

**Ex. 3.** ***A certain sum of money amounts to Rs. 678 in 2 years and to Rs. 736.50 in*** $3\frac{1}{2}$ ***years. Find the sum and the rate of interest.***

**Sol.** (Principal) + (S.I. for $3\frac{1}{2}$ years) = Rs. 736.50.

(Principal) + (S.I. for 2 years) = Rs. 678.

On subtraction, S.I. for $1\frac{1}{2}$ years = Rs. 58.50.

$\therefore$ S.I. for 2 years = Rs. $\left(\frac{58.50 \times 2 \times 2}{3}\right)$ = Rs. 78.

*Principal = Rs. (678 – 78) = Rs. 600.*

*Also, rate* $= \frac{100 \times 78}{600 \times 2} = 6\frac{1}{2}$ *% per annum.*

**Ex. 4.** ***A merchant borrowed Rs. 2500 from two money lenders. For one loan he paid 12% p.a. and for the other 14% per annum. The total interest paid for one year was Rs. 326. How much did he borrow at each rate ?***

**Sol.** Suppose he borrowed Rs. $x$ at 12% and Rs. $(2500-x)$ at 14% per annum.

Then, S.I. on Rs. $x$ = Rs. $\left(\frac{x \times 12 \times 1}{100}\right)$ = Rs. $\left(\frac{3x}{25}\right)$

And, S.I. on Rs. $(2500-x)$ = Rs. $\left[\frac{(2500-x) \times 14 \times 1}{100}\right]$

= Rs. $\left[\frac{17500-7x}{50}\right]$

$\therefore \left(\frac{3x}{25} + \frac{17500-7x}{50}\right) = 326$ *or* $x = 1200$.

*Thus, he borrowed Rs. 1200 at 12% and Rs. 1300 at 14%.*

**Ex. 5.** ***Simple interest on a certain sum is $\frac{9}{16}$ of the sum. Find the rate percent and time, if both are equal.***

**Sol.** Let the sum be Re. 1. Then, S.I. = Re $\frac{9}{16}$.

Now, if rate = $r$% per annum and time = $r$ years, then

$$\frac{1 \times r \times r}{100} = \frac{9}{16} \text{ or } r^2 = \frac{900}{16} \text{ or } r = \frac{30}{4} = \frac{15}{2}.$$

$\therefore$ *Rate* = $7\frac{1}{2}$ *% per annum and Time* = $7\frac{1}{2}$ *years.*

**Ex. 6.** ***At what rate percent per annum will a sum of money double itself in 12 years?***

**Sol.** Let the principal be Rs. $P$ and the rate be $R$% per annum. Since it doubles in 12 years, so Simple interest is Rs. $P$.

$$\therefore P = \frac{P \times R \times 12}{100} \text{ or } 12R = 100 \text{ or } R = \frac{100}{12} = 8\frac{1}{3}$$

*Hence, rate* = $8\frac{1}{3}$ *% per annum.*

**Ex. 7.** ***What annual instalment will discharge a debt of Rs. 2210 due in 4 years at 7% simple interest?***

**Sol.** Let the annual payment be Rs. $x$. Since the first instalment is to be paid after 1 year, the sum of amounts of Rs. $x$ for 3 years, of Rs. $x$ for 2 years, of Rs. $x$ for 1 year and Rs. $x$, gives the amount of debt.

Now, amount of Rs. $x$ for 3 years

= Rs. $\left[x + \frac{x \times 7 \times 3}{100}\right]$ = Rs. $\left(\frac{121x}{100}\right)$

Amount of Rs. $x$ for 2 years

= Rs. $\left[x + \frac{x \times 7 \times 2}{100}\right]$ = Rs. $\left(\frac{57x}{50}\right)$

Amount of Rs. $x$ for 1 year

$$= \text{Rs.}\left[x + \frac{x \times 7 \times 1}{100}\right] = \text{Rs.}\left(\frac{107x}{100}\right)$$

$$\therefore \left(\frac{121x}{100} + \frac{57x}{50} + \frac{107x}{100} + x\right) = 2210 \text{ or } x = 500.$$

*Hence, an annual payment of Rs. 500 will discharge the debt.*

**Ex. 8.** ***A sum was put at simple interest at a certain rate for 2 years. Had it been put at 3% higher rate, it would have fetched Rs. 72 more. Find the sum.***

**Sol.** Let the sum be Rs. 100 and original rate $r\%$ p.a.

Then, S.I. at $r\% = \text{Rs.}\left(\frac{100 \times r \times 2}{100}\right) = \text{Rs. } 2r$

S.I. at $(r+3)\% = \text{Rs.}\left[\frac{100 \times (r+3) \times 2}{100}\right] = \text{Rs. } (2r+6).$

Difference in S.I.'s = Rs. $(2r + 6 - 2r)$ = Rs. 6.

If the diff. in S.I. is Rs. 6, sum = Rs. 100.

*If the diff. in S.I. is Rs. 72, sum =* $\textit{Rs.}\left(\frac{100}{6} \times 72\right) = \textit{Rs. } 1200.$

**Ex. 9.** ***A mahajan lends out Rs. 9 on the condition that the loan is payable in 10 months by 10 equal instalments of Re. 1 each. Find the rate percent simple interest.***

**Sol.** Since the borrower pays a monthly instalment of Re. 1 to the lender, it means the lender keeps his Re. 1 for 9 months, Re. 1 for 8 months, Re. 1 for 7 months and so on. On the other hand, the borrower keeps his Rs. 9 for 10 months. Therefore,

[Rs. 10 + S.I. on Re.1 for $(9+8+7+6+5+4+3+2+1)$ months]

= (Rs. 9 + S.I. on Rs. 9 for 10 months)

$\therefore$ (Rs. 10 + S.I. on Re. 1 for 45 months)

= (Rs. 9 + S.I. on Re. 1 for 90 months)

or S.I. on Re. 1 for 45 months = Re. 1

$\therefore$ *S.I. on Rs. 100 for 1 year =* $\textit{Rs.}\left(\frac{100 \times 12 \times 1}{45}\right) = 26\frac{2}{3}\%.$

## EXERCISE 18A (Subjective)

1. Find the unknown term in the following :
   (i) Principal = Rs. 212.50, Rate = 15%, Time = 4 years.
   Simple interest = ?
   (ii) Principal = Rs. 1000, Amount = Rs. 1015,
   Time : From April, 6 to June, 18; Rate = ?
   (iii) Principal = Rs. 8500, Amount = Rs. 15767.50,

Rate = $9\frac{1}{2}$% p.a., Time = ?

**2.** At what rate percent per annum simple interest will Rs. 750 amount to Rs. 900 in 3 years ?

**3.** A man deposits Rs. 4200 in a bank at 8% per annum and Rs. 1400 in a post office at 6% per annum. Find the rate of interest for the whole sum.
**Hint.** *The sum of interests on Rs. 4200 at 8% p.a. for 1 year and that on Rs. 1400 at 6% p.a. is the S.I. on Rs. 5600 for 1 year. Find rate.*

**4.** The simple interest on a certain sum amounts to $\frac{3}{8}$ of the whole sum at the end of 6 years 3 months. Find rate percent per annum.

**5.** Divide Rs. 9300 into two parts such that the simple interest on one part for 3 years at $4\frac{1}{2}$ % per annum may be equal to that on the other part for $3\frac{1}{2}$ years at 5% per annum.
**Hint.** *Let one part be x. Then, another part* = (9300 – *x*).

$$\therefore \left[x\times 3\times\frac{9}{2}\times\frac{1}{100}\right]=\left[(9300-x)\times\frac{7}{2}\times\frac{5}{100}\right], \textit{find } x.$$

**6.** A man had Rs. 2000, part of which he lent at 5% and the rest at 4%. The whole interest received was Rs. 96. How much did he lend at 4% ?

**7.** If Rs. 5600 amount to Rs. 6678 in $3\frac{1}{2}$ years. What will Rs. 9400 amount to in $5\frac{1}{4}$ years at the same rate percent per annum.

**Hint.** *Rate* = $\left(\frac{100\times 1078\times 2}{5600\times 7}\right)=5\frac{1}{2}$ *% p.a.*

$\therefore$ *S.I. on Rs. 9400 for* $5\frac{1}{4}$ *years*

$$= \text{Rs.}\left(\frac{9400}{100}\times\frac{21}{4}\times\frac{11}{2}\right)=\text{Rs. } 2714.25.$$

**8.** A sum of money lent out at simple interest amounts to Rs. 2990 in 2 years and to Rs. 3282.50 in $3\frac{1}{2}$ years. Find the sum of money and the rate of interest.

**9.** A father left a will of Rs. 9000 to be divided between his two sons aged 12 years and 14 years such that they may get equal amount when each attains the age of 18 years. If the money is reckoned at 10% per annum simple interest, find how much each gets at the time of the will.
**Hint.** *Suppose the younger one gets Rs. x and elder gets Rs.* (5000 – *x*) *at the time of will. Then,*

*Amount of Rs. x for 6 yrs. = Amt. of Rs. (5000 – x) for 4 yrs.*

**10.** Divide Rs. 18102 in three parts so that their amounts after 1, 2, 3 years at 10% per annum may all be equal.

**Hint.** *Let amount in each case be Rs. 100. Then,*

$$\text{Sum in 1st case} = \text{Rs.}\left(\frac{100}{110}\times 100\right) = \text{Rs.}\frac{1000}{11}$$

$$\text{Sum in 2nd case} = \text{Rs.}\left(\frac{100}{120}\times 100\right) = \text{Rs.}\frac{250}{3}$$

$$\text{Sum in 3rd case} = \left(\frac{100}{130}\times 100\right) = \text{Rs.}\frac{1000}{13}$$

$$\therefore \text{Ratio of Principal} = \left(\frac{1000}{11} : \frac{250}{3} : \frac{1000}{13}\right) \text{ or } (156 : 143 : 132)$$

**11.** A man promises his wife a birthday present, giving her each year a number of rupees equal to the number of years in her age. If her birthday falls on Aug. 8th, what sum must be placed at simple interest at 7% on Jan. 1st, before she is 42, in order to raise the required sum ?

**Hint.** *Let the sum be Rs. 100.*

$$\text{Time from 1st Jan. to Aug. 8th} = 219 \text{ days} = \frac{3}{5} \text{ years.}$$

$$\text{S.I. on Rs. 100 for } \frac{3}{5} \text{ years at } 7\% = \text{Rs.}\frac{21}{5}.$$

$$\text{If reqd. money is Rs.}\frac{21}{5}, \text{ sum} = \text{Rs. } 100.$$

$$\text{If reqd. money is Rs. 42, sum} = \text{Rs.}\left(100\times\frac{5}{21}\times 42\right) = \text{Rs. } 1000.$$

**12.** What quarterly payment will discharge a debt of Rs. 2120 due in one year at 16% per annum simple interest ?

**Hint.** *Let each instalment be of Rs. x. Then sum of the amount of Rs. x for 9 months, the amount of Rs. x for 6 months, the amount of Rs. x for 3 months and x gives Rs. 2120. Find x.*

## ANSWERS (Exercise 18A)

**1.** (i) Rs. 127.50 (ii) $7\frac{1}{2}$ % p.a. (iii) 9 years

**2.** $6\frac{2}{3}$ % **3.** $7\frac{1}{2}$ % **4.** 6% **5.** Rs. 5250, Rs. 4050

**6.** Rs. 400 **7.** Rs. 12114.25 **8.** Rs. 2600, $7\frac{1}{2}$ %

**9.** Rs. 4200, Rs. 4800 **10.** Rs. 6552, Rs. 6006, Rs. 5544

**11.** Rs. 1000 **12.** Rs. 500

## EXERCISE 18B (Objective Type Questions)

**1.** The simple interest on Rs. 500 for 6 years at 5% p.a. is :

(*a*) Rs. 250 (*b*) Rs. 150
(*c*) Rs. 140 (*d*) Rs. 120 **(Clerical Grade, 1991)**

**2.** A man will get Rs. 87 as simple interest on Rs. 725 at 4% per annum in :

(*a*) 3 years (*b*) $3\frac{1}{2}$ years
(*c*) 4 years (*d*) 5 years **(Clerical Grade, 1991)**

**3.** The simple interest on a certain sum for 3 years at $9\frac{1}{2}$ % p.a. amounts to Rs. 159.60. The sum is :

(*a*) Rs. 480 (*b*) Rs. 560
(*c*) Rs. 720 (*d*) Rs. 640

**4.** Interest on a certain sum of money for $2\frac{1}{3}$ years at $3\frac{3}{4}$ % per annum is Rs. 210. The sum is :

(*a*) Rs. 2800 (*b*) Rs. 1580
(*c*) Rs. 2400 (*d*) none of these

**(Railways, 1989)**

**5.** At simple interest, a sum doubles after 20 years. The rate of interest per annum is :

(*a*) 5% (*b*) 10%
(*c*) 20% (*d*) Data inadequate

**(Bank P.O. 1990)**

**6.** A sum of money amounts to Rs. 767 in 3 years and Rs. 806 in 4 years. The sum is :

(*a*) Rs. 600 (*b*) Rs. 650
(*c*) Rs. 675 (*d*) Rs. 700

**7.** A certain sum of money at S.I. amounts to Rs. 1012 in $2\frac{1}{2}$ years and to Rs. 1067.20 in 4 years. The rate of interest per annum is :

(*a*) 2.5% (*b*) 3%
(*c*) 4% (*d*) 5% **(Central Excise, 1989)**

**8.** A sum of money, put at simple interest, trebles itself in 15 years. The rate percent per annum is :

(*a*) $13\frac{1}{3}$ % (*b*) $16\frac{2}{3}$ %
(*c*) $12\frac{2}{3}$ % (*d*) 20%

**9.** In how many years will a sum of money double itself at 12% p.a. ?
(*a*) 6 yrs. 9 mths. (*b*) 8 yrs. 3 mths.
(*c*) 7 yrs. 6 mths. (*d*) 8 yrs. 6 mths.

**10.** The simple interest at $x$% for $x$ years will be Rs. $x$ on a sum of :
(*a*) Rs. $x$ (*b*) Rs. $100x$
(*c*) Rs. $\left(\frac{100}{x}\right)$ (*d*) Rs. $\left(\frac{100}{x^2}\right)$

**11.** A certain sum of money at simple interest amounts to Rs. 1260 in 2 years and to Rs. 1350 in 5 years. The rate percent per annum is :
(*a*) 2.5% (*b*) 3.75%
(*c*) 5% (*d*) 7.5% **(Central Excise, 1988)**

**12.** If Rs. 64 amount to Rs. 78.40 in 3 years, what will Rs. 86 amount to in 4 years; the rate of interest being the same in both the cases ?
(*a*) Rs. 108.60 (*b*) Rs. 111.80
(*c*) Rs. 118.40 (*d*) Rs. 121.20

**13.** The simple interest on a sum of money is (1/9) of the principal and the number of years is equal to the rate percent per annum. The rate percent per annum is :
(*a*) 3 (*b*) $\frac{1}{3}$
(*c*) $3\frac{1}{3}$ (*d*) $\frac{3}{10}$ **(Clerical Grade, 1991)**

**14.** Rs. 1,200 amounts to Rs. 1,632 in 4 years at a certain rate of simple interest. If the rate of interest is increased by 1%, it would amount to how much ?
(*a*) Rs. 1635 (*b*) Rs. 1644
(*c*) Rs. 1670 (*d*) Rs. 1680 **(Bank P.O. 1991)**

**15.** A sum of Rs. 10 is lent out to be returned in 11 monthly instalments of Re. 1 each, interest being simple. The rate of interest is :
(*a*) 10% (*b*) 11%
(*c*) $9\frac{1}{11}$ % (*d*) $21\frac{9}{11}$ %

**16.** *A* lent Rs. 600 to *B* for 2 years and Rs. 150 to *C* for 4 years and received altogether from both Rs. 90 as simple interest. The rate of interest is :
(*a*) 12% (*b*) 10%
(*c*) 5% (*d*) 4% **(Railways, 1988)**

**17.** *A* borrowed Rs. 5000 from *B* at simple interest. After 4 years, *B* received Rs. 1000 more than the amount given to *B* on loan. The rate of interest was :
(*a*) 5% (*b*) 25%
(*c*) 20% (*d*) 4% **(BSRB Exam, 1991)**

**18.** A sum of money doubles itself in 5 years. It will become 4 times of itself in :

(*a*) 10 years (*b*) 12 years

(*c*) 15 years (*d*) 20 years **(Clerical Grade, 1991)**

**19.** If the interest on Rs. 1200 be more than the interest on Rs. 1000 by Rs. 50 in 3 years, the rate percent is :

(*a*) $10\frac{1}{3}$ % (*b*) $6\frac{2}{3}$ %

(*c*) $8\frac{1}{3}$ % (*d*) $9\frac{2}{3}$ %

**20.** A sum was put at simple interest at a certain rate for 2 years. Had it been put at 1% higher rate, it would have fetched Rs. 24 more. The sum is :

(*a*) Rs. 600 (*b*) Rs. 800

(*c*) Rs. 1200 (*d*) Rs. 480

**21.** What annual payment will discharge a debt of Rs. 850 due in 5 years, the rate being 8% p.a. ?

(*a*) Rs. 166.40 (*b*) Rs. 120

(*c*) Rs. 100 (*d*) Rs. 65.60

**22.** The simple interest on a sum of money will be Rs. 600 after 10 years. If the principal is trebled after 5 years, the total interest at the end of 10 years will be :

(*a*) Rs. 600 (*b*) Rs. 900

(*c*) Rs. 1200 (*d*) data inadequate

**(Bank P.O. 1987)**

**23.** A sum of Rs. 1550 is lent out into two parts, one at 8% and another one at 6%. If the total annual income is Rs. 106, the money lent at 8% is :

(*a*) Rs. 900 (*b*) Rs. 840

(*c*) Rs. 720 (*d*) Rs. 650

**24.** A man lends Rs. 10000 in 4 parts. If he gets 8% on Rs. 2000; $7\frac{1}{2}$ % on Rs. 4000 and $8\frac{1}{2}$ % on Rs. 1400; what percent must he get for the remainder, if his average interest is 8.13% ?

(*a*) 7% (*b*) 9%

(*c*) $9\frac{1}{4}$ % (*d*) $10\frac{1}{2}$ %

**25.** A man invested $\frac{1}{3}$ of his capital at 7%, $\frac{1}{4}$ at 8% and the remainder at 10%. If his annual income is Rs. 561, the capital is :

(*a*) Rs. 5400 (*b*) Rs. 6600

(*c*) Rs. 7200 (*d*) Rs. 6000

**26.** A money lender finds that due to a fall in the rate of interest from 8% to $7\frac{3}{4}$ %, his yearly income diminishes by Rs. 61.50. His capital is :

(*a*) Rs. 22400 (*b*) Rs. 23800
(*c*) Rs. 24600 (*d*) Rs. 26000

**27.** A sum of Rs. 2600 is lent out in two parts in such a way that the interest on one part at 10% for 5 years is equal to that on another part at 9% for 6 years. The sum lent out at 10% is :

(*a*) Rs. 1250 (*b*) Rs. 1350
(*c*) Rs. 1150 (*d*) Rs. 1450

**28.** The simple interest on Rs. 200 for 7 months at 5 paise per rupee per month is :

(*a*) Rs. 70 (*b*) Rs. 7
(*c*) Rs. 35 (*d*) Rs. 30.50

**29.** If 1 Re. produces Rs. 9 as interest in 60 years at simple interest, the rate percent is :

(*a*) 30 (*b*) 15
(*c*) $13\frac{1}{3}$ (*d*) 9 **(Clerical Grade, 1991)**

**30.** The sum of money that will give Re. 1 as interest per day at 5% per annum simple interest is :

(*a*) Rs. 3650 (*b*) Rs. 36500
(*c*) Rs. 730 (*d*) Rs. 7300 **(Clerical Grade, 1991)**

## HINTS & SOLUTIONS (Exercise 18B)

1. Simple interest = Rs. $\left(\frac{500 \times 5 \times 6}{100}\right)$ = Rs. 150.

2. Required time = $\left(\frac{87 \times 100}{725 \times 4}\right)$ years = 3 years.

3. Sum = Rs. $\left(\frac{100 \times 15960}{3} \times \frac{2}{19} \times \frac{1}{100}\right)$ = Rs. 560.

4. Sum = Rs. $\left(\frac{210 \times 100 \times 3 \times 4}{7 \times 15}\right)$ = Rs. 2400.

5. Let the sum be Rs. $x$. Then, S.I. = Rs. $x$.

$\therefore$ Rate = $\left(\frac{100 \times x}{x \times 20}\right)$% = 5%.

6. S.I. for 1 year = Rs. (806 – 767) = Rs. 39.
S.I. for 3 years = Rs. (39 × 3) = Rs. 117.
$\therefore$ Sum = Rs. (767 – 117) = Rs. 650.

7. S.I. for $1\frac{1}{2}$ years = Rs. (1067.20 – 1012) = Rs. 55.20

S.I. for $2\frac{1}{2}$ years = Rs. $\left(55.20\times\frac{2}{3}\times\frac{5}{2}\right)$ = *Rs.* 92

$\therefore$ Principal = Rs. (1012 – 92) = Rs. 920.

$\therefore$ Rate % = $\left(\frac{100\times 92\times 2}{920\times 5}\right)\%$ = 4%.

**8.** Let the sum be Rs. $x$. Then, S.I. = Rs. $2x$.

$\therefore$ Rate = $\left(\frac{100\times 2x}{x\times 15}\right) = 13\frac{1}{3}$ %.

**9.** Let the sum be Rs. $x$. Then, S.I. = Rs. $x$.

$\therefore$ Time = $\left(\frac{100\times x}{x\times 12}\right)$ years = 8 yrs. 3 mths.

**10.** Principal = Rs. $\left(\frac{100\times x}{x\times x}\right)$ = Rs. $\left(\frac{100}{x}\right)$.

**11.** S.I. for 3 years = Rs. (1350 – 1260) = Rs. 90.

S.I. for 2 years = Rs. $\{(90/3)\times 2\}$ = Rs. 60.

$\therefore$ Principal = Rs. (1260 – 60) = Rs. 1200.

Hence, rate = $\left(\frac{100\times 60}{1200\times 2}\right)\%$ = 2.5%

**12.** Rate = $\left(\frac{14.40\times 100}{64\times 3}\right)\% = 7\frac{1}{2}$ % p.a.

S.I. on Rs. 86 for 4 years

= Rs. $\left(\frac{86\times 15\times 2}{100\times 2}\right)$ = Rs. 25.80.

$\therefore$ Amount = Rs. (86 + 25.80) = Rs. 111.80.

**13.** Let the sum be Rs. $x$. Then, S.I. = Rs. (1/9) $x$.

Let the rate be $r$% per annum and the time be $r$ years. Then,

$\frac{x}{9} = \frac{x\times r\times r}{100}$ or $r = \sqrt{\frac{100\times 1}{9}} = \frac{10}{3} = 3\frac{1}{3}$ %.

**14.** Rate = $\left(\frac{432\times 100}{1200\times 4}\right)\%$ = 9%

New rate = (9 + 1)% = 10%.

S.I. at new rate = Rs. $\left(\frac{1200\times 10\times 4}{100}\right)$ = Rs. 480.

Amount at new rate = Rs. (1200 + 480) = Rs. 1680.

**15.** Rs. 10 + S.I. on Rs. 10 for 11 months

= Rs. 11 + S.I. on Re. 1 for 55 months

or Rs. 10 + S.I. on Re. 1 for 110 months
= Rs. 11 + S.I. on Re. 1 for 55 months
or S.I. on Re. 1 for 55 months = Re. 1.

$\therefore$ Rate $= \left(\frac{100 \times 1}{1 \times 55} \times 12\right)\% = 21\frac{9}{11}\%$.

16. $\left(\frac{600 \times x \times 2}{100}\right) + \left(\frac{150 \times x \times 4}{100}\right) = 90$ or $x = 5\%$.

17. Rate % $= \left(\frac{100 \times 1000}{5000 \times 4}\right)\% = 5\%$.

18. Let the sum be Rs. $x$. Then,
S.I. for 5 years = Rs. $x$.

$\therefore$ Rate $= \left(\frac{100 \times x}{x \times 5}\right) = 20\%$.

Further, sum = Rs. $x$ ; S.I. = Rs. $3x$ and rate = 20%.

$\therefore$ Time $= \left(\frac{100 \times 3x}{x \times 20}\right) = 15$ years.

19. $\left(\frac{1200 \times r \times 3}{100}\right) - \left(\frac{1000 \times r \times 3}{100}\right) = 50$ or $r = 8\frac{1}{3}\%$.

20. Let sum be Rs. $x$ and rates be $r\%$ & $(r + 1)\%$

Then, $\frac{x \times (r+1) \times 2}{100} - \frac{x \times r \times 2}{100} = 24$ or $x =$ Rs. 1200.

21. Let annual instalment be Rs. $x$. Then,

$$\left[x + \left(\frac{x \times 4 \times 8}{100}\right)\right] + \left[x + \left(\frac{x \times 3 \times 8}{100}\right)\right] + \left[x + \left(\frac{x \times 2 \times 8}{100}\right)\right] + \left[x + \left(\frac{x \times 1 \times 8}{100}\right)\right] + x = 580$$

or $\frac{33x}{25} + \frac{31x}{25} + \frac{29x}{25} + \frac{27x}{25} + x = 580$ or $x = 100$.

22. Let the sum be Rs. $x$.
S.I. = Rs. 600, Time = 10 years

$\therefore$ Rate $= \left(\frac{600 \times 100}{x \times 10}\right)\% = \left(\frac{6000}{x}\right)\%$ per annum.

S.I. for first 5 years = Rs. $\left(\frac{x \times 5 \times 6000}{100 \times x}\right)$ = Rs. 300.

S.I. for last 5 years = Rs. $\left(\frac{3x \times 5 \times 6000}{100 \times x}\right)$ = Rs.900.

Hence, total interest at the end of 10 yrs. = Rs. 1200.

**23.** Average rate $= \left(\frac{106}{1550} \times 100\right)\% = \frac{212}{31}\%.$

Use allegation rule :

Ratio of amounts $= \frac{26}{31} : \frac{36}{31} = 13 : 18.$

Hence, money at 8% = Rs. $\left(1550 \times \frac{13}{31}\right)$ = Rs. 650.

**24.** $\left(\frac{2000 \times 8 \times 1}{100}\right) + \left(4000 \times \frac{15}{2} \times \frac{1}{100}\right) + \left(1400 \times \frac{17}{2} \times \frac{1}{100}\right)$

$+ \left(\frac{2600 \times x \times 1}{100}\right) = \left(\frac{10000 \times 8.13 \times 1}{100}\right)$

or $(160 + 300 + 119 + 26x = 813)$ or $26x = 234$ or $x = 9\%$.

**25.** Let the capital be Rs. $x$. Then,

$\left(\frac{x}{3} \times \frac{7}{100} \times 1\right) + \left(\frac{x}{4} \times \frac{8}{100} \times 1\right) + \left(\frac{5x}{12} \times \frac{10}{100} \times 1\right) = 561$

or $x = 6600$.

**26.** Let the capital be Rs. $x$. Then,

$\left(\frac{x \times 8 \times 1}{100}\right) - \left(x \times \frac{31}{4} \times \frac{1}{100}\right) = 61.50$

or $x\left(\frac{2}{25} - \frac{31}{400}\right) = 61.50$ or $x = 24600$.

**27.** Let the money at 10% be Rs. $x$. Then,

$\left(\frac{x \times 10 \times 5}{100}\right) = \left[\frac{(2600 - x) \times 9 \times 6}{100}\right]$

or $50x = (2600 \times 54) - 54x$

or $x = \left(\frac{2600 \times 54}{104}\right)$ = Rs. 1350.

**28.** S.I. on Re. 1 for 1 month = 5 paise = Re. 0.05

S.I. on Rs. 200 for 1 month = Rs. $(200 \times 0.05)$ = Rs. 10.00.

$\therefore$ S.I. on Rs. 200 for 7 months = Rs. 70.

**29.** Rate $= \left(\frac{100 \times 9}{1 \times 60}\right) = 15\%.$

**30.** Sum = Rs. $\left(\frac{365 \times 100}{1 \times 5}\right)$ = Rs. 7300.

## ANSWERS (Exercise 18B)

| | | | | | |
|---|---|---|---|---|---|
| 1. (*b*) | 2. (*a*) | 3. (*b*) | 4. (*c*) | 5. (*a*) | 6. (*b*) |
| 7. (*c*) | 8. (*a*) | 9. (*b*) | 10. (*c*) | 11. (*a*) | 12. (*b*) |
| 13. (*c*) | 14. (*d*) | 15. (*d*) | 16. (*c*) | 17. (*a*) | 18. (*c*) |
| 19. (*c*) | 20. (*c*) | 21. (*c*) | 22. (*c*) | 23. (*d*) | 24. (*b*) |
| 25. (*b*) | 26. (*c*) | 27. (*b*) | 28. (*a*) | 29. (*b*) | 30. (*d*) |

# 19

# COMPOUND INTEREST

*Sometimes it so happens that the borrower and the lender agree to fix up a certain unit of time, say* **quarterly** *or* ***half yearly*** *or* ***yearly*** *to settle the previous account. In such cases, the amount after first unit of time becomes the principal for the second unit, the amount after second unit becomes the principal for the third unit and so on. After a certain period,* **the difference between the amount and the money borrowed is called the Compound Interest (C.I.) for that period.**

**FORMULAE.**

**(i)** If *Principal* = Rs. P, *Time* = t years and *Rate* = *r*% p.a. and compound interest reckoned annually, then

$$\textbf{Amount after t years} = P\left\{1+\frac{r}{100}\right\}$$

**(ii)** If *Principal* = Rs. P, *time* = t years and *Rate* = *r*% p.a. and compound interest reckoned half yearly, then

*Rate* = (*r*/2) % per half year and *time* = (2 t) half years.

$$\textbf{Amount after t years} = P\left\{1+\frac{r}{2\times 100}\right\}^{2t}$$

**(iii)** If *Principal* = Rs. P, *Time* = t years and *Rate* = *r*% p.a. and compound interest reckoned quarterly, then

*Rate* = (*r*/4) % per quarter, *time* = (4t) quarters

$$\textbf{Amount after t years} = P\left\{1+\frac{r}{4\times 100}\right\}^{4t}$$

**(iv)** If *Principal* = Rs. P, *time* = n years,

*Rate* = $r_1$% for 1st year; $r_2$% for 2nd year; $r_3$% for 3rd year and so on and in the last $r_n$% for nth year. Then,

**Amount after n years**

$$= P\left\{1+\frac{r_1}{100}\right\}\left\{1+\frac{r_2}{100}\right\}\cdots\cdots\left\{1+\frac{r_n}{100}\right\}.$$

**(v)** If *Principal* = Rs. P, *Rate* = *r*% p.a. and *the time is the fraction of a year* then the rate is considered as that fraction of the year.

For example, if the time is $4\frac{3}{5}$ years, then

$$\text{Amount} = P\left[\left(1+\frac{r}{100}\right)^4\left(1+\frac{\frac{3}{5}r}{100}\right)\right].$$

(vi) Suppose a sum of Rs. 100 is borrowed at $r\%$ per annum compound interest reckoned quarterly or half-yearly, then the compound interest for 1 year is called the *Effective Annual Rate.*

(vii) *Present worth* of a sum of Rs. $x$ due $t$ years hence at $r\%$ p.a. compound interest is

$$\text{Rs.}\left\{\frac{x}{(1+r/100)^t}\right\}.$$

**Solved Problems.**

**Ex. 1.** ***Find the compound interest (reckoned yearly) on :***

(i) Rs. 1600 at $7\frac{1}{4}$ % per annum for 2 years.

(ii) Rs. 2000 at 6% per annum for $1\frac{1}{4}$ years.

(iii) Rs. 2400 at 10% per annum for 2 years 4 months.

(iv) Rs. 2500 for 3 years at 10% for first year, 12% for the second year and 15% for the third year per annum.

**Sol.** (i) $\text{Amount} = \text{Rs.}\left[1600\times\left(1+\frac{29}{4\times100}\right)^2\right]$

$= \text{Rs.}\left(1600\times\frac{429}{400}\times\frac{429}{400}\right)$

$= \text{Rs. } 1840.41$

*C.I. = Rs. (1840.41 – 1600) = Rs. 240.41*

(ii) $\text{Amount} = \text{Rs.}\left[2000\times\left(1+\frac{6}{100}\right)\left(1+\frac{\frac{1}{4}\times6}{100}\right)\right]$

$= \text{Rs.}\left[2000\times\frac{53}{50}\times\frac{203}{200}\right] = \text{Rs. } 2151.80$

*C.I. = Rs. (2151.80 – 2000) = Rs. 151.80.*

(iii) Time = 2 years 4 months = $2\frac{1}{3}$ years.

$$\text{Amount} = \text{Rs.}\left[2400\times\left(1+\frac{10}{100}\right)^2\left(1+\frac{\frac{1}{3}\times10}{100}\right)\right]$$

$= \text{Rs.}\left[2400 \times \frac{11}{10} \times \frac{11}{10} \times \frac{31}{30}\right] = \text{Rs. } 3000.80.$

*C.I. = Rs. (3000.80 – 2400) = Rs. 600.80*

(iv) Amount $= \text{Rs.}\left[2500 \times \left(1 + \frac{10}{100}\right) \times \left(1 + \frac{12}{100}\right) \times \left(1 + \frac{15}{100}\right)\right]$

$= \text{Rs.}\left(2500 \times \frac{11}{10} \times \frac{28}{25} \times \frac{23}{20}\right) = \text{Rs. } 3542.$

*C.I. = Rs. (3542 – 2500) = Rs. 1042.*

**Ex. 2.** ***Find the compound interest on Rs. 4000 for 9 months at 6% per annum, being given that the interest is reckoned (i) quarterly (ii) half yearly and (iii) yearly.***

**Sol.** (i) Principal = Rs. 4000, Time = 9 months = 3 quarters, Rate = 6%

per annum $= \frac{6}{4}$ or $\frac{3}{2}$ % per quarter.

$\therefore \text{Amount} = \text{Rs.}\left\{4000 \times \left(1 + \frac{\frac{3}{2}}{100}\right)^3\right\}$

$= \text{Rs.}\left\{4000 \times \frac{203}{200} \times \frac{203}{200} \times \frac{203}{200}\right\}$

$= \text{Rs. } (4182.71)$

*C.I. = Rs. (4182.71 – 4000) = Rs. 182.71*

(ii) Principal = Rs. 4000, Time = 9 months = $1\frac{1}{2}$ half-years.

Rate = 6% per annum = 3% per half-yearly.

$\therefore \text{Amount} = \text{Rs.}\left\{4000 \times \left(1 + \frac{3}{100}\right)\left(1 + \frac{\frac{3}{2}}{100}\right)\right\}$

$= \text{Rs.}\left\{4000 \times \frac{103}{100} \times \frac{203}{200}\right\} = \text{Rs. } 4181.80$

*C.I. = Rs. (4181.80 – 4000) = Rs. 181.80*

(iii) Principal = Rs. 4000, Time = 9 months = $\frac{3}{4}$ years.

Rate = 6% per annum.

$\therefore \text{Amount} = \text{Rs.}\left[4000 \times \left(1 + \frac{\frac{3}{4} \times 6}{100}\right)\right] = \text{Rs. } 4180$

*C.I. = Rs. (4180 – 4000) = Rs. 180.*

**Ex. 3.** ***If the compound interest on a certain sum of money for $2\frac{1}{2}$ years at 5% per annum be Rs. 104.05, what would be the simple interest ?***

**Sol.** Let the Principal be Rs. 100.

Amount of Rs. 100 for $2\frac{1}{2}$ years at 5% C.I.

$$= \text{Rs.}\left[100 \times \left(1+\frac{5}{100}\right)^2 \times \left(1+\frac{\frac{5}{2}}{100}\right)\right]$$

$$= \text{Rs.}\left(100 \times \frac{21}{20} \times \frac{21}{20} \times \frac{41}{40}\right) = \text{Rs.}\left(\frac{18081}{160}\right)$$

C.I. on Rs. 100 for $2\frac{1}{2}$ years

$$= \text{Rs.}\left(\frac{18081}{160} - 100\right) = \text{Rs.}\left(\frac{2081}{160}\right)$$

If C.I. is Rs. $\frac{2081}{160}$, principal = Rs. 100.

If C.I. is Rs. 104.05, principal

$$= \text{Rs.}\ \frac{100 \times 160}{2081} \times 104.05 = \text{Rs. } 800.$$

Now, Principal = Rs. 800, Rate = 5% p.a., Time = $\frac{5}{2}$ years.

$$\therefore \text{ S.I.} = \text{Rs.}\left(\frac{800 \times 5 \times 5}{100 \times 2}\right) = \text{Rs. } 100.$$

**Ex. 4.** ***The difference between the compound interest and the simple interest on a certain sum at $7\frac{1}{2}$ % per annum for 3 years is Rs. 110.70. Find the sum.***

**Sol.** Let the sum be Rs. 100.

Amount of Rs. 100 for 3 years at $7\frac{1}{2}$ % p.a. compound interest

$$= \text{Rs.}\left[100 \times \left(\frac{15}{2 \times 100}\right)^3\right]$$

$$= \text{Rs.}\left(100 \times \frac{43}{40} \times \frac{43}{40} \times \frac{43}{40}\right) = \text{Rs.}\left(\frac{79507}{640}\right)$$

$$\therefore \text{ C.I.} = \text{Rs.}\left(\frac{79507}{640} - 100\right) = \text{Rs.}\left(\frac{15507}{640}\right).$$

S.I. on Rs. 100 for 3 years at $7\frac{1}{2}$ %

$$= \text{Rs.} \left( \frac{100 \times 3 \times 15}{2 \times 100} \right) = \text{Rs.} \left( \frac{45}{2} \right)$$

Difference between C.I. and S.I.

$$= \text{Rs.} \left( \frac{15507}{640} - \frac{45}{2} \right) = \text{Rs.} \left( \frac{1107}{640} \right)$$

If the difference between C.I. and S.I. is Rs. $\frac{1107}{640}$,

sum = Rs. 100

If the difference between C.I. and S.I. is Rs. 110.70, sum

$$= \text{Rs.} \left( \frac{100 \times 640}{1107} \times 110.70 \right) = \text{Rs. } 6400.$$

**Ex. 5.** ***Find the difference between the compound interests on Rs. 5000 for*** $1\frac{1}{4}$ ***years at 8% per annum, according as the interest is payable yearly or half yearly.***

**Sol. Case I. Yearly Payment :**

Principal = Rs. 5000, Time = $1\frac{1}{4}$ years, Rate = 8%

$$\text{Amount} = \text{Rs.} \left[ 5000 \times \left( 1 + \frac{8}{100} \right) \left( 1 + \frac{\frac{1}{4} \times 8}{100} \right) \right]$$

$$= \text{Rs.} \left( 5000 \times \frac{27}{25} \times \frac{51}{50} \right) = \text{Rs. } 5508$$

*C.I. = Rs. (5508 – 5000) = Rs. 508.*

**Case II. Half Yearly Payment :**

Principal = Rs. 5000, Time = $2\frac{1}{2}$ half-years, Rate = 4% half yearly.

$$\text{Amount} = \text{Rs.} \left[ 5000 \times \left( 1 + \frac{4}{100} \right)^2 \left( 1 + \frac{2}{100} \right) \right]$$

$$= \text{Rs.} \left[ 5000 \times \frac{26}{25} \times \frac{26}{25} \times \frac{51}{50} \right] = \text{Rs.} \left( \frac{137904}{25} \right)$$

$$\text{C.I.} = \text{Rs.} \left( \frac{137904}{25} - 5000 \right) = \text{Rs.} \left( \frac{12904}{25} \right)$$

= Rs. 516.16.

*Difference in C.I.'s = Rs. (516.16 − 508) = Rs. 8.16.*

**Ex. 6.** ***On a certain sum of money the compound interest for 2 years is Rs. 282.15 and the simple interest is Rs. 270. Find the sum and the rate percent per annum.***

**Sol.** S.I. for 2 years = Rs. 270

S.I. for 1 year = Rs. $\frac{270}{2}$ = Rs. 135.

Now, (C.I. for 2 years) − (S.I. for 2 years) = Rs. [282.15 − 270]
= Rs. 12.15.

This is nothing but the interest on simple interest for 1 year.

∴ S.I. on Rs. 135 for 1 year = Rs. 12.15

Hence, Rate $= \left(\frac{12.15}{135} \times 100\right)\% = 9\%$

Also, Principal = Rs. $\frac{100 \times 270}{9 \times 2}$ = Rs. 1500.

**Ex. 7.** ***Find the present value of Rs. 4813 due 3 years hence at $6\frac{1}{4}\%$ per annum compound interest.***

**Sol.** Amount of Rs. 100 for 3 years at $6\frac{1}{4}\%$ C.I.

$$= \text{Rs.}\left[100\left(1+\frac{25}{4 \times 100}\right)^3\right] = \text{Rs.}\left(\frac{120325}{1024}\right).$$

Now present value of Rs. $\left(\frac{120325}{1024}\right)$ = Rs.100.

Present value of Rs. 4813 = Rs. $\left(\frac{100 \times 1024 \times 4813}{120325}\right)$ = Rs. 4096.

**Ex. 8.** ***What annual payment will discharge a debt of Rs. 8116 due in 3 years at 8% per annum compound interest?***

**Sol.** Let Rs. $x$ be the amount of each instalment. Then the instalments of Rs. $x$ are paid at the end of 1 year, 2 years and 3 years respectively.

Present values of these instalments are

$$\frac{x}{\left(1+\frac{8}{100}\right)}, \frac{x}{\left(1+\frac{8}{100}\right)^2}, \frac{x}{\left(1+\frac{8}{100}\right)^3}$$

Total present values of these instalments.

$$= \text{Rs. } x\left[\frac{25}{27}+\frac{625}{729}+\frac{15625}{19683}\right] = \text{Rs.}\left[\frac{50725x}{19683}\right].$$

Also, present value of Rs. 8116 due 3 years hence

$$= \text{Rs.}\left[\frac{8116}{\left(1+\frac{8}{100}\right)^3}\right] = \text{Rs.}\left[\frac{8116 \times 15625}{19683}\right]$$

$$\therefore \frac{50725x}{19683} = \frac{8116 \times 15625}{19683} \text{ or } x = 2500.$$

**Ex. 9.** ***A sum of money is borrowed and paid back in two equal annual instalments of Rs. 7290, allowing 8% Compound interest. What was the sum borrowed ?***

**Sol.** Sum borrowed

= (Present value of Rs. 7290 due 1 year hence)

+ (Present value of Rs. 7290 due 2 years hence)

$$= \text{Rs.}\left[\frac{7290}{\left(1+\frac{8}{100}\right)} + \frac{7290}{\left(1+\frac{8}{100}\right)^2}\right]$$

= *Rs. (6750 + 6250) = Rs. 13000.*

**Ex. 10.** ***At what rate percent Compound interest per annum will Rs. 640 amount to Rs. 795.07 in 3 years ?***

**Sol.** Let the rate be $r$% per annum. Then,

$$640 \times \left\{1+\frac{r}{100}\right\}^3 = 795.07$$

$$\text{or } \left(1+\frac{r}{100}\right)^3 = \left(\frac{79507}{64000}\right) = \left(\frac{43}{40}\right)^3$$

$$\text{or } \left(1+\frac{r}{100}\right) = \frac{43}{40} \text{ or } r = \left(\frac{3}{40} \times 100\right) = 7\frac{1}{2}.$$

Hence, rate = $7\frac{1}{2}$ % per annum.

**Ex. 11.** ***The Compound interest on Rs. 30000 at 7% per annum for a certain time is Rs. 4347. Find the time.***

**Sol.** Let the time be $t$ years. Then,

$$30000 \times \left(1+\frac{7}{100}\right)^t = 30000 + 4347$$

$$\text{or } \left(\frac{107}{100}\right)^t = \left(\frac{34347}{30000}\right) = \left(\frac{11449}{10000}\right) = \left(\frac{107}{100}\right)^3$$

$\therefore$ $t = 3$. *i.e.*, the time is 3 years.

**Ex. 12.** ***A sum of Rs. 13360 is borrowed at $8\frac{3}{4}$% compound interest and paid back in 3 equal instalments. What is the value of each instalment ?***

**Sol.** Let the value of each instalment be Rs. $x$.

Then, (Present value of Rs. $x$ due 1 year hence)

+ (Present value of Rs. $x$ due 2 years hence) = Rs. 13360.

$$\text{or}\quad \frac{x}{\left(1+\frac{35}{4\times 100}\right)}+\frac{x}{\left(1+\frac{35}{4\times 100}\right)^2}=13360$$

$$\text{or}\quad \frac{80x}{87}+\frac{6400x}{7569}=13360 \quad\text{or}\quad x=7569$$

∴ Value of each instalment = Rs. 7569.

## EXERCISE 19A (Subjective)

1. *Find the compound interest on :*

(i) Rs. 16000 at $9\frac{3}{4}$ % per annum for 2 years.

(ii) Rs. 12800 at $7\frac{1}{2}$ % per annum for $2\frac{1}{2}$ years.

(iii) 20480 at $6\frac{1}{4}$ % per annum for 2 years 73 days.

2. *Find compound interest on :*

(i) Rs. 8000 in $1\frac{1}{2}$ years at 10% per annum compound interest payable half yearly.

**Hint.** *Time = 3 half-years, Rate = 5% per half-year.*

(ii) Rs. 12500 in $1\frac{1}{4}$ years at 8% per annum Compound interest payable half yearly.

**Hint.** *Time = $2\frac{1}{2}$ half years, Rate = 4% per half year.*

$$\therefore \textit{Amount} = \text{Rs. } 12500\times\left(1+\frac{4}{100}\right)^2\times\left(1+\frac{2}{100}\right).$$

3. I lent Rs. 16000 on simple interest for 3 years at $7\frac{1}{2}$ % per annum. How much more should have I gained, had I given it at Compound interest at the same rate and for the same time ?

**Hint.** *Find the difference between C.I. and S.I.*

4. If the difference between the simple interest and the compound interest on a certain sum of money for 3 years at 5% per annum is Rs. 122, find the sum.

5. A sum of money put at Compound interest amounts to Rs. 2809 in 2 years and to Rs. 2977.54 in 3 years.
Find the rate of interest and the sum.
**Hint.** *S.I. on Rs. 2809 for 1 year = Rs. 168.54*

$$\therefore \text{Rate} = \left(\frac{168.54}{2809} \times 100\right) = 6\%. \text{ Now, find the sum.}$$

6. A sum of money is put at compound interest for 2 years at 20% per annum. It would fetch Rs. 482 more, if the interest were payable half yearly, than if it were payable yearly. Find the sum.
**Hint.** *Let the sum be Rs. 100.*

*The difference between C.I. on Rs. 100 when payable half yearly and that payable yearly, when calculated is Rs.* $\left(\frac{241}{100}\right)$,

*Now, if the difference in C.I.'s is Rs.* $\frac{241}{100}$, *sum = Rs. 100*

$\therefore$ *If the difference in C.I.'s is Rs. 482, sum*

$$= \text{Rs}\left(100 \times \frac{100}{241} \times 482\right) = \text{Rs. } 20000.$$

7. *A* sum of Rs. 8448 is to be divided between *A* and *B* who are respectively 18 and 19 years old, in such a way that if their shares be invested at 6.25% per annum compound interest, they shall receive equal amounts on attaining the age of 21 years. Find the present share of each and how much each will receive at the age of 21.
**Hint.** *Let the present shares of A and B be Rs. x and Rs.* (8448 – *x*) *respectively. Then*

$$x \times \left(1 + \frac{25}{4 \times 100}\right)^3 = (8448 - x)\left(1 + \frac{25}{4 \times 100}\right)^2, \text{find } x.$$

8. The Compound interest on a certain sum of money for 2 years is Rs. 330 and the simple interest on the same sum for the same period and at the same rate is Rs. 320. Find the sum and the rate percent per annum.
**Hint.** *S.I. for 1 year = Rs. 160.*

$\therefore$ *S.I. on Rs. 160 for 1 year = Rs. (330 – 320) = Rs. 10*

$$\text{So, rate} = \left(\frac{100 \times 10}{160 \times 1}\right) = 6\frac{1}{4}\%.$$

*Now, S.I. = Rs. 320. Time = 2 years, Rate =* $6\frac{1}{4}$%. *Find time.*

**9.** A money lender borrows money at 4% per annum and pays the interest at the end of the year. He lends it at 6% per annum payable half yearly and receives the interest at the end of the year. In this way, he gains Rs. 104.50 a year. How much money does he borrow ?

**Hint.** *Suppose the money borrowed is Rs. 100. Now, find the difference of C.I. on Rs. 100 at 4% for 1 year (paid annually) and the C.I. on Rs. 100 at 6% for1 year (reckoned half yearly). This difference is Rs.* $\left(\frac{209}{100}\right)$.

*Now, if the difference is Rs.* $\frac{209}{100}$, *sum = Rs. 100*

∴ *If the difference is Rs. 104.50, sum*

$$= Rs. \left(\frac{100 \times 100}{209} \times 104.50\right) = Rs. 5000.$$

**10.** A man borrowed Rs. 1261 at 5% compound interest. Show that by making equal annual repayments of Rs. 463.05, he can clear off his debt in 3 years.

**Hint.** *Let Rs. x be the value of each repayment.*

*Then, the sum of the present values of Rs. x due 1 year hence, Rs. x due 2 years hence and Rs. x due 3 years hence must be equal to Rs. 1261.*

*or* $\frac{x}{\left(1+\frac{5}{100}\right)} + \frac{x}{\left(1+\frac{5}{100}\right)^2} + \frac{x}{\left(1+\frac{5}{100}\right)^3} = 1261$, *find x.*

*The value of x obtained is 463.05.*

**11.** A tree increases annually by $\frac{1}{8}$ th of its height. By how much will it increase after $2\frac{1}{2}$ years, if it stands today 8 m. high ?

**Hint.** *Increase %* $= \left(\frac{1}{8} \times 100\right) = 12\frac{1}{2}\%$

∴ *Height after* $2\frac{1}{2}$ *years*

$$= 8 \times \left\{1 + \frac{25}{2 \times 100}\right\}^2 \times \left\{1 + \frac{25}{4 \times 100}\right\} = 10.75 \text{ m.}$$

**12.** A machine depreciates in value in a year by 2% of its value at the beginning of the year. If the value of a machine be Rs. 625000, what is its depreciated value after 3 years ?

**Hint.** *Depreciated value* $= Rs. 625000 \times \left(1 + \frac{2}{100}\right)^3$.

**13.** What is the nominal rate percent per annum, when interest is payable half yearly that would give an effective rate of 8% per annum ?

**Hint.** $100 \times \left(1 + \frac{R}{100 \times 2}\right)^2 = 108$, *find R.*

**14.** Find the effective rate percent per annum equivalent to a nominal rate of 10% per annum, interest payable half yearly.

**Hint.** *Find C.I. on Rs. 100 at 5% half yearly for 2 half years.*

## ANSWERS (Exercise 19A)

1. (i) Rs. 3272.10 (ii) Rs. 2546.70 (iii) Rs. 2929
2. (i) Rs. 1261 (ii) Rs. 1370.40
3. Rs. 276.75
4. Rs. 16000
5. 6%, Rs. 2500
6. Rs. 20000
7. Rs. 4096, Rs. 4352
8. Rs. 2560, $6\frac{1}{4}$ % p.a.
9. Rs. 5000
11. 10.75 m.
12. Rs. 588245
13. 7.846%
14. $10\frac{1}{4}$ %

## EXERCISE 19B (Objective Type Questions)

**1.** The compound interest on Rs. 450 at $6\frac{2}{3}$ % per annum for $2\frac{3}{4}$ years is :

(*a*) Rs. 105 (*b*) Rs. 87.60
(*c*) Rs. 94.60 (*d*) Rs. 82.50

**2.** Rs. 800 at 5% per annum compound interest amount to Rs. 882 in :

(*a*) 4 years (*b*) 3 years
(*c*) 2 years (*d*) 1 year **(Clerical Grade 1991)**

**3.** A sum amounts to Rs. 1352 in 2 years at 4% compound interest. The sum is :

(*a*) Rs. 1300 (*b*) Rs. 1200
(*c*) Rs. 1250 (*d*) Rs. 1260

**4.** At what rate percent compound interest will Rs. 625 amount to Rs. 900 in 2 years ?

(*a*) 10% (*b*) 15%
(*c*) 20% (*d*) 25%

**5.** The compound interest on Rs. 1250 for $1\frac{1}{2}$ years at 8 percent per annum, interest payable half yearly is :

(*a*) Rs. 153 (*b*) Rs. 156.08
(*c*) Rs. 162.08 (*d*) none of these

**6.** The compound interest on a certain sum of money for 2 years at 10% per annum is Rs. 420. The simple interest on the same sum at the same rate and same time will be :

(*a*) Rs. 350 (*b*) Rs. 375

(*c*) Rs. 380 (*d*) Rs. 400 **(Clerical Grade 1991)**

7. The difference between simple interest and the compound interest on Rs. 800 at 5% per annum for 1 year is :
(*a*) nil (*b*) Rs. 41
(*c*) Rs. 56.50 (*d*) Rs. 81

8. The difference between simple interest and the compound interest on Rs. 600 for 1 year at 10% per annum, reckoned half yearly is :
(*a*) nil (*b*) Rs. 1.50
(*c*) Rs. 4.40 (*d*) Rs. 6.50

9. The difference of compound interests on Rs. 800 for 1 year at 20% per annum when compounded half yearly and quarterly is :
(*a*) nil (*b*) Rs. 6.60
(*c*) Rs. 4.40 (*d*) Rs. 2.50

10. The simple interest on a sum of money for 2 years is Rs. 100 and the compound interest on the same sum at the same rate for the same time is Rs. 104. The rate of interest is :
(*a*) 2% p.a. (*b*) $3\frac{11}{13}$ % p.a.
(*c*) 4% p.a. (*d*) 8% p.a.

11. The simple interest on a sum of money for 3 years is Rs. 240 and the compound interest on the same sum, at the same rate for 2 year is Rs. 170. The rate of interest is :
(*a*) $12\frac{1}{2}$ % (*b*) $29\frac{1}{6}$ %
(*c*) $5\frac{5}{17}$ % (*d*) 8%

12. The simple interest on a certain sum for 2 years at 10% per annum is Rs. 90. The corresponding compound interest is :
(*a*) Rs. 99 (*b*) Rs. 95.60
(*c*) Rs. 94.50 (*d*) Rs. 108

13. The difference between simple interest and the compound interest on a certain sum of money for 2 years at 10% is Rs. 8. The sum is :
(*a*) Rs. 1600 (*b*) Rs. 800
(*c*) Rs. 1200 (*d*) Rs. 640

14. The value of a machine depreciates every year at the rate of 10% on its value at the beginning of that year. If the present value of the machine is Rs. 729, its worth 3 years ago was :
(*a*) Rs. 947.10 (*b*) Rs. 750.87
(*c*) Rs. 800 (*d*) Rs. 1000

15. If the compound interest on a certain sum for 2 years at 5% per annum is Rs. 41, the simple interest is :

(*a*) Rs. 37.50 (*b*) Rs. 31
(*c*) Rs. 40 (*d*) Rs. 35

**16.** A sum is invested at compound interest payable annually. The interest in two successive years was Rs. 225 and Rs. 236.25. The sum is :
(*a*) Rs. 4000 (*b*) Rs. 4500
(*c*) Rs. 4250 (*d*) Rs. 4100

**17.** A loan was repaid in two annual instalments of Rs. 112 each. If the rate of interest be 10% per annum compounded annually, the sum borrowed was :
(*a*) Rs. 200 (*b*) Rs. 220
(*c*) Rs. 210 (*d*) Rs. 217.80

**18.** A sum of money placed at compound interest doubles itself in 5 years. It will amount to eight times itself in :
(*a*) 15 years (*b*) 20 years
(*c*) 12 years (*d*) 10 years

**19.** A sum of Rs. 550 was taken as a loan. This is to be repaid in two equal annual instalments. If the rate of interest be 20% compounded annually, then the value of each instalment is :
(*a*) Rs. 396 (*b*) Rs. 350
(*c*) Rs. 360 (*d*) Rs. 421

**20.** A man borrowed Rs. 4000 from a bank at $7\frac{1}{2}$ compound interest. At the end of every year, he pays Rs. 1500 as part repayment of loan and interest. How much does he still owe to the bank after 3 such instalments ?
(*a*) Rs. 123.25 (*b*) Rs. 125
(*c*) Rs. 400 (*d*) Rs. 469.18

**21.** The difference between the compound interest and simple interest on a certain sum at 5% for 2 years is Rs. 1.50. The sum is :
(*a*) Rs. 600 (*b*) Rs. 500
(*c*) Rs. 400 (*d*) Rs. 300 **(Bank P.O. 1987)**

**22.** S.I. on a sum at 4% p.a. for 2 years is Rs. 80. The C.I. on the same sum for the same period is :
(*a*) Rs. 1081.60 (*b*) Rs. 81.60
(*c*) Rs. 160 (*d*) none of these
**(Asstt. Grade 1987)**

## HINTS & SOLUTIONS

**1.** Amount. = Rs. $\left[450 \times \left(1 + \frac{20}{3 \times 100}\right)^2 \left(1 + \frac{3}{4} \times \frac{20}{3 \times 100}\right)\right]$

$$= \text{Rs.}\left[450 \times \frac{16}{15} \times \frac{16}{15} \times \frac{21}{20}\right] = \text{Rs. } 537.60$$

$\therefore$ C.I. = Rs. (537.60 – 450) = Rs. 87.60.

**2.** Let time be $t$ years.

$$882 = 800\left(1 + \frac{5}{100}\right)^t \Rightarrow \frac{882}{800} = \left(\frac{21}{20}\right)^t$$

$$\Rightarrow \left(\frac{21}{20}\right)^2 = \left(\frac{21}{20}\right)^t \Rightarrow t = 2 \text{ years.}$$

**3.** Let the sum be P. Then,

$$1352 = P\left(1 + \frac{4}{100}\right)^2 \Rightarrow 1352 = P \times \frac{26}{25} \times \frac{26}{25}$$

$$\Rightarrow \frac{1352 \times 25 \times 25}{26 \times 26} = P \Rightarrow P = Rs.\ 1250.$$

**4.** Let rate be $r$. Then,

$$900 = 625\left(1 + \frac{r}{100}\right)^2 \Rightarrow \frac{900}{625} = \left(1 + \frac{r}{100}\right)^2$$

$$\Rightarrow \left(\frac{6}{5}\right)^2 = \left(1 + \frac{r}{100}\right)^2 \Rightarrow \left(\frac{6}{5}\right) = \left(1 + \frac{r}{100}\right) \Rightarrow r = 20\%.$$

**5.** $\text{Amount} = \text{Rs.}\left[1250\left(1 + \frac{4}{100}\right)^3\right] = \text{Rs. } 1406.08$

$\therefore$ C.I. = Rs. (1406.08 – 1250) = Rs. 156.08.

**6.** Let principal be P. Then,

$$P\left(1 + \frac{10}{100}\right)^2 - P = 420 \Rightarrow P = \text{Rs. } 2000$$

$$\text{S.I.} = \text{Rs.}\left(\frac{2000 \times 10 \times 2}{100}\right) = Rs.\ 400$$

**7.** $\text{S.I.} = \text{Rs.}\left(\frac{800 \times 5 \times 1}{100}\right) = \text{Rs. } 40.$

$$\text{C.I.} = \text{Rs.}\left[\left\{800\left(1 + \frac{5}{100}\right)\right\} - 800\right] = \text{Rs. } 40.$$

$\therefore$ Difference between S.I. and C.I. is nil.

**8.** $\text{S.I.} = \text{Rs.}\left(\frac{600 \times 10 \times 1}{100}\right) = \text{Rs. } 60.$

$$\text{C.I.} = \text{Rs.}\left[\left\{600\left(1+\frac{5}{100}\right)^2\right\}-600\right] = \text{Rs. } 61.50.$$

$\therefore$ Difference = Rs. 1.50.

**9.** C.I. when reckoned half yearly

$$= \text{Rs.}\left[\left\{800\left(1+\frac{10}{100}\right)^2\right\}-800\right] = \text{Rs. } 168.$$

C.I. when reckoned quarterly

$$= \text{Rs.}\left[\left\{800\left(1+\frac{5}{100}\right)^4\right\}-800\right] = \text{Rs. } 172.40.$$

$\therefore$ Difference = Rs. 4.40.

**10.** S.I. for 1 year $= \text{Rs.}\left(\frac{100}{2}\right) = \text{Rs. } 50.$

Difference between S.I. and C.I. for 1 year = Rs. 4

$\therefore$ Interest on Rs. 50 for 1 year = Rs. 4.

Hence, rate $= \left(\frac{4}{50}\times 100\right)\% = 8\%.$

**11.** S.I. for 1 year $= \text{Rs.}\left(\frac{240}{3}\right) = \text{Rs. } 80.$

S.I. for 2 years = Rs. 160.

Diff. in S.I. and C.I. for 2 years = Rs. (170 – 160) = Rs. 10.

Now, interest on Rs. 80 for 1 year = Rs. 10.

$\therefore$ Rate $= \left(\frac{10}{80}\times 100\right) = 12\frac{1}{2}\ \%.$

**12.** Sum $= \text{Rs.}\left(\frac{100\times 90}{2\times 10}\right) = \text{Rs. } 450.$

$$\text{C.I.} = \text{Rs.}\left[450\left(1+\frac{10}{100}\right)^2 - 450\right] = \textit{Rs}.\ 94.50.$$

**13.** Let the sum be Rs. 100. Then,

$$\text{S.I.} = \text{Rs.}\left(\frac{100\times 10\times 2}{100}\right) = \text{Rs. } 20.$$

$$\text{C.I.} = \text{Rs.}\left\{100\left(1+\frac{10}{100}\right)^2 - 100\right\} = \text{Rs. } 21.$$

(C.I.) = (S.I.) = Re. 1.

Now, 1 : 8 :: 100 : $x$

$$\therefore x = \left(\frac{8 \times 100}{1}\right) = \text{Rs. } 800.$$

**14.** $$P\left(1 - \frac{10}{100}\right)^3 = 729$$

$$\therefore P = Rs. \left(\frac{729 \times 10 \times 10 \times 10}{9 \times 9 \times 9}\right) = Rs.\ 1000.$$

**15.** S.I. on Re. 1 = Re. $\left(\frac{1 \times 2 \times 5}{100}\right) = \text{Re. } \frac{1}{10}.$

C.I. on Re. 1 = Re. $\left[1\left(1 + \frac{5}{100}\right)^2 - 1\right] = \text{Re. } \frac{41}{400}.$

$$\frac{\text{S.I.}}{\text{C.I.}} = \left(\frac{1}{10} \times \frac{400}{41}\right) = \frac{40}{41}.$$

$$\text{S.I.} = \left(\frac{40}{41} \times C.I.\right) = Rs. \left(\frac{40}{41} \times 41\right) = Rs.\ 40.$$

**16.** S.I. on Rs. 225 for 1 year = Rs. 11.25.

$$\therefore \text{Rate} = \left(\frac{100 \times 1125}{225 \times 1 \times 100}\right)\% = 5\%.$$

Hence, sum = Rs. $\left(\frac{100 \times 225}{5 \times 1}\right) = Rs.\ 4500.$

**17.** Principal = (P.W. of Rs. 121 due 1 year hence) + (P.W. of Rs. 121 due 2 years hence).

$$= \text{Rs.} \left[\frac{121}{\left(1 + \frac{10}{100}\right)} + \frac{121}{\left(1 + \frac{10}{100}\right)^2}\right] = Rs.\ 210.$$

**18.** Let the Principal be Rs. $P$ and rate be $r\%$ . Then amount after 5 years = Rs. $2P$

$$\therefore 2P = P\left(1 + \frac{r}{100}\right)^5$$

$$\Rightarrow 2 = \left(\frac{100 + r}{100}\right)^5 \quad \text{...(i)}$$

Let, the Principal be 8 times in $t$ years. Then,

$$8P = P\left(1+\frac{r}{100}\right)^t$$

$$\Rightarrow 8 = \left(1+\frac{r}{100}\right)^t$$

$$\Rightarrow 2^3 = \left(1+\frac{r}{100}\right)^t \Rightarrow \left(\frac{100+r}{100}\right)^{15} = \left(\frac{100+r}{100}\right)^t \text{ [using (i)]}$$

$\Rightarrow t = 15$ *years.*

**19.** Let the value of each instalment be Rs. $x$.

Then, $$\left\{\frac{x}{\left(1+\frac{20}{100}\right)} + \frac{x}{\left(1+\frac{20}{100}\right)^2}\right\} = 550$$

or $\frac{5x}{6} + \frac{25x}{36} = 550$ *or* $x = 360$.

**20.** Balance = Rs. $$\left[\left\{4000 \times \left(1+\frac{15}{2\times 100}\right)^3\right\} - \left\{1500 \times \left(1+\frac{15}{2\times 100}\right)^2 + 1500 \times \left(1+\frac{15}{2\times 100}\right) + 1500\right\}\right]$$

= Rs. 123.25.

**21.** Let the sum be Rs. 100. Then,

S.I. = Rs. $\left(\frac{100\times 5\times 2}{100}\right)$ = Rs. 10

C.I. = Rs. $\left[\left\{100 \times \left(1+\frac{5}{100}\right)^2\right\} - 100\right]$ = Rs. $\frac{41}{4}$

Difference between C.I. and S.I. = Rs. $\left(\frac{41}{4} - 10\right)$ = Re. 0.25

$\therefore$ 0.25 : 1.50 :: 100 : $x$

$\Rightarrow x = \frac{1.50\times 100}{0.25}$ = *Rs.* 600.

**22.** Principal = Rs. $\left(\frac{100\times 80}{4\times 2}\right)$ = Rs.1000

C.I. = Rs. $\left[\left\{1000 \times \left(1+\frac{4}{100}\right)^2\right\} - 1000\right] \cong$ *Rs.* 81.60.

## ANSWERS (Exercise 19B)

| | | | | | |
|---|---|---|---|---|---|
| **1.** (*b*) | **2.** (*c*) | **3.** (*c*) | **4.** (*c*) | **5.** (*b*) | **6.** (*d*) |
| **7.** (*a*) | **8.** (*b*) | **9.** (*c*) | **10.** (*d*) | **11.** (*a*) | **12.** (*c*) |
| **13.** (*b*) | **14.** (*d*) | **15.** (*c*) | **16.** (*b*) | **17.** (*c*) | **18.** (*a*) |
| **19.** (*c*) | **20.** (*a*) | **21.** (*a*) | **22.** (*b*) | | |

# 20

# AREA OF PLANE FIGURES

Either a circle or a figure bounded by three or more than three straight lines is called a *plane figure*.

The space enclosed by a figure is called its *area*.

**Quadrilateral :** *A plane figure bounded by four straight lines is called a quadrilateral.*

**Various Types of Quadrilaterals :**

(i) **Rectangle :** *A quadrilateral whose opposite sides are equal and all angles are at right angles.*
A line joining two opposite non-adjacent corners of a quadrilateral is called its **diagonal**. The diagonals of a rectangle are equal.

(ii) **Square :** *A rectangle having all its sides equal is called a square.*

(iii) **Parallelogram :** *A quadrilateral whose opposite sides are equal and parallel is called a parallelogram.*

(iv) **Rhombus :** *A parallelogram having all the sides equal is called a rhombus.*
Diagonals of a rhombus are not equal and they bisect each other at right angles.

(v) **Trapezium :** *A quadrilateral having one pair of opposite sides equal, is called a trapezium.*

(vi) **Kite :** *A quadrilateral having two pairs of adjacent sides equal, is called a kite.*

**Triangle :** *A plane figure bounded by three straight lines is called a triangle.*

(i) **Equilateral Triangle :** *A triangle with all sides equal, is called an equilateral triangle.*

(ii) **Isosceles Triangle :** *A triangle with two sides equal is called an isosceles triangle.*

(iii) **Scalene triangle :** *A triangle with all sides unequal is called a scalene triangle.*

(iv) **Right Angled Triangle :** *A triangle having one of the angles equal to 90° is called a right angled triangle.*
The side opposite to the right angle of a triangle is called its **hypotenuse.**

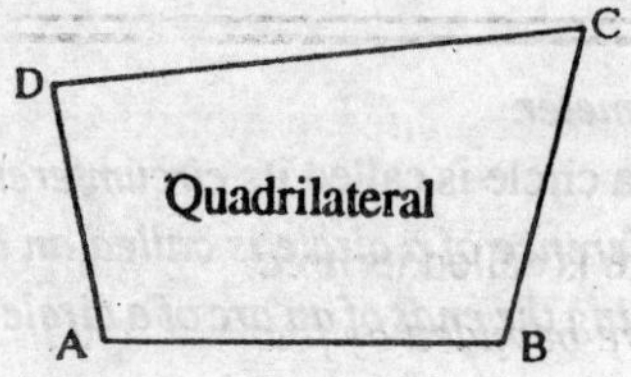
D
C
Quadrilateral
A
B

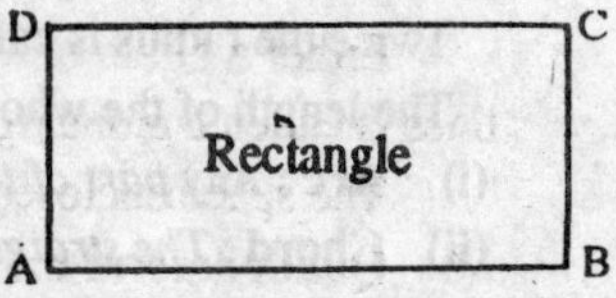
D
C
Rectangle
A
B

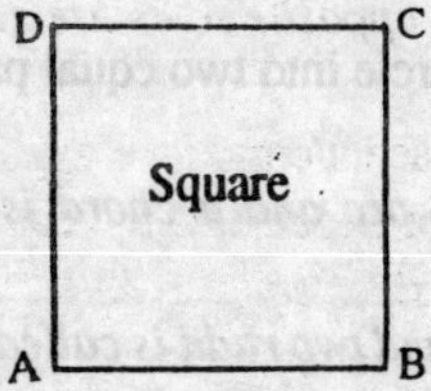
D
C
Square
A
B

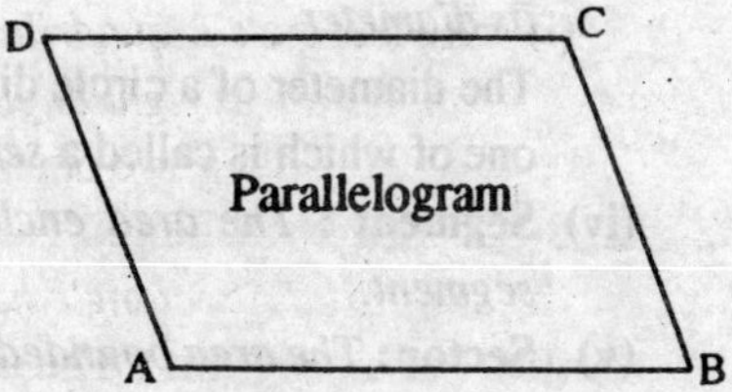
D
C
Parallelogram
A
B

D
C
Rhombus
A
B

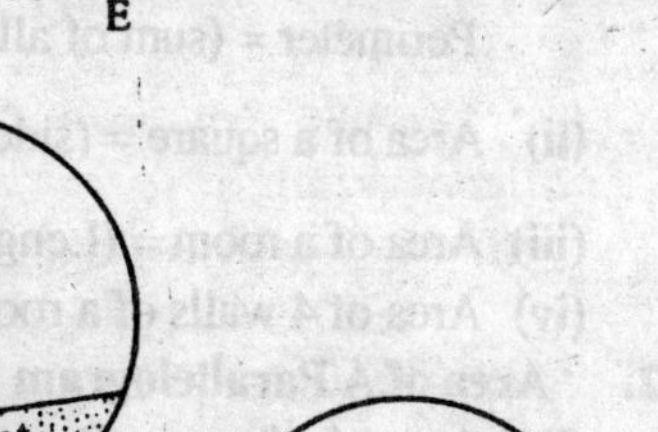
D
C
A
B
E

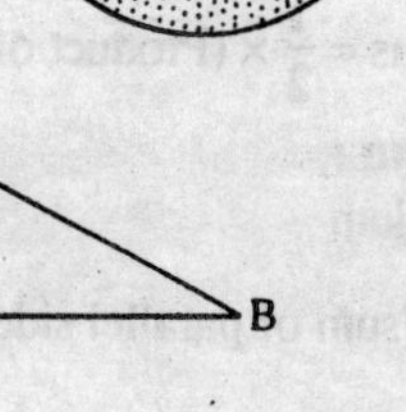
O
Chord
Segment

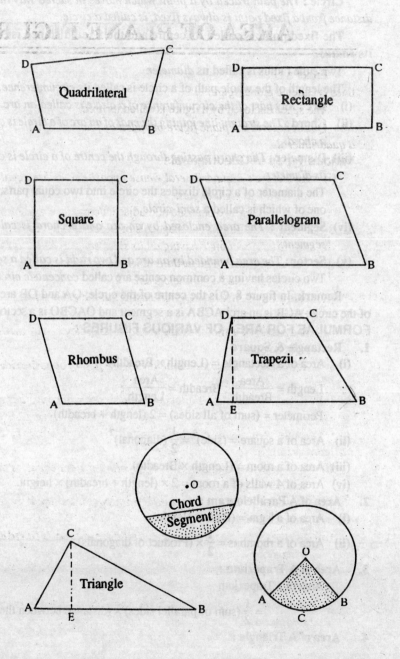
C
Triangle
A
B
E

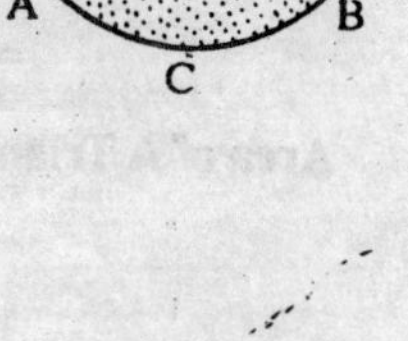
O
A
B
C

**In a right angled triangle :**

$$(Hypotenuse)^2 = (Base)^2 + (Perpendicular)^2.$$

**Circle :** *The path traced by a point which moves in such a way that its distance from a fixed point is always fixed, is called a circle.*

The fixed point is called its centre and the fixed distance is called its *radius.*

Twice the radius is called its *diameter.*

The length of the whole path of a circle is called its *circumference.*

(i) **Arc :** *Any part of the circumference of a circle is called an arc.*

(ii) **Chord :** *The straight line joining the ends of an arc of a circle is called a chord.*

(iii) **Diameter :** *The chord passing through the centre of a circle is called its diameter.*

The diameter of a circle divides the circle into two equal parts, each one of which is called a *semi-circle.*

(iv) **Segment :** *The area enclosed by an arc and a chord is called a segment.*

(v) **Sector :** *The area bounded by an arc and two radii is called a sector.*

Two circles having a common centre are called *concentric circles.*

**Remark.** In figure 8, O is the centre of the circle; OA and OB are radii of the circle; ACB is an arc; ACBA is a segment and OACBO is a sector.

## FORMULAE FOR AREA OF VARIOUS FIGURES :

**1. Rectangle & Square :**

(i) Area of a Rectangle = (Length × Breadth).

$$\text{Length} = \frac{\text{Area}}{\text{Breadth}}; \ \text{Breadth} = \frac{\text{Area}}{\text{Length}};$$

Perimeter = (sum of all sides) = 2 (length + breadth).

(ii) Area of a square $= (\text{side})^2 = \frac{1}{2}(\text{diagonal})^2$.

(iii) Area of a room = (Length × Breadth).

(iv) Area of 4 walls of a room = 2 × (length + breadth) × height.

**2. Area of A Parallelogram :**

(i) Area of a ∥ gm.= (Base × Height).

(ii) Area of a rhombus $= \frac{1}{2} \times$ (Product of diagonals).

**3. Area of A Trapezium :**

Area of a Trapezium

$$= \frac{1}{2}(\text{sum of parallel sides}) \times (\text{distance between them}).$$

**4. Area of A Triangle :**

(i) Area of a triangle $= \left(\frac{1}{2} \times \text{Base} \times \text{Height}\right)$.

(ii) Area of an equilateral triangle $= \frac{\sqrt{3}}{4} \times (\text{side})^2$.

(iii) If the lengths of the sides of a triangle be a, b and c and if $s = \frac{a+b+c}{2}$, then

Area of Triangle $= \sqrt{s(s-a)(s-b)(s-c)}$.

4. **Circle :**

(i) Area of a circle $= \pi r^2$

(ii) Circumference of a circle $= 2\pi r$.

(iii) Arc AB $= \frac{2\pi r\theta}{360}$ where $\theta^\circ = \angle AOB$.

(iv) Area of sector ACBO $= \frac{\pi r^2 \theta}{360} = \frac{1}{2} \times (arc\ AB) \times r$.

## Solved Examples :

**Ex. 1.** ***Find the area and perimeter of a rectangular plot whose length is 24.5 metres and breadth is 16.8 metres.***

**Sol.** Length = 24.5 m., Breadth = 16.8 m.

$\therefore$ Area $= (24.5 \times 16.8)\ m^2 = 411.6$ sq. metres.

*Perimeter* $= 2 \times (24.5 + 16.8)\ m = 82.6\ m$.

**Ex. 2.** ***Find the area of a rectangle whose one side is 3 metres and the diagonal is 5 metres.***

**Sol.** Another side $= \sqrt{(5)^2 - (3)^2} = \sqrt{16} = 4\ m$.

$\therefore$ Area $= (3 \times 4)$ *sq. m. = 12 sq. metres.*

**Ex. 3.** ***Find the area of a square whose diagonal is 2.9 metres long.***

**Sol.** *Area of the square*

$$= \frac{1}{2} \times (\text{diagonal})^2 = \left(\frac{1}{2} \times 2.9 \times 2.9\right) = 4.205 \text{ sq. metres.}$$

**Ex. 4.** ***Find the area of a parallelogram whose base is 35 metres and altitude 18 metres.***

**Sol.** Area of the parallelogram

= Base × Height = (35 × 18) = 630 sq. metres.

**Ex. 5.** ***Find the area of a rhombus one side of which measures 20 cm. and one diagonal 24 cm.***

**Sol.** Since the diagonals of a rhombus bisect at right angles, so one side and half of each of the diagonals form a right angled triangle. In this right angled triangle,

one side = 20 cms; another side = (24/2) = 12 cms.

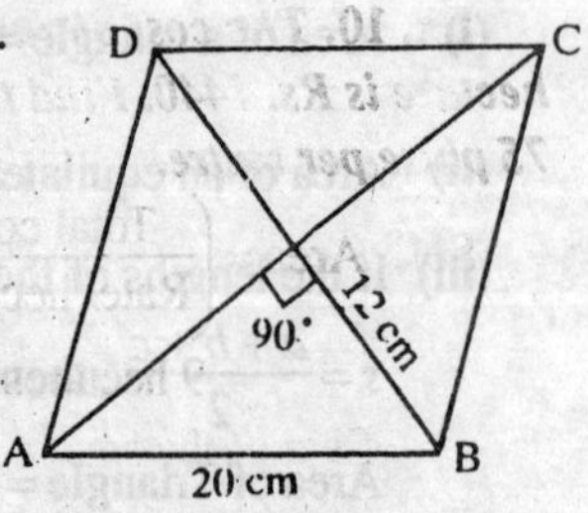


Third side = $\sqrt{(20)^2 - (12)^2} = \sqrt{256}$ cms.

= 16 cms.

∴ Half of another diagonal = 16 cms.

or another diagonal = 32 cms.

*Hence, area of rhombus* = $\left(\frac{1}{2} \times 24 \times 32\right)$

= 384 sq. cms.

**Ex. 6.** ***Find the area of a triangle in which base is 1.5 m. and height is 75 cm.***

**Sol.** Area of the triangle = $\frac{1}{2} \times$ Base $\times$ Height

$= \left(\frac{1}{2} \times 150 \times 75\right)$ *sq. cm.* = *5625 sq. cm.*

**Ex. 7.** ***Find the area of an equilateral triangle each of whose sides measures 12 cm.***

**Sol.** Area of the triangle = $\frac{\sqrt{3}}{4} \times (\text{side})^2$

$= \left(\frac{\sqrt{3}}{4} \times 12 \times 12\right) = 36\sqrt{3}$ sq. cm.

= *(36 × 1.73) = 62.28 sq. cm.*

**Ex. 8.** ***Find the area of a triangle whose one angle is 90°, the hypotenuse is 9 metres and the base is 6.5 metres.***

**Sol.** Altitude of the triangle = $\sqrt{9^2 - (6.5)^2} = \sqrt{38.75}$ *m*

= 6.2 metres.

∴ Area of the triangle = $\frac{1}{2} \times$ base $\times$ altitude

$= \left(\frac{1}{2} \times 6.5 \times 6.2\right)$ sq. m = 20.15 sq. m.

**Ex. 9.** ***Find the area of a triangle with two sides equal, each being 5.1 metres and the third side 4.6 metres.***

**Sol.** Let $a$ = 5.1 metres, $b$ = 5.1 metres and $c$ = 4.6 metres.

Then, $s = \left(\frac{a+b+c}{2}\right) = \left(\frac{5.1 + 5.1 + 4.6}{2}\right)$ = 7.4 metres.

∴ $(s - a) = 2.3$ ; $(s - b) = 2.3$ and $(s - c) = 2.8$ m.

∴ Area = $\sqrt{s(s-a)(s-b)(s-c)}$

$= \sqrt{7.4 \times 2.3 \times 2.3 \times 2.8} = \sqrt{109.6088}$ = *10.46 sq. m.*

**Ex. 10.** ***The cost of cultivating a square field at the rate of Rs. 160 per hectare is Rs. 1440. Find the cost of putting a fence around it at the rate of 75 paise per metre.***

**Sol.** Area $= \left(\frac{\text{Total cost}}{\text{Rate/hectare}}\right)$ Hectares $= \frac{1440}{160}$ hectares

$= 9$ hectares $= 90000$ sq. metres.

Side of the square $= \sqrt{90000} = 300$ metres.

Perimeter of the square $= 4 \times 300$ or 1200 metres.

*Cost of fencing* $=$ *Rs.* $\left(\frac{1200 \times 75}{100}\right) =$ *Rs. 900.*

**Ex. 11.** ***A rectangular courtyard 3.78 metres long and 5.25 metres broad is to be paved exactly with square tiles, all of the same size. What is the largest size of such a tile ? Also find the number of tiles.***

**Sol.** Largest size of tile = H.C.F. of 378 cm. and 525 cm.

= 21 cms.

Area of courtyard $= (378 \times 525)$ sq. cm.

Area of a tile $= (21 \times 21)$ sq. cm.

*Number of tiles* $= \left(\frac{378 \times 525}{21 \times 21}\right) = 450.$

**Ex. 12.** ***A lawn is in the form of a rectangle having its sides in the ratio 2 : 3. The area of the lawn is $\frac{1}{6}$ hectare. Find the length and the breadth of the lawn.***

**Sol.** Let the breadth be $2x$ metres and length $3x$ metres.

Then $(2x \times 3x) = \left(\frac{1}{6} \times 10000\right)$ [$\because$ 1 hectare = 10000 sq. m.]

or $x = \sqrt{\frac{10000}{6 \times 6}} = \frac{100}{6} = \frac{50}{3}.$

$\therefore$ *Breadth* $= \frac{2 \times 50}{3} = 33\frac{1}{3}$ *metres,*

*Length* $= \frac{3 \times 50}{3} = 50$ *metres.*

**Ex. 13.** ***A room is 13 metres long, 9 metres broad and 10 metres high. Find the cost of carpeting the room with a carpet 75 cm. broad at the rate of Rs. 2.40 per metre. What will be the cost of painting the four walls of the room at Rs. 4.65 per sq. metre, it being given that the doors and windows occupy 40 sq. metres ?***

**Sol.** Area of floor $= (13 \times 9) = 117$ sq. metres.

Area of carpet $= 117$ sq. metres.

Breadth of carpet = 75 cms. = $\frac{3}{4}$ metres.

Length of carpet $= \left(\frac{117 \times 4}{3}\right)$ metres = 156 metres.

Cost of carpet = Rs. $(156 \times 2.40)$ = Rs. 374.40

Area of 4 walls $= [2(13+9) \times 10] = 440$ sq. metres

Area to be painted = Rs. $(440 - 40) = 400$ sq. metres.

*Cost of painting = Rs. $(400 \times 4.65)$ = Rs. 1860.*

Ex. 14. ***The cost of papering the walls of a room 12 metres long at the rate of 45 paise per square metre is Rs. 113.40 and the cost of matting the floor at the rate of 35 paise per square metre is Rs. 37.80. Find the height of the room.***

**Sol.** Area of floor $= \left(\frac{\text{Total Cost}}{\text{Rate/sq .m}}\right)$ Sq. metres.

$= \left(\frac{3780}{35}\right) = 108\ m^2$

$\therefore$ Breadth of the room $= \frac{\text{Area of floor}}{\text{Length of the room}}$

$= \left(\frac{108}{12}\right)$ metres = 9 metres.

Now, area of 4 walls $= \left(\frac{\text{Total cost of Papering}}{\text{Rate/Sq. metre}}\right)$

$= \left(\frac{11340}{35}\right)$ sq. metres = 252 sq. m.

Let the height of room be $h$ metres.

Then, $2 \times (12+9) \times h = 252$

$\therefore$ *Height,* $h = \frac{252}{2 \times 21} = 6$ *metres.*

Ex. 15. ***A room is half as long again as it is broad. The cost of carpeting the room at Rs. 3.25 per sq. metre is Rs. 175.50 and the cost of papering the walls at Rs. 1.40 per sq. metre is Rs. 240.80. If a door and 2 windows occupy 8 sq. metres, find the dimensions of the room.***

**Sol.** Area of room $= \left(\frac{175.50}{3.25}\right) = 54$ sq. metres.

Now let breadth be $x$ metres, then length $= \frac{3}{2}x$ metres.

$$\therefore \text{ Area} = \left(x \times \frac{3}{2}x\right) \text{ sq. metres} = \frac{3x^2}{2} \text{ sq. metres.}$$

$$\text{So, } \frac{3}{2}x^2 = 54 \text{ or } x = \sqrt{\frac{54 \times 2}{3}} = 6.$$

$$\therefore \text{ Breadth} = 6 \text{ metres, Length} = \left(\frac{3}{2} \times 6\right) = 9 \text{ metres.}$$

$$\text{Now, papered area} = \left(\frac{240.80}{1.40}\right) = 172 \text{ sq. metres.}$$

Area of a door and two windows = 8 sq. metres.

Total area of 4 walls = (172 + 8) = 180 sq. metres.

$$\therefore 2\,(9 + 6) \times \textit{Height} = 180 \text{ or } \textit{Height} = \left(\frac{180}{30}\right) = 6 \text{ metres.}$$

**Ex. 16.** ***A rectangular grassy plot is 112 metres by 78 metres. It has a gravel path 2.5 metres wide all round it on the inside. Find the area of the path and the cost of constructing it at 72 paise per 1000 sq. cm.***

**Sol.** Area of grassy plot

= (112 × 78) = 8736 sq. m.

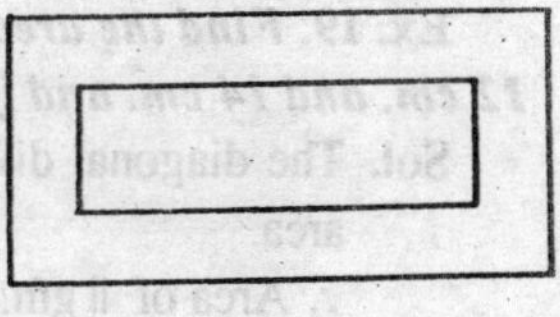

Area of plot excluding the path

= [(112 − 5) × (78 − 5)] sq. m.

= [(107 × 73)] = 7811 sq. m.

Area of path = (8734 − 7811) = 925 sq. m.

= (925 × 100 × 100) sq. cm.

Cost on 1000 sq. cm. = 72 paise

$$\textit{Cost on } (925 \times 100 \times 100) \textit{ sq. cm.} = \textit{Rs.} \left(\frac{72}{1000} \times \frac{925 \times 100 \times 100}{100}\right)$$

$$= \textit{Rs. } 6660.$$

**Ex. 17.** ***A rectangular lawn 60 metres by 40 metres has two roads each 5 metres wide, running in the middle of it, one parallel to length and the other parallel to the breadth. Find the cost of gravelling them at 60 paise per square metre.***

**Sol.** Area of road *ABCD*

= (60 × 5) = 300 sq. metres.

Area of road *EFGH*

= (40 × 5) = 200 sq. metres

Clearly, area *JKLM* is common to both of the above roads.

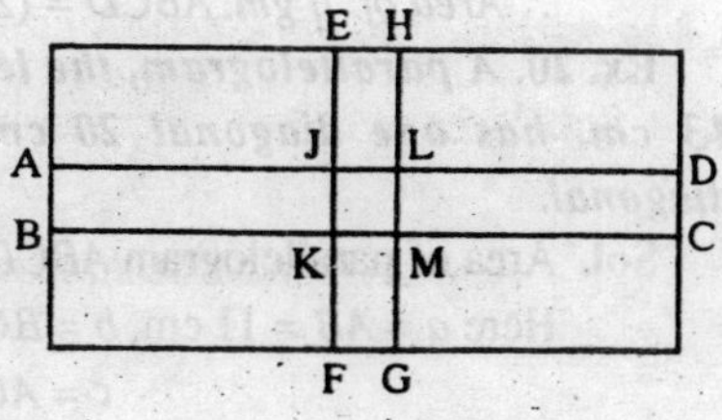


Now, area *JKLM*

= (5 × 5) sq. m. = 25 sq. metres.

$\therefore$ Area of the path to be gravelled.

$= (300 + 200 - 25) = 475$ sq. metres

Hence, the cost of gravelling the path

$= \text{Rs.} \left(\frac{60}{100} \times 475\right) = \text{Rs. } 285.$

**Ex. 18.** ***The base of a triangular field is three times its altitude. If the cost of cultivating the field at Rs. 24.60 per hectare is Rs. 332. 10, find its base and height.***

**Sol.** Area of the field $= \frac{332.10}{24.60} = 13.5$ hectares.

$= (13.5 \times 10000) = 135000$ sq. m.

Let, altitude be $x$ metres. Then, base $= 3x$ metres.

$\text{Area} = \left(\frac{1}{2} \times \text{base} \times \text{altitude}\right) = \left(\frac{1}{2} \times x \times 3x\right) = \frac{3x^2}{2}$ sq. m.

$\therefore \frac{3x^2}{2} = 135000$ or $x = \sqrt{\frac{135000 \times 2}{3}} = 300.$

*Hence, altitude = 300 metres and base = 900 metres.*

**Ex. 19.** ***Find the area of a parallelogram; if its two adjacent sides are 12 cm. and 14 cm. and if the diagonal connecting the ends is 18 cm.***

**Sol.** The diagonal divides the parallelogram into two triangles of equal area.

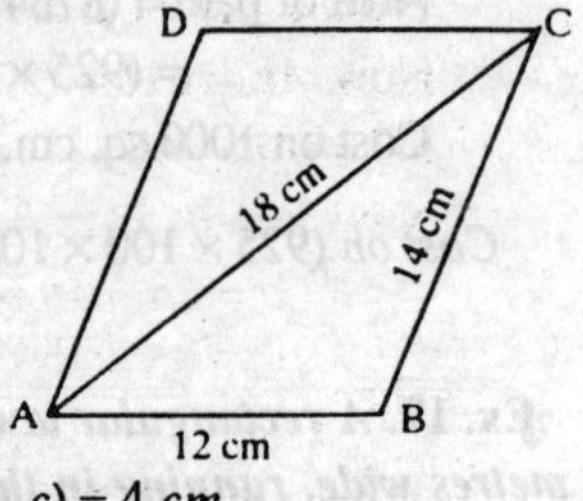


$\therefore$ Area of $\parallel$ gm. $ABCD$

$= 2 \times (\text{area of } \Delta ABC).$

Let $AB = a = 12$ cm., $BC = b = 14$ cm.

and $AC = c = 18$ cm.

Then, $s = \frac{1}{2}(a + b + c) = 22$ cm.

$\therefore$ $(s - a) = 10$ *cm*, $(s - b) = 8$ *cm* & $(s - c) = 4$ *cm*.

So, area of $\Delta ABC = \sqrt{s(s-a)(s-b)(s-c)}$

$= \sqrt{22 \times 10 \times 8 \times 4} = \sqrt{7040} = 83.97$sq. cm.

$\therefore$ *Area of // gm. ABCD = (2 × 83.97) sq. cm. = 167.94 sq. cm.*

**Ex. 20.** ***A parallelogram, the lengths of whose sides are 11 cm. and 13 cm. has one diagonal 20 cm. long. Find the length of another diagonal.***

**Sol.** Area of parallelogram $ABCD = 2\ \Delta ABC$

Here $a = AB = 11$ cm, $b = BC = 13$ cm.,

$c = AC = 20$ cms.

$$s = \left(\frac{11+13+20}{2}\right) = 22 \text{ cm.}$$

Area of $\Delta ABC = \sqrt{(22 \times 11 \times 9 \times 2)} = 66$ sq. cm.

Since the diagonal of a parallelogram divides it into two triangles of equal area,

$\therefore \Delta ADC = \Delta ABC = 66$ sq. cm.

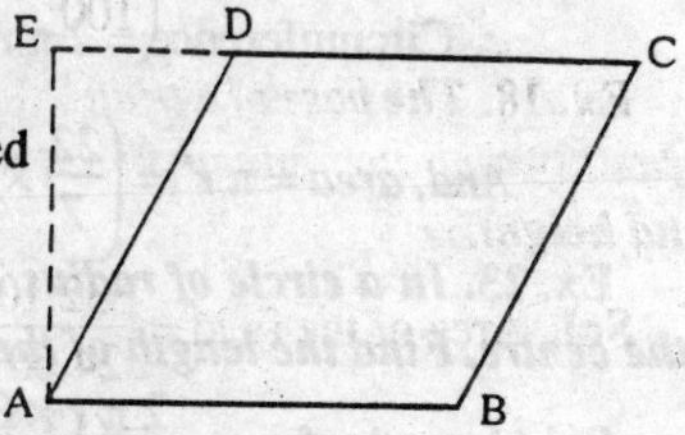


From $A$ draw $AE \perp CD$ produced

Then, $\frac{1}{2} \times CD \times AE = 66$

or $AE = \frac{66 \times 2}{11} = 12$ cm.

$\therefore \quad DE = \sqrt{(AD^2 - AE^2)} = \sqrt{(13)^2 - (12)^2} = 5$ cm.

Now, $CE = CD + DE = (11 + 5)$ cm. $= 16$ cm.

$\therefore AC = \sqrt{CE^2 + AE^2} = \sqrt{(12)^2 + (16)^2} = \sqrt{400} = 20$ *cm.*

**Ex. 21.** ***A field is in the form of a trapezium whose parallel sides are 120 metres and 75 metres and the non-parallel sides are 105 metres and 72 metres.Find the cost of ploughing the field at the rate of 60 paise per square metre.***

**Sol.** Let $ABCD$ be the field in which :

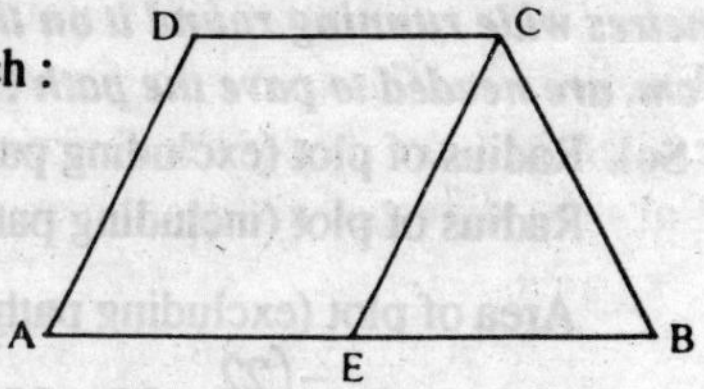


$AB = 120$ m., $CD = 75$ m.,

$AD = 72$ m. and $BC = 105$ m.

From $C$ draw $CE \parallel AD$.

Now, $AE = CD = 75$ m.

$BE = (AB - AE) = (120 - 75) = 45$ m. and $CE = AD = 72$ m.

Now, in $\Delta BCE$, let

$$a = BE = 45\ m,\ b = BC = 105\ m \text{ and } c = CE = 72\ m.$$

Then, $s = \left(\frac{45 + 105 + 72}{2}\right) = 111$ metres.

$\therefore$ Area of $\Delta$ BCE $= \sqrt{111 \times 66 \times 6 \times 39} = 1309.26$ sq. metres.

So, $\frac{1}{2} \times 45 \times \text{Height} = 1309.26$

or $\quad \text{Height} = \frac{1309.26 \times 2}{45} = 58.19$ m.

$\therefore$ Area ABCD $= \left(\frac{1}{2} \times (120 + 75) \times 58.19\right)$ sq. metres

$= 5672.55$ sq. metres.

*Hence, the cost of ploughing the field*

$$= Rs. \left(\frac{5672.55 \times 60}{100}\right) = Rs.\ 3403.53.$$

**Ex. 22.** ***Find the circumference and the area of a circle of diameter 98 cm.***

**Sol.** Diameter of the circle = 98 cm.

$\therefore$ Its radius = 49 cm.

$\therefore$ Circumference $= 2\pi r = \left(2 \times \frac{22}{7} \times 49\right)$ cm. = 3.08 metres.

*And, area* $= \pi r^2 = \left(\frac{22}{7} \times 49 \times 49\right)$ *sq. cm. = 7546 sq. cm.*

**Ex. 23.** ***In a circle of radius 28 cm., an arc subtends an angle of 72° at the centre. Find the length of the arc and the area of the sector so formed.***

**Sol.** Length of arc $= \frac{2\pi r \times \theta}{360} = \left(2 \times \frac{22}{7} \times 28 \times \frac{72}{360}\right)$ cm.

$= 35.2$ cm.

Area of the sector $= \frac{\pi r^2 \theta}{360} = \left(\frac{22}{7} \times 28 \times 28 \times \frac{72}{360}\right)$ sq. cm.

$= 492.8$ sq. cm.

**Ex. 24.** ***A circular grassy plot of land 70 metres in diameter has a path 7 metres wide running round it on the outside. How many stones 25 cm. by 11 cm. are needed to pave the path ?***

**Sol.** Radius of plot (excluding path) = 35 metres.

Radius of plot (including path) = 42 metres.

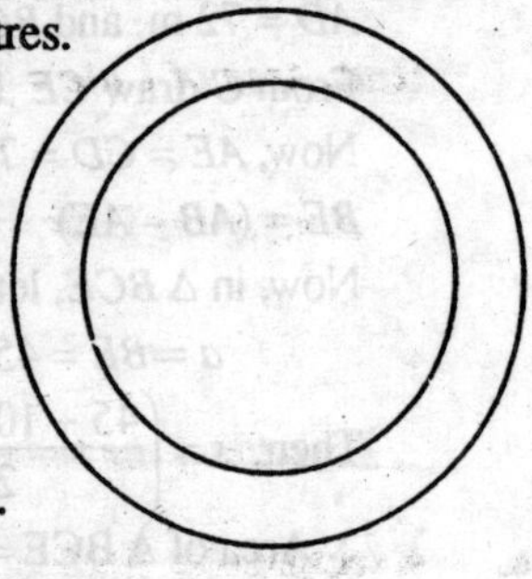

Area of plot (excluding path)

$= \left(\frac{22}{7} \times 35 \times 35\right)$ sq. metres

$= 3850$ sq. metres.

Area of plot (including path)

$= \left(\frac{22}{7} \times 42 \times 42\right)$ sq. metres.

$= 5544$ sq. metres.

Area of path = (5544 – 3850) = 1694 sq. metres.

$= (1694 \times 100 \times 100)$ sq. cms.

Area of one stone $= (25 \times 11)$ sq. cms.

*Number of stones* $= \frac{1694 \times 100 \times 100}{25 \times 11} = 61600.$

**Ex. 25.** ***The diameter of the driving wheel of a bus is 140 cm. How many revolutions per minute must the wheel make in order to keep a speed of 66 km. per hour ?***

**Sol.** Distance covered by wheel in 1 minute

$$= \frac{66 \times 1000 \times 100}{60} = \text{cms.} = 110000 \text{ cms.}$$

$$\text{Circumference of wheel} = \left(2 \times \frac{22}{7} \times 70\right) \text{cms.} = 440 \text{ cms.}$$

$$\textit{Number of revolutions in 1 minute} = \left(\frac{110000}{440}\right) = 250.$$

## EXERCISE 20A (Subjective)

1. Calculate the area of a rectangular field whose :
   (i) Length = 13.5 m, breadth = 8 m;
   (ii) Length = 66 m and perimeter = 242 m;
   (iii) One side = 12 m and diagonal = 13 m.
2. Find the area of a square field, the length of whose diagonal is 36 metres.
3. The cost of fencing a rectangular field at Rs. 3.50 per metre is Rs. 595. If the length of the field be 60 metres, find the cost of levelling it at 50 paise per square metre.
4. The length and breadth of a rectangular field are in the ratio 5 : 3. If the cost of cultivating the field at 25 paise per square metre is Rs. 6000, find the dimensions of the field.
5. A room 15 m long requires 7500 tiles, each 15 cm. by 12 cm, to cover the entire floor. Find the breadth of the room.

   **Hint.** *Area of 1 tile* $= \left(\frac{15}{100} \times \frac{12}{100}\right)$ *sq. m.*

   $$\textit{Area of the floor} = \left(7500 \times \frac{15}{100} \times \frac{12}{100}\right) \textit{sq. m.}$$
   $$= 135 \textit{ sq. m.}$$

   $$\therefore \textit{Breadth of the room} = \frac{\text{Area}}{\text{Length}} = \left(\frac{135}{15}\right) m = 9 \, m.$$

6. A rectangular field is 125 m long and 68 m broad. A path of uniform width of 3 metres runs round the field inside it. Fina the area of the path.
7. Find the length of the diagonal of a square of area 200 square centimetres.

   **Hint.** $\frac{1}{2} \times (\textit{diagonal})^2 = 200.$

8. A footpath of uniform width runs round the inside of a rectangular field 38 metres long and 32 metres wide. If the area of the path be 600 sq. metres, find its width.

   **Hint.** *Let the width of the path be x metres.*
   *Now, area of the field* = $(38 \times 32)$ *sq. m.* = *1216 sq. m.*
   $\therefore (1216 - 600) = (38 - 2x)(32 - 2x)$ .

9. The length and breadth of a rectangular field are in the ratio 7 : 4. A path 4 metre wide running all round outside it has an area of 416 square metres. Find the length and breadth of the field.
10. The dimensions of a room are 12.5 metres by 9 metres by 7 metres. There are 2 doors and 4 windows in the room; each door measures 2.5 metres by 1.2 metres and each window 1.5 metres by 1 metre. Find the cost of painting the walls at Rs. 3.50 per square metre.
**Hint.** *Area to be painted*
$= [2 \times (12.5 + 9) \times 7] - [(2 \times 2.5 \times 1.2) + (4 \times 1.5 \times 1)]$ *sq. m.*
11. The area of a square field is 8 hectares. How long would a man take to cross it diagonally by walking at the rate of 4 km. per hour.
12. The length of a hall is 20 metres and its width is 16 metres. The sum of the areas of the floor and the flat roof is equal to sum of the areas of the four walls. Find the height of the hall.
**Hint.** $2\ (20 \times 16) = 2\ (20 + 16) \times h$. *Find h.*
13. The length and breadth of a room are in the ratio 4 : 3 and its height is 5.5 metres. The cost of decorating its walls at Rs. 6.60 per square metre is Rs. 5082. Find the length and breadth of the room.
**Hint.** *Area of 4 walls* $= \left(\frac{5082}{6.60}\right)$ *sq. m.* $= 770$ *sq. m.*
$2\ (4x + 3x) \times 5.5 = 770$. *Find x.*
*Length* $= 4x$ and *Breadth* $= 3x$ *in metres.*
14. The perimeter of a rectangular field is 0.6 km. and its length is twice its breadth. Determine the area of the field in hectares.
**Hint.** *Let breadth* $= x$ *metres & length* $= 2x$ *metres.*
*Then,* $2\ (x + 2x) = (0.6 \times 1000) \Rightarrow x = 100$ *m.*
15. If the length of each side of a square be doubled, show that its area becomes 4 times.
**Hint.** *Let side* $= x$ *metres. Then, area* $= x^2$ *sq. m.*
*New side* $= (2x)$ *metres. New area* $= 4x^2$ *sq. m.*
16. A rectangular park is 35 metres by 18.5 metres. In the middle, there is a circular flower bed 7 metres in diameter. Find the cost of planting grass in the remaining part at 60 paise per square metre.
**Hint.** *Remaining area* $= \left[(35 \times 18.5) - \left(\frac{22}{7} \times 3.5 \times 3.5\right)\right]$ *sq. m.*
17. The base of a triangular field is three times its height. If the cost of cultivating the field at Rs. 36.72 per hectare is Rs. 495.72, find its base and height.
18. Find the area of a quadrilateral piece of ground ABCD in which $AB = 143$ metres, $BC = 154$ metres, $AC = 165$ metres, $CD = 60$ metres and $AD = 125$ metres.

**Hint.** *Area of Quad. ABCD*
$= (Area\ of\ \Delta ABC + Area\ of\ \Delta ACD)$.

**19.** Find the area of a parallelogram whose two adjacent sides are 130 metres and 140 metres and one of the diagonals is 150 metres long. Find also the cost of gravelling it at the rate of 10 paise per square metre.

**20.** Find the area of a quadrilateral piece of ground one of whose diagonals is 50 metres long and the lengths of perpendiculars from the other two vertices are 29 and 21 metres respectively.

**Hint.** $Area = \frac{1}{2} \times 50\ [29 + 21]$.

**21.** A carpet is laid on a rectangular floor of a drawing room in a house, measuring 8 metres by 5 metres. There is a boarder of constant width around the carpet and the area of the boarder is 12 square metres. What is the width of the boarder.

**Hint.** *Let the width of boarder be* $(x/2)$ *metres. Then,*
$40 - \{(8 - x)(5 - x)\} = 12$ *or* $x^2 - 13x + 12 = 0$
*or* $(x - 1)(x - 12) = 0$ *So,* $x = 1$ *or* $x = 12$.
*But* $x = 12$ *is not possible, so* $x = 1$.

**22.** The two parallel sides of a trapezium measure 58 metres and 42 metres respectively. The other two sides are equal, each being 17 metres. Find its area.

**23.** The area of a parallelogram is 338 sq. metres. If its altitude is twice the corresponding base, determine the base and the altitude.

**24.** Find the area of a rhombus one side of which measures 20 cm. and one of whose diagonals is 24 cms.

**25.** The cost of ploughing trapezoid field at the rate of Rs. 1.35 per square metre is Rs. 421.20. The difference between the parallel sides is 8 metres and the perpendicular distance between them is 24 metres. Find the length of parallel sides.

**26.** Find the area of a triangle in which :
(i) Base = 36.8 cm. and Height = 7.5 cm;
(ii) each side measures 8 cm;
(iii) $a = 25$ cm, $b = 17$ cm, $c = 12$ cm.

**27.** The area of a circle is 38.5 square metres. Find its circumference.

**Hint.** $\pi r^2 = 38.5 \Rightarrow r^2 = \left(38.5 \times \frac{7}{22}\right) = (12.25)$.
$\therefore r = \sqrt{12.25} = 3.5\ m$.
$Circumference = \left(2 \times \frac{22}{7} \times 3.5\right) m$.

**28.** The circumference of a circle is 6.6 metres. Find its area.

**Hint.** $2\pi r = 6.6 \Rightarrow r = \left(6.6 \times \frac{7}{22} \times \frac{1}{2}\right) m. = 1.05\ m.$ *Find* $\pi r^2$.

**29.** The length of an arc subtending an angle of 72° is 22 cm. Find the radius of the circle.

**Hint.** *Length of the arc* $= \left(2 \times \frac{22}{7} \times r \times \frac{72}{360}\right)$,

$\therefore \frac{44r}{35} = 22$, *find r.*

**30.** The radius of a circle is 35 cm. Find the area of a sector enclosed by two radii and an arc 44 cm. in length.

**Hint.** $44 = \left(2 \times \frac{22}{7} \times 35 \times \frac{\theta}{360}\right)$ or $\theta = 72°$.

*Area of sector* $= \left(\frac{22}{7} \times \frac{72}{360} \times 35 \times 35\right) = 770$ *sq. cm.*

**31.** The area of a triangular plate of which the base and the altitude are 33 cm. and 14 cm. respectively is to be reduced to one third by drilling a circular hole through it. Calculate the diameter of the hole.

**Hint.** *Area of plate* $= \left(\frac{1}{2} \times 33 \times 14\right)$ *sq. cm.* $= 231$ *sq. cm.*

*Reduced area* $= \left(\frac{1}{3} \times 231\right)$ *sq. cm.* $= 77$ *sq. cm.*

*Area of hole* $= (231 - 77) = 154$ *sq. cm.*

$\pi r^2 = 154$, *find r.*

**32.** Fill in the blanks :

(i) Area of four walls of a room = (...?...) × Height × (...).

(ii) Height of a room $= \frac{\text{(Area of 4 walls)}}{(\ ...+...\ )} \times \frac{1}{(\ ...\ )}$.

(iii) Area of a triangle $= \frac{1}{2} \times$ (..........) × (..........).

(iv) Area of an equilateral triangle $= \frac{\sqrt{3}}{4}$ (..........).

(v) Area of a Trapezium

$= \frac{1}{2} \times$ (Perp. distance between parallel sides) × (..........).

(vi) Area of a rhombus $= \frac{1}{2} \times$ (.........).

(vii) Area of a parallelogram = (.........) × ( .......... ).

(viii) Area of a sector with central angle θ° = (Area of circle) × (..........).

(ix) Length of an arc making a central angle θ° = (..........) $\times \frac{\theta}{360}$

(x) In a square, $\frac{1}{2}$ (diagonal)$^2$ = (...........).

(xi) Length of an arc = (Area of sector) × (..........).

(xii) Circumference of a circle = $\frac{\text{(Area of the circle)}}{(.............)}$.

## ANSWERS (Exercise 20A)

| | | |
|---|---|---|
| **1.** (*i*) 108 sq. m. | (*ii*) 3630 sq. m. | (*iii*) 60 sq. m. |
| **2.** 648 sq. m. | **3.** Rs. 750 | **4.** 200 m. by 120 m. |
| **5.** 9 metres | **6.** 1122 sq. m. | **7.** 20 cm. |
| **8.** 5 metres | **9.** 28 m, 16 m. | **10.** Rs. 1011.50 |
| **11.** 6 min. | **12.** 8.89 metres | **13.** 40 m, 30 m. |
| **14.** 200 m, 100 m | **16.** Rs. 365.40 | **17.** 900 m, 300 m |
| **18.** 13336.5 sq. m. | **19.** 16800 sq. m, Rs. 1680 | **20.** 1250 sq. m. |
| **21.** 50 cm. | **22.** 750 sq. m. | **23.** 13 m, 26 m. |
| **24.** 384 sq. cm | **25.** 17 m, 9 m. | |
| **26.** (*i*) 138 sq. cm. | (*ii*) 27.68 sq. cm. | (*iii*) 90 sq. cm. |
| **27.** 22 metres | **28.** 3.465 sq. m. | **29.** 17.5 cm. |
| **30.** 770 sq. cm. | **31.** 14 cm. | |

**32.** (*i*) (length + breadth), 2 (*ii*) length, breadth, 2

(*iii*) Base × Height (*iv*) (side)$^2$

(*v*) Sum of parallel sides (*vi*) product of diagonals

(*vii*) (Base × Height of base from opposite side)

(*viii*) $\frac{\theta}{360}$ (*ix*) Circumference

(*x*) Area of the square (*xi*) $\frac{2}{\text{(radius)}}$ (*xii*) $\frac{\pi}{2}$

## EXERCISE 20B (Objective Type Questions)

1. One side of a rectangular field is 4 metres and its diagonal is 5 metres. The area of the field is :

   (a) $12 \text{ m}^2$ (b) $20 \text{ m}^2$

   (c) $15 \text{ m}^2$ (d) $4\sqrt{5} \text{ m}^2$

2. A lawn in the form of a rectangle is half as long again as it is broad. The area of the lawn is $\frac{2}{3}$ hectares. The length of the lawn is :

   (a) 100 metres (b) $33\frac{1}{3}$ metres

   (c) $66\frac{2}{3}$ metres (d) $\left(\frac{100}{\sqrt{3}}\right)$ metres

3. The length of a rectangle is increased by 60%. By what percent would the width have to be decreased to maintain the same area ?

   (a) $37\frac{1}{2}\%$ (b) 60%

   (c) 75% (d) 120% **(Bank P.O. 1988)**

4. The perimeter of a rectangle is 82 m and its area is $400 \text{ m}^2$. The breadth of the rectangle is :

   (a) 14 m (b) 16 m

   (c) 18 m (d) 12 m

5. The width of a rectangular hall is $\frac{3}{4}$ of its length. If the area of the hall is $300 \text{ m}^2$, then the difference between its length and width is :

   (a) 3 m (b) 4 m

   (c) 5 m (d) 15 m **(Bank P.O. 1990)**

6. A room $8 \text{ m} \times 6 \text{ m}$ is to be carpeted by a carpet 2 m wide. The length of carpet required is :

   (a) 12 m (b) 36 m

   (c) 24 m (d) 48 m

   **(Railway Recruitment, 1990)**

7. The dimensions of the floor of a rectangular hall are $4 \text{ m} \times 3 \text{ m}$. The floor of the hall is to be tiled fully with $8 \text{ cm} \times 6 \text{ cm}$ rectangular tiles without breaking tiles to smaller sizes. The number of tiles required is :

   (a) 4800 (b) 2600

   (c) 2500 (d) 2400 **(C.D.S. 1991)**

8. The length of a rectangle is twice its breadth. If its length is decreased by 5 cm and breadth is increased by 5 cm, the area of rectangle is increased by $75 \text{ cm}^2$. Therefore, the length of the rectangle is :

(*a*) 20 cm. (*b*) 30 cm.
(*c*) 40 cm. (*d*) 50 cm. **(C.D.S. 1991)**

**9.** A 5 m wide lawn is cultivated all along the outside of a rectangular plot measuring 90 m × 40 m. The total area of the lawn is :
(*a*) 1200 $m^2$ (*b*) 1300 $m^2$
(*c*) 1350 $m^2$ (*d*) 1400 $m^2$ **(C.D.S. 1991)**

**10.** The length of a plot of land is 4 times its breadth. A playground measuring 1200 $m^2$ occupies one-third of the total area of the plot. What is the length of the plot, in metres ?
(*a*) 90 (*b*) 80
(*c*) 60 (*d*) none of these
**(Bank P.O. 1990)**

**11.** The length and breadth of a playground are 36 m and 21 m respectively. Flagstaffs are required to be fixed on all along the boundary at a distance 3 m apart. The number of flagstaffs will be :
(*a*) 37 (*b*) 38
(*c*) 39 (*d*) 40
**(I. Tax & Central Excise 1989)**

**12.** If the width of a rectangle is 2 m less than its length, and its perimeter is 32 m, the area of the rectangle is :
(*a*) 224 $m^2$ (*b*) 108 $m^2$
(*c*) 99 $m^2$ (*d*) 63 $m^2$

**13.** If the length of a rectangle is increased by $12\frac{1}{2}$ % and the width increased by $6\frac{1}{4}$ %, the area of the rectangle will :
(*a*) increase by 19.53% (*b*) decrease by $6\frac{1}{4}$ %
(*c*) increase by $19\frac{3}{4}$ % (*d*) increase by 19.92%

**14.** A room 5m × 4m is to be carpeted leaving a margin of 25 cm from each wall. If the cost of the carpet is Rs. 80 per $m^2$, the cost of carpeting the room will be :
(*a*) Rs. 1440 (*b*) Rs. 1260
(*c*) Rs. 1228 (*d*) Rs. 1192

**15.** A rectangular carpet has an area of 120 $m^2$ and a perimeter of 46 m. The length of its diagonal is :
(*a*) 15 m (*b*) 16 m
(*c*) 17 m (*d*) 20 m
**(Railway Recruitment 1991)**

16. A rectangle is having 15 cm as its length and 150 $cm^2$ as its area. Its area is increased to $1\frac{1}{3}$ times the original area by increasing only its length. Its new perimeter is :
(*a*) 50 cm (*b*) 60 cm
(*c*) 70 cm (*d*) 80 cm **(Bank P.O. 1989)**

17. Length of a room is 6 m longer than its breadth. If the area of the room is 72 $m^2$, its breadth will be :
(*a*) 12 m (*b*) 6 m
(*c*) 8 m (*d*) 10 m

18. The length of the longest rod which can be laid across a floor of a rectangular room 12 m in length and 5 m in breadth will be :
(*a*) 17 m (*b*) 7 m
(*c*) 2.4 m (*d*) 13 m

19. A man drives 4 km. distance to go around a rectangular park. If the area of the rectangle is 0.75 sq. km, the difference between the length and the breadth of the rectangle is :
(*a*) 10.25 km (*b*) 0.5 km
(*c*) 1 km (*d*) 2.75 km

20. A man walked 20 m to cross a rectangular field diagonally. If the length of the field is 16 m, the breadth of the field is :
(*a*) 4 m (*b*) 16 m
(*c*) 12 m (*d*) cannot be determined
**(Railway Recruitment 1991)**

21. If only the length of a rectangular plot is reduced to $\frac{2}{3}$ rd of its original length, the ratio of original area to reduced area is :
(*a*) 2 : 3 (*b*) 3 : 2
(*c*) 1 : 2 (*d*) none of these
**(Railway Recruitment 1991)**

22. The length and breadth of a rectangular piece of land are in the ratio of 5 : 3. The owner spent Rs. 3000 for surrounding it from all the sides at the rate of Rs. 7.50 per metre. The difference between length & breadth is :
(*a*) 50 m (*b*) 100 m
(*c*) 200 m (*d*) 150 m **(BSRB Exam. 1991)**

23. A rectangular lawn 60 m by 40 m has two roads, each 5 m wide, running in the middle of it, one parallel to length and the other parallel to breadth. The cost of gravelling the roads at 60 p. per $m^2$ is :
(*a*) Rs. 300 (*b*) Rs. 285
(*c*) Rs. 250 (*d*) Rs. 265

**24.** The sides of a rectangular park are in the ratio 3 : 2 and its area is 3750 $m^2$. The cost of fencing it at 50 paise per metre is :
(*a*) Rs. 312.50 (*b*) Rs. 375
(*c*) Rs. 187.50 (*d*) Rs. 125

**25.** The cost of carpeting a room 15 m long with a carpet 75 cm wide at 30 paise per metre is Rs. 36. The breadth of the room is :
(*a*) 8 m (*b*) 12 m
(*c*) 9 m (*d*) 6 m

**26.** The diagonal of a square is 3.2 m. Its area is :
(*a*) 10.24 $m^2$ (*b*) 2.56 $m^2$
(*c*) 3.41 $m^2$ (*d*) 5.12 $m^2$

**27.** Area of a square field is $\frac{1}{2}$ hectare. The diagonal of the square is :
(*a*) 50 m (*b*) 100 m
(*c*) 250 m (*d*) $50\sqrt{2}$ m

**28.** If the side of a square be increased by 4 cm, the area increases by 60 sq. cms. The side of the square is :
(*a*) 12 cm (*b*) 13 cm
(*c*) 14 cm (*d*) none of these

**29.** The cost of cultivating a square field at the rate of Rs. 160 per hectare is Rs. 1440. The cost of fencing it at 75 paise per m is :
(*a*) Rs. 900 (*b*) Rs. 1800
(*c*) Rs. 360 (*d*) Rs. 810

**30.** The length and breadth of a square are increased by 40% and 30% respectively. The area of resulting rectangle exceeds the area of the square by :
(*a*) 42% (*b*) 62%
(*c*) 82% (*d*) none of these
**(I. Tax & Central Excise 1988)**

**31.** If the side of a square be increased by 50%, the percent increase in area is :
(*a*) 50 (*b*) 100
(*c*) 125 (*d*) 150 **(N.D.A. Exam. 1987)**

**32.** The length of a square is increased by 40% while breadth is decreased by 40%. The ratio of area of the resulting rectangle so formed to that of the original square is :
(*a*) 25 : 21 (*b*) 21 : 25
(*c*) 16 : 15 (*d*) 15 : 16
**(I. Tax & Central Excise 1989)**

**33.** The ratio of areas of two squares, one having double its diagonal than the other, is :

(*a*) 2 : 1 (*b*) 3 : 1
(*c*) 3 : 2 (*d*) 4 : 1

**34.** Of the two square fields, the area of one is 1 hectare, while the other one is broader by 1%. The difference in areas is :
(*a*) 101 $m^2$ (*b*) 201 $m^2$
(*c*) 100 $m^2$ (*d*) 200 $m^2$

**35.** Of all the rectangles, the square has
(*a*) smallest area for a given perimeter
(*b*) largest area for a given perimeter
(*c*) largest perimeter for a given area
(*d*) none of the above

**36.** The area of a rectangle is thrice that of a square. Length of the rectangle is 40 cm and breadth of the rectangle is $\frac{3}{2}$ times that of the side of the square. The side of the square in cms is :
(*a*) 60 (*b*) 20
(*c*) 30 (*d*) 15 **(Bank P.O. 1989)**

**37.** If the ratio of areas of two squares is 9 : 1, the ratio of their perimeters is :
(*a*) 9 : 1 (*b*) 3 : 4
(*c*) 3 : 1 (*d*) 1 : 3 **(Asstt. Grade 1990)**

**38.** The length and breadth of a room are 10 m 75 cm and 8 m 25 cm respectively. The floor is to be paved with square tiles of the largest possible size. The size of the tiles is :
(*a*) 25 cm × 25 cm (*b*) 50 cm × 50 cm
(*c*) 20 cm × 20 cm (*d*) 30 cm × 30 cm
**(Hotel Management 1991)**

**39.** Area of four walls of a room is 77 $m^2$. The length and breadth of the room are 7.5 m and 3.5 m respectively. The height of the room is :
(*a*) 7.7 m (*b*) 3.5 m
(*c*) 6.77 m (*d*) 5.4 m

**40.** Area of four walls of a room is 168 $m^2$. The breadth and height of the room are 8 m and 6 m respectively. The length of the room is :
(*a*) 14 m (*b*) 12 m
(*c*) 3.5 m (*d*) 6 m

**41.** The cost of papering four walls of a room is Rs. 48. Each one of length, breadth and height of another room is double that of the room. The cost of papering the walls of this new room is :
(*a*) Rs. 96 (*b*) Rs. 192
(*c*) Rs. 384 (*d*) Rs. 288

**42.** One side of a parallelogram is 14 cms. Its distance from the opposite side is 16 cms. The area of the parallelogram is :

(*a*) 112 $cm^2$ (*b*) 224 $cm^2$
(*c*) 56 $\pi$ $cm^2$ (*d*) 210 $cm^2$

**43.** One side of a rhombus is 10 cms and one of its diagonals is 12 cms The area of the rhombus is :

(*a*) 120 $cm^2$ (*b*) 60 $cm^2$
(*c*) 80 $cm^2$ (*d*) 96 $cm^2$

**44.** If the perimeter of a rhombus is $4a$ and lengths of the diagonals are $x$ and $y$, then its area is :

(*a*) $a(x+y)$ (*b*) $x^2+y^2$
(*c*) $xy$ (*d*) $\frac{1}{2}xy$ **(N.D.A. 1990)**

**45.** In a rhombus whose area is 144 $cm^2$ one of its diagonals is twice as long as the other. The lengths of its diagonals are :

(*a*) 24 cm, 48 cm (*b*) 12 cm, 24 cm
(*c*) $6\sqrt{2}$ cm, $12\sqrt{2}$ cm (*d*) 6 cm, 12 cm **(C.D.S. 1989)**

**46.** The two parallel sides of a trapezium are 1 m and 2 m respectively. The perpendicular distance between them is 6 m. The area of trapezium is :

(*a*) 9 $m^2$ (*b*) 12 $m^2$
(*c*) 6 $m^2$ (*d*) 18 $m^2$

**47.** The cross section of a canal is a trapezium in shape. If the canal is 10 m wide at the top and 6 m wide at the bottom and the area of cross section is 640 $m^2$, the length of canal is :

(*a*) 40 m (*b*) 80 m
(*c*) 160 m (*d*) 384 m

**48.** The area of a trapezium is 384 $cm^2$. If its parallel sides are in the ratio 3 : 5 and the perpendicular distance between them be 12 cm, the smaller of parallel sides is :

(*a*) 16 cm (*b*) 24 cm
(*c*) 32 cm (*d*) 40 cm

**49.** If each of the dimensions of a rectangle is increased by 100%, its area is increased by :

(*a*) 100% (*b*) 200%
(*c*) 300% (*d*) 400%

**50.** The length of the rectangular floor is twice its width. If the length of a diagonal is $9\sqrt{5}$ m, then perimeter of the rectangle is :

(*a*) 27 m (*b*) 54 m
(*c*) 81 m (*d*) 162 m

51. The length of a rectangular plot is 144 m and its area is same as that of a square plot with one of its sides being 84 m. The width of the plot is :
(*a*) 7 m (*b*) 49 m
(*c*) 14 m (*d*) data inadequate

52. The perimeters of both, a square and a rectangle are each equal to 48 m and the difference between their areas is 4 $m^2$. The breadth of the rectangle is :
(*a*) 10 m (*b*) 12 m
(*c*) 14 m (*d*) none of these

53. The length of a rectangle is doubled while its breadth is halved. The percentage change in area is :
(*a*) 50 sq. units (*b*) 75 sq. units
(*c*) no change (*d*) none of these

54. The area of a parallelogram is 72 $cm^2$ and its altitude is twice the corresponding base. Then the length of the base is :
(*a*) 3 cm (*b*) 6 cm
(*c*) 12 cm (*d*) none of these

55. The perimeter of a rhombus is 52 m while its longer diagonal is 24 m. Its other diagonal is :
(*a*) 5 m (*b*) 10 m
(*c*) 20 m (*d*) 28 m

56. A tin sheet is in the form of a rhombus whose side is 5 cm and one of its diagonals is 8 cm. Then the cost of painting the sheet at the rate of Rs. 3.50 per $cm^2$ on both of its sides is :
(*a*) Rs. 84 (*b*) Rs. 140
(*c*) Rs. 168 (*d*) none of these

57. The side of an equilateral triangle is 8 cms. Its area is :
(*a*) $4\sqrt{3}$ $cm^2$ (*b*) $16\sqrt{3}$ $cm^2$
(*c*) 21.3 $cm^2$ (*d*) 64 $cm^2$

58. The area of an equilateral triangle is $4\sqrt{3}$ $cm^2$. The length of each side of the triangle is :
(*a*) 4 cms (*b*) 3 cms
(*c*) $\frac{\sqrt{3}}{4}$ cms (*d*) $\frac{4}{\sqrt{3}}$ cms

59. The altitude of an equilateral triangle of side $2\sqrt{3}$ cms is :
(*a*) $\frac{\sqrt{3}}{2}$ cms (*b*) $\frac{1}{2}$ cms
(*c*) $\frac{\sqrt{3}}{4}$ cms (*d*) 3 cms

**60.** The front of a tent has the form of a triangle and has a base of length 1.5 m and a height of 2 m. How much material is required to make a cover for this front ?

(*a*) 3.5 $m^2$ (*b*) 1.5 $m^2$

(*c*) 3.0 $m^2$ (*d*) 1.3 $m^2$

**61.** The sides of a triangle are 13 m, 14 m and 15 m. The area of the triangle is :

(*a*) 2730 $m^2$ (*b*) 52.2 $m^2$

(*c*) 84 $m^2$ (*d*) 106 $m^2$

**62.** One side of a right angled triangular scarf is 80 cm and its longer side is 1 m. Its cost at the rate of Rs. 5 per $m^2$ will be :

(*a*) Rs. 1.20 (*b*) Re. 1.00

(*c*) Rs. 1.25 (*d*) Rs. 1.15 **(N.D.A. 1990)**

**63.** Heights of two similar triangles are 3 cm and 4 cm respectively. Their areas will be in the ratio :

(*a*) 4 : 9 (*b*) 9 : 4

(*c*) 16 : 9 (*d*) 9 : 16

**64.** If the area of an equilateral triangle is $36\sqrt{3}$ $cm^2$, the perimeter of the triangle is :

(*a*) 18 cm (*b*) 24 cm

(*c*) 30 cm (*d*) 36 cm

**65.** The area of a right angled triangle is 432 $cm^2$. If the hypotenuse of the triangle is $30\sqrt{2}$ cm, its perimeter will be nearly :

(*a*) $72\sqrt{2}$ cm (*b*) 108 cm

(*c*) 84 cm (*d*) none of these

**66.** The legs of a right triangle are in the ratio of 1 : 2 and its area is 36. The hypotenuse of the triangle is :

(*a*) 3 (*b*) $\sqrt{5}$

(*c*) $\sqrt{3}$ (*d*) $6\sqrt{5}$

**(Railway Recruitment 1991)**

**67.** If the sides of a triangle are doubled, its area

(*a*) remains same (*b*) becomes double

(*c*) becomes 3 times (*d*) becomes 4 times

**(Railway Recruitment 1991)**

**68.** Each side of an equilateral triangle is increased by 1.5%. The percentage increase in its area is :

(*a*) 1.5% (*b*) 3%

(*c*) 4.5% (*d*) 5.7%

**(Railway Recruitment 1991)**

**69.** The perimeters of an equilateral triangle and a square are same. The ratio of their areas is

(*a*) 1 (*b*) more than 1

(*c*) less than 1 (*d*) may be any of these

**70.** If the area of a triangle with base $x$ is equal to the area of a square with side $x$, then the altitude of the triangle is :

(*a*) $\frac{x}{2}$ (*b*) $x$

(*c*) $2x$ (*d*) $3x$

**(I. Tax & Central Excise 1988)**

**71.** The perimeter of an isosceles triangle is equal to 14 cm; the lateral side is to the base in the ratio 5 to 4. The area, in $cm^2$, of the triangle is :

(*a*) $\frac{1}{2}\sqrt{21}$ (*b*) $\sqrt{21}$

(*c*) $\frac{3}{2}\sqrt{21}$ (*d*) $2\sqrt{21}$ **(C.D.S. Exam 1989)**

**72.** If a square and a rhombus stand on the same base, then the ratio of areas of square and rhombus is :

(*a*) greater than 1 (*b*) equal to 1

(*c*) equal to $\frac{1}{4}$ (*d*) equal to $\frac{1}{2}$

**(N.D.A. Exam. 1990)**

**73.** The circumference of a circle is 352 m. Its area is :

(*a*) 9856 $m^2$ (*b*) 8956 $m^2$

(*c*) 6589 $m^2$ (*d*) 5986 $m^2$ **(N.D.A. Exam. 1990)**

**74.** The area of a circle is 38.5 $cm^2$. Its circumference is :

(*a*) 6.20 cms (*b*) 11 cms

(*c*) 22 cms (*d*) 121 cms

**75.** The difference between the circumference and radius of a circle is 37 cms. The area of the circle is :

(*a*) 148 $cm^2$ (*b*) 111 $cm^2$

(*c*) 154 $cm^2$ (*d*) 259 $cm^2$

**76.** If the diameter of a circle is doubled, its area :

(*a*) is doubled (*b*) is trebled

(*c*) is quadrupled (*d*) becomes 6 times

**77.** The radius of a circle has been reduced from 9 cms. to 7 cms. The approximate percentage decrease in area is :

(*a*) 31.5% (*b*) 39.5%

(*c*) 34.5% (*d*) 65.5%

**78.** If the radius of a circle be reduced by 50%, its area is reduced by :

(*a*) 25% (*b*) 75%
(*c*) 50% (*d*) 100%

**79.** The area of a circular park is 13.86 hectares. The cost of fencing it at the rate of 20 paise per metre is :
(*a*) Rs. 277.20 (*b*) Rs. 264
(*c*) Rs. 324 (*d*) Rs. 198

**80.** The radius of a circle is increased so that its circumference increases by 5%. The area of the circle will increase by :
(*a*) 10% (*b*) 10.25%
(*c*) 8.75% (*d*) 10.5%

**81.** The diameter of a wheel is 2 cm. It rolls forward covering 10 revolutions. The distance travelled by it is :
(*a*) 3.14 cms. (*b*) 62.8 cms.
(*c*) 31.4 cms. (*d*) 125.6 cms.

**(Railway Recruitment 1990)**

**82.** A circular disc of area $0.49\pi\ m^2$ rolls down a length of 1.76 kms. The number of revolutions it makes is :
(*a*) 300 (*b*) 400
(*c*) 600 (*d*) 4000

**83.** The radius of a wheel is 1.4 dm. How many times does it revolve during a journey of 0.66 kms. ?
(*a*) 375 (*b*) 750
(*c*) 1500 (*d*) 3000

**84.** A wheel makes 100 revolutions in covering a distance of 88 km. The diameter of the wheel is :
(*a*) 240 metres (*b*) 400 metres
(*c*) 280 metres (*d*) 140 metres

**85.** If the wheel of the engine of a train $4\frac{2}{7}$ metres in circumference makes seven revolutions in 4 seconds, the speed of the train in km/hr is :
(*a*) 35 (*b*) 32
(*c*) 27 (*d*) 20 **(Clerk's Grade Exam 1991)**

**86.** The area of a sector of a circle of radius 5 cms formed by an arc of length 3.5 cms, is :
(*a*) 35 $cm^2$ (*b*) 17.5 $cm^2$
(*c*) 8.75 $cm^2$ (*d*) 55 $cm^2$

**87.** The length of a minute hand on a wall clock is 7 cms. The area swept by the minute hand in 30 minutes, is :
(*a*) 147 $cm^2$ (*b*) 210 $cm^2$
(*c*) 154 $cm^2$ (*d*) 77 $cm^2$

**88.** The area of the sector of a circle, whose radius is 12 m and whose angle at the centre is 42°, is :

(*a*) 26.4 $m^2$ (*b*) 39.6 $m^2$

(*c*) 52.8 $m^2$ (*d*) 79.2 $m^2$

**89.** The length of a rope by which a buffalo must be tethered in order that she may be able to graze an area of 9856 $m^2$, is :

(*a*) 56 m (*b*) 64 m

(*c*) 88 m (*d*) 168 m

**(I. Tax & Central Excise 1989)**

**90.** A circular wire of radius 42 cm is cut and bent in the form of a rectangle whose sides are in ratio 6 : 5. The smaller side of the rectangle is :

(*a*) 30 cm (*b*) 60 cm

(*c*) 72 cm (*d*) 132 cm

**(I. Tax & Central Excise 1989)**

**91.** The perimeters of a circular field and square field are equal. If the area of square field is 12100 $m^2$, the area of circular field will be :

(*a*) 15500 $m^2$ (*b*) 15400 $m^2$

(*c*) 15200 $m^2$ (*d*) 15300 $m^2$

**92.** The perimeter of a semi circle of 56 cm diameter will be :

(*a*) 144 cm (*b*) 232 cm

(*c*) 154 cm (*d*) 116 cm **(Bank P.O. 1989)**

**93.** Four circular cardboard pieces each of radius 7 cm are placed in such a way that each piece touches two other pieces. The area of the space enclosed by 4 pieces is :

(*a*) 38 $cm^2$ (*b*) 42 $cm^2$

(*c*) 45 $cm^2$ (*d*) none of these

**94.** The area of the largest circle that can be drawn inside a square of side 14 m, is :

(*a*) 98 $m^2$ (*b*) 202 $m^2$

(*c*) 120 $cm^2$ (*d*) 154 $m^2$

**95.** A square is inscribed in a circle of radius 8 cm. The area of the portion between the circle and the square is :

(*a*) 48.27 $cm^2$ (*b*) 73.14 $cm^2$

(*c*) 169.14 $cm^2$ (*d*) 88.26 $cm^2$

**96.** From a rectangular sheet of metal 10 cm × 8 cm, a circular sheet of maximum possible area is cut. The area of the remaining sheet is :

(*a*) 29. 71 $cm^2$ (*b*) 20.66 $cm^2$

(*c*) 12.82 $cm^2$ (*d*) 8.66 $cm^2$

**97.** If a piece of wire 20 cm long is bent into an arc of a circle subtending an angle of 60° at the centre, then the radius of the circle (in cms) is :

(*a*) $\frac{\pi}{120}$ (*b*) $\frac{\pi}{60}$

(*c*) $\frac{120}{\pi}$ (*d*) $\frac{60}{\pi}$ **(N.D.A. Exam. 1990)**

**98.** A plot of land is in the shape of a right angled isosceles triangle. The length of hypotenuse is $50\sqrt{2}$ m. The cost of fencing is Rs. 3 per metre. The cost of fencing the plot will be :

(*a*) less than Rs. 300 (*b*) less than Rs. 400

(*c*) more than Rs. 500 (*d*) more than Rs. 600

**(C.D.S. Exam. 1991)**

**99.** There are two concentric circles with outer circle of radius 24 cm. If the area of inner circle is one third of area between the circles, the ratio of the circumferences of the two circles is :

(*a*) 4 : 1 (*b*) 3 : 1

(*c*) 2 : 1 (*d*) 8 : 1

**100.** The radius of the circle inscribed in an equilateral triangle is 2 cm. The altitude of the triangle is :

(*a*) $4\sqrt{3}$ cm (*b*) 6 cm

(*c*) $6\sqrt{3}$ cm (*d*) 6.5 cm

## HINTS & SOLUTIONS (Exercise 20B)

1. $l^2 + b^2 = (\text{diagonal})^2$ or $b^2 = [(\text{diagonal})^2 - l^2]$.

$\therefore b^2 = (5^2 - 4^2) = 9$ and therefore $b = 3$ m.

*Hence, the area of the field* $= (4 \times 3)\ m^2 = 12\ m^2$.

2. Let breadth $= x$ metres. Then, length $= \frac{3}{2}x$ metres.

$\therefore x \times \frac{3}{2}x = \frac{2}{3} \times 10000$ or $x^2 = \left(\frac{4}{9} \times 10000\right)$. So, $x = \frac{200}{3}$.

$\therefore$ *Length* $= \left(\frac{3}{2} \times \frac{200}{3}\right) = 100$ *metres.*

3. Let the original length be $x$ metres & breadth be $y$ metres.
Let, new breadth $= z$ metres.

$\therefore$ (160% of $x$) $z = xy$ or $z = xy \times \frac{100}{160x} = \frac{5}{8}y$.

Decrease $= \left(y - \frac{5}{8}y\right) = \frac{3}{8}y$.

$\therefore$ *Decrease percent* $= \left(\frac{3}{8}y \times \frac{1}{y} \times 100\right)\% = 37\frac{1}{2}\%$.

**4.** Let length = $x$ metres and breadth = $y$ metres.

Then, $2(x+y) = 82 \Rightarrow x + y = 41 \Rightarrow x = (41 - y)$.

$\therefore xy = 400 \Rightarrow (41 - y)\, y = 400$.

or $y^2 - 41y + 400 = 0$

or $(y - 16)(y - 25) = 0$

$\therefore y = 16$ or $y = 25$.

Hence, breadth = 16 m.

**5.** Let length = $x$ metres. Then, breadth = $\frac{3}{4}x$ metres.

$\therefore x \times \frac{3}{4}x = 300 \Rightarrow x^2 = \frac{300 \times 4}{3} = 400$ or $x = 20$.

$\therefore$ *(length – breadth)* $= \left(20 - \frac{3}{4} \times 20\right) = 5$ *metres.*

**6.** Length of the carpet $= \left(\frac{8 \times 6}{2}\right)$ m = 24 m.

**7.** Area of the floor = $(400 \times 300)\ cm^2$.

Area of one tile = $(8 \times 6)\ cm^2$.

*Number of tiles* $= \frac{400 \times 300}{8 \times 6} = 2500$.

**8.** Let breadth = $x$ cm and length = $2x$ cm.

Then, $(2x - 5)(x + 5) - x \times 2x = 75$.

$\therefore 2x^2 + 5x - 25 - 2x^2 = 75$ or $5x = 50$ or $x = 10$.

*Hence, length = 20 cm.*

**9.** Area of the lawn

$= [(100 \times 50) - (90 \times 40)]\ m^2 = 1400\ m^2$.

**10.** $x \times 4x = 3600 \Rightarrow x^2 = 900 \Rightarrow x = 30$.

$\therefore$ *Length* $= (4 \times 30)$ *m* $= 120$ *m.*

**11.** Number of flagstaffs

$$= \left[\left(\frac{36}{3} + 1\right) + \left(\frac{36}{3} + 1\right) + \left(\frac{21}{3} - 1\right) + \left(\frac{21}{3} - 1\right)\right] = 38.$$

**12.** Let length = $x$ metres and breadth = $(x - 2)$ metres.

$\therefore$ Perimeter $= 2[x + (x - 2)] = (4x - 4)$.

So, $4x - 4 = 32 \Rightarrow x = 9$.

$\therefore$ *Length = 9 m, breadth = 7 m. Hence, area = 63* $m^2$.

**13.** Let length = $x$ metres & breadth = $y$ metres.

Then, area = $(xy)\ m^2$.

New length = $112\frac{1}{2}$ % of $x$ & new breadth = $106\frac{1}{4}$ % of $y$.

$\therefore$ New area $= \frac{225}{200}x \times \frac{425}{400}y = \left(\frac{153}{128}xy\right)m^2$.

Increase in area $= \left(\frac{153}{128}xy - xy\right) = \left(\frac{25}{128}xy\right)m^2$.

$\therefore$ *Increase percent* $= \left(\frac{25}{128}xy \times \frac{1}{xy} \times 100\right)\% = 19.53\ \%$.

14. Area of the carpet $= (4.5 \times 3.5)\ m^2$.
*Cost of the carpet = Rs. $(80 \times 4.5 \times 3.5)$ = Rs. 1260.*

15. $2(x + y) = 46 \Rightarrow x + y = 23 \Rightarrow y = 23 - x$.
$\therefore x(23 - x) = 120 \Rightarrow x^2 - 23x + 120 = 0 \Rightarrow (x - 15)(x - 8) = 0$.
$\therefore$ length = 15 m, breadth = 8 m.
*So, diagonal* $= \sqrt{(15)^2 + (8)^2} = 17$ *m.*

16. Original length = 15 cm & breadth $= \frac{150}{15} = 10$ cm.

New area $= \left(150 \times \frac{4}{3}\right)m^2 = 200\ m^2$.

$\therefore$ New length $= \frac{\text{New area}}{\text{Original breadth}} = \frac{200}{10}$ cm = 20 cm.

*New perimeter = 2 (20 + 10) cm = 60 cm.*

17. $x(x - 6) = 72 \Rightarrow x^2 - 6x - 72 = 0$
$\Rightarrow (x - 12)(x + 6) = 0$
$\Rightarrow x = 12$ (neglecting $x = -6$)
$\therefore$ *Breadth = (12 – 6) m = 6 m.*

18. Length of the rod $= \sqrt{(12)^2 + (5)^2} = 13$ m.

19. $2(x + y) = 4 \Rightarrow x + y = 2$ & $xy = 0.75 = \frac{3}{4}$.

$\therefore (x - y)^2 = (x + y)^2 - 4xy = \left(4 - 4 \times \frac{3}{4}\right) = 1$.

*So,* $(x - y) = 1$ *km.*

20. Breadth $= \sqrt{(20)^2 - (16)^2} = \sqrt{144} = 12$ m.

21. Let length = $x$ & breadth = $y$.

New length $= \frac{2}{3}x$ & breadth $= y$.

$\therefore \dfrac{\textit{Original area}}{\textit{Reduced area}} = \dfrac{xy}{\frac{2}{3}xy} = \dfrac{3}{2}.$

**22.** Perimeter of the field $= \dfrac{3000}{7.50} = 400$ m.

$\therefore 2(5x + 3x) = 400 \Rightarrow x = 25.$

So, length = 125 m & breadth = 75 m.

*Difference between length & breadth = (125 − 75) m = 50 m.*

**23.** Area of cross roads $= (60 \times 5 + 40 \times 5 - 5 \times 5) = 475\ m^2$.

$\therefore$ *Cost of gravelling the roads* $= Rs.\left(475 \times \dfrac{60}{100}\right) = Rs.\ 285.$

**24.** $3x \times 2x = 3750 \Rightarrow x^2 = 625 \Rightarrow x = 25.$

$\therefore$ Length = 75 m & breadth = 50 m.

Perimeter $= [2 \times (75 + 50)]$ m = 250 m.

$\therefore$ *Cost of fencing* $= Rs.\left(250 \times \dfrac{1}{2}\right) = Rs.\ 125.$

**25.** Length of carpet $= \dfrac{3600}{30}$ m = 120 m.

Area of the carpet $= \left(120 \times \dfrac{75}{100}\right) m^2 = 90\ m^2.$

$\therefore$ Area of the room $= 90\ m^2$.

*So, breadth of the room* $= \left(\dfrac{90}{15}\right) m = 6\ m.$

**26.** Area $= \dfrac{1}{2} \times (\text{diagonal})^2 = \left(\dfrac{1}{2} \times 3.2 \times 3.2\right) m^2 = 5.12\ m^2.$

**27.** $\dfrac{1}{2} \times (\textit{diagonal})^2 = \dfrac{1}{2} \times 10000 \Rightarrow \text{diagonal} = \sqrt{10000} = 100$ m.

**28.** $(x + 4)^2 - x^2 = 60 \Rightarrow x^2 + 16 + 8x - x^2 = 60 \Rightarrow x = 5.5$ cm.

**29.** Area $= \dfrac{1440}{160} = 9$ hectares $= (9 \times 10000)\ m^2.$

Side = 3 × 100 = 300 m. So, perimeter = 1200 m.

$\therefore$ *Cost of fencing* $= Rs.\left(1200 \times \dfrac{3}{4}\right) = Rs.\ 900.$

**30.** Let the original side be $x$. Then, area $= x^2$.

Then, new area $= \left(\dfrac{140}{100}x \times \dfrac{130}{100}x\right) = \dfrac{91}{50}x^2.$

$\therefore$ Change in area $= \left(\frac{91}{50}x^2 - x^2\right) = \frac{41}{50}x^2$.

*Increase percent* $= \left(\frac{41}{50}x^2 \times \frac{1}{x^2} \times 100\right)\% = 82\%$.

**31.** Let the side of the square be 100 m.

*Then, increase in area* $= \left[\frac{(150)^2 - (100)^2}{100 \times 100} \times 100\right]\% = 125\%$.

**32.** Let each side of the square be 100 m.

New area = $(140 \times 60)\ m^2$.

$\therefore \frac{\text{New area}}{\text{Original area}} = \frac{140 \times 60}{100 \times 100} = \frac{21}{25}$.

**33.** Let their diagonals be $2d$ and $d$ respectively.

*Ratio of their areas* $= \frac{\frac{1}{2} \times (2d)^2}{\frac{1}{2} \times d^2} = \frac{4}{1}$.

**34.** Area of the first square = $10000\ m^2$.

Side of this square = $\sqrt{10000} = 100$ m.

Side of the new square = 101 m.

*Difference in areas* $= [(101)^2 - (100)^2]\ m^2 = 201\ m^2$.

**35.** Of all the rectangles, the square has the largest area for a given perimeter.

**36.** Length of the rectangle = 40 cm.

Let the side of the square be $x$ cm.

Then, breadth of the rectangle $= \frac{3}{2}x$ cm.

$40 \times \frac{3}{2}x = 3x^2 \Rightarrow x = 20$.

$\therefore$ *Side of the square = 20 cm.*

**37.** $\frac{x^2}{y^2} = \frac{9}{1} \Rightarrow \frac{x}{y} = \sqrt{\frac{9}{1}} = \frac{3}{1} \Rightarrow \frac{4x}{4y} = \frac{12}{4} = \frac{3}{1}$.

$\therefore$ *Ratio of perimeters = 3 : 1.*

**38.** H.C.F. of 1075 and 825 is 25.

$\therefore$ *Size of each tile = 25 cm × 25 cm.*

**39.** $2(7.5 + 3.5) \times h = 77 \Rightarrow h = \frac{77}{2 \times 11} = 3.5$ m.

**40.** $2(x + 8) \times 6 = 168 \Rightarrow x + 8 = 14 \Rightarrow x = 6$ m.

**41.** Cost of papering $[2(l + b) \times h]\ m^2$ = Rs. 48.

$\therefore$ Cost of papering $[2(2l+2b)\times 2h]$ m$^2$ *i.e.* $4[2(l+b)\times h]$
$= Rs. (48\times 4) = Rs. 192.$

**42.** Area of the parallelogram $= 14\times 16 = 224$ cm$^2$.

**43.** $(10)^2 = (6)^2 + x^2 \Rightarrow x = 8.$
$\therefore$ Another diagonal $= 16$ cm.

*So, area of the rhombus* $= \left(\frac{1}{2}\times 16\times 12\right) cm^2 = 96\ cm^2.$

**44.** Area of rhombus $= \frac{1}{2}xy.$

**45.** Let its diagonals be $x$ cm & $2x$ cm. Then,
$\frac{1}{2}\times x\times 2x = 144 \Rightarrow x^2 = 144$ or $x = 12.$
$\therefore$ *Lengths of diagonals are 12 cm, 24 cm.*

**46.** Area $= \left[\frac{1}{2}(1+2)\times 6\right]$ m$^2$ $= 9$ m$^2$.

**47.** Let the length of canal be $x$ m.
*Then,* $\frac{1}{2}(10+6)\times x = 640 \Rightarrow x = \frac{640\times 2}{16} = 80$ m.

**48.** $\frac{1}{2}(3x+5x)\times 12 = 384 \Rightarrow x = \frac{384\times 2}{8\times 12} = 8.$
$\therefore$ *Smaller side = 24 cm.*

**49.** Let length $= x$ & breadth $= y$. Then, area $= xy$.
New length $= 2x$ & new breadth $= 2y$. So, area $= 4xy$.
$\therefore$ *Increase percent* $= \left(\frac{3xy}{xy}\times 100\right)\% = 300\%.$

**50.** Let breadth $= x$ metres and length $= 2x$ metres.
Then, $x^2 + (2x)^2 = (9\sqrt{5})^2 \Rightarrow 5x^2 = 405 \Rightarrow x = \sqrt{81} = 9.$
$\therefore$ *Perimeter* $= 2(18+9)\ m = 54\ m.$

**51.** Let the width of the plot be $x$ metres.
*Then,* $144\times x = 84\times 84 \Rightarrow x = \frac{84\times 84}{144} = 49\ m.$

**52.** Let the length of rectangle $= x$ metres & its breadth $= y$ m.
Also, let the side of the square be $z$ metres.
Then, $2(x+y) = 4z = 48 \Rightarrow x+y = 24$ and $z = 12.$
Also, $z^2 - xy = 4 \Rightarrow xy = z^2 - 4 = 144 - 4 = 140.$
So, $(x-y)^2 = (x+y)^2 - 4xy = 576 - 560 = 16.$
$\therefore x - y = 4$ *and* $x + y = 24$. *So,* $2y = 20$ *or* $y = 10\ m.$

**53.** Let the length be $x$ metres and breadth $y$ metres.

New length = $2x$ metres and new breadth = $\frac{1}{2}y$.

$\therefore$ *New area* $= 2x \times \frac{1}{2} y = xy =$ *old area.*

**54.** Let base = $x$ cm & altitude = $2x$ cm.

$\therefore x \times 2x = 72 \Rightarrow x^2 = 36$ or $x = 6$ cm.

*Hence, base = 6 cm.*

**55.** Side of rhombus = $52 \div 4 = 13$ m.

$(13)^2 = (12)^2 + x^2 \Rightarrow x^2 = 25 \Rightarrow x = 5$ m.

$\therefore$ *Another diagonal = 10 m.*

**56.** Let another diagonal = $2x$ cm.

Then, $x^2 + 4^2 = 5^2 \Rightarrow x = \sqrt{(25-16)} = 3$ cm.

$\therefore$ Area of both the sides $= 2 \times \left(\frac{1}{2} \times 6 \times 8\right) \text{cm}^2 = 48 \text{ cm}^2$.

*So, cost of painting the sheet = Rs. (3.50 × 48) = Rs. 168.*

**57.** Area $= \frac{\sqrt{3}}{4} \times (8)^2 = 16\sqrt{3} \text{ cm}^2$.

**58.** $\frac{\sqrt{3}}{4} \times a^2 = 4\sqrt{3} \Rightarrow a^2 = 16$ or $a = 4$ cm.

**59.** Area $= \frac{\sqrt{3}}{4} \times 2\sqrt{3} = \frac{3}{2} \text{ cm}^2$.

$\frac{1}{2} \times 2\sqrt{3} \times h = \frac{3}{2} \Rightarrow h = \frac{3}{2} \times \frac{2}{2\sqrt{3}} = \frac{\sqrt{3}}{2}$ *cm.*

**60.** Material required $= \left(\frac{1}{2} \times 1.5 \times 2\right) \text{m}^2 = 1.5 \text{ m}^2$.

**61.** $s = \frac{1}{2}(13 + 14 + 15)$ m = 21 m.

$(s-a) = 8$ m, $(s-b) = 7$ m and $(s-c) = 6$ m.

$\therefore$ *Area* $= \sqrt{21 \times 8 \times 7 \times 6} = 84 \, m^2$.

**62.** Let another side = $x$ cm.

Then, $(80)^2 + x^2 = (100)^2 \Rightarrow x^2 = 3600 \Rightarrow x = 60$ cm.

$\therefore$ Area of the scarf $= \left(\frac{1}{2} \times \frac{60}{100} \times \frac{80}{100}\right) \text{m}^2 = \frac{6}{25} \text{ m}^2$.

*So, its cost = Rs.* $\left(\frac{6}{25} \times 5\right)$ *= Rs. 1.20.*

**63.** Their areas are in the ratio $3^2 : 4^2$ *i.e.* 9 : 16.

**64.** $\frac{\sqrt{3}}{4} a^2 = 36\sqrt{3} \Rightarrow a^2 = 144 \Rightarrow a = 12$.

$\therefore$ *Its perimeter* $= (12 \times 3)$ *cm* $= 36$ *cm.*

**65.** Let its sides be $x$ cm & $y$ cm.

Then, $\frac{1}{2} xy = 432 \Rightarrow xy = 864.$

And, $x^2 + y^2 = (30\sqrt{2})^2 = 1800.$

$\therefore (x+y)^2 = (x^2 + y^2) + 2xy = 1800 + 2 \times 864 = 3528$

or $x + y = \sqrt{3528} = \sqrt{4 \times 441 \times 2} = 42\sqrt{2}$

$\therefore$ *Perimeter* $= 42\sqrt{2} + 30\sqrt{2} = 72\sqrt{2}$ *cm.*

**66.** $\frac{1}{2} \times x \times 2x = 36 \Rightarrow x = 6.$

$\therefore$ *Hypotenuse* $= \sqrt{6^2 + (12)^2} = \sqrt{180} = 6\sqrt{5}.$

**67.** Let the original sides be $a, b, c$ then $s = \frac{1}{2}(a + b + c).$

Area of this triangle $= \sqrt{s(s-a)(s-b)(s-c)}$

For new triangle, the sides are $2a, 2b, 2c$ & $S = 2s$.

$\therefore$ Area of new triangle $= \sqrt{S(S-2a)(S-2b)(S-2c)}$

$= \sqrt{2s(2s-2a)(2s-2b)(2s-2c)} = \sqrt{16\, s(s-a)(s-b)(s-c)}$

$= 4\sqrt{s(s-a)(s-b)(s-c)} = 4 \times$ *(area of original triangle).*

**68.** Let original length of each side $= a$.

Then, area $= \frac{\sqrt{3}}{4} a^2 = A$

New area $= \frac{\sqrt{3}}{4}\left[\left(\frac{101.5}{100} a\right)^2\right] = \frac{\sqrt{3}}{4}\left(\frac{20.3}{20}\right) a^2 = \left(\frac{20.3}{20}\right) A.$

*Increase in area* $= \left(\frac{0.3}{20} A \times \frac{1}{A} \times 100\right)\% = 1.5\%.$

**69.** Let the side of each triangle be $x$ and the side of each square be $y$ and let the perimeter of each be $P$.

Then, $3x = P \Rightarrow x = \frac{P}{3}$ and $4y = P \Rightarrow y = \frac{P}{4}.$

$$\textit{Ratio of their areas} = \frac{\frac{\sqrt{3}}{4} \times \left(\frac{P}{3}\right)^2}{\left(\frac{P}{4}\right)^2} = \frac{4}{3\sqrt{3}} < 1.$$

**70.** $\frac{1}{2} \times x \times h = x^2 \Rightarrow h = 2x.$

**71.** Let lateral side $= 5x$ & base $= 4x$.

Then, $5x + 5x + 4x = 14 \Rightarrow x = 1.$

$\therefore$ The sides are 5 cm, 5 cm, 4 cm.

Now, $h^2 = 5^2 - 2^2 \Rightarrow h = \sqrt{21}$.

$\therefore$ *Area* $= \left(\frac{1}{2} \times 4 \times \sqrt{21}\right) cm^2 = 2\sqrt{21}\ cm^2$.

**72.** $\frac{\text{Area of square}}{\text{Area of rhombus}} = \frac{\text{base} \times \text{base}}{\text{base} \times \text{height}} = \frac{a \times a}{a \times h} = \frac{a}{h} > 1$, *since* $a > h$.

**73.** $2\pi r = 352 \Rightarrow r = \frac{176}{\pi} = \frac{176}{22} \times 7 = 56$ m.

$\therefore$ *Area* $= \pi r^2 = \left(\frac{22}{7} \times 56 \times 56\right) m^2 = 9856\ m^2$.

**74.** $\pi r^2 = 38.5 \Rightarrow r^2 = \left(\frac{38.5}{22} \times 7\right) = \frac{49}{4} \Rightarrow r = \frac{7}{2}$.

$\therefore$ *Circumference* $= 2\pi r = \left(2 \times \frac{22}{7} \times \frac{7}{2}\right) = 22$ *cm*.

**75.** $2\pi r - r = 37 \Rightarrow r(2\pi - 1) = 37 \Rightarrow r\left(\frac{44}{7} - 1\right) = 37 \Rightarrow r = 7$.

$\therefore$ *Area* $= \pi r^2 = \left(\frac{22}{7} \times 7 \times 7\right) cm^2 = 154\ cm^2$.

**76.** Let diameter $= d$. Then, radius $= \frac{d}{2}$.

$\therefore$ Area $= \pi\left(\frac{d}{2}\right)^2 = \frac{\pi d^2}{4}$.

New diameter $= 2d \Rightarrow$ New radius $= d$.

$\therefore$ New area $= \pi d^2 = 4$ (old area).

*So, the area is quadrupled.*

**77.** Original area $= [\pi \times (9)^2]$ cm$^2$; New area $= [\pi \times (7)^2]$ cm$^2$.

$\therefore$ *Decrease in area* $= \left[\frac{\pi \times (9)^2 - \pi(7)^2}{\pi \times (9)^2} \times 100\right]\%$

$= \left(\frac{\pi \times (9^2 - 7^2)}{81\pi} \times 100\right)\% = \left(\frac{32 \times 100}{81}\right)\% = 39.5\%$.

**78.** Let radius $= x$. New radius $= \left(\frac{50}{100}x\right) = \frac{1}{2}x$.

*Area reduced* $= \left[\frac{\pi x^2 - \pi\left(\frac{x^2}{4}\right)}{\pi x^2} \times 100\right]\% = \left(\frac{3}{4} \times 100\right)\% = 75\%$.

**79.** $\pi r^2 = 13.86 \times 10000 \Rightarrow r^2 = \left(\frac{13.86 \times 10000 \times 7}{22}\right) \Rightarrow r = 210$ m.

Length of fence $= 2\pi r = \left(2 \times \frac{22}{7} \times 210\right)$ m $= 1320$ m.

$\therefore$ *Cost of fencing = Rs.* $\left(1320 \times \frac{20}{100}\right)$ *= Rs. 264.*

**80.** $C = 2\pi r$ & $\frac{105}{100} C = 2\pi R$.

$\therefore r = \frac{C}{2\pi}$ and $R = \frac{21C}{40\pi}$.

Original area $= \pi \times \left(\frac{C}{2\pi}\right)^2 = \frac{\pi C^2}{4\pi^2} = \frac{C^2}{4\pi}$.

New area $= \pi \times \left(\frac{21C}{40\pi}\right)^2 = \frac{441\pi C^2}{1600\pi^2} = \frac{441C^2}{1600\pi}$.

Increase in area $= \frac{441C^2}{1600\pi} - \frac{C^2}{4\pi} = \frac{41C^2}{1600\pi}$

$\therefore$ *Increase percent* $= \left(\frac{41C^2}{1600\pi} \times \frac{4\pi}{C^2} \times 100\right) = 10.25\%$.

**81.** Distance covered in 1 revolution $= 2\pi \times (1)^2 = \frac{44}{7}$ cm.

*Distance covered in 10 revolutions* $= \left(\frac{44}{7} \times 10\right)$ *cm = 62.8 cm.*

**82.** $\pi r^2 = 0.49\pi \Rightarrow r = 0.7$ m.

$\therefore$ Circumference $= \left(2 \times \frac{22}{7} \times 0.7\right)$ m $= 4.4$ m.

Number of revolutions $= \left(\frac{1.76 \times 1000}{4.4}\right) = 400$.

**83.** Circumference $= \left(2 \times \frac{22}{7} \times 14\right)$ cm $= 88$ cm.

*Number of revolutions* $= \frac{0.66 \times 1000 \times 100}{88} = 750$.

**84.** Distance travelled in 1 revolution $= \frac{88 \times 1000}{100} = 880$ m.

$\therefore 2\pi r = 880 \Rightarrow 2r = \dfrac{880}{\pi} = \dfrac{880 \times 7}{22} = 280$ *m.*

**85.** Distance covered in 4 seconds $= \left(\dfrac{30}{7} \times 7\right)$ m = 30 m.

*Distance covered in* $60 \times 60$ *seconds* $= \left(\dfrac{30}{4} \times 60 \times 60\right)$ m = 27000 *m*

= 27 *km/hr.*

**86.** Area of sector $= \left(\dfrac{1}{2} \times r^2 \times \dfrac{\text{arc}}{r}\right) = \left(\dfrac{1}{2} \times 5 \times 3.5\right) \text{cm}^2 = 8.75 \text{ cm}^2.$

**87.** Angle formed in 30 min. $= \left(\dfrac{360}{60} \times 30\right) = 180°.$

*Area swept* $= \dfrac{\pi r^2 \theta}{360} = \left(\dfrac{22}{7} \times 7 \times 7 \times \dfrac{180}{360}\right) = 77 \text{ cm}^2.$

**88.** Area of sector $= \dfrac{\pi r^2 \theta}{360} = \left(\dfrac{22}{7} \times 144 \times \dfrac{42}{360}\right) \text{m}^2 = 52.8 \text{ m}^2.$

**89.** $\pi r^2 = 9856 \Rightarrow r^2 = \left(9856 \times \dfrac{7}{22}\right) = (448 \times 7) \Rightarrow r = 56$ m.

**90.** $2(6x + 5x) = \left(2 \times \dfrac{22}{7} \times 42\right) \Rightarrow x = 12.$

$\therefore$ *Smaller side* $= 5x = 60$ cm..

**91.** $2\pi r = 4x$, where $x^2 = 12100$. So, $x = 110$.

$\therefore 2\pi r = 440 \Rightarrow r = \dfrac{440 \times 7}{2 \times 22} = 70$ m.

*So, area of the circular field* $= \left(\dfrac{22}{7} \times 70 \times 70\right) m^2 = 15400 \ m^2.$

**92.** Perimeter $= (\pi r + 2r)$ cm $= \left(\dfrac{22}{7} \times 28 + 56\right)$ cm = 144 cm.

**93.** Required area $= \left[(14)^2 - 4 \times \dfrac{\pi \times (7)^2 \times 90}{360}\right] \text{cm}^2 = 42 \text{ cm}^2.$

**94.** Clearly, the radius of such a circle is 7 m.

Area of such a circle $= \left(\dfrac{22}{7} \times 7 \times 7\right) \text{m}^2 = 154 \text{ m}^2.$

**95.** Diagonal of such a square = 16 cm.

$\therefore$ Required area $= \left[\dfrac{22}{7} \times (8)^2 - \dfrac{1}{2} \times 16 \times 16\right] \text{cm}^2 = 73.14 \text{ cm}^2.$

**96.** Radius of the circular sheet = 4 cm.

$\therefore$ Area of remaining sheet $= \left(10 \times 8 - \frac{22}{7} \times 4 \times 4\right) cm^2 = 29.71\ cm^2$.

**97.** $\frac{2 \pi r \times 60}{360} = 20 \Rightarrow r = \frac{60}{\pi}$ cm.

**98.** Let each of the equal sides be $a$ metres long.
Then, $a^2 + a^2 = (50\sqrt{2})^2 = 5000 \Rightarrow a^2 = 2500 \Rightarrow a = 50$.
$\therefore$ Perimeter of the triangle
$= (50 + 50 + 50\sqrt{2}) = 100 + 50 \times 1.4146 = 170.73$ m.
$\therefore$ *Cost of fencing* = *Rs.* (170.73 × 3) = *Rs. 512.19.*

**99.** $\pi r^2 = \frac{1}{3} \times [\pi \times (24)^2 - \pi r^2]$

$4 \pi r^2 = \pi \times 24 \times 24 \Rightarrow r^2 = \frac{24 \times 24}{4} = 144 \Rightarrow r = 12$.

$\therefore$ *Ratio of circumferences* $= \frac{2 \times \pi \times 24}{2 \times \pi \times 12} = \frac{2}{1}$.

**100.** The incentre of an equilateral triangle is its centroid also. The altitude is the same as the length of median. But, the centroid divides the median in the ratio 2 : 1.
So, the lower length from the centre is 2 & upper length is 4.
$\therefore$ *Altitude* = *(4 + 2) cm = 6 cm.*

## ANSWERS (Exercise 20B)

| | | | | | |
|---|---|---|---|---|---|
| **1.** (*a*) | **2.** (*a*) | **3.** (*a*) | **4.** (*b*) | **5.** (*c*) | **6.** (*c*) |
| **7.** (*c*) | **8.** (*a*) | **9.** (*d*) | **10.** (*d*) | **11.** (*b*) | **12.** (*d*) |
| **13.** (*a*) | **14.** (*b*) | **15.** (*c*) | **16.** (*b*) | **17.** (*b*) | **18.** (*d*) |
| **19.** (*c*) | **20.** (*c*) | **21.** (*b*) | **22.** (*a*) | **23.** (*b*) | **24.** (*d*) |
| **25.** (*d*) | **26.** (*d*) | **27.** (*b*) | **28.** (*d*) | **29.** (*a*) | **30.** (*c*) |
| **31.** (*c*) | **32.** (*b*) | **33.** (*d*) | **34.** (*b*) | **35.** (*b*) | **36.** (*b*) |
| **37.** (*c*) | **38.** (*a*) | **39.** (*b*) | **40.** (*d*) | **41.** (*b*) | **42.** (*b*) |
| **43.** (*d*) | **44.** (*d*) | **45.** (*b*) | **46.** (*a*) | **47.** (*b*) | **48.** (*b*) |
| **49.** (*c*) | **50.** (*b*) | **51.** (*b*) | **52.** (*a*) | **53.** (*c*) | **54.** (*b*) |
| **55.** (*b*) | **56.** (*c*) | **57.** (*b*) | **58.** (*a*) | **59.** (*a*) | **60.** (*b*) |
| **61.** (*c*) | **62.** (*a*) | **63.** (*d*) | **64.** (*d*) | **65.** (*a*) | **66.** (*d*) |
| **67.** (*d*) | **68.** (*a*) | **69.** (*c*) | **70.** (*c*) | **71.** (*d*) | **72.** (*a*) |
| **73.** (*a*) | **74.** (*c*) | **75.** (*c*) | **76.** (*c*) | **77.** (*b*) | **78.** (*b*) |
| **79.** (*b*) | **80.** (*b*) | **81.** (*b*) | **82.** (*b*) | **83.** (*b*) | **84.** (*c*) |
| **85.** (*c*) | **86.** (*c*) | **87.** (*d*) | **88.** (*c*) | **89.** (*a*) | **90.** (*b*) |
| **91.** (*b*) | **92.** (*a*) | **93.** (b) | **94.** (d) | **95.** (b) | **96.** (a) |
| **97.** (*d*) | **98.** (*c*) | **99.** (c) | **100.** (b) | | |

# 21

# VOLUME AND AREA OF SOLID FIGURES

*Bodies occupying space are called* **Solids.** *The space occupied by a solid body is called its* **volume.** *The solid bodies occur in various shapes. In this chapter, we will discuss about finding the areas and volumes of some special type of solid bodies.*

**1. A cuboid and a Cube :** The solids like wooden boxes, tea containers, tiles etc. which have six faces, each of which is a rectangle, are called **cuboides.**

A cuboid in which every face is a square is called a **cube.** Length of each face of a cube is called its edge.

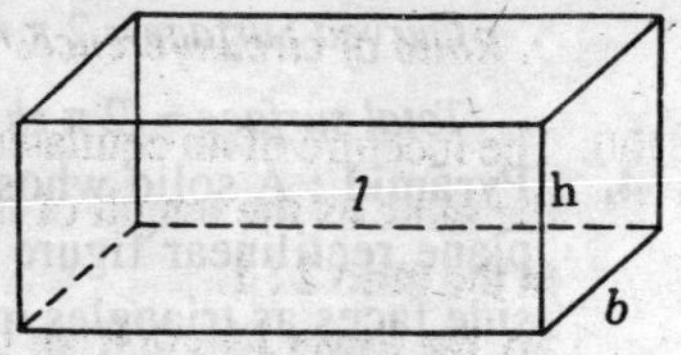


**CUBOID**
*Fig. 1*

**Formulae :**

(1) **CUBOID**

*Volume of a Cuboid*

$= (Length \times Breadth \times Height)$

$= (l \times b \times h)$ *cubic units.*

*Diagonal of a Cuboid* $= \sqrt{(l^2 + b^2 + h^2)}$ .

*Whole surface of a Cuboid* $= 2\ (lb + bh + lh)$ *sq. units.*

(2) **CUBE**

Volume of a Cube = $(\text{Edge})^3 = l^3$ cubic units

Diagonal of a Cube = $\sqrt{3} \times (\text{Edge})$

Whole surface of a Cube = $6 \times (\text{Edge})^2$ sq. units.

**2. Prism :** A right prism is a solid in which the two ends are congruent parallel figures and the side faces are rectangles. Fig. 2 shows a right prism with triangular ends. The total area of side faces of a prism is called the lateral surface of the prism.

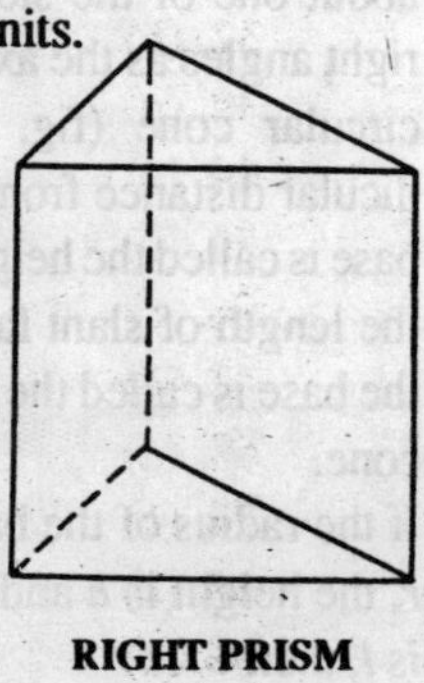

**RIGHT PRISM**
*Fig. 2*

**Formulae :**

*Volume of Prism*

$= (Area\ of\ the\ base) \times (Height)$

*Lateral Surface* = *(Perimeter of the base)* × *(Height).*

3. **Cylinder :** The solid generated by the revolution of a rectangle about one of its sides as axis is called a cylinder. For example, 4 kg. Dalda ghee container is in the form of a cylinder. Fig. 3 shows the shape of a cylinder.

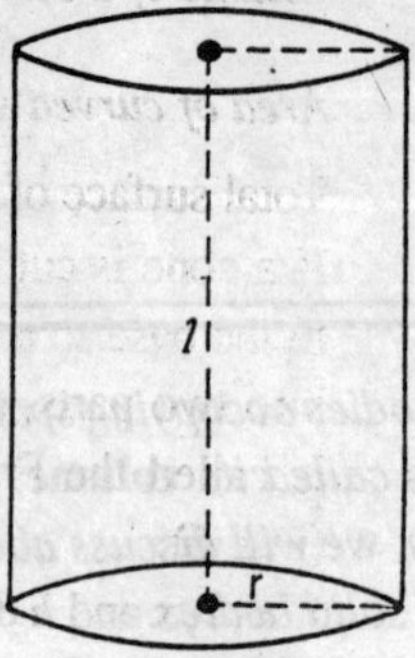

**CYLINDER**
*Fig. 3*

**Formulae :**

*The volume of a cylinder, the radius of whose base is r and whose length (or height) is h, is given by*

$V = \pi r^2 h$

= (Area of the base) × (Height)

*Curved Surface* $= 2\pi rh$.

*Total surface* $= (2\pi rh + 2\pi r^2)$.

4. **Pyramid :** A solid whose base is a plane rectilinear figure having the side faces as triangles meeting at a common vertex is called a pyramid (fig. 4). When the base of a pyramid is a triangle, the pyramid is called a tetrahedron.

Volume of a Pyramid

$= \frac{1}{2} \times$ (Area of base) × Height

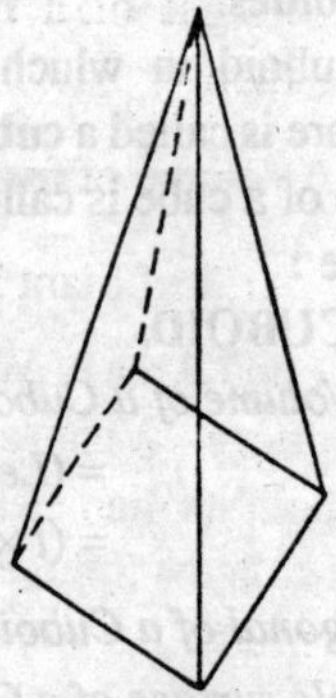
**PYRAMID**
*Fig. 4*

5. **Cone :** The solid generated by the revolution of a right angled triangle about one of the sides containing the right angles as the axis is called a right circular cone (fig. 5). The perpendicular distance from the vertex to the base is called the height of the cone and the length of slant face from vertex to the base is called the slant height of the cone.

If the radius of the base of a cone is $r$, the height is $h$ and the slant height is $l$, then

$l^2 = h^2 + r^2$ or $l = \sqrt{(h^2 + r^2)}$.

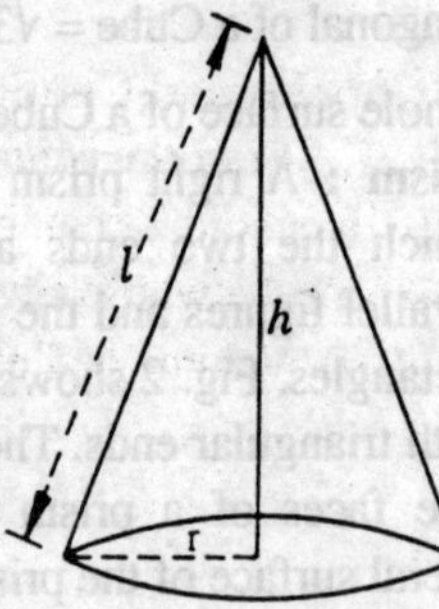

**CONE**
*Fig. 5*

**Formulae :**

*Volume of a cone* $= V = \frac{1}{3}\pi r^2 h$

*Area of curved surface of a cone* $= \pi rl = \pi r\sqrt{(r^2 + h^2)}$.

Total surface of a cone $= \pi rl + \pi r^2$

If a cone is cut by a plane parallel to the base so as to divide the cone into two parts, then the lower part is called the Frustum of the cone (Fig. 6).

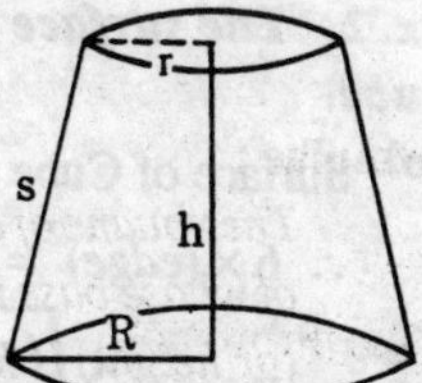


**FRUSTUM OF A CONE**
*Fig. 6*

If $R$, $r$, $s$ and $h$ denote the radius of bigger end, radius of smaller end, slant height and the perpendicular height of a frustum respectively, then

Volume of frustum $= \frac{\pi h}{3}[R^2 + r^2 + Rr]$.

Area of Slant Surface of frustum $= \pi(R + r)s$.

and Area of Whole Surface $= \pi(R^2 + r^2 + Rs + rs)$.

6. **Sphere.** When a semi circle moves about its diameter, the solid generated is called a Sphere (Fig. 7).

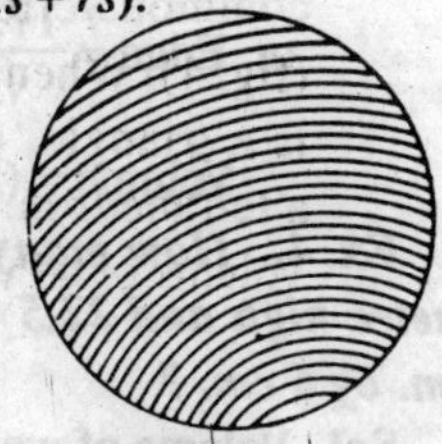

**SPHERE**
*Fig. 7*

**Formulae :**

Volume of a Sphere $= \frac{4}{3}\pi r^3$

Curved surface of a Sphere $= 4\pi r^2$.

## Solved Problems.

1. **Cuboids and Cubes.**

**Ex. 1.** ***Find the volume, surface area and diagonal of :***

(i) a cuboid 22 cm. by 12 cm. by 7.5 cm.

(ii) a cube whose edge is 18 cm. long.

**Sol. (i)** Volume of Cuboid

$= (22 \times 12 \times 7.5)$ cu.cm $= 1980$ cu.cm.

Surface of Cuboid $= 2\,(lb + bh + lh)$

$= 2 \times (22 \times 12 + 12 \times 7.5 + 22 \times 7.5)$ sq. cm.

$= 1038$ sq. cm.

Diagonal of Cuboid $= \sqrt{(l^2 + b^2 + h^2)}$

$= \sqrt{[(22)^2 + (12)^2 + (7.5)^2]}$ cm.

$= \sqrt{(684.25)}$ cm. $= 26.15$ cm.

(ii) Volume of Cube = $(\text{Edge})^3 = (18 \times 18 \times 18)$ cu. cm.

$= 5832$ cu. cm.

Surface of Cube $= 6 \times (\text{Edge})^2 = (6 \times 18 \times 18)$ sq. cm.

$= 1944$ sq. cm.

Diagonal of Cube $= \sqrt{3} \times (Edge) = (1.732 \times 18) cm. = 31.176$ cm.

Ex. 2. ***The surface of a cube is 552.96 sq. cm. Find the volume of the cube.***

Sol. Surface of Cube $= 6 \times (\text{Edge})^2$.

$\therefore\ 6 \times (edge)^2 = 552.26$

or Edge $= \sqrt{\left(\frac{552.96}{6}\right)} = \sqrt{(92.16)}$ cm. $= 9.6$ cm.

$\therefore$ *Volume of Cube* $= (\text{Edge})^3 = (9.6)^3$ *cu. cm.*

$= 884.736$ *cu. cm.*

Ex. 3. ***A cube has a diagonal 17.32 cm. long. Find the volume of the cube.***

Sol. Diagonal of the cube $= \sqrt{3} \times$ (edge).

$\therefore\ \sqrt{3} \times (\text{Edge}) = 17.32$

or Edge $= \frac{17.32}{\sqrt{3}}$ cm. $= \frac{17.32}{1.732} = 10$ cm.

$\therefore$ Volume $= (\text{Edge})^3 = (10)^3$ cu. cm. $= 1000$ cu.cm.

Ex. 4. ***How many bricks will be required for a wall 8 metre long, 6 metre high and 22.5 cm. thick, if each brick measures 25 cm. by 11.25 cm. by 6 cm. ?***

Sol. Volume of wall $= (800 \times 600 \times 22.5)$ cu. cm.

Volume of a brick $= (25 \times 11.25 \times 6)$ cu. cm.

$\therefore$ Number of bricks $= \frac{\text{Volume of the wall}}{\text{Volume of a brick}}$

$= \left(\frac{800 \times 600 \times 22.5}{25 \times 11.25 \times 6}\right) = 6400.$

Ex. 5. ***An open rectangular cistern when measured from out side is 1 m. 35 cm. long; 1 m. 8 cm. broad and 90 cm. deep, and is made of iron 2.5 cm. thick. Find (i) the capacity of the cistern, (ii) the volume of the iron used and (iii) the total surface area of the cistern.***

Sol. **External Dimensions :**

Length = 135 cm., Breadth = 108 cm., Depth = 90 cm.

**Internal Dimensions :**

Length = 130 cm., Breadth = 103 cm., Depth = 87.5 cm.

$\therefore$ (i) Capacity = Internal volume $= (130 \times 103 \times 87.5)$ cu. cm.

$= 1171625$ cu. cm.

(ii) Volume of iron = (outer volume) – (Internal volume)

$= [(135 \times 108 \times 90) - (1171625)]$ cu. cm.

$= (1312200 - 1171625)$ cu. cm.

$= 140575$ cu. cm.

(iii) External area = Area of 4 sides + Area of the base

$= [\{2(135 + 108) \times 90\} + \{135 \times 108\}]$ sq. cm.

$= (43740 + 14580)$ sq. cm. $= 58320$ sq. cm.

Internal area $= [\{2(130 + 103) \times 87.5\} + \{130 \times 103\}]$ sq. cm.

$= [(40775 + 13390)]$ sq. cm. $= 54165$ sq. cm.

Area at the top = area between the outer and inner rectangles

$= [(135 \times 108) - (130 \times 103)]$ sq. cm.

$= (14580 - 13390)$ sq. cm. $= 1190$ sq. cm.

$\therefore$ The total surface area of the cistern

$= (58320 + 54165 + 1190)$ sq. cm.

$= 113675$ sq. cm.

**Ex. 6.** ***Half cubic metre of gold sheet is extended by hammering so as to cover an area of 1 hectare. Find the thickness of the gold.***

**Sol.** Volume of Sheet $= \frac{1}{2}$ cu. metre $= \left(\frac{1}{2} \times 100 \times 100 \times 100\right)$ cu. cm.

Area of Sheet = 1 hectare = 10000 sq. metres

$= (10000 \times 100 \times 100)$ sq. cm.

Thickness $= \frac{\text{Volume}}{\text{Area}} = \frac{1 \times 100 \times 100 \times 100}{2 \times 10000 \times 100 \times 100}$ cm.

$= \frac{1}{200}$ *cm.* $= 0.005$ *cm.*

**Ex. 7.** ***In a shower 5 cm. of rain falls. Find in cubic metres the volume of water that falls on 2 hectares of ground.***

**Sol.** Depth of rain = 5 cm. $= \frac{5}{100}$ *i.e.* $\frac{1}{20}$ metres.

Area of ground = 2 hectares = 20000 sq. metres.

Volume of water $= 20000 \times \frac{1}{20}$ cu. metres = 1000 cu. metres.

**Ex. 8.** ***The water in a rectangular reservoir having a base 80 metres by 60 metres is 6.5 metres deep. In what time can the water be emptied by a pipe of which the cross section is a square of side 20 cm., if the water runs through the pipe at the rate of 15 km. per hour ?***

**Sol.** Volume of water $= (80 \times 60 \times 6.5)$ cu. m. $= 31200$ cu. m.

Area of cross section of the pipe

$$= \left(\frac{20}{100} \times \frac{20}{100}\right) \text{sq. m} = \frac{1}{25} \text{ sq. m.}$$

Volume of water emptied in 1 hour

$$= \left(\frac{15 \times 1000 \times 1}{25}\right) = 600 \text{ cu. m.}$$

*Time taken to empty the reservior* $= \frac{31200}{600}$ *hrs.= 52 hrs.*

**Ex. 9.** ***The area of a playground is 5600 sq. metres. Find the cost of covering it with gravel 1 cm. deep, if the gravel costs Rs. 2.80 per cubic metre.***

**Sol.** Volume of gravel $= \left(5600 \times \frac{1}{100}\right)$ cu. metres = 56 cu. m.

*Cost of gravelling = Rs. (56 × 2.80) = Rs. 156.80.*

**Ex. 10.** ***A rectangular plot of area 43560 $m^2$ has length and breadth in the ratio 5 : 2. A gravel path 5 m. wide runs outside the plot close to its four sides. If it costs Rs. 590 to gravel the path at 25 paise per cu. m., find to what depth is the path gravelled.***

**Sol.** Volume of path $= \frac{\text{Total Cost}}{\text{Rate}} = \frac{590 \times 100}{25} = 2360$ cu. m.

Let the length and breadth of the plot be $5x$ and $2x$ metres respectively so that area is $10x^2$ sq. m.

But, area $= 43560\ m^2$.

$\therefore\ 10x^2 = 43560$ or $x = \sqrt{\left(\frac{43560}{10}\right)} = 66.$

So, length = 66 × 5 or 330 m. and Breadth = 66 × 2 or 132 m.

Area of plot = 330 × 132 or 43560 sq. m.

Area of plot including path = (330 + 10) × (132 + 10) sq. m.

= 340 × 142 or 48280 sq. m.

Area of path = (48280 – 43560) sq. m. = 4720 sq. m.

*Depth of path* $= \frac{Volume}{Area} = \frac{2360}{4720} = \frac{1}{2}$ *m.*

**2. Problems on Volume and Areas of Pyramids and Prisms.**

**Ex. 11.** ***The base of a prism is a triangle of which the sides are 17 cm., 25 cm. and 28 cm. respectively. The volume of the prism is 4200 cu. cm. What is the height ? Find the lateral area also.***

**Sol.** Volume of prism = (Area of base) × (Height) .

Let $a$ = 17 cm., $b$ = 25 cm., $c$ = 28 cm.

Then, $S = \left(\frac{17 + 25 + 28}{2}\right) = 35$ cm.

$\therefore\ (S - a) = 18$ cm., $(S - b) = 10$ cm. and $(S - c) = 7$ cm.

So, Area of base = $\sqrt{(35 \times 18 \times 10 \times 7)}$ sq. cm. = 210 sq. cm.

So, Height of prism = $\frac{\text{Volume}}{\text{(Area of base)}} = \frac{4200}{210}$ cm. = 20 cm.

Lateral Surface = (Perimeter of base) × (Height)

= [(17 + 25 + 28) × 20] *sq. cm.* = *1400 sq. cm.*

**Ex. 12.** ***A right pyramid stands on a rectangular base, whose sides are 24 cm. and 16 cm. If each of the slant edges is 17 cm., find the height and the volume of the pyramid.***

**Sol.** Height of pyramid = $\sqrt{[(17)^2 - (8)^2]}$ = 15 cm.

Volume = $\frac{1}{8}$ × Area of base × Height

= $\left(\frac{1}{8} \times 24 \times 16 \times 15\right)$ cu. cm.

= *1920 cu. cm.*

**3. Problems on Volume and Area of Cylinders.**

**Ex. 13.** ***Find the volume, the curved surface and the whole surface of a cylinder of length 80 cm. and the diameter of whose base is 7 cm.***

**Sol.** Length of cylinder = 80 cm. and Radius of base = 3.5 cm.

∴ Volume = $\pi r^2 h = \left(\frac{22}{7} \times 3.5 \times 3.5 \times 80\right)$ cu. cm. = 3080 cu. cm.

Curved Surface = $2\pi rh = \left(2 \times \frac{22}{7} \times 3.5 \times 80\right)$ sq. cm.

= 1760 sq. cm.

Total Surface = $(2\pi rh + 2\pi r^2)$

= (1760) + $\left(2 \times \frac{22}{7} \times 3.5 \times 3.5\right)$ sq. cm.

= *(1760 + 77) sq. cm.* = *1837 sq. cm.*

**Ex. 14.** ***The thickness of a hollow cylinder is 3.5 cm. and its outside diameter is 36.4 cm. Find the cost of painting its surface at 5 paise per square centimetre, if the cylinder is 70 cm. long.***

**Sol.** Outside radius = 18.2 cm.

Inside radius = (18.9 – 3.5) cm. = 14.7 cm.

Area of outside curved surface = $\left(2 \times \frac{22}{7} \times 18.2 \times 70\right)$

= 8008 sq. cm.

Area of inside curved surface = $\left(2 \times \frac{22}{7} \times 14.7 \times 70\right)$ = 6468 sq. cm.

Areas of two end circular rings

$$= 2 \times \frac{22}{7} \times [(18.2)^2 - (14.7)^2] = 723.8 \text{ sq. cm.}$$

Total area to be painted = (8008 + 6468 + 723.8) sq. cm.

= 15199.8 sq. cm.

$$\textit{Cost of painting} = Rs. \left(\frac{15199.80 \times 5}{100}\right) = Rs.\ 759.99.$$

**Ex. 15.** ***Find the weight of a lead pipe 3.5 metres long, if the external diameter of the pipe is 2.4 cm. and the thickness of the lead is 2 m m. and 1 c.c. of lead weights 11.4 gms.***

**Sol.** External radius of the pipe = 1.2 cm.

Internal radius of the pipe = (1.2 – 0.2) cm. = 1 cm.

$$\text{External volume} = \left(\frac{22}{7} \times 1.2 \times 1.2 \times 3.5 \times 100\right) = 1584 \text{ cu. cm.}$$

$$\text{Internal volume} = \left(\frac{22}{7} \times 1 \times 1 \times 3.5 \times 100\right) \text{cu. cm.} = 1100 \text{ cu. cm.}$$

Volume of lead = (External volume) – (Internal volume)

= (1584 – 1100) cu. cm. = 484 cu. cm.

$$\textit{Weight of the pipe} = \left(\frac{484 \times 11.4}{1000}\right) kg. = 5.5176\ kg.$$

**Ex. 16.** ***A solid iron rectangular block of dimensions 4.4 m., 2.6 m. and 1 m. is cast into a hollow cylinderical pipe of internal radius 30 cm. and thickness 5 cm. Determine the length of the pipe.***

**Sol.** Volume of iron block = (440 × 260 × 100) cu. cm.

Internal radius of pipe = 30 cm.

External radius of the pipe = (30 + 5) or 35 cm.

Let the length of the pipe be *h* cm.

$$\text{Then, external volume} = \frac{22}{7} \times 35 \times 35 \times h = (3850\ h) \text{ cu. cm.}$$

$$\text{and internal volume} = \left(\frac{22}{7} \times 30 \times 30 \times h\right) = \left(\frac{19800}{7} h\right) \text{cu. cm.}$$

$$\text{Volume of pipe} = \left[\left(3850 - \frac{19800}{7}\right) h\right] \text{cu. cm.} = \left(\frac{7150}{7} h\right) \text{cu. cm.}$$

$$\therefore \frac{7150}{7} h = 440 \times 260 \times 100$$

$$\text{or } h = \frac{440 \times 260 \times 100 \times 7}{7150 \times 100} = 112 \text{ m}$$

**Ex. 17.** ***A well with 11.2 metres inside diameter is dug 8 metres deep. Earth taken out of it has been evenly spread all round it to a width of 7 metres to form an embankment. Find the height of the embankment.***

**Sol.** Here the embankment is a circular path 7 metres wide round the well and the radius of the well is 5.6 metres.

Radius of well with embankment = (5.6 + 7) or 12.6 metres.

$$\text{Volume of earth taken out} = \left(\frac{22}{7} \times 5.6 \times 5.6 \times 8\right) = 788.48 \text{ cu.m}$$

Area of embankment

$$= \left[\left(\frac{22}{7} \times 12.6 \times 12.6\right) - \left(\frac{22}{7} \times 5.6 \times 5.6\right)\right] \text{sq. m.} = 400.4 \text{ sq. m.}$$

$$\textit{Height of embankment} = \frac{\textit{Volume}}{\textit{Area}} = \frac{788.48}{400.4} = 1.97 \textit{ metres.}$$

**4. Problems on Volume and Area of Cones.**

**Ex. 18.** ***The height of a cone is 16 cm. and the diameter of its base is 24 cm. Find the slant height, volume, area of curved surface and the total surface of the cone.***

**Sol.** Height of the cone = 16 cm., Radius of the base = 12 cm.

$$\therefore \text{Slant height} = \sqrt{[(\text{Height})^2 + (\text{Radius})^2]}$$

$$= \sqrt{[(16)^2 + (12)^2]} = 20 \text{ cm.}$$

$$\text{Volume} = \frac{1}{3}\pi r^2 h = \left(\frac{1}{3} \times \frac{22}{7} \times 12 \times 12 \times 16\right) \text{cu. cm.}$$

$$= \frac{16896}{7} \text{ cu. cm.}$$

$$= 2413.71 \text{ cu. cm.}$$

$$\text{Area of curved surface} = \pi rl = \left(\frac{22}{7} \times 12 \times 20\right) \text{sq. cm.}$$

$$= \frac{5280}{7} \text{ sq. cm.} = 754.28 \text{ sq. cm.}$$

$$\text{Total Surface} = (\pi rl + \pi r^2)$$

$$= \left[(754.28) + \left(\frac{22}{7} \times 12 \times 12\right)\right] \text{sq. cm.}$$

$$= (754.28 + 452.57) \textit{ sq. cm.} = 1206.85 \textit{ sq. cm.}$$

**Ex. 19.** ***From a solid right circular cylinder with height 10 cm. and radius of the base 6 cm., a right circular cone of the same height and base is removed. Find the volume of the remaining solid.***

**Sol.** Volume of cylinder $= \left(\frac{22}{7} \times 6 \times 6 \times 10\right)$ cu. cm.

$= 1131.42$ cu. cm.

Volume of cone $= \left(\frac{1}{3} \times \frac{22}{7} \times 6 \times 6 \times 10\right)$ cu. cm. $= 377.14$ cu. cm.

Volume of remaining solid $= (1131.42 - 377.14)$ cu. cm.

$= 755.28$ cu. cm.

**Ex. 20.** ***A conical vessel whose internal radius is 10 cm. and height 48 cm. is full of water. Find the volume of water. If this water is poured into a cylindrical vessel with internal radius 20 cm., find the height to which the water rises in it.***

**Sol.** Volume of water $= \left(\frac{1}{3} \times \frac{22}{7} \times 10 \times 10 \times 48\right) = \left(\frac{35200}{7}\right)$ cu. cm.

Let the height to which water rises be $h$ cm.

Internal radius of cylindrical vessel = 20 cm.

Volume of water in this vessel $= \frac{22}{7} \times 20 \times 20 \times h$ cu. cm.

$= \left(\frac{8800}{7} h\right)$ cu. cm.

$\therefore \frac{8800}{7} h = \frac{35200}{7}$ or $h = \frac{35200 \times 7}{7 \times 8800} = 4$ cm.

*Height of water level = 4 cm.*

**Ex. 21.** ***Find what length of canvas 1.5 metres in width is required to make a conical tent 9 metres in diameter and 6.3 metres in slant height. Also find the cost of the canvas at the rate of Rs. 2.15 per metre.***

**Sol.** Area of curved surface $= \pi rl = \left(\frac{22}{7} \times 4.5 \times 6.3\right)$ sq. metres

$= 89.1$ sq. metres.

$\therefore$ Area of canvas required = 89.1 sq. metres.

Width of canvas = 1.5 metres.

Length of canvas $= \frac{89.1}{1.5}$ metres $= 59.4$ metres.

*Cost of canvas = Rs. (59.4 × 2.15) = Rs. 127.71.*

**Ex. 22.** ***A right angled triangle of which the sides containing the right angle are 6.3 cm. and 10 cm. in length, is made to turn round on the longer side. Find the volume of the cone thus formed. Also find the surface of the cone.***

**Sol.** The height of the cone formed by revolving the given triangle is 10 cm. and radius of the base is 6.3 cm.

$\therefore$ Volume of the cone $= \left(\frac{1}{3} \times \frac{22}{7} \times 6.3 \times 6.3 \times 10\right)$ cu. cm.

$= 415.8$ cu. cm.

Slant height $= \sqrt{[(10)^2 + (6.3)^2]} = 11.62$ cm.

$\therefore$ *Curved surface* $= \left(\frac{22}{7} \times 6.3 \times 11.62\right)$ *sq. cm.* $= 230.076$ *sq. cm.*

**Ex. 23.** ***A tent is in the form of a right circular cylinder surmounted by a cone. The diameter of cylinder is 24 m. The height of the cylindrical portion is 11 m. while the vertex of the cone is 16 m. above the ground. Find the area of the canvas required for the tent.***

**Sol. Cylindrical portion.**

Area of curved surface =

$2\pi rh = \left(2 \times \frac{22}{7} \times 12 \times 11\right) = \frac{5808}{7}$ sq. m.

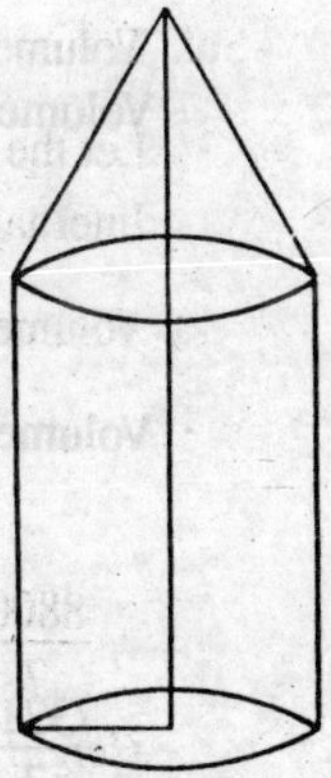

**Conical portion.**

Height = (16 – 11) or 5 m.

Radius of base = 12 m.

$\therefore$ Slant height

$= \sqrt{[(5)^2 + (12)^2]}$ m. = 13 m.

Area of curved surface

$= \pi rl = \left(\frac{22}{7} \times 12 \times 13\right)$ sq. m. $= \frac{3432}{7}$ sq.m.

*Total area of the canvas*

$= \left(\frac{5808}{7} + \frac{3432}{7}\right)$ *sq.m* $= 1320$ *sq. m.*

**Ex. 24.** ***The slant height of the frustum of a cone is 20 cm. and the height of the frustum is 16 cm. The radius of the smaller circle is 8 cm. Find (i) the volume of the frustum, (ii) the total area of the surface of the frustum and (iii) the height of the complete cone of which the frustum is a part.***

**Sol.** Let us denote the radius of smaller circle, radius of bigger circle, the height of frustum and the slant height by $r, R, h$ and $s$ respectively.

Then, $r = 8$ cm, $h = 16$ cm and $s = 20$ cm.

But $s = \sqrt{[h^2 + (R - r)^2]} = \sqrt{[(16)^2 + (R - 8)^2]}$

or $20 = \sqrt{(R^2 + 320 - 16R)}$ or $R^2 + 320 - 16R = 400$

or $R^2 - 16R - 80 = 0$ or $(R - 20)(R + 4) = 0$

$\therefore R = 20$ or $R = -4$. But, $R = -4$ is not possible.

So, the radius of bigger circle is 20 cm.

Now,

(i) Volume of frustum $= \frac{\pi h}{3}(R^2 + r^2 + Rr)$

$= \left[\left(\frac{1}{3} \times \frac{22}{7} \times 16\right)(400 + 64 + 160)\right]$ cu. cm. $= \frac{73216}{7}$ cu. cm.

(ii) Total area of surface of frustum

$= \pi (R^2 + r^2 + Rs + rs)$

$= \frac{22}{7}(400 + 64 + 160 + 160) = 2464$ sq. cm.

(iii) Let the height of the cone be $H$ cm.

Then, volume of cone $= \left(\frac{1}{3} \times \frac{22}{7} \times 20 \times 20 \times H\right)$ cu. cm. $= \left(\frac{8800}{21} H\right)$ cm

Volume of (cone – frustum) $= \left[\frac{1}{3} \times \frac{22}{7} \times 8 \times 8 \times (H - 16)\right]$ cu. cm.

$= \left(\frac{1408}{21} \times (H - 16)\right)$ cu. cm.

Volume of frustum $= \left[\frac{8800}{21} H\right] - \left[\frac{1408}{21} \times (H - 16)\right]$ cu. cm.

$= \left(\frac{8800\,H - 1408\,H + 22528}{21}\right)$ cu. cm.

$\therefore \frac{73216}{7} = \frac{7392\,H + 22528}{21}$ or $H = 26.66$ cm.

$\therefore$ *Height of cone = 26.66 cm.*

**5. Problems on Volume and Area of sphere.**

**Ex. 25.** ***Find the surface and volume of a sphere whose radius is 2.1 metres.***

**Sol.** Volume of sphere $= \frac{4}{3}\pi r^3$

$= \left(\frac{4}{3} \times \frac{22}{7} \times 2.1 \times 2.1 \times 2.1\right)$ cu. metres

$= 38.808$ cu. metres.

surface of sphere $= 4\pi r^2$

$= \left(4 \times \frac{22}{7} \times 2.1 \times 2.1\right)$ sq. metres,

$= 55.44$ *sq. metres.*

**Ex. 26.** ***A hemisphere of lead of radius 7 cm. is cast into a right circular cone of height 49 cm. Find the radius of the base.***

**Sol.** Volume of hemisphere $= \left(\frac{1}{2} \times \frac{4}{3} \times \frac{22}{7} \times 7 \times 7 \times 7\right)$ cu. cm.

$$= \frac{2156}{3} \text{ cu. cm.}$$

Let the radius of the base of cone be $r$ cm.

Then, $\frac{1}{3} \times \frac{22}{7} \times r^2 \times 49 = \frac{2156}{3}$ or $r^2 = \frac{2156 \times 7 \times 3}{3 \times 22 \times 49} = 14.$

$\therefore$ *Radius* $= \sqrt{(14)} = 3.74$ *cm.*

**Ex. 27.** ***A spherical shell of metal has an outer radius of 9 cm. and inner radius of 8 cm. If the metal costs Rs. 1.80 per cu. cm., find the cost of shell.***

**Sol.** Volume of metal $= \left\{\frac{4}{3}\pi (9)^3 - \frac{4}{3}\pi (8)^3\right\}$ cu. cm.

$$= \frac{4}{3}\pi (9^3 - 8^3) \text{ cu. cm.}$$

$$= \left(\frac{4}{3} \times \frac{22}{7} \times 217\right) \text{cu. cm.} = \frac{2728}{3} \text{ cu. cm.}$$

$$\textit{Cost of metal} = \text{Rs.} \left(\frac{2728}{3} \times 1.80\right) = \text{Rs. } 1636.80.$$

**Ex. 28.** ***A metal sphere of diameter 42 cm. is dropped into a cylindrical vessel, which is partly filled with water. The diameter of the vessel is 1.68 metres. If the sphere is completely submerged, find by how much the surface of water will rise.***

**Sol.** Radius of the sphere = 21 cm.

Volume of sphere $= \left(\frac{4}{3} \times \frac{22}{7} \times 21 \times 21 \times 21\right)$ cu. cm.

$= 38808$ cu. cm.

Volume of water displaced by sphere = 38808 cu. cm.

Let the water rise by $h$ cm.

Then, $\frac{22}{7} \times 84 \times 84 \times h = 38808$

*or* $h = \frac{38808 \times 7}{22 \times 84 \times 84}$ *cm.* $= \frac{7}{4}$ *cm.* $= 1.75$ *cm.*

**Ex. 29.** ***A solid is composed of a cylinder with hemispherical ends. If the whole length of the solid is 9 metres and its diameter 3 metres, find the cost of polishing its surface at the rate of Rs. 1.05 per sq. metre.***

**Sol.** Length of cylindrical portion = (9 – 3) or 6 metres.

Radius of cylinder $= \frac{3}{2}$ metres.

Curved surface of cylindrical portion

$$= \left(2 \times \frac{22}{7} \times \frac{3}{2} \times 6\right) \text{ sq. metres} = \frac{396}{7} \text{ sq. metres.}$$

Curved surface of two hemispheres each of radius $\frac{3}{2}$ metres.

$$= 2\left(2 \times \frac{22}{7} \times \frac{3}{2} \times \frac{3}{2}\right) \text{ sq. metres} = \frac{198}{7} \text{ sq. m.}$$

Total area to be polished

$$= \left(\frac{396}{7} + \frac{198}{7}\right) \text{ sq. metres} = \left(\frac{584}{7}\right) \text{ sq. metres.}$$

$$\textit{Cost of polishing} = Rs. \left(\frac{594}{7} \times 1.05\right) = Rs.\ 89.10.$$

## EXERCISE 21A (Subjective)

1. Find the volume of a cuboid 90 metres by 50 metres by 75 cm.
2. Find the volume and surface area of a cube, whose each edge measures 25 cm.
3. Find the volume of a cube whose surface area is 150 square metres.

   **Hint.** $6 \times (edge)^2 = 150 \Rightarrow edge = 5$ *m.*
4. Find the volume of a cube whose diagonal is $10\sqrt{3}$ metres.

   **Hint.** $\sqrt{3} \times edge = 10\sqrt{3} \Rightarrow edge = 10$ *m.*
5. Find the length of the longest pole that can be put in a room 10 metres by 8 metres by 5 metres.

   **Hint.** *Length of longest pole = diagonal* $= \sqrt{l^2 + b^2 + h^2}$.
6. A closed rectangular box has inner dimensions 24 cm. by 12 cm. by 10 cm. Calculate its capacity and the area of tin foil needed to line its inner surface.

   **Hint.** *Capacity = Volume* = $(24 \times 12 \times 10)$ *sq. cm.*

   *Area of tin foil needed = Total surface area*

   $= [2\,(24 \times 12 + 12 \times 10 + 24 \times 10)]$ *sq. cm.*
7. A wall of length 25 metres, width 60 cm. and height 2 metres is to be constructed by using bricks, each of dimensions 20 cm. by 12 cm. by 8 cm. How many bricks will be needed ?

   **Hint.** *Number of bricks* $= \left(\frac{2500 \times 60 \times 200}{20 \times 12 \times 8}\right)$
8. A certain cube of wood was bought for Rs. 768. If the wood costs Rs. 1500 per cubic metre, find the length of each edge of the cube.

   **Hint.** *Volume of the cube* $= \left(\frac{768}{1500} \times 100 \times 100\right.$ *cu. cm.*

$\therefore$ *Length of each edge* $= (512000)^{1/3} = (80 \times 80 \times 80)^{1/3} = 80$ *cm.*

**9.** A room 5 metres high is half as long again as it is broad and its volume is 480 cubic metres. Find the length and breadth of the room.

**Hint.** *Let breadth* $= x$. *Then, length* $= \frac{3}{2} x$ *metres.*

$\therefore \frac{3}{2} x \times x \times 5 = 480.$

**10.** A river 2 metres deep and 45 metres wide is flowing at the rate of 3 km per hour. Find how much water runs into the sea per minute.

**Hint.** *Distance covered by water in 1 min* $= \left(\frac{3000}{60}\right) m = 50$ *m.*

*Water that runs in 1 min.* $= (50 \times 45 \times 2)$ *cu. m.*

**11.** Find the volume of an iron rod which is 7 cm long and whose diameter is 1 cm.

**12.** Calculate the curved surface area, the total surface area and the volume of a cylinder with base radius 14 cm and height 60 cm.

**13.** Water flows at 10 km per hour through a pipe with cross section a circle of radius 35 cm, into a cistern of dimensions 25 m by 12 m by 10 m. By how much will the water level rise in the cistern in 24 minutes ?

**Hint.**

*Volume flown in 24 min.* $= \left(\frac{22}{7} \times \frac{35}{100} \times \frac{35}{100} \times \frac{10000}{60} \times 24\right)$ *cu. m.*

$= 1540$ *cu. m*

*Initial volume of cistern* $= (25 \times 12 \times 10)$ *cu. m.* $= 3000$ *cu. m.*

New level $= \left(\frac{4540}{25 \times 12}\right) = 15.13$ *m.*

$\therefore$ *Rise in level* $= (15.13 - 10)$ *m* $= 5.13$ *m.*

**14.** The internal diameter of an iron pipe is 6 cm and the length is 2.8 metres. If the thickness of the metal be 5 mm and 1 cu. cm of iron weighs 8 gm, find the weight of the pipe.

**Hint.** *Volume of the metal in the pipe*

$= \frac{22}{7} \times 300 \times [(3.5)^2 - (3)^2]$ *cu. cm.* $= 28600$ *cu. cm.*

$\therefore$ *Weight of the pipe* $= \left(28600 \times \frac{8}{1000}\right)$ *kg.*

**15.** A cylindrical bucket 28 cm in diameter and 72 cm high is full of water. The water is emptied into a rectangular tank 66 cm long and 28 cm wide. Find the height of water level in the tank.

**16.** Find the volume of a conical tent which is 27 metres high and the radius of whose base is 7 metres.

**17.** Find the volume of a cone the diameter of whose base is 21 cm and the slant height is 37.5 cm.

**Hint.** $h = \sqrt{(37.5)^2 - (10.5)^2} = 36 cm.$

*Now, volume* $= \frac{1}{3}\pi r^2 h.$

**18.** It is required to make a hollow cone 24 cm high whose base radius is 7 cm. Find the area of the sheet metal required including the base. Also, find the capacity of this cone.

**Hint.** *Metal required* $= (\pi rl + \pi r^2)$ *sq. cm.*,

*Where* $r = 7$ *cm and* $l = \sqrt{(24)^2 + (7)^2} = 25$ *cm.*

**19.** A cone of height 7 cm and base radius 3 cm is carved from a rectangular block of wood 10 cm × 5 cm × 2 cm. Calculate the percentage of wood wasted.

**Hint.** *From 100 cu. cm, volume carved*

$$= \left(\frac{1}{3} \times \frac{22}{7} \times 3 \times 3 \times 7\right) \text{cu. cm.} = 66 \text{ cu. cm.}$$

**20.** In a hollow cylinder made of iron, the volume of iron is 1430 cu. cm. If the length of the cylinder be 35 cm and its external diameter be 14 cm., find the thickness of the cylinder.

**Hint.** *Volume of iron* $= \frac{22}{7} \times 35 \times [(7)^2 - r^2]$ *cu. cm.* $= (5390 - 110r^2).$

$\therefore (5390 - 110r^2) = 1430.$ *Find r.*

**21.** The area of the curved surface of a cylinder is 4400 $cm^2$ and the circumference of its base is 110 cm. Find the height and the volume of the cylinder.

**22.** A powder tin has a square base with side 8 cm and height 13 cm. Another is cylindrical with radius of its base 7 cm and height 15 cm. Find the difference in their capacities.

**23.** How many cubic metres of the earth must be dug out to sink a well 21 metres deep and 6 metres in diameter ? Find the cost of plastering the inner surface of the well at Rs. 9.50 per sq. metre.

**Hint.** *Volume of earth dug out = Volume of the well.*

*Inner surface of the well = Curved surface of the well.*

**24.** The radii of two cylinders are in the ratio 2 : 3 and their heights are in the ratio 5 : 3. Calculate the ratio of their volumes and the ratio of their curved surfaces.

**Hint.** *Let their radii be* 2r, 3r *and heights be* 5h, 3h.

*Then,* $\frac{V_1}{V_2} = \frac{\pi(2r)^2 \times (5h)}{\pi(3r)^2 \times (3h)} = \frac{20}{27}$ & $\frac{S_1}{S_2} = \frac{2\pi \times 2r \times 5h}{2\pi \times 3r \times 3h} = \frac{10}{9}.$

**25.** A solid cylinder has a total surface area of 231 square cm. Its curved surface area is (2/3) of the total surface area. Find the volume of the cylinder.

**Hint.** $2\pi rh + 2\pi r^2 = 231$ *and* $2\pi rh = \left(\frac{2}{3}\times 231\right)$.

**26.** The sum of the radius of the base and the height of a solid cylinder is 37 m. If the total surface area of the cylinder be 1628 sq. m., find its volume.

**Hint.** $r + h = 37$ *and* $2\pi r(r+h) = 1628$.

**27.** The radius and height of a right circular cone are in the ratio 5 : 12 and its volume is 2512 cu. cm. Find the slant height, radius and curved surface area of the cone. (Take $\pi = 3.14$).

**Hint.** $\frac{1}{3}\times 3.14\times(25x^2)\times(12x) = 2512 \Rightarrow x = 2.$

$\therefore$ *Radius* = 10 *cm & height* = 24 *cm.*

$\therefore l = \sqrt{r^2+h^2} = \sqrt{(10)^2+(24)^2} = 26$ *cm.*

**28.** From a solid right circular cylinder with height 12 cm and radius of the base 7 cm, a right circular cone of the same height and base is removed. Find the volume of the remaining solid.

**Hint.** *Remaining solid*

$$= \left(\frac{22}{7}\times 7\times 7\times 12 - \frac{1}{3}\times\frac{22}{7}\times 7\times 7\times 12\right) \text{ cu. cm.}$$

**29.** If the radii of the ends of a bucket 45 cm. high are 28 cm and 7 cm, determine its capacity and the surface area.

**Hint.** $r = 7$ *cm*, $R = 28$ *cm* and $h = 45$ *cm.*

$l = \sqrt{h^2 + (R-r)^2} = 49.6$ *cm.*

*Capacity = Volume of the frustum*

$$= \frac{1}{3}\pi h(R^2 + r^2 + Rr) = 48510 \text{ cu. cm.}$$

Surface area of the bucket = $[\pi l(R+r) + \pi r^2]$ sg. cm. = 5610 sq. cm.

**30.** Find the volume and surface area of a sphere of radius 2.1 cm.

**31.** Find the surface area of a sphere whose volume is 310464 cu. cm.

**32.** Find the volume of a solid sphere whose surface area is 636 sq. cm.

**33.** The diameter of a copper sphere is 6 cm. The sphere is melted and drawn into a long wire of uniform circular cross section. If the length of the wire is 36 cm, find its radius.

**Hint.** $\frac{4}{3}\times\pi\times(3)^3 = \pi r^2\times 36$. *Find r.*

**34.** A metallic sphere of radius 21 cm is dropped into a cylindrical vessel, which is partially filled with water. The diameter of the vessel is 1.68

metres. If the sphere is completely submerged, find by how much the surface of water will rise.

**Hint.** *Volume of sphere* $= \left(\frac{4}{3}\times\frac{22}{7}\times 21\times 21\times 21\right) = 38808$ *cu. cm.*

$\therefore \frac{22}{7}\times 84\times 84\times h = 38808.$ *Find h.*

**35.** Find the weight of an iron shell, the external and internal diameters of which are 13 cm and 10 cm respectively, if 1 cu. cm of iron weighs 8 gms.

**Hint.** *Volume of iron* $= \frac{4}{3}\pi\left\{\left(\frac{13}{2}\right)^3 - (5)^3\right\}$

$= \left(\frac{4}{3}\times\frac{22}{7}\times\frac{1197}{8}\right)$ *cu. cm.* $= 627$ *cu. cm.*

$\therefore$ *Weight of iron* $= \left(\frac{627\times 8}{1000}\right)$ *kg.*

**36.** A cylinder of length 1 metre and diameter 15 cm is melted down and cast into spheres of diameter 5 cm. How many spheres can be made ?

**Hint.** *Number of spheres* $= \left(\frac{\pi\times 7.5\times 7.5\times 100}{\frac{4}{3}\times\pi\times 2.5\times 2.5\times 2.5}\right)$

**37.** How many lead shots each 0.3 cm in diameter can be made from a cuboid of dimension 9 cm by 11 cm by 12 cm ?

**Hint.** *Number of lead shots* $= \frac{\text{Volume of cuboid}}{\text{Volume of 1 lead shot}}$.

**38.** A spherical ball of radius 3 cm is melted and recast into three spherical balls. The radii of two of these balls are 1.5 cm and 2 cm. Find the radius of the third ball.

**Hint.** *Volume of 3rd ball* $= \frac{4}{3}\pi(3)^3 - \left\{\frac{4}{3}\pi\left(\frac{3}{2}\right)^3 + \frac{4}{3}\pi(2)^3\right\} = \frac{125}{6}\pi.$

$\therefore \frac{4}{3}\pi r^3 = \frac{125}{6}\pi \Rightarrow r = \frac{5}{2}.$

**39.** A cone, a hemisphere and a cylinder stand on equal bases and have the same height. Show that their volumes are in the ratio 1 : 2 : 3.

**Hint.** *Ratio of volumes*

$= \frac{1}{3}\pi r^3 : \frac{2}{3}\pi r^3 : \frac{3}{3}\pi r^3 = 1 : 2 : 3.$

**40.** The largest sphere is carved out of a cube of side 7 cm. Find the volume of the sphere (Take $\pi = 3.14$).

**Hint.** *Diameter of the sphere = 7 cm.*

**41.** The bottom of a tent is in the form of a cylinder of 28 metre diameter and 8 metre height, surmounted by a cone of equal base and 10.5 m in height. Find the cost of canvas at Rs. 2.50 per square metre and the capacity of the tent.

**Hint.** *Area of canvas* $= [2\pi \times 14 \times 8 + \pi \times 14 \times \sqrt{(14)^2 + (10.5)^2}]$

*Capacity of the tent* $= \{\pi \times (14)^2 \times 8\} + \left\{\frac{1}{3}\pi \times (14)^2 \times 10.5\right\}$ *cu. m.*

## ANSWERS (Exercise 21A)

**1.** 3375 cu. m. **2.** 15625 cu. cm., 3750 sq. cm.
**3.** 125 cu. m. **4.** 1000 cu. m.
**5.** 13 m. **6.** 2880 cu. cm., 1296 sq. cm.
**7.** 15625 **8.** 80 cm **9.** 12 m, 8 m
**10.** 4500 cu. m. **11.** 5.5 cu. cm.
**12.** 5280 sq. cm., 6512 sq.cm., 36960 cu. cm.
**13.** 5.13 m **14.** 228.8 kg. **15.** 24 cm
**16.** 1386 cu. m **17.** 4158 cu. cm. **18.** 704 sq. cm.
**19.** 34% **20.** 1 cm
**21.** 40 cm, 38500 cu. cm. **22.** 1478 cu. cm. **23.** 594 cu. m, Rs. 3762
**24.** 20 : 27, 10 : 9 **25.** 269.5 cu. cm. **26.** 4620 cu. m.
**27.** 24 cm, 10 cm, 816.4 sq. cm. **28.** 1232 cu. cm.
**29.** 48510 cu. cm., 5610 sq. cm. **30.** 388.08 cu. cm., 55.44 sq. cm.
**31.** 22176 sq. cm. **32.** 1437.3 cu. cm. **33.** 1 cm
**34.** 1.75 cm **35.** 5.016 kg. **36.** 270
**37.** 84000 **38.** 2.5 cm.
**40.** 179.6 cu. cm. **41.** Rs. 4435, 7084 cu. m.

## EXERCISE 21B (Objective Type Questions)

**1.** If the length, breadth and height of a cuboid are 2m, 2m and 1 m respectively, then its surface area (in $m^2$) is :
(*a*) 8 (*b*) 12
(*c*) 16 (*d*) 24 **(N.D.A. Exam. 1990)**

**2.** The diagonal of a cuboid 22 cm by 12 cm by 7.5 cm is :
(*a*) 13.83 cm (*b*) 6.04 cm
(*c*) 24.25 cm (*d*) 26.15 cm

**3.** Sum of the length, width and depth of a cuboid is *s* and its diagonal is *d*. Its surface area is :
(*a*) $s^2$ (*b*) $d^2$

(c) $s^2 - d^2$ (d) $s^2 + d^2$ **(C.D.S. Exam. 1989)**

4. The area of the cardboard (in $cm^2$) needed to make a box of size 25 cm × 15 cm × 8 cm will be :
(a) 390 (b) 1000
(c) 1390 (d) 2780

5. A wooden box of dimensions 8 m × 7 m × 6 m is to carry rectangular boxes of dimensions 8 cm × 7 cm × 6 cm. The maximum number of boxes that can be carried in the wooden box, is :
(a) 980,00,00 (b) 750,00,00
(c) 100,00,00 (d) 120,00,00
**(C.D.S. Exam. 1991)**

6. The length of the longest rod that can be placed in a room 30 m long, 24 m broad and 18 m high, is :
(a) 30 m (b) $15\sqrt{2}$ m
(c) 60 m (d) $30\sqrt{2}$ m
**(C.D.S. Exam. 1991)**

7. A tank 3 m long, 2 m wide and 1.5 m deep is dug in a field 22 m long and 14 m wide. If the earth dug out is evenly spread out over the field, the level of the field will rise by nearly :
(a) 0.299 cm (b) 0.29 mm
(c) 2.98 cm (d) 4.15 cm

8. If the length, breadth and height of a rectangular parallelopiped are in the ratio 6 : 5 : 4 and if total surface area is 33,300 $m^2$ then the length, breadth and height of a parallelopiped (in cm) respectively are :
(a) 90, 85, 60 (b) 90, 75, 70
(c) 85, 75, 60 (d) 90, 75, 60
**(N.D.A. Exam. 1990)**

9. $\frac{1}{2}$ $m^3$ gold sheet is extended by hammering so as to cover an area of 1 hectare. The thickness of the sheet is :
(a) 0.5 cm (b) 0.05 cm
(c) 0.005 cm (d) 0.0005 cm

10. In a shower, 5 cm of rain falls. The volume of water that falls on 2 hectares of ground is :
(a) 100 $m^3$ (b) 1000 $m^3$
(c) 10000 $m^3$ (d) 10 $m^3$

11. A river 2 metres deep and 45 metres wide is flowing at the rate of 3 km/hr. the amount of water that runs into the sea per minute is :
(a) 4500 $m^3$ (b) 27000 $m^3$
(c) 3000 $m^3$ (d) 2700 $m^3$

**12.** The dimensions of an open box are 52 cms, 40 cms and 29 cms. Its thickness is 2 cms. If 1 $cm^3$ of metal used in the box weighs 0.5 gms, the weight of the box is :

(*a*) 8.56 kg (*b*) 7.76 kg
(*c*) 7.756 kg (*d*) 6.832 kg

**13.** The surface of a cube is 1176 $cm^2$. The volume of this cube is :

(*a*) 7056 $cm^3$ (*b*) 4704 $cm^3$
(*c*) 2744 $cm^3$ (*d*) 3528 $cm^3$

**14.** The length of diagonal of a cube is $(14 \times \sqrt{3})$ cm. The surface area of the cube is :

(*a*) 588 $cm^2$ (*b*) 1176 $cm^2$
(*c*) 339.08 $cm^2$ (*d*) 294 $cm^2$

**15.** The volume of a cube is 125 $cm^3$. The surface area of the cube is :

(*a*) 625 $cm^2$ (*b*) 125 $cm^2$
(*c*) 150 $cm^2$ (*d*) 100 $cm^2$

**16.** If each side of a cube is doubled, then its volume :

(*a*) is doubled (*b*) becomes 4 times
(*c*) becomes 6 times (*d*) becomes 8 times

**17.** The length of the longest rod that can fit in a cubical room of 4 m side, is :

(*a*) 8.66 m (*b*) 5.196 m
(*c*) 6.928 m (*d*) 7.264 m

**18.** Three cubes whose edges are 3 cm, 4 cm and 5 cm respectively are melted without any loss of metal into a single cube. The edge of new cube is :

(*a*) 12 cm (*b*) 10 cm
(*c*) 9 cm (*d*) 6 cm

**19.** A metal sheet 27 cm long, 8 cm broad and 1 cm thick is melted into a cube. The difference between the surface areas of two solids will be :

(*a*) 284 $cm^2$ (*b*) 296 $cm^2$
(*c*) 286 $cm^2$ (*d*) 300 $cm^2$ **(C.D.S. Exam. 1991)**

**20.** The sum of sides of a cube and a rectangular parallelopiped are same. The ratio of the volume of the cube to that of the parallelopiped will be :

(*a*) 1 (*b*) greater than 1
(*c*) less than 1 (*d*) may be any of these

**21.** The surface area of a cube is 600 $cm^2$. The length of its diagonal is :

(*a*) $10\sqrt{2}$ cm (*b*) $10\sqrt{3}$ cm

(*c*) $\frac{10}{\sqrt{2}}$ cm (*d*) $\frac{10}{\sqrt{3}}$ cm

**22.** The percentage increase in surface area of a cube when each side is doubled, is :

(*a*) 25% (*b*) 50%

(*c*) 150% (*d*) 300%

**23.** The number of small cubes with edges of 10 cm that can be accomodated in a cubical box of 1 metre edge is :

(*a*) 100 (*b*) 1000

(*c*) 10 (*d*) 10000

**24.** The length of a cylinder is 80 cm and the diameter of its base is 7 cm. The whole surface of the cylinder is :

(*a*) 1837 $cm^2$ (*b*) 1760 $cm^2$

(*c*) 3080 $cm^2$ (*d*) 1942 $cm^2$

**25.** A cylindrical tower is 5 m in diameter and 14 m high. The cost of white washing its curved surface at 50 paise per $m^2$ is :

(*a*) Rs. 90 (*b*) Rs. 97

(*c*) Rs. 100 (*d*) Rs. 110

**26.** The height of a cylinder is 14 cm and its curved surface is 264 $cm^2$. The radius of its base is :

(*a*) 3 cm (*b*) 4 cm

(*c*) 2.4 cm (*d*) 12.4 cm

**27.** A hollow garden roller 63 cm wide with a girth of 440 cm is made of iron 4 cm thick. The volume of iron is :

(*a*) 56372 $cm^3$ (*b*) 58752 $cm^3$

(*c*) 54982 $cm^3$ (*d*) 57636 $cm^3$

**28.** If 1 $cm^3$ cast iron weighs 21 gms, then weight of a cast iron pipe of length 1 m with a bore of 3 cm and in which thickness of metal is 1 cm, is :

(*a*) 21 kg (*b*) 24.2 kg

(*c*) 26.4 kg (*d*) 18.6 kg

**29.** A well with 14 metres inside diameter is dug 8 metres deep. Earth taken out of it has been evenly spread all around it to a width of 21 metres to form an embankment. The height of embankment is :

(*a*) 43 cm (*b*) 47.6 cm

(*c*) 53.3 cm (*d*) 41 cm

**30.** Two circular cylinders of equal volume have their heights in the ratio 1 : 2. The ratio of their radii is :

(*a*) $1 : \sqrt{2}$ (*b*) $\sqrt{2} : 1$

(*c*) 1 : 2 (*d*) 1 : 4

**31.** The ratio of total surface area to lateral surface area of a cylinder whose radius is 80 cm and height is 20 cm is :

(*a*) 2 : 1 (*b*) 3 : 1
(*c*) 4 : 1 (*d*) 5 : 1

**32.** The radii of two cylinders are in the ratio of 2 : 3 and their heights are in the ratio 5 : 3. The ratio of their volumes is :

(*a*) 27 : 20 (*b*) 20 : 27
(*c*) 4 : 9 (*d*) 9 : 4

**(Hotel Management 1991)**

**33.** If the diameter of the base of a closed right circular cylinder is equal to its height *h*, then its whole surface area is :

(*a*) $2\pi h^2$ (*b*) $\frac{4}{3}\pi h^2$
(*c*) $\frac{3}{2}\pi h^2$ (*d*) $\pi h^2$ **(N.D.A. Exam. 1990)**

**34.** The diameter of the base of a right circular cone is 4 cm and its perpendicular height is $2\sqrt{3}$. The slant height of the cone is :

(*a*) 5 cm (*b*) 4 cm
(*c*) $4\sqrt{3}$ cm (*d*) 3 cm

**35.** The radius of base of a right circular cone is 6 cm and its slant height is 28 cm. The curved surface of the cone is :

(*a*) 268 $cm^2$ (*b*) 658 $cm^2$
(*c*) 462 $cm^2$ (*d*) 528 $cm^2$

**36.** The area of the base of a right circular cone is 154 $cm^2$ and its height is 14 cm. The curved surface of cone is :

(*a*) $(154 \times \sqrt{5})$ $cm^2$ (*b*) 11 $cm^2$
(*c*) $(154 \times \sqrt{7})$ $cm^2$ (*d*) 5324 $cm^2$

**37.** If the height of a cone is increased by 100%, then its volume is increased by :

(*a*) 100% (*b*) 200%
(*c*) 300% (*d*) 400%

**38.** If a right circular cone of vertical height 24 cm has a volume of 1232 $cm^3$, then the area of its curved surface in $cm^2$ is :

(*a*) 1254 (*b*) 704
(*c*) 550 (*d*) 154 **(N.D.A. Exam. 1990)**

**39.** If the volumes of two cones are in ratio 1 : 4 and their diameters are in ratio 4 : 5, then the ratio of their heights is :

(*a*) 1 : 5 (*b*) 5 : 4
(*c*) 5 : 16 (*d*) 25 : 64 **(N.D.A. Exam. 1990)**

40. A cylinder and a cone have same height and same radius of the base. The ratio between the volumes of the cylinder and cone is :
(a) 1 : 3 (b) 3 : 1
(c) 1 : 2 (d) 2 : 1

41. A right cylinder and a right circular cone have same radius and same volume. The ratio of height of the cylinder to that of the cone is :
(a) 3 : 5 (b) 2 :5
(c) 3 : 1 (d) 1 : 3 **(C.D.S. Exam. 1991)**

42. A right circular cone is cut off at the middle of its height and parallel to base. Call smaller cone thus formed *A* and remaining part *B*. Then :
(a) $Vol\ A < Vol\ B$ (b) $Vol\ A = Vol\ B$
(c) $Vol\ A > Vol\ B$ (d) $Vol\ A = \frac{Vol\ B}{2}$
**(C.D.S. Exam. 1991)**

43. A reservoir is in the shape of a frustum of a right circular cone. It is 8 m across at the top and 4 m across the bottom. It is 6 m deep. Its capacity is :
(a) 176 $m^3$ (b) 196 $m^3$
(c) 200 $m^3$ (d) 110 $m^3$ **(C.D.S. Exam. 1991)**

44. A solid consists of a circular cylinder with exact fitting right circular cone placed on the top. The height of the cone is *h*. If total volume of the solid is 3 times the volume of cone, then the height of circular cylinder is :
(a) $2h$ (b) $4h$
(c) $\frac{3h}{2}$ (d) $\frac{2h}{3}$ **(C.D.S. Exam. 1991)**

45. A cone of height 7 cm and base radius 3 cm is carved from a rectangular block of wood (10 cm × 5 cm × 2 cm). The percentage of wood wasted is :
(a) 34% (b) 66%
(c) 46% (d) 54%

46. A right cylindrical vessel is full with water. How many right cones having same diameter and height as those of right cylineder will be needed to store that water ?
(a) 2 (b) 3
(c) 4 (d) 5 **(C.D.S. Exam. 1991)**

47. A cylindrical piece of metal of radius 2 cm and height 6 cm is shaped into a cone of same radius. The height of cone is :
(a) 18 cm (b) 14 cm
(c) 12 cm (d) 8 cm
**(Railway Recruitment 1991)**

48. The curved surface of a sphere is 1386 $cm^2$. Its volume is :

(*a*) 2772 $cm^3$ (*b*) 4158 $cm^3$

(*c*) 4851 $cm^3$ (*d*) 5544 $cm^3$

**49.** The volume of a hemi-sphere is 19404 $cm^3$. The radius of the hemi-sphere is :

(*a*) 10.5 cm (*b*) 21 cm

(*c*) 42 cm (*d*) 17.5 cm

**50.** The volume of a sphere is $\frac{88}{21}(14)^3$ $cm^3$. The curved surface of this sphere is :

(*a*) 2424 $cm^2$ (*b*) 2446 $cm^2$

(*c*) 2464 $cm^2$ (*d*) 2484 $cm^2$

**51.** A hemispherical bowl has inner diameter 42 cm. The quantity of liquid that the bowl can hold (in $cm^3$) is :

(*a*) $\frac{2}{3}\times\frac{22}{7}\times(21)^3$ (*b*) $\frac{2}{3}\times\frac{7}{22}\times(21)^3$

(*c*) $\frac{3}{2}\times\frac{22}{7}\times(21)^3$ (*d*) $\frac{2}{3}\times\frac{22}{7}\times(42)^3$

**(C.D.S. Exam. 1991)**

**52.** Six spherical balls of radius *r* are melted and cast into a cylindrical rod of metal of same radius. The height of rod will be :

(*a*) 4 *r* (*b*) 6 *r*

(*c*) 8 *r* (*d*) 12 *r*

**53.** The amount of water contained in a sphere of radius 3 cm is poured into a cube of side 5 cm. The height upto which water will rise in the cube is :

(*a*) 2 cm (*b*) $\frac{3\pi}{4}$ cm

(*c*) 0.5 π cm (*d*) 1.44 π cm

**54.** If the radii of two spheres are in ratio 1 : 2, then ratio of their surface areas will be

(*a*) 1 : 2 (*b*) 1 : 4

(*c*) 1 : 8 (*d*) 1 : 16 **(N.D.A. Exam. 1990)**

**55.** The number of solid spheres, each of diameter 6 cms, that could be moulded to form a solid metal cylinder of height 45 cms and diameter 4 cms, is :

(*a*) 3 (*b*) 4

(*c*) 5 (*d*) 6 **(N.D.A. Exam. 1990)**

**56.** The areas of two spheres are in the ratio 1 : 4. The ratio of their volumes is :

(*a*) 1 : 4 (*b*) $1:2\sqrt{2}$
(*c*) 1 : 8 (*d*) 1 : 64

**57.** If a solid sphere of radius 10 cms is moulded into 8 spherical solid balls of equal radius, then surface area of each ball (in $cm^2$) is :
(*a*) $100\pi$ (*b*) $75\pi$
(*c*) $60\pi$ (*d*) $50\pi$ **(N.D.A. Exam. 1990)**

**58.** A spherical ball of lead, 3 cm in diameter is melted and recast into 3 spherical balls. The diameter of two of these are $1\frac{1}{2}$ cm and 2 cm respectively. The diameter of third ball is :
(*a*) 2.66 cm (*b*) 2.5 cm
(*c*) 3 cm (*d*) 3.5 cm

**59.** The number of solid spheres of radius $\frac{1}{4}$ cm, which may be formed from a solid sphere of radius 4 cms is :
(*a*) 4096 (*b*) 4964
(*c*) 6904 (*d*) 9640

**60.** If the volume and surface area of a sphere are numerically the same, then its radius is :
(*a*) 1 unit (*b*) 2 units
(*c*) 3 units (*d*) 4 units **(N.D.A. Exam. 1990)**

**61.** A solid lead ball of 7 cm radius was melted and then drawn into a wire of 0.2 cm diameter. The length of wire will be :
(*a*) 458.43 m (*b*) 457.33 m
(*c*) 468.26 m (*d*) 437.29 m

**62.** How many bullets can be made out of a cube of lead whose edge measures 22 cm, each bullet being 2 cm in diameter ?
(*a*) 5324 (*b*) 2662
(*c*) 1347 (*d*) 2541

**63.** A cylindrical vessel 60 cm in diameter is partially filled with water. A sphere, 30 cm in diameter is gently dropped into the vessel. To what further height will water in the cylinder rise ?
(*a*) 15 cm (*b*) 30 cm (*c*) 40 cm
(*d*) cannot be determined, since height of cylinder has not been given

**64.** The material of a cone is converted into the shape of a cylinder of equal radius. If the height of the cylinder is 5 cm, the height of cone is :
(*a*) 10 cm (*b*) 15 cm
(*c*) 18 cm (*d*) 20 cm

**65.** The radii of two cylinders are in ratio 2 : 3 and their heights in ratio 5 : 3. Their volumes will be in ratio :

(*a*) 4 : 9 (*b*) 27 : 20
(*c*) 20 : 27 (*d*) 9 : 4

**(Clerks' Grade Exam. 1991)**

**66.** For a given volume, which of the following has minimum surface area ?
(*a*) cube (*b*) cone
(*c*) sphere (*d*) cylinder

**67.** If the volumes of two cubes are in the ratio 8 : 1, the ratio of their edges is :
(*a*) 8 : 1 (*b*) 4 : 1
(*c*) 2 : 1 (*d*) 3 : 1

**68.** The radius of a wire is decreased to one third. If volume remains same, length will increase :
(*a*) 1 time (*b*) 6 times
(*c*) 3 times (*d*) 9 times

**(Railway Recruitment 1990)**

## HINTS & SOLUTIONS (Exercise 21B)

**1.** Surface area $= 2\,(lb + bh + lh) = 2\,(2 \times 2 + 2 \times 1 + 2 \times 1)\ \text{m}^2 = 16\ \text{m}^2$.

**2.** diagonal $= \sqrt{l^2 + b^2 + h^2}$

$$= \sqrt{(22)^2 + (12)^2 + (7.5)^2} = \sqrt{684.25} = 26.15\ \text{cm}.$$

**3.** $l + b + h = s$ and $\sqrt{l^2 + b^2 + h^2} = d$. So, $(l^2 + b^2 + h^2) = d^2$.

$$\therefore (l + b + h)^2 = s^2 \Rightarrow (l^2 + b^2 + h^2) + 2\,(lb + bh + lh) = s^2$$

$$\Rightarrow d^2 + 2\,(lb + bh + lh) = s^2$$

$$\Rightarrow 2\,(lb + bh + lh) = (s^2 - d^2).$$

$\therefore$ *Surface area* $= (s^2 - d^2)$.

**4.** Area needed $= 2\,(lb + bh + lh)$

$$= 2\,[(25 \times 15) + (15 \times 8) + (25 \times 8)]\ \text{cm}^2$$

$$= 1390\ \text{cm}^2.$$

**5.** Number of boxes $= \dfrac{800 \times 700 \times 600}{8 \times 7 \times 6} = 1000000.$

**6.** Length of the longest rod = length of diagonal

$$= \sqrt{(l^2 + b^2 + h^2)} = \sqrt{(30)^2 + (24)^2 + (18)^2} = \sqrt{1800} = 30\sqrt{2}\ \text{m}.$$

**7.** Volume of earth dug out $= (3 \times 2\ 1.5)\ \text{m}^3 = 9\ \text{m}^3$.

Area over which earth is spread $= [(22 \times 14) - (3 \times 2)]\ \text{m}^2 = 302\ \text{m}^2$.

$\therefore$ Increase in level $= \frac{9}{302} m = \frac{9 \times 100}{302} cm = 2.98 \, cm.$

8. Let the length, breadth and height be $6x$, $5x$ and $4x$ metres respectively.

Then, $2 \times [6x \times 5x + 5x \times 4x + 6x \times 4x] = 33300$.

$\therefore 148x^2 = 33300$ or $x^2 = 225$ or $x = 15$.

So, length = 90 m, breadth = 75 m and height = 60 m.

9. Thickness $= \left( \frac{\frac{1}{2}}{10000} \right) m = \frac{1}{2} \times \frac{100}{10000}$ cm = 0.005 cm

10. Volume $= \left( 2 \times 10000 \times \frac{5}{100} \right) m^3 = 1000 \, m^3$.

11. Length of water column flown in 1 min $= \frac{3 \times 1000}{60} = 50$ m.

$\therefore$ Volume $= (50 \times 2 \times 45) \, m^3 = 4500 \, m^3$.

12. Volume of metal $= (52 \times 40 \times 29 - 48 \times 36 \times 27) \, cm^3 = 13664 \, cm^3$.

$\therefore$ Weight of metal $= \left( 13664 \times 0.5 \times \frac{1}{1000} \right) kg = 6.832 \, kg$

13. $6a^2 = 1176 \Rightarrow a^2 = 196 \Rightarrow a = 14$.

$\therefore$ *Volume* $= a^3 = (14 \times 14 \times 14) \, cm^3 = 2744 \, cm^3$.

14. $\sqrt{a^2 + a^2 + a^2} = 14 \times \sqrt{3} \Rightarrow \sqrt{3}\, a = 14 \times \sqrt{3} \Rightarrow a = 14$.

$\therefore$ *Surface area* $= 6 \times a^2 = (6 \times 14 \times 14) \, cm^2 = 1176 \, cm^2$.

15. $a^3 = 125 \Rightarrow a = 5$. So, surface area $= 6a^2 = 6 \times 25 = 150 \, cm^2$.

16. Let original length of edge $= a$. Then, volume $= a^3$.

New edge $= 2a$; New volume $= (2a)^3 = 8a^3$.

$\therefore$ *The volume becomes 8 times the original volume.*

17. Length of longest rod $= \sqrt{4^2 + 4^2 + 4^2} = \sqrt{48} = 6.928 \, m$.

18. Volume of new cube $= (3^3 + 4^3 + 5^3) \, cm^3 = 216 \, cm^3$.

*Edge of this cube* $= (216)^{1/3} = (6 \times 6 \times 6)^{1/3} = 6 \, cm$.

19. Volume of new cube formed $= (27 \times 8 \times 1) \, cm^3 = 216 \, cm^3$.

Edge of this cube $= (216)^{1/3} = (6 \times 6 \times 6)^{1/3} = 6$ cm.

$\therefore$ Surface area of this cube $= 6a^2 = 6 \times 6 \times 6 = 216 \, cm^2$.

Surface area of given cuboid

$$= 2\,(27 \times 8 + 8 \times 1 + 27 \times 1)\ cm^2 = 502\ cm^2.$$

$\therefore$ *Difference between the surface areas* $= (502 - 216)\ cm^2 = 286\ cm^2.$

**21.** $6\,a^2 = 600 \Rightarrow 3\,a^2 = 300 \Rightarrow \sqrt{3\,a^2} = \sqrt{300} = 10\sqrt{3}.$

$\therefore$ *Diagonal* $= 10\sqrt{3}$ *cm.*

**22.** Let the edge of the cube $= a$. Then, surface area $= 6\,a^2$.

New edge $= 2a$. So, new surface area $= 6 \times (2a)^2 = 24\,a^2$.

*Increase* $= \left(\dfrac{18\,a^2}{6\,a^2} \times 100\right)\% = 300\%.$

**23.** Number of Cubes $= \dfrac{100 \times 100 \times 100}{10 \times 10 \times 10} = 1000.$

**24.** Whole surface $= 2\,\pi\,rh + 2\,\pi\,r^2$

$$= \left(2 \times \frac{22}{7} \times \frac{7}{2} \times 80 + 2 \times \frac{22}{7} \times \frac{49}{4}\right) = 1837\ cm^2.$$

**25.** Curved surface $= 2\,\pi\,rh = \left(2 \times \dfrac{22}{7} \times \dfrac{5}{2} \times 14\right) = 220\ m^2.$

Cost of white washing = Rs. $\left(220 \times \dfrac{1}{2}\right) = Rs.\ 110.$

**26.** $2 \times \dfrac{22}{7} \times r \times 14 = 264 \Rightarrow r = \dfrac{264}{88} = 3\ cm.$

**27.** Circumference $= 440$ cm $\Rightarrow 2\,\pi\,r = 440$ or $r = \left(\dfrac{440}{2 \times 22} \times 7\right) = 70$ cm.

Inner radius $= (70 - 4)$ cm $= 66$ cm.

*Volume of iron*

$$= \pi\,[(70)^2 - (66)^2] \times 63 = \left(\frac{22}{7} \times 136 \times 4 \times 63\right) cm^3 = 58752\ cm^3.$$

**28.** External radius = 2.5 cm; Internal radius = 1.5 cm.

$\therefore$ Volume of metal $= [\pi \times (2.5)^2 \times 100 - \pi \times (1.5)^2 \times 100]\ cm^3$

$$= \frac{22}{7} \times 100 \times [(2.5)^2 - (1.5)^2]\ cm^3 = \frac{8800}{7}\ cm^3.$$

*Weight of the metal* $= \left(\dfrac{8800}{7} \times 21 \times \dfrac{1}{1000}\right) kg. = 26.4\ kg.$

**29.** Volume of earth dug $= \left(\dfrac{22}{7} \times 7 \times 7 \times 8\right) = 1232\ m^3.$

Area of embankment $= \frac{22}{7} \times \{(28)^2 - 7^2\}\ m^2 = 2310\ m^2$.

*Height of embankment* $= \left(\frac{1232}{2310} \times 100\right) cm = 53.3\ cm.$

**30.** Let their heights be $h$ & $2h$ & radii $r$ and $R$.

Then, $\pi r^2 h = \pi R^2 (2h) \Rightarrow \frac{r^2}{R^2} = \frac{2}{1}$ or $\frac{r}{R} = \frac{\sqrt{2}}{1}$.

**31.** $\frac{\text{Total surface area}}{\text{Lateral surface area}} = \frac{2\pi rh + 2\pi r^2}{2\pi rh} = \frac{2\pi r(h+r)}{2\pi rh}$

$= \left(\frac{h+r}{h}\right) = \left(\frac{20+80}{20}\right) = \frac{5}{1}.$

**32.** Let the radii of the cylinders be $2r$ and $3r$ respectively and their heights be $5h$ and $3h$ respectively. Then,

ratio of their volumes $= \frac{\pi (2r)^2 \times 5h}{\pi (3r)^2 \times 3h} = \frac{20}{27}.$

**33.** Radius $= \frac{1}{2} h$ & height $= h$.

Whole surface $= 2\pi \times \left(\frac{1}{2} h\right) \times h + 2\pi \times \left(\frac{1}{2} h\right)^2 = \frac{3}{2} \pi h^2.$

**34.** Slant height $= \sqrt{h^2 + r^2} = \sqrt{(2\sqrt{3})^2 + 2^2} = 4\ cm.$

**35.** Curved surface $= \pi rl = \left(\frac{22}{7} \times 6 \times 28\right) cm^2 = 528\ cm^2.$

**36.** $\pi r^2 = 154 \Rightarrow r^2 = \left(\frac{154}{22} \times 7\right) \Rightarrow r = 7.$

$\therefore$ slant height $= \sqrt{h^2 + r^2} = \sqrt{(14)^2 + (7)^2} = 7\sqrt{5}$ cm.

*Curved surface* $= \left(\frac{22}{7} \times 7 \times 7\sqrt{5}\right) cm^2 = 154\sqrt{5}\ cm^2.$

**37.** Let the height be $h$ & radius be $r$.
New height $= 2h$.

$\therefore$ *Change in volume* $= \left[\frac{\frac{1}{3}\pi r^2 (2h) - \frac{1}{3}\pi r^2 h}{\frac{1}{3}\pi r^2 h} \times 100\right] \% = 100\%.$

**38.** $\frac{1}{3} \times \frac{22}{7} \times r^2 \times 24 = 1232 \Rightarrow r^2 = \left(\frac{1232 \times 3 \times 7}{22 \times 24}\right) \Rightarrow r = 7.$

Slant height $= \sqrt{(24)^2 + (7)^2} = 25$ cm.

$\therefore$ *Curved surface* $= \left(\frac{22}{7} \times 7 \times 25\right) cm^2 = 550\ cm^2.$

**39.** Let the diameters of the bases of the cones be $4r$ and $5r$. Let their heights be $h$ and $H$.

Then, $\dfrac{\frac{1}{3}\pi \times \left(\frac{4r}{2}\right)^2 \times h}{\frac{1}{3}\pi \times \left(\frac{5r}{2}\right)^2 \times H} = \frac{1}{4} \Rightarrow \frac{16}{25} \times \frac{h}{H} = \frac{1}{4}.$

$\therefore \frac{h}{H} = \left(\frac{1}{4} \times \frac{25}{16}\right) = \frac{25}{64}.$

**40.** $\dfrac{\text{Volume of the cylinder}}{\text{Volume of the cone}} = \dfrac{\pi r^2 h}{\frac{1}{3}\pi r^2 h} = \frac{3}{1}.$

**41.** Let $r$ be radius of each. Let $h$ & $H$ be the heights of the cylinder and cone respectively. Then,

$\pi r^2 h = \frac{1}{3}\pi r^2 H \Rightarrow \frac{h}{H} = \frac{1}{3}.$

**42.** Let $h$ be the height of the cone and $r$ be the radius of its base.

Radius of base of $A = \frac{1}{2} r$, Height of $A = \frac{1}{2} h$.

Volume of $A = \frac{1}{3}\pi \times \left(\frac{1}{2} r\right)^2 \times \left(\frac{1}{2} h\right) = \frac{1}{24}\pi r^2 h.$

Volume of $B = \left(\frac{1}{3}\pi r^2 h - \frac{1}{24}\pi r^2 h\right) = \frac{7}{24}\pi r^2 h.$

$\therefore$ *Vol. A < Vol. B.*

**43.** Volume $= \frac{\pi h}{3}[R^2 + r^2 + Rr]$, where $R = 4$ m, $r = 2$ m, $h = 6$ m.

$= \left\{\frac{22}{7} \times \frac{1}{3} \times 6 \times (16 + 4 + 8)\right\} m^3 = 176\ m^3.$

**44.** Let $r$ be the radius of the base of each.

Then, volume of the cone $= \frac{1}{3}\pi r^2 h.$

Total volume $= \pi r^2 h.$

$\therefore$ Volume of cylinder $= \left(\pi r^2 h - \frac{1}{3}\pi r^2 h\right) = \frac{2}{3}\pi r^2 h.$

Let $H$ be the height of the cylinder.

*Then,* $\pi r^2 H = \frac{2}{3}\pi r^2 h \Rightarrow H = \frac{2}{3}h.$

45. Total volume $= (10 \times 5 \times 2)\ cm^3 = 100\ cm^3.$

Volume carved $= \left(\frac{1}{3} \times \frac{22}{7} \times 3 \times 3 \times 7\right) cm^3 = 66\ cm^3.$

Wood wasted $= (100 - 66)\ cm^3 = 34\ cm^3.$

$\therefore$ *Percentage of wood wasted* $= \left(\frac{34}{100} \times 100\right)\% = 34\%.$

46. Volume of 1 cylinder $= \pi r^2 h.$

Volume of 1 cone $= \frac{1}{3}\pi r^2 h.$

*Number of cones* $= \frac{\pi r^2 h}{\frac{1}{3}\pi r^2 h} = 3.$

47. $\frac{1}{3}\pi \times (2)^2 \times h = \pi \times (2)^2 \times 6 \Rightarrow h = 18\ cm.$

48. $4\pi r^2 = 1386 \Rightarrow r^2 = \left(\frac{1386}{4 \times 22} \times 7\right) = \frac{441}{4} \Rightarrow r = \frac{21}{2}.$

$\therefore$ *Volume of sphere* $= \left(\frac{4}{3} \times \frac{22}{7} \times \frac{21}{2} \times \frac{21}{2} \times \frac{21}{2}\right) cm^3 = 4851\ cm^3.$

49. $\frac{2}{3} \times \frac{22}{7} \times r^3 = 19404 \Rightarrow r^3 = \frac{19404 \times 3 \times 7}{2 \times 22} = 9261.$

$\therefore r = (9261)^{1/3} = (21 \times 21 \times 21)^{1/3} = 21\ cm.$

50. $\frac{4}{3}\pi r^3 = \frac{48 \times (14)^3}{21} \Rightarrow r^3 = \left(\frac{88 \times 14 \times 14 \times 14}{21} \times \frac{3}{4} \times \frac{7}{22}\right).$

$\therefore r = 14$ cm.

*So, curved surface of the sphere* $= \left(4 \times \frac{22}{7} \times 14 \times 14\right) cm^2 = 2464\ cm^2.$

51. Capacity of the bowl $= \left[\frac{2}{3} \times \frac{22}{7} \times (21)^3\right] cm^3.$

52. Let the height of the rod be $H$. Then,

$6 \times \frac{4}{3} \pi r^3 = \pi r^2 H \Rightarrow H = 8r.$

**53.** Let the required height be *h* cm.

Then, $5 \times 5 \times h = \frac{4}{3} \pi \times (3)^3$

$\therefore h = \frac{36\pi}{25} = 1.44\pi.$

**54.** Let the radii be *r* and 2*r*.

*Ratio of surface areas* $= \frac{4\pi r^2}{4\pi (2r)^2} = \frac{1}{4}.$

**55.** Volume of 1 sphere $= \frac{4}{3} \pi \times (3)^3 = 36\pi.$

Volume of 1 cylinder $= \pi \times (2)^2 \times 45 = 180\pi.$

$\therefore$ *Number of spheres* $= \frac{180\pi}{36\pi} = 5.$

**56.** Let *r* and *R* be the radii of the spheres. Then,

$$\frac{4\pi r^2}{4\pi R^2} = \frac{1}{4} \Rightarrow \left(\frac{r}{R}\right)^2 = \left(\frac{1}{2}\right)^2 \Rightarrow \frac{r}{R} = \frac{1}{2} \Rightarrow \frac{r^3}{R^3} = \frac{1}{8}.$$

$\therefore$ *Ratio of volumes* $= \frac{\frac{4}{3}\pi r^3}{\frac{4}{3}\pi R^3} = \frac{r^3}{R^3} = \frac{1}{8}.$

**57.** Let *r* be the radius of each moulded sphere.

Then, $8 \times \frac{4}{3} \pi r^3 = \frac{4}{3} \pi \times (10)^3 \Rightarrow (2r)^3 = (10)^3 \Rightarrow 2r = 10.$

$\therefore r = 5$ cm.

*So, the surface area of each ball* $= [4\pi \times (25)]\ cm^2 = (100\pi)\ cm^2.$

**58.** Let the radius of the third ball be *r* cm.

Then, $\frac{4}{3} \pi r^3 = \frac{4}{3} \pi \times \left[\left(\frac{3}{2}\right)^3 - \left\{\left(\frac{3}{4}\right)^3 + (1)^3\right\}\right]$

$\therefore r^3 = \frac{125}{64}$ or $r = \frac{5}{4}.$

*So, the diameter of the third ball* $= \frac{5}{2}$ *cm* = *2.5 cm.*

**59.** Number of spheres $= \dfrac{\frac{4}{3} \times \pi \times 4 \times 4 \times 4}{\frac{4}{3} \times \pi \times \frac{1}{4} \times \frac{1}{4} \times \frac{1}{4}} = 4096.$

**60.** Let $r$ be the radius of the sphere.

*Then,* $\frac{4}{3}\pi r^3 = 4\pi r^2 \Rightarrow r = 3$ *units.*

**61.** Let the length of the wire be $x$ cm.

Then, $\frac{4}{3} \times \pi \times 7 \times 7 \times 7 = \pi \times 0.1 \times 0.1 \times x$

$\therefore x = \frac{4}{3} \times \frac{7 \times 7 \times 7}{0.1 \times 0.1}\ cm = \frac{457333}{100}\ m = 457.33\ m.$

**62.** Volume of cube = $(22 \times 22 \times 22)\ cm^3$.

Volume of 1 bullet = $\left(\frac{4}{3} \times \frac{22}{7} \times 1 \times 1 \times 1\right) cm^3$.

$\therefore$ *Number of bullets* $= \left(\dfrac{22 \times 22 \times 22 \times 3 \times 7}{4 \times 22}\right) = 2541.$

**63.** Let $H$ and $h$ be the heights of water level before and after dropping the sphere into it.

Then, $[\pi \times (30)^2 \times H] - [\pi \times (30)^2 \times h] = \frac{4}{3}\pi \times (30)^3$

*or* $\pi \times 900 \times (H - h) = \frac{4}{3}\pi \times 27000$ or $(H - h) = 40$ *cm.*

**64.** Let $h$ be the height of the cone and $r$ be the radius of the base of each of the cone & cylinder.

Then, $\pi r^2 \times 5 = \frac{1}{3}\pi r^2 h \Rightarrow h = 15$ *cm.*

**65.** Let their radii be $2x$ & $3x$ respectively and their heights be $5h$ & $3h$ respectively. Then, ratio of volumes $= \dfrac{\pi \times (2x)^2 \times 5h}{\pi \times (3x)^2 \times 3h} = \dfrac{20}{27}.$

**67.** Let the edges be $a$ & $b$ respectively.

*Then,* $\frac{a^3}{b^3} = \frac{8}{1} \Rightarrow \left(\frac{a}{b}\right)^3 = (2)^3 \Rightarrow \frac{a}{b} = \frac{2}{1}.$

**68.** Let radius = $R$ and length = $h$, volume = $\pi R^2 h$.

New radius $= \frac{1}{3} R$. Let new length $= H$.

Volume $= \pi \times \left(\frac{1}{3} R\right)^2 \times H = \frac{\pi R^2 H}{9}$.

$\therefore \pi R^2 h = \frac{\pi R^2 H}{9}$ or $H = 9h$.

## ANSWERS (Exercise 21B)

| | | | | | |
|---|---|---|---|---|---|
| **1.** (*c*) | **2.** (*d*) | **3.** (*c*) | **4.** (*c*) | **5.** (*c*) | **6.** (*d*) |
| **7.** (*c*) | **8.** (*d*) | **9.** (*c*) | **10.** (*b*) | **11.** (*a*) | **12.** (*d*) |
| **13.** (*c*) | **14.** (*b*) | **15.** (*c*) | **16.** (*d*) | **17.** (*c*) | **18.** (*d*) |
| **19.** (*c*) | **20.** (*c*) | **21.** (*b*) | **22.** (*d*) | **23.** (*b*) | **24.** (*a*) |
| **25.** (*d*) | **26.** (*a*) | **27.** (*b*) | **28.** (*c*) | **29.** (*c*) | **30.** (*b*) |
| **31.** (*d*) | **32.** (*b*) | **33.** (*c*) | **34.** (*b*) | **35.** (*d*) | **36.** (*a*) |
| **37.** (*a*) | **38.** (*c*) | **39.** (*d*) | **40.** (*b*) | **41.** (*d*) | **42.** (*a*) |
| **43.** (*a*) | **44.** (*d*) | **45.** (*a*) | **46.** (*b*) | **47.** (*a*) | **48.** (*c*) |
| **49.** (*b*) | **50.** (*c*) | **51.** (*a*) | **52.** (*c*) | **53.** (*d*) | **54.** (*b*) |
| **55.** (*c*) | **56.** (*c*) | **57.** (*a*) | **58.** (*b*) | **59.** (*a*) | **60.** (*c*) |
| **61.** (*b*) | **62.** (*d*) | **63.** (*c*) | **64.** (*b*) | **65.** (*c*) | **67.** (*c*) |
| **68.** (*d*). | | | | | |

# 22

# RACES & GAMES OF SKILL

**Races :** *A contest of speed in running, riding, driving, sailing or rowing is called a* **race.** *The ground or path on which contests are made is called a* **race course.** *The point from which a race begins is known as a* **starting point.** *The point set to bound a race is called a* **winning post or a goal.** *The person who first reaches the winning post is called a* **winner.** *If all the persons contesting a race reach the goal exactly at the same time, then the race is said to be a* **dead–heat race.**

Suppose *A* and *B* are two contestants in a race. If, before the start of the race, *A* is at the starting point and *B* is ahead of *A* by 15 metres. Then *A* is said to give *B*, a start of 15 metres. To cover a race of 200 metres in this case, *A* will have to cover a distance of 200 metres and *B* will have to cover (200–15) or 185 metres only.

In a 100 metres race, '*A* can give *B* 15 metres' or '*A* can give *B*, a start of 15 metres' or '*A* beats *B* by 15 metres', means that while *A* runs 100 metres, *B* runs (100–15) or 85 metres.

**Games.** '*A game of 100* means that the person among the contestants who scores 100 points first is the winner.' If *A* scores 100 points, while *B* scores only 80 points, then we say that *A* can give *B* 20 points.

**Solved Examples**

**Ex. 1.** ***In a km. race A beats B by 35 metres or 7 seconds. Find A's time over the course.***

**Sol.** Here *B* runs 35 metres in 7 seconds.

$\therefore$ *B* s time over the course $= \left(\frac{7}{35} \times 1000\right)$ sec. = 200 seconds.

*So A's time over the course* $= (200 - 7)$ *sec* = *193 seconds*

= 3 *minutes 13 seconds.*

**Ex. 2.** ***A runs $1\frac{2}{3}$ times as fast as B. If A gives B a start of 80 metres how far must the winning post be so that A and B might reach it at the same time ?***

**Sol.** The rates of *A* and *B* are as 5 : 3

i.e., in a race of 5 metres, *A* gains 2 metres over *B*.

2 metres are gained by *A* in a race of 5 metres.

80 metres will be gained by *A* in a race of

$\left(\frac{5}{2} \times 80\right)$ metres = 200 metres.

∴ *Winning post is 200 metres away from the starting point.*

**Ex. 3.** ***A, B and C are the three contestants in a km. race. If A can give B a start of 40 metres and A can give C a start of 64 metres., how many metres start can B give C ?***

**Sol.** While *A* covers 1000 metres, *B* covers (1000 – 40) or 960 metres and *C* covers (1000 – 64) or 936 metres.

Now when *B* covers 960 metres, *C* covers 936 metres.

∴ When *B* covers 1000 metres, *C* covers

$$\left(\frac{936}{960} \times 1000\right) \text{ metres} = 975 \text{ metres.}$$

*So, B gives C a start of (1000 – 975) or 25 metres.*

**Ex. 4.** ***A can run a km. in 3 min. 10 sec. and B in 3 min. 20 sec. By what distance can A beat B ?***

**Sol.** *A* beats *B* by 10 seconds.

Distance covered by *B* in 200 sec. = 1000 metres.

Distance covered by *B* in 10 sec. = $\frac{1000}{200} \times 10 = 50$ metres.

∴ *A beats B by 50 metres.*

**Ex. 5.** ***In a 100 metres race, A runs at 6 km. per hour. If A gives B a start of 4 metres and still beats him by 12 seconds, what is the speed of B?***

**Sol.** Time taken by *A* to cover 100 metres.

$$= \left(\frac{60 \times 60}{6000} \times 100\right) \text{ sec.} = 60 \text{ sec.}$$

∴ *B* covers (100 – 4) or 96 metres in (60 + 12) or 72 sec.

*Hence, speed of B* = $\left(\frac{96 \times 60 \times 60}{72 \times 1000}\right)$ *km./hr. = 4.8 km./hr.*

**Ex. 6.** ***A can run a kilometre in 4 minutes 50 seconds and B in 5 minutes. How many metre's start can A give B in a km. race so that the race may end in a dead heat ?***

**Sol.** Time taken by *A* to run 1 km. = 4 mts. 50 sec. = 290 sec.

Time taken by *B* to run 1 km. = 5 mts. = 300 sec.

∴ *A* can give *B*, a start of (300 – 290) or 10 sec.

Now, in 300 Seconds, *B* runs 1000 metres.

∴ In 10 seconds, *B* runs $\left(\frac{1000}{300} \times 10\right)$ metres = $33\frac{1}{3}$ metres.

*So, A can give B a start of* $33\frac{1}{3}$ *metres.*

**Ex. 7.** ***In a kilometre race, if A gives B, a start of 40 metres, then A wins***

***by 19 seconds, but if A gives B, a start of 30 seconds, then B wins by 40 metres. Find the time taken by each to run a kilometre.***

**Sol.** Suppose that the time taken by $A$ and $B$ to run 1 km is $x$ and $y$ seconds respectively.

When $A$ gives $B$, a start of 40 metres, then $A$ has to run 1000 metres, while $B$ has to run only 960 metres.

Time taken by $A$ to run 1000 m = $x$ sec.

$$\text{Time taken by } B \text{ to run 960 metres} = \left(\frac{y}{1000} \times 960\right) \text{ sec.}$$

$$= \left(\frac{24}{25} y\right) \text{ sec.}$$

$$\text{Clearly, } \frac{24}{25} y - x = 19 \text{ or } 24y - 25x = 475 \qquad \text{...(i)}$$

Again, $A$ gives $B$, a start of 30 seconds, then $B$ runs for $y$ seconds, while $A$ runs for $(y - 30)$ seconds.

Now, in $x$ seconds, $A$ covers 1000 metres

$$\therefore \text{ In } (y-30) \text{ sec., } A \text{ will cover } \left[\frac{1000}{x} \times (y-30)\right] \text{ metres.}$$

$$\text{So, } 1000 - \frac{1000 \times (y-30)}{x} = 40 \text{ or } 25y - 24x = 750 \qquad \text{...(ii)}$$

Solving (i) and (ii) we get, $x = 125$ & $y = 150$.

∴ *Time taken by A to run 1 km.* = *125* sec.

*Time taken by B to run 1 km.* = 150 sec.

**Ex. 8.** ***A and B run a km. and A wins by 1 minute. A and C run a km and A wins by 375 metres. B and C run a km and B wins by 30 seconds. Find the time taken by each to run a km.***

**Sol.** Since $A$ beats $B$ by 60 seconds and $B$ beats $C$ by 30 seconds, so $A$ beats $C$ by 90 seconds. But, it being given that $A$ beats $C$ by 375 metres. So, it means that $C$ covers 375 metres in 90 seconds.

$$\therefore \textit{ Time taken by C to cover 1 km.} = \left(\frac{90}{375} \times 1000\right) \text{ sec.}$$

$$= 240 \text{ seconds.}$$

*Time taken by A to cover 1 km. = (240 – 90) sec. = 150 seconds.*

*Time taken by B to cover 1 km. = (240 – 30) sec. = 210 seconds.*

**Ex. 9.** ***A and B run a race. A has a start of 50 metres and A sets off 6 minutes before B, at the rate of 10 km an hour. How soon will B overtake A, if his rate of running is 12 km an hour.***

$$\textbf{Sol. } \text{Distance run by } A \text{ in 6 minutes} = \left(\frac{10}{60} \times 6\right) \text{ km} = 1 \text{ km}$$

Thus, $A$ has a start of (1000 + 50) or 1050 metres.

So, in order to overtake $A$, $B$ has to gain 1050 metres.

Now, 2000 metres is gained by $B$ in 60 minutes.

So, 1050 metres will be gained by $B$ in

$$\left(\frac{60}{2000} \times 1050\right) \text{ mts.} = 31 \text{ min. } 30 \text{ sec.}$$

**Ex. 10.** ***In a race of 600 metres, A can beat B by 60 metres and in a race of 500 metres, B can beat C by 50 metres. By how many metres will A beat C in a race of 400 metres ?***

**Sol.** Clearly, if $A$ runs 600 metres, $B$ runs = 540 metres.

$\therefore$ If $A$ runs 400 metres, $B$ runs $= \left(\frac{540}{600} \times 400\right) = 360$ metres.

Again, when $B$ runs 500 metres, $C$ runs = 450 metres.

$\therefore$ When $B$ runs 360 metres, $C$ runs $= \left(\frac{450}{500} \times 360\right)$ metres.

$= 324$ metres.

*So, A beats C by (400 – 324) or 76 metres.*

**Ex. 11.** ***A can give, B 20 metres and C 25 metres in a 100 metres race, while B can give C one second over the course. How long does each take to run 100 metres ?***

**Sol.** It is clear that, when $A$ runs 100 metres, then $B$ runs 80 metres and $C$ runs 75 metres.

$\therefore$ When $B$ runs 100 metres, $C$ runs $= \left(\frac{75}{80} \times 100\right)$

$= \left(\frac{375}{4}\right)$ metres.

Thus, $B$ beats $C$ by $\left(100 - \frac{375}{4}\right) = \frac{25}{4}$ metres.

But, it being given that $B$ beats $C$ by 1 second.

$\therefore$ $\frac{25}{4}$ metres are covered by $C$ in 1 second.

So, 100 metres are covered by $B$ in $\left(\frac{4}{25} \times 100\right) = 16$ seconds.

$\therefore$ 100 metres are covered by $B$ in (16 – 1) or 15 seconds.

Now 80 metres are covered by $B$ in $\left(\frac{15}{100} \times 80\right)$ or 12 seconds.

$\therefore$ *100 metres are covered by A in 12 seconds.*

**Games of Skill.**

**Ex. 12.** ***At a game of billiards, A can give B 10 points in 60 and he can give C 15 in 60. How many can B give C in a game of 90 ?***

**Sol.** If $A$ scores 60 points, then $B$ scores 50 points.

If $A$ scores 60 points, then $C$ scores 45 Points

Now, when $B$ scores 50 points, $C$ scores 45 points.

$\therefore$ When $B$ scores 90 points, $C$ scores $\left(\frac{45}{50} \times 90\right) = 81$ points.

*Hence, B can give C, 9 points in a game of 90.*

**Ex. 13.** ***In a game of billiards, A can give B 12 points in 60 and A can give C 10 in 90. How many can C give B in a game of 70 ?***

**Sol.** If $A$ scores 60 points, then $B$ scores 48 points.

Also, if $A$ scores 90 points, then $C$ scores 80 points.

$\therefore$ If $A$ scores 60 points, then $C$ scores $\left(\frac{80}{90} \times 60\right) = \left(\frac{160}{3}\right)$ points.

Now, when $C$ scores $\frac{160}{3}$ points, $B$ scores 48 points

$\therefore$ *When C scores 70 points, B scores* $\left(\frac{48 \times 3}{160} \times 70\right) = 63$ *points.*

**Ex. 14.** ***In a game, A can give B 20 points, A can give C 32 points and B can give C 15 points. How many points make the game ?***

**Sol.** Suppose $x$ points make the game.

Clearly, when $A$ scores $x$ points, $B$ scores $(x - 20)$ points and $C$ scores $(x - 32)$ points.

Now, when $B$ scores $x$ points, $C$ scores $(x - 15)$ points.

When $B$ scores $(x - 20)$ points, $C$ scores $\left[\frac{(x-15)}{x} \times (x-20)\right]$ points.

$\therefore \frac{(x-15)(x-20)}{x} = (x - 32)$ or $x = 100$.

Hence, 100 points make the game.

***Running Round A Circle.***

**Ex. 15.** ***Two men A and B run a 4 km race on a course 250 metres round. If their rates be 5 : 4, how often does the winner pass the other ?***

**Sol.** Clearly, when $A$ makes 5 rounds, then $B$ makes 4 rounds

Distance covered by $A$ in 5 rounds $\left(\frac{5 \times 250}{1000}\right)$ km. $= \left(\frac{5}{4}\right)$ km.

Distance covered by $B$ in 4 rounds $= \left(\frac{4 \times 250}{1000}\right)$ km. $= 1$ km.

It is clear that $A$ passes $B$ each time, when $A$ makes 5 rounds.

In other words after covering $\frac{5}{4}$ km. each time, $A$ passes $B$.

$\therefore$ In covering $\frac{5}{4}$ km., $A$ passes $B$ 1 time.

So, in covering 4 km. $A$ passes $B$

$\left(\frac{1 \times 4}{5} \times 4\right) = 3\frac{1}{5}$ *times. i.e., 3 times.*

**Ex. 16.** ***Two men, A and B, walk round a circle 1200 metres in circumference. A walks at the rate of 150 metres and B at the rate of 80 metres per minute. If they both start at the same time from the same point and walk in the same direction, when will they be together again for the first time and when will they be together again at the starting point for the first time ?***

**Sol.** A and B will be together again for the first time when A has gained one complete round on B.

Now, A gains 70 metres on B in 1 min.

$\therefore$ A gains 1200 metres on B in $\left(\frac{1}{70} \times 1200\right) = 17\frac{1}{7}$ min.

Thus, A and B will be together for the first time in $17\frac{1}{7}$ min.

Again, time taken by A to make 1 round $= \left(\frac{1200}{150}\right)$ min. = 8 min.

*Time taken by B to make 1 round* $= \left(\frac{1200}{80}\right)$ min = 15 min.

*Thus A will be at the starting point after each interval of 8 min. and B after each interval of 15 minutes. So, they will be together at the starting point after 120 minutes. (L.C.M. of 8 & 15). Thus, A and B will be together at the starting point for the first time after 2 hours.*

**Ex. 17.** ***Three men A, B and C walk round a circle 1760 metres in circumference at the rates of 160 metres, 120 metres and 105 metres per minute respectively. If they all start together and walk in the same direction, when will they be first together again ?***

**Sol.** To gain 40 metres over B, A takes 1 min.

To gain 1760 metres over B, A takes $\left(\frac{1}{40} \times 1760\right) = 44$ min.

Again, to gain 55 metres over C, A takes 1 min.

$\therefore$ To gain 1760 metres over C, A takes $\left(\frac{1}{55} \times 1760\right)$ min. = 32 min.

Thus, A and B will be together after 44 minutes, while A and C are together again after 32 minutes.

Now, L.C.M. of 44 & 32 is 352.

So, A, B & C are first together again after 352 minutes or 5 hrs. 32 minutes.

## EXERCISE 22A (Subjective).

1. In a km race A beats B by 40 metres or 8 seconds. What is A's time over the course ?

**Hint.** *B takes 8 seconds to run 40 metres.*

**2.** *A* can beat *B* by 25 metres in a 250 metres race and *B* can beat *C* by 20 metres in a 500 metres race. By how much can *A* beat *C* in a km race ?

**3.** *A* can run 440 metres in 51 seconds and *B* in 55 seconds. By how many seconds will *B* win if he has 40 metres start ?

**Hint.** *Time taken by B to cover 400 metres* $= \left(\frac{55}{440} \times 400\right) = 50$ *sec.*

**4.** In a race of 800 metres, if *A* gives *B* a start of 5 metres, *A* beats *B* by 6 seconds. If *A* gives *B* a start of 10 seconds, *B* wins by $26\frac{2}{3}$ metres. Find the time taken by each to run the race.

**5.** Two men *A* and *B* ran 100 metres in 11.25 sec. and 12.5 sec. respectively. If they started together, find how far was *B* from the finishing line, when *A* completed his 100 metres.

**6.** *A* can walk 3 km while *B* walks 5 km. Also, *C* can walk 6 km., while *A* walks 3.5 km. What start can *C* give *B* in a 3 km. walk ?

**Hint.** *When A walks 3 km., then C walks* $= \left(\frac{6}{3.5} \times 3\right)$ *km.* $= \frac{36}{7}$ *km.*

*Now, when C walks* $\frac{36}{7}$ *km., then B walks 5 km.*

*When C walks 3 km., then B walks* $\left(\frac{5 \times 7}{36} \times 3\right)$ *km* $= \frac{35}{12}$ *km.*

$\therefore$ *C can give B a start of* $\left(3 - \frac{35}{12}\right)$ *km.*

**7.** *A* can run a km in half minute less time than *B*. In a km race *B* gets a start of 100 metres and loses by 25 metres. Find the time *A* and *B* take to run a km.

**8.** In a 500 metre's race *B* gives *A*, a start of 160 metres. The ratio of the speeds of *A* and *B* is 2 : 3. Who wins, and by how much ?

**9.** *A* can beat *B* by 15 metres when he gives *B* a start of 5 metres in a race of 800 metres. If *B* gives *C* a start of 16 metres in a race of 500 metres, *C* wins by 4 metres. Who will win and by how much, if *A* gives *C* a start of 61 metres on a race of 600 metres ?

**10.** At a game of billiards *A* can give *B* 6 points in 60 and he can give *C* 13 points in 65. How many points can *B* give *C* in a game of 55 ?

**11.** *A* can give *B* 15 points. *A* can give *C* 22 points, and *B* can give *C* 10 points. How many points make the game ?

**12.** In a game of 100 points *A* can give *B* 20 points and *C* 28 points. How many points can *B* give *C* in a game of 100 points ?

**13.** In running a 2 km race on a course $\frac{1}{4}$ km. round, *A* overlaps *B* at the middle of 6th round. By what distance will $A$ win ?

**Hint.** *When A runs $5\frac{1}{2}$ rounds, then B runs $4\frac{1}{2}$ rounds.*

*i.e., (A's speed : B's speed)* $= \frac{11}{2} : \frac{9}{2}$ or $11 : 9$

*Thus when A runs 11 km., B runs 9 km.*

*So, when A runs 2 km; B runs* $= \left(\frac{9}{11} \times 2\right)$ *km.*

**14.** Starting from the same point, two persons, $A$ and $B$ run round a circular track, half a km. in length, at 3 km. and 2.5 km. per hour respectively. Find, when they will meet again if :–

(i) they run in the same direction.

(ii) they run in opposite directions.

**Hint.** *While running in opposite directions, they will meet when total distance travelled by them is $\frac{1}{2}$ km. Clearly, in 1 hr. they are (3 + 2.5) or 5.5 km. apart.*

## ANSWERS (Exercise 22A)

**1.** 3 min. 12 sec. **2.** 136 metres **3.** 1 second

**4.** 100 sec., $106\frac{2}{3}$ sec. **5.** 10 metres **6.** $83\frac{1}{3}$ metres

**7.** $3\frac{1}{2}$ min., $4\frac{1}{2}$ min. **8.** $B$ wins by $6\frac{2}{3}$ metres

**9.** $A$ wins by 60 metres **10.** 5 points **11.** 50

**12.** 10 points **13.** $\frac{4}{11}$ km. **14.** (i) 1 hour (ii) $5\frac{5}{11}$ minutes.

## EXERCISE 22B (OBJECTIVE TYPE QUESTIONS)

**1.** A can run 100 metres in 27 seconds and $B$ in 30 seconds. A will beat $B$ by

(*a*) 9 metres (*b*) 10 metres

(*c*) $11\frac{1}{8}$ metres (*d*) 12 metres

**2.** A can run a kilometer in 4 min. 54 sec. and $B$ in 5 min. How many metres start can $A$ give $B$ in a km. race so that the race may end in a dead heat ?

(*a*) 20 metres (*b*) 16 metres

(*c*) 18 metres (*d*) 14.5 metres

**3.** In a 300 metres race $A$ beats $B$ by 15 metres or 5 seconds. $A$'s time over the course is

(*a*) 100 seconds (*b*) 95 seconds

(*c*) 105 seconds (*d*) 90 seconds

4. In a 100 metres race, *A* can beat *B* by 25 metres and *B* can beat *C* by 4 metres. In the same race, *A* can beat *C* by :

(*a*) 29 metres (*b*) 21 metres
(*c*) 28 metres (*d*) 26 metres

5. In a 100 metres race *A* can give *B* 10 metres and *C* 28 metres. In the same race, *B* can give *C* :

(*a*) 18 metres (*b*) 20 metres
(*c*) 27 metres (*d*) 9 metres

6. *A* can run 20 metres while *B* runs 25 metres. In a km. race *B* beats *A* by

(*a*) 250 metres (*b*) 225 metres
(*c*) 200 metres (*d*) 125 metres

7. *A* runs $1\frac{3}{4}$ times as fast as *B*. If *A* gives *B* a start of 60 metres, how far must the winning post be in order that *A* and *B* reach it at the same time ?

(*a*) 105 metres (*b*) 80 metres
(*c*) 140 metres (*d*) 45 metres

8. In a 500 metres race, the ratio of speeds of two contestants *A* and *B* is 3 : 4. *A* has a start of 140 metres. Then, *A* wins by :

(*a*) 60 metres (*b*) 40 metres
(*c*) 20 metres (*d*) 10 metres

9. In a 100 metres race, *A* beats *B* by 10 metres and *C* by 1 metre. In a race of 180 metres, *B* will beat *C* by :

(*a*) 5.4 metres (*b*) 4.5 metres
(*c*) 5 metres (*d*) 6 metres

10. *A* can beat *B* by 31 metres and *C* by 18 metres in a race of 200 metres. In a race of 350 metres *C* will beat *B* by :

(*a*) 22.75 metres (*b*) 25 metres
(*c*) $7\frac{4}{7}$ metres (*d*) 19.5 metres

11. *A* and *B* take part in a 100 metres race. *A* runs at 5 km. per hour. *A* gives *B* a start of 8 metres and still beats him by 8 seconds. Speed of *B* is :

(*a*) 5.15 km./hr. (*b*) 4.14 km./hr.
(*c*) 4.25 km./hr. (*d*) 4.4 km./hr.

12. In a game of 100 points, *A* can give *B* 20 points and *C* 28 points. Then, *B* can give *C* :

(*a*) 8 points (*b*) 10 points
(*c*) 14 points (*d*) 40 points

13. At a game of billiards, *A* can give *B* 15 points in 60 and *A* can give *C* 20 in 60. How many can *B* give *C* in a game of 90 ?

(*a*) 30 points (*b*) 20 points
(*c*) 10 points (*d*) 12 points

## SOLUTIONS (Exercise 22B)

1. Distance covered by $B$ in 3 seconds

$= \left(\frac{100}{30} \times 3\right)$ metres = 10 metres.

$\therefore$ *A beats B by 10 metres.*

2. Distance covered by $B$ in 6 seconds

$= \left(\frac{1000}{300} \times 6\right)$ metres = 20 metres.

Thus, $A$ beats $B$ by 20 metres.

*So, for a dead heat race, A must give B a start of 20 metres.*

3. 15 metres are covered by $B$ in 5 seconds

300 metres are covered by $B$ in $\left(\frac{5}{15} \times 300\right) = 100$ seconds.

$\therefore$ *Time taken by A = (100 – 5) = 95 seconds.*

4. $A : B = 100 : 75$ and $B : C = 100 : 96$.

$\therefore A : C = \frac{A}{B} \times \frac{B}{C} = \frac{100}{75} \times \frac{100}{96} = \frac{100}{72} = 100 : 72$.

*So, A beats C by (100 –72)= 28 metres.*

5. $A : B : C = 100 : 90 : 72$

$\therefore B : C = \frac{90}{72} = \frac{\left(90 \times \frac{100}{90}\right)}{\left(72 \times \frac{100}{90}\right)} = \frac{100}{80} = (100 : 80)$

*So, B can give C 20 metres.*

6. In a 25 metres race, $B$ beats $A$ by 5 metres

*In a km. race B beats A by* $\left(\frac{5}{25} \times 1000\right) = 200$ *metres.*

7. Ratio of rates of $A$ and $B = 7 : 4$

i.e., 3 metres are gained by $A$ in a race of 7 metres

$\therefore$ *60 metres are gained by A in a race of* $\left(\frac{7}{3} \times 60\right) = 140$ *metres.*

8. To reach the winning post $A$ will have to cover a distance of (500 – 140) i.e., 360 metres

While $A$ covers 3 metres, $B$ covers 4 metres

While, $A$ covers 360 metres, $B$ covers $\left(\frac{4}{3} \times 360\right) = 480$ metres.

So, $A$ reaches the winning post while $B$ remains 20 metres behind.

$\therefore$ *A wins by 20 metres.*

9. $A : B : C = 100 : 90 : 87$

$\therefore \frac{B}{C} = \frac{90}{87} = \frac{90 \times 2}{87 \times 2} = \frac{180}{174}$.

So, while $B$ covers 180 metres, $C$ covers = 174 metres.

$\therefore$ *B beats C by 6 metres.*

10. $A : B : C = 200 : 169 : 182$.

$\therefore \frac{C}{B} = \frac{182}{169} = \frac{182 \times \left(\frac{350}{182}\right)}{169 \times \left(\frac{350}{182}\right)} = \frac{350}{325}$.

So, while $C$ covers 350 metres, $B$ covers 325 metres.

$\therefore$ *C beats B by 25 metres.*

11. $A$'s speed = $\left(5 \times \frac{5}{18}\right)$ metres/sec. = $\frac{25}{18}$ metres/sec.

Time taken by $A$ to cover 100 metres

$= \left(100 \times \frac{18}{25}\right)$ sec. = 72 sec.

$\therefore$ $B$ covers 92 metres in (72 + 8) or 80 sec.

*B's speed* = $\left(\frac{92}{80} \times \frac{18}{5}\right)$ *km./hr. = 4.14 km./hr.*

12. $A : B : C = 100 : 80 : 72$

$\therefore B : C = \frac{80}{72} = \frac{10}{9} = \frac{100}{90}$.

Thus, if $B$ scores 100, $C$ scores 90

$\therefore$ *B can give C 10 points.*

13. $A : B : C = 60 : 45 : 40$

$\therefore B : C = \frac{45}{40} = \frac{9}{8} = \frac{9 \times 10}{8 \times 10} = \frac{90}{80}$.

So, if $B$ scores 90, then $C$ scores 80.

$\therefore$ *B can give C 10 points in a game of 90.*

## ANSWERS (Exercise 22B)

| | | | | | |
|---|---|---|---|---|---|
| **1.** (*b*) | **2.** (*a*) | **3.** (*b*) | **4.** (*c*) | **5.** (*b*) | **6.** (*c*) |
| **7.** (*c*) | **8.** (*c*) | **9.** (*d*) | **10.** (*b*) | **11.** (*b*) | **12.** (*b*) |
| **13.** (*c*) | | | | | |

# 23

# CALENDAR

Under this heading we mainly deal with finding the day of the week on a particular given date. The process of finding it lies in obtaining the number of odd days.

*The number of days more than the complete number of weeks in a given period, are called* **odd days.**

**Leap & Ordinary Year.** *Every year which is divisible by 4 such as 1992, is called a leap year. Every 4th century is a leap year but no other century is a leap year viz.* 400, 800, 1200, 1600 are all leap years, but none of 700, 900, 1100 etc. is a leap year.

An ordinary year has 365 days *i.e.* (52 weeks + 1 day).

A leap year has 366 days *i.e.* (52 weeks + 2 days).

*An ordinary year has 1 odd day and a leap year has 2 odd days.*

A century *i.e.* 100 years has 76 ordinary years and 24 leap years.

$\therefore$ 100 years = 76 ordinary years + 24 leap years

= [(76 × 52) weeks + 76 days]

+ (24 × 52) weeks + 48 days]

= (5217 weeks + 5 days) = 5 odd days.

*i.e.*, 100 years contain 5 odd days.

200 years contain 10 and therefore 3 odd days.

300 years contain 15 and therefore 1 odd day.

400 years contain (20 + 1) and therefore 0 odd day.

Similary, the years 800, 1200, 1600 etc. contain no odd day.

We count days according to number of odd days.

Sunday for 0 odd day, Monday for 1 odd day and so on.

## Solved Problems.

**Ex. 1.** ***Find the day of the week on***

(i) 16th July, 1776 .

(ii) 12th January, 1979.

**Sol.** (i) 16th July, 1776 means

(1775 years + 6 months + 16 days)

Now, 1600 years have 0 odd days.

100 years have 5 odd days.

75 years contain 18 leap years & 57 ordinary years and therefore

(36 + 57) or 93 or 2 odd days.

∴ 1775 years give 0 + 5 + 2 = 7 and so 0 odd day.

Also number of days from 1st Jan., 1776 to 16th July, 1776

| Jan. | Feb. | March | April | May | June | July |
|---|---|---|---|---|---|---|
| 31 + | 29 + | 31 + | 30 + | 31 + | 30 + | 16 |

= 198 days = 28 weeks + 2 days = 2 odd day.

∴ Total number of odd days = 0 + 2 = 2.

*Hence the day on 16th July, 1776 was 'Tuesday'.*

(ii) 12th January, 1979 means, '(1978 years + 12 days)'

Now 1600 years have 0 odd days

300 years have 15 or 1 odd day

78 years have

{19 leap years + 59 ordinary years
= (38 + 59) or 97 odd days or 6 odd days

12 days of January has 5 odd days

Total number of odd days : 0 + 1 + 6 + 5 = 12 *or* 5 *odd days.*

*So, the day was 'Friday'.*

**Ex. 2.** ***On what dates of August 1980 did Monday fall ?***

**Sol.** First find the day on 1st August, 1980.

1st August, 1980 means, '(1979 years + 7 months + 1 day)'.

Now 1600 years contain 0 odd day.

300 years contain 15 or 1 odd day.

{19 leap years + 60 ordinary years
= 38 + 60 or 98 or 0 odd day}

Thus 1979 years contain 0 + 1 + 0 = 1 *odd day.*

Number of days from Jan., 1980 upto 1st Aug., 1980.

| Jan. | Feb. | March | April | May | June | July | Aug. |
|---|---|---|---|---|---|---|---|
| 31 + | 29 + | 31 + | 30 + | 31 + | 30 + | 31 + | 1 |

= 214 days = 30 weeks + 4 days = 4 odd days.

Total number of odd days = 1 + 4 = 5.

So, on 1st Aug., 1980, it was 'Friday'.

So, 1st Monday in August, 1980 lies on 4th August.

∴ *Monday falls on 4th, 11th, 18th, & 25th in August, 1980.*

**Ex. 3.** ***Prove that the calendar for 1990 will serve for 2001 also.***

**Sol.** In order that the calendar for 1990 and 2001 be the same, 1st January of both the years must be on the same day of the week. For this, the total number of odd days between 31st Dec. 1989 and 31st Dec. 2000 must be zero.

**Odd days are as under :**

| Year | 1990 | 1991 | 1992 | 1993 | 1994 | 1995 |
|---|---|---|---|---|---|---|
| *odd days* | *1* | *1* | *2* | *1* | *1* | *1* |
| **Year** | 1996 | 1997 | 1998 | 1999 | 2000 (leap) | |
| *odd days* | 2 | *1* | *1* | *1* | 2 | |

∴ Total number of odd days = 14 days *i.e.* 0 odd days.

*Hence the result follows.*

**Ex. 4.** ***Prove that the last day of a century can not be either Tuesday, Thursday or Saturday.***

**Sol.** Ist Century, *i.e.* 100 years contain 76 ordinary years & 24 leap years and therefore, (76 + 48) or 124 odd days or 5 odd days.

∴ The last day of 1st century is 'Friday'.

Two Centuries, *i.e.* 200 years contain 152 ordinary years & 48 leap years and therefore (152 + 96) or 248 or 3 odd days.

∴ The last day of 2nd century is 'Wednesday'.

Three Centuries, *i.e.* 300 years contain 228 ordinary years & 72 leap years and therefore, (228 + 144) or 372 or 1 odd day.

∴ The last day of third century is 'Monday'.

Four Centuries, *i.e.* 400 years contain 303 ordinary years & 97 leap years and therefore, (303 + 194) or 497 or 0 odd day.

∴ The last day of 4th century is 'Sunday'.

*Since the order is continually kept in successive cycles, we see that the last day of a century can not be Tuesday, Thursday or Saturday.*

**Ex. 5.** ***Prove that any date in March is the same day of the week as the corresponding date in November of that year.***

**Sol.** In order to prove the required result, we have to show that the total number of odd days between last day of February and last day of October is zero.

Number of days between these dates are :

March April May June July Aug. Sept. Oct.

31 + 30 + 31 + 30 + 31 + 31 + 30 + 31

= 245 days = 35 weeks = 0 odd day.

*Hence, the result follows.*

## Exercise 23A (Subjective)

1. Find the day of the week on :
   (*i*) 1st January, 1901 (*ii*) 15th August, 1947
   (*iii*) 28th April, 1973 (*iv*) 31st October, 1984
   (*v*) 14th March, 1993 (*vi*) 27th December, 1985
2. On what dates of December, 1984 did Sunday fall ?

**Hint.** *Find the day on 1st December 1984 which was Saturday. So Sunday fell on 2nd and therefore on 9th, 16th, 23rd & 30th.*

3. Prove that the calender for 1993 will serve for 1999 also.

4. Today is 3rd November. The day of the week is sunday. Last year was a leap year. What will be the day of the week on this date after 2 years ?

**Hints.** *Clearly, none of the next 2 years is a leap year. Each year gives 1 odd day. So, the required day is 2 days beyond Sunday i.e. it is Tuesday.*

5. If the first day of the year 1979 was Monday, what day of the week must have been on 1st January, 1986 ?

**Hint.** *Total number of odd days from 1st January 1979 to 1st January 1986 is 2. So, required day is 2 days beyond Monday.*

## ANSWERS (Exercise 23A)

1. (i) Tuesday (ii) Friday (iii) Saturday
(iv) Friday (v) Wednesday (vi) Sunday
2. 2nd, 9th, 16th, 23rd & 30th 4. Tuesday 5. Wednesday

## EXERCISE 23B (Objective Type Questions)

1. January 1, 1992 was a Wednesday. What day of the week will it be on January 1, 1993 ?
(*a*) Monday (*b*) Tuesday
(*c*) Sunday (*d*) Friday

2. On January 12, 1980, it was Saturday. The day of the week on January 12, 1979 was :
(*a*) Saturday (*b*) Friday
(*c*) Sunday (*d*) Thursday

3. On July, 2, 1985, it was Wednesday. The day of the week on July 2, 1984 was :
(*a*) Wednesday (*b*) Tuesday
(*c*) Monday (*d*) Thursday

4. Monday falls on 4th April, 1988. What was the day on 3rd November, 1987 ?
(*a*) Monday (*b*) Sunday
(*c*) Tuesday (*d*) Wednesday

5. Today is Friday. After 62 days it will be :
(*a*) Friday (*b*) Thursday
(*c*) Saturday (*d*) Monday

6. Smt. Indira Gandhi died on 31st October, 1984. The day of the week was:
(*a*) Monday (*b*) Tuesday
(*c*) Wednesday (*d*) Friday

7. The number of odd days in a leap year is :

(*a*) 1 (*b*) 2
(*c*) 3 (*d*) 4

**8.** The year next to 1988 having the same calendar as that of 1988 is :
(*a*) 1990 (*b*) 1992
(*c*) 1993 (*d*) 1995

**9.** The year next to 1991 having the same calendar as that of 1990 is :
(*a*) 1998 (*b*) 2001
(*c*) 2002 (*d*) 2003

**10.** Today is 1st August. The day of the week is Monday. this is a leap year. The day of the week on this day after 3 years will be :
(*a*) Wednesday (*b*) Thursday
(*c*) Friday (*d*) Saturday

**11.** How many days are there from 2nd January 1993 to 15th March 1993 :
(*a*) 72 (*b*) 73
(*c*) 74 (*d*) 71

**12.** The first republic day of India was celebrated on 26th January, 1950. It was :
(*a*) Monday (*b*) Tuesday
(*c*) Thursday (*d*) Friday

**13.** P.V. Narsimha Rao was elected party leader on 29th May, 1991. What was the day of the week ?
(*a*) Tuesday (*b*) Friday
(*c*) Wednesday (*d*) Sunday

## SOLUTIONS (Exercise 23B)

**1.** 1992 being a leap year, it has 2 odd days. So, the first day of the year 1993 will be two days beyond Wednesday. *i.e.* it will be Friday.

**2.** The year 1979 being an ordinary year, it has 1 odd day.
So, the day on 12th January 1980 is one day beyond the day on 12th January, 1979.
But, January 12, 1980 being Saturday
∴ January 12, 1979 was Friday.

**3.** The year 1984 being a leap year, it has 2 odd days.
So, the day on 2nd July, 1985 is two days beyond the day on 2nd July, 1984.
But, 2nd July 1985 was Wednesday.
∴ 2nd July, 1984 was Monday.

**4.** Counting the number of days after 3rd November, 1987 we have :

| | Nov. | | Dec. | | Jan. | | Feb. | | March | | April |
|---|---|---|---|---|---|---|---|---|---|---|---|
| days | 27 | + | 31 | + | 31 | + | 29 | + | 31 | + | 4 |

= 153 days containing 6 odd days

*i.e.*,(7 – 6) = 1 day beyond the day on 4th April, 1988.

So, the day was Tuesday.

5. Each day of the week is repeated after 7 days.

∴ After 63 days, it would be Friday.

So, after 62 days, it would be Thursday.

6. 1600 years contain 0 odd day; 300 years contain 1 odd day.

Also, 83 years contain 20 leap years and 63 ordinary years and therefore (40 + 0) odd days *i.e.*,5 odd days.

∴ 1983 years contain (0 + 1 + 5) *i.e.*, 6 odd days.

Number of days from Jan., 1984 to 31st. Oct. 1984

= (31 + 29 + 31 + 30 + 31 + 30 + 31 + 31 + 30 + 31) = 305 days

= 4 odd days

∴ Total number of odd days = 6 + 4 = 3 odd days.

So, 31st Oct., 1984 was Wednesday.

7. A leap year has (52 weeks + 2 days). So, the number of odd days in a leap year is 2.

8. Starting with 1988, we go on counting the number of odd days till the sum is divisible by 7

| | | | | | | |
|---|---|---|---|---|---|---|
| Years → | 1988 | 1989 | 1990 | 1991 | 1992 | |
| Odd days → | 2 | 1 | 1 | 1 | 2 | = 7 *i.e.* 0 odd day |

∴ Calendar for 1993 is the same as that of 1988.

9. We go on counting the odd days from 1991 onwards till the sum is divisible by 7. The number of such days are 14 upto the year 2001.

So, the calendar for 1991 will be repeated in the year 2002.

10. This being a leap year none of the next 3 years is a leap year. So, the day of the week will be 3 days beyond Monday *i.e.*, it will be Thursday.

11. Jan. Feb. March

30 + 28 + 15 = 73 days.

12. 1600 years have 0 odd day and 300 years have 1 odd day.

49 years contain 12 leap years and 37 ordinary years and therefore (24 + 37) odd days *i.e.*, 5 odd days

*i.e.*, 1949 years contain (0 + 1 + 5) or 6 odd days.

26 days of January contain 5 odd days.

Total odd days = (6 + 5) = 11 or 4 odd days.

So, the day was Thursday.

13. Try youself. It was Wednesday

## ANSWERS (Exercise 23B)

| | | | | | |
|---|---|---|---|---|---|
| 1. (*d*) | 2. (*b*) | 3. (*c*) | 4. (*c*) | 5. (*b*) | 6. (*c*) |
| 7. (*b*) | 8. (*c*) | 9. (*c*) | 10. (*b*) | 11. (*b*) | 12. (*c*) |
| 13. (*c*) | | | | | |

# 24

# CLOCKS

The face or dial of a clock of watch is a circle whose circumference is divided into 60 equal parts, called *minute spaces*. The clock has two hands, the smaller one, called *hour hand* (or short hand) goes over 5 minute spaces whilst the larger one, called *minute hand* or (long hand) passes over 60 minute spaces.

In 60 minutes, the minute hand gains 55 minutes on the hour hand. It may be noted that :

(i) In every hour, both the hand coincide once.

(ii) When the two hands are at right angle, they are 15 minute spaces apart. This happens twice in every hour.

(iii) When the hands are in opposite directions, they are 30 minute spaces apart. This happens once in every hour.

(iv) The hands are in the same straight line when they are coincident or opposite to each other.

**Too Fast & Too Slow** : If a clock or watch indicates 9.15, when the correct time is 9, it is said to be 15 minutes too fast. On the other hand, if it indicates 8.45, when the correct time is 9, it is said to be 15 minutes too slow.

## Solved Examples.

**Ex. 1.** ***At what time between 3 and 4 O'clock are the hands of a clock together ?***

**Sol.** At 3 O'clock, the hour hand is at 3 and the minute hand is at 12. i.e., they are 15 min. spaces apart. To be together, the minute hand must gain 15 min. over the hour hand.

Now 55 min. are gained in 60 min.

$\therefore$ 15 min. will be gained in $\left(\frac{60}{55} \times 15\right)$ min. $= 16\frac{4}{11}$ min.

*So, the hands will coincide at* $16\frac{4}{11}$ *min. past 3.*

**Ex. 2.** ***At what time between 4 and 5 O'clock will the hand of clock be at right angle ?***

**Sol.** At 4 O'clock, the minute hand will be 20 min. spaces behind the hour hand. Now, when the two hands are at right angle, they are 15 min. spaces apart. So, there are two cases :

**Case I. When the min. hand is 15 min. spaces behind the hour hand.**

**To be in this position, the min. hand will have to gain (20 – 15) = 5 min. spaces.**

**Now, 55 min. spaces are gained in 60 min.**

**$\therefore$ 5 min. spaces are gained in $\left(\frac{60}{55} \times 5\right)$ min. = $5\frac{5}{11}$ min.**

**$\therefore$ They are at right angle at $5\frac{5}{11}$ min. past 5.**

**Case II. When the min. hand is 15 min. spaces ahead of the hour hand.**

**To be in this position, the min. hand will have to gain (20 + 15) = 35 min. spaces.**

**Now, 55 min. spaces are gained in 60 min.**

**$\therefore$ 35 min. spaces will be gained in $\left(\frac{60}{55} \times 35\right)$ min. = $38\frac{2}{11}$ min.**

***So, they are at right angle at $38\frac{2}{11}$ min. past 4.***

**Ex. 3.** ***Find at what time between 8 and 9 O'clock will the hands of a clock be in the same straight line but not together.***

**Sol. At 8 O'clock, the hour hand is at 8 and the min. hand is at 12 i.e., the two hands are 20 min. spaces apart. To be in the same straight line but not together they will be 30 min. spaces apart. So, the min. hand will have to gain (30 – 20) = 10 min. spaces over the hour hand.**

**Now, 55 min. are gained in 60 min.**

**10 min. will be gained in $\left(\frac{60}{55} \times 10\right)$ min. = $10\frac{10}{11}$ min.**

***$\therefore$ The hand will be at right angle but not together at $50\frac{10}{11}$ min. past 8.***

**Ex. 4.** ***At what time between 5 and 6 are the hands of a clock 3 minutes apart ?***

**Sol. At 5 O'clock, the minute hand is 25 minute spaces apart.**

**Case I.** ***Minute hand is 3 minute spaces behind the hour hand.***

**In this case, the minute hand has to gain (25 – 3) i.e., 22 minute spaces.**

**Now, 55 min. are gained in 60 min.**

**22 min. are gained in $\left(\frac{60}{55} \times 22\right)$ min. = 24 min.**

**$\therefore$ The hands will be 3 minutes apart at 24 min. past 5.**

**Case II.** ***Minute hand is 3 minute spaces ahead of the hour hand.***

In this case, the minute hand has to gain (25 + 3) i.e., 28 minute spaces.

Now, 55 min. are gained in 60 min.

28 min. are gained in $\left(\frac{60}{55} \times 28\right) = 31\frac{5}{11}$ min.

∴ *The hands will be 3 minutes apart at $31\frac{5}{11}$ min. past 5.*

**Ex. 5.** ***The minute hand of a clock overtakes the hour hand at intervals of 65 minutes of correct time. How much a day does the clock gain or lose ?***

**Sol.** In a correct clock, the minute hand gains 55 minute spaces over the hour hand in 60 minutes. To be together again, the minute hand must gain 60 minutes over the hour hand.

Now, 55 min. are gained in 60 min.

∴ 60 min. are gained in $\left(\frac{60}{55} \times 60\right)$ min. $= 65\frac{5}{11}$ min.

But, they are together after 65 minutes.

∴ Gain in 65 minutes $= \left(65\frac{5}{11} - 65\right) = \frac{5}{11}$ min.

*Gain in 24 hrs.* $= \left(\frac{5}{11} \times \frac{60 \times 24}{65}\right)$ *min.* $= 10\frac{10}{143}$ *min.*

**Ex. 6.** ***A watch which gains uniformly, is 5 min. slow at 8 O'clock in the morning on Sunday, and is 5 minutes 48 seconds fast at 8 p.m. on following Sunday. When was it correct ?***

**Sol.** Time from 8 a.m. on Sunday to 8 p.m. on following Sunday = 7 days 12 hours = 180 hours.

Thus, the watch gains $\left(5 + 5\frac{4}{5}\right)$ min. or $\frac{54}{5}$ min. in 180 hours

Now, $\frac{54}{5}$ min. are gained in 180 hours.

∴ 5 min. are gained in $\left(\frac{180 \times 5}{54} \times 5\right)$ hours.

= 83 hrs. 20 min. = 3 days 11 hrs. 20 min.

Thus, the watch is correct 3 days 11 hrs. 20 min. after 8 a.m. on Sunday i.e., *it will be correct at 20 min. past 7 p.m. on Wednesday.*

**Ex. 7.** ***A clock is set right at 8 a.m. The clock gains 10 minutes in 24 hours. What will be the true time when the clock indicates 1 p.m. on the following day ?***

**Sol.** Time from 8 a.m. on a day to 1 p.m. on the following day is 29 hrs.

Now, 24 hrs. 10 min. of this clock are the same as 24 hours of the correct clock.

i.e., $\frac{145}{6}$ hrs. of this clock = 24 hrs. of correct clock.

29 hrs. of this clock = $\left(\frac{24 \times 6}{145} \times 29\right)$ hrs. of correct clock.

= 28 hrs. 48 min. of correct clock.

*So, the correct time is 28 hrs. 48 min. after 8 a.m. or 48 min. past 12.*

**Ex. 8.** ***A clock is set right at 5 a.m. The clock loses 16 min. in 24 hours. What will be the true time when the clock indicates 10 p.m. on the 4th day ?***

**Sol.** Time from 5 a.m. on a day to 10 p.m. on 4th day is 89 hours.

Now, 23 hrs. 44 min. of this clock are the same as 24 hours of the correct clock.

i.e., $\frac{356}{15}$ hrs. of this clock = 24 hrs. of correct clock.

$\therefore$ 89 hrs. of this clock = $\left(\frac{24 \times 15}{356} \times 89\right)$ hrs. of correct clock.

= 90 hrs. of correct clock.

*So, the correct time is 11 p.m.*

## EXERCISE 24 (OBJECTIVE TYPE QUESTIONS)

**1.** At what time between 5 and 6 are the hands of a clock coincident ?

(*a*) 22 minutes past 5 (*b*) 30 minutes past 5

(*c*) $22\frac{8}{11}$ minutes past 5 (*d*) $27\frac{3}{11}$ minutes past 5

**2.** At what time between 9 and 10 will the hands of a watch be together ?

(*a*) 45 minutes past 9 (*b*) 50 minutes past 9

(*c*) $49\frac{1}{11}$ minutes past 9 (*d*) $48\frac{2}{11}$ minutes past 9

**3.** At what time between 7 and 8 will the hands of a clock be in the same straight line, but not together ?

(*a*) 5 minutes past 7 (*b*) $5\frac{2}{11}$ minutes past 7

(*c*) $5\frac{3}{11}$ minutes past 7 (*d*) $5\frac{5}{11}$ minutes past 7

**4.** At what time between 4 and 5 will the hands of a watch point in opposite directions ?

(*a*) 45 minutes past 4 (*b*) 40 minutes past 4

(*c*) $50\frac{4}{11}$ minutes past 4 (*d*) $54\frac{6}{11}$ minutes past 4

5. At what time between 5.30 and 6 will the hands of a clock be at right angles ?

(a) $43\frac{5}{11}$ minutes past 5 (b) $43\frac{7}{11}$ minutes past 5

(c) 40 minutes past 5 (d) 45 minutes past 5

6. A watch, which gains uniformly, is 2 min. slow at noon on Monday, and is 4 min. 48 seconds fast at 2 p.m. on the following Monday. When was it correct ?

(a) 2 p.m. on Tuesday (b) 2 p.m. on Wednesday

(c) 3 p.m. on Thursday (d) 1 p.m. on Friday

7. A watch which gains 5 seconds in 3 minutes was set right at 7 a.m. In the afternoon of the same day, when the watch indicated quarter past 4 O'clock, the true time is :

(a) $59\frac{7}{12}$ minutes past 3 (b) 4 p.m.

(c) $58\frac{7}{11}$ minutes past 3 (d) $2\frac{3}{11}$ minutes past 4

8. How much does a watch gain or lose per day, if its hands coincide every 64 minutes ?

(a) loses 96 minutes (b) loses 90 minutes

(c) loses $36\frac{5}{11}$ minutes (d) loses $32\frac{8}{11}$ minutes

9. How many times do the hands of a clock coincide in a day ?

(a) 24 (b) 20

(c) 21 (d) 22

10. How many times do the hands of a clock point towards each other in a day ?

(a) 24 (b) 20

(c) 12 (d) 22

11. How many times are the hands of a clock at right angles in a day ?

(a) 24 (b) 48

(c) 22 (d) 44

12. How many times in a day, are the hands of a clock straight ?

(a) 24 (b) 48

(c) 22 (d) 44

13. At what angle the hands of a clock are inclined at 15 minutes past 5 ?

(a) $72\frac{1}{2}^\circ$ (b) $67\frac{1}{2}^\circ$

(c) $58\frac{1}{2}^\circ$ (d) $64^\circ$

# SOLUTIONS (Exercise 24)

1. At 5 O'clock, the minute hand is 25 minute spaces apart.
To be coincident, it must gain 25 minute spaces.
Now, 55 minutes are gained in 60 minutes.
25 minutes will be gained in $\left(\frac{60}{55} \times 25\right)$ min. or $27\frac{3}{11}$ min.
*So, the hands are coincident at $27\frac{3}{11}$ min. past 5.*

2. To be together between 9 and 10, the minute hand has to gain 45 minute spaces.
Now, 55 min. spaces are gained in 60 minutes.
$\therefore$ 45 min. spaces are gained in $\left(\frac{60}{55} \times 45\right)$ min. or $49\frac{1}{11}$ min.
*So, the hands are together at $49\frac{1}{11}$ min. past 9.*

3. When the hands are in the same straight line, but not together, they are 30 min. spaces apart.
At 7 O'clock, they are 25 min. spaces apart.
So, the minute hand has to gain only 5 min. spaces
Now, 55 min. spaces are gained in 60 min.
5 min. spaces are gained in $\left(\frac{60}{55} \times 5\right)$ min. or $5\frac{5}{11}$ min.
$\therefore$ *The hands are in the same straight line, but not together at $5\frac{5}{11}$ min. past 7.*

4. At 4 O'clock, the hands are 20 min. spaces apart.
To be in opposite directions, they must be 30 min. spaces apart.
So, the min. hand has to gain 50 min. spaces.
Now, 55 min. spaces are gained in 60 min.
50 min. spaces are gained in $\left(\frac{60}{55} \times 50\right)$ min. or $54\frac{6}{11}$ min.
$\therefore$ *The hands are in opposite direction at $54\frac{6}{11}$ min. past 4.*

5. At 5 O'clock, the hands are 25 min. spaces apart. To be at right angles and that too between 5.30 and 6, the min. hand has to gain (25 + 15) or 40 min. spaces.
Now, 55 min. spaces are gained in 60 min.
$\therefore$ 40 min. spaces are gained in $\left(\frac{60}{55} \times 40\right)$ min. or $43\frac{7}{11}$ min.

*So, the hands are at right angles at $43\frac{7}{11}$ min. past 5.*

6. Time from Monday noon to 2 p.m. on following Monday

$$= 7 \text{ days } 2 \text{ hours} = 170 \text{ hours.}$$

The watch gains $\left(2 + 4\frac{4}{5}\right)$ or $\frac{34}{5}$ min. in 170 hours.

$\therefore$ It will gain 2 min. in $\left(\frac{170 \times 5}{34} \times 2\right)$ hrs. = 50 hrs = 2 days 2hrs.

*So, the watch is correct 2 days 2 hours after Monday noon. i.e., at 2 p.m. on Wednesday.*

7. Time from 7 a.m. to quarter past 4

$$= 9 \text{ hours } 15 \text{ min.} = 555 \text{ min.}$$

Now, $\frac{37}{12}$ min. of this watch = 3 min. of the correct watch.

555 min. of this watch $= \left(\frac{3 \times 12}{37} \times 555\right)$ min.

$= \left(\frac{3 \times 12}{37} \times \frac{555}{60}\right)$ hrs. = 9 hrs. of the correct watch.

*Correct time is 9 hours after 7 a.m. i.e., 4 p.m.*

8. 55 min. spaces are covered in 60 min.

60 min. spaces are covered in $\left(\frac{60}{55} \times 60\right)$ min. or $65\frac{5}{11}$ min.

$\therefore$ Loss in 64 min. $= \left(65\frac{5}{11} - 64\right) = \frac{16}{11}$ min.

*Loss in* $(24 \times 60)$ *min.* $= \left(\frac{16}{11 \times 64} \times 24 \times 60\right)$ *min.* $= 32\frac{8}{11}$ *min.*

9. The hands of a clock coincide 11 times in every 12 hours (because between 11 and 1, they coincide only once, at 12 O'clock). So, the hands coincide 22 times in a day.

10. The hands of a clock point towards each other 11 times in every 12 hours (because between 5 and 7, at 6 O'clock only they point towards each other).

*So, in a day the hands point towards each other 22 times.*

11. In 12 hours, they are at right angles 22 times (because two positions 3 O'clock and 9 O'clock are common).

*So, in a day they are at right angles 44 times.*

12. The hands coincide or are in opposite direction (22 + 22) i.e., 44 times in a day.

13. At 15 minutes past 5, the minute hand is at 3 and hour hand slightly advanced from 5. Angle between their 3rd and 5th position is 60°.

**Angle through which hour hand shifts in 15 mts. is $\left(15 \times \frac{1}{2}\right)^\circ = 7\frac{1}{2}^\circ$.**

*$\therefore$ Required angle* $= \left(60 + 7\frac{1}{2}\right) = 67\frac{1}{2}^\circ$.

## ANSWERS (Exercise 24)

| | | | | | |
|---|---|---|---|---|---|
| 1. (*d*) | 2. (*c*) | 3. (*d*) | 4. (*d*) | 5. (*b*) | 6. (*b*) |
| 7. (*b*) | 8. (*d*) | 9. (*d*) | 10. (*d*) | 11. (*d*) | 12. (*d*) |
| 13. (*b*) | | | | | |

# STOCK & SHARES

**Stock.** In order to meet the expanses of a certain plan, the Government of India sometimes raises a loan from the public at a certain fixed rate of interest. Bonds or Promisery Notes each of a fixed value are issued for sale to the public.

If a man purchases a bond of Rs. 100 at which 5% interest has been fixed by the Government, then the holder of such a bond is said to have, 'a Rs. 100 stock at 5%'. Here Rs. 100 is called the Face value of the stock. Usually, a period is fixed for the repayment of the loan i.e., the stock matures at a fixed date only. Now, if a person holding a stock is in need of the money before the date of maturity of stock, he can sell the bond or bonds to some other person, whereby the claim of interest is transferred to that person.

Stocks are sold and bought in the open market through brokers at stocks exchanges. The broker's charge is usually called 'brokerage'

**Remarks**

(i) When stock is purchased, brokerage is added to cost price.

(ii) When stock is sold, brokerage is subtracted from selling price.

The selling price of a Rs. 100 stock is said to be at par, above par (or at a premium) and below par (or at a discount), according as the selling price of this stock is Rs. 100 exactly, more than Rs. 100 and less than Rs. 100 respectively.

**Remark.** 'By a Rs. 700, 6% stock at 97', we mean a stock whose face value is Rs. 700, the market price of a Rs. 100 stock is Rs. 97 and the annual interest on this stock is 5% of the face value.

## Solved Problems

**Ex. 1.** *Find the cost of :*

(i) Rs. 9100, $8\frac{3}{4}\%$ stock at 92.

(ii) Rs. 8500, $9\frac{1}{2}\%$ stock at 6 permium.

(iii) Rs. 7200, 10% stock at 7 discount.

(iv) Rs. 6400, 8% stock at par $\left(\text{brokerage } \frac{1}{8}\%\right)$.

**Sol.** (i) Cost of Rs. 100 stock = Rs. 92

Cost of Rs. 9100 stock = Rs. $\left(\frac{92}{100} \times 9100\right)$ = Rs. 8372.

**(ii)** Cost of Rs. 100 stock = Rs. (100 + 6) = Rs. 106.

Cost of Rs. 8500 stock = Rs. $\left(\frac{106}{100} \times 8500\right)$ = Rs. 9010.

**(iii)** Cost of Rs. 100 stock = Rs. (100 – 7) = Rs. 93

Cost of Rs. 7200 stock = Rs. $\left(\frac{93}{100} \times 7200\right)$ Rs. = 6696.

**(iv)** C.P. of Rs. 100 stock = Rs. $\left(100 + \frac{1}{8}\right)$ = Rs. $\frac{801}{8}$.

C.P. of Rs. 6400 stock = Rs. $\left(\frac{801 \times 6400}{8 \times 100}\right)$ = Rs. 6408.

**Ex. 2.** ***Find the cash required to purchase Rs. 1600, $8\frac{1}{2}$ % stock at 105 (brokerage $\frac{1}{2}$ %).***

**Sol.** Cash required for purchasing Rs. 100 stock

$$= \text{Rs.}\left(105 + \frac{1}{2}\right) = \text{Rs.}\left(\frac{211}{2}\right).$$

Cash required for purchasing Rs. 1600 stock

$$= \text{Rs.}\left(\frac{211 \times 1600}{2 \times 100}\right) = \text{Rs. } 1688.$$

**Ex. 3.** ***Find the cash realized by selling Rs. 2400, $5\frac{1}{2}$ % stock at 5 premium (brokerage $\frac{1}{4}$ %).***

**Sol.** By selling Rs. 100 stock, cash realized

$$= \text{Rs.}\left(105 - \frac{1}{4}\right) = \text{Rs.}\left(\frac{419}{4}\right)$$

By selling Rs. 2400 stock, cash realized

$$= \text{Rs.}\left(\frac{419 \times 2400}{4 \times 100}\right) = \text{Rs. } 2514.$$

**Ex. 4.** ***How much $4\frac{1}{2}$% stock at 95 can be pruchased by investing Rs. 1905, (brokerage $\frac{1}{4}$%) ?***

**Sol.** By investing Rs. $\left(95 + \frac{1}{4}\right)$, stock purchased = Rs. 100.

By investing Rs. 1905, stock purchased

$$= \text{Rs.}\left(\frac{100 \times 4 \times 1905}{381}\right) = \text{Rs. } 2000.$$

**Ex. 5.** ***What is the annual income derived from Rs. 1800, 5% stock at 104 ?***

**Sol.** Income from Rs. 100 stock = Rs. 5

*Income from Rs. 1800 stock* = Rs. $\left(\frac{5}{100} \times 1800\right)$ = Rs. 90.

**Ex. 6.** ***What is the annual income by investing Rs. 3000 in 6% stock at 120 ?***

**Sol.** On investing Rs. 120, income = Rs. 6

*On investing Rs. 3000, income* = Rs. $\left(\frac{6}{120} \times 3000\right)$ = Rs. 150 .

**Ex. 7.** ***Find the annual income derived by investing Rs. 770 in $4\frac{1}{2}$% stock at 96 $\left(\text{brokerage } \frac{1}{4}\%\right)$.***

**Sol.** On investing Rs. $\left(96 + \frac{1}{4}\right)$, income = Rs. $\frac{11}{2}$.

On investing Rs. 100, income = Rs. $\left(\frac{11 \times 4 \times 100}{2 \times 385}\right)$ = Rs. $5\frac{5}{7}$.

$\therefore$ Rate = $5\frac{5}{7}$ %.

**Ex. 9.** ***Find the market value of a $5\frac{1}{4}$% stock, in which an income of Rs. 756 is derived by investing Rs. 14976, brokerage being $\frac{1}{4}$ %.***

**Sol.** For an income of Rs. 756, investment = Rs. 14976

For an income of Rs. $\frac{21}{4}$, investment = Rs. $\left(\frac{14976}{756} \times \frac{21}{4}\right)$ = Rs. 104

$\therefore$ For a Rs. 100 stock, investment = Rs. 104

*Hence, market value* = Rs. $\left(104 - \frac{1}{4}\right)$ = Rs. 103.75.

**Ex. 10.** ***Which is the better investment, $5\frac{1}{2}$% stock at 102 or $4\frac{3}{4}$% stock at 96 ?***

**Sol.** Let the investment be Rs. (102 × 96) in each case.

**Case I.** $5\frac{1}{2}$% stock at 102.

Income from this stock = Rs. $\left(\frac{11 \times 102 \times 96}{2 \times 102}\right)$ = Rs. 528.

**Case II.** $4\frac{3}{4}$% stock at 96.

Income from this stock = Rs. $\left(\frac{19 \times 102 \times 96}{4 \times 96}\right)$ = Rs. 484.50

*Thus, by investing the same amount, the income from $5\frac{1}{2}$% stock at 102 is more. So, the investment in this stock is better.*

**Ex. 11.** *How much money must I invest in $6\frac{2}{3}$ % stock at 10 premium to secure an annual income of Rs. 600 ?*

**Sol.** For an income of Rs. $\left(\frac{20}{3}\right)$, investment = Rs. 110

∴ For an income of Rs. 600, investment

$$= \text{Rs.} \left(\frac{110 \times 3 \times 600}{20}\right) = \text{Rs. } 9900.$$

**Ex. 12.** *A person has Rs. 16500 stock in 3%. He sells it out at $101\frac{1}{8}$ and invests the proceeds in 4% railway debentures at $131\frac{7}{8}$. Find the change in his income, a brokerage of $\frac{1}{8}$ % being charged on each transaction.*

**Sol. Case I.** 3% stock at $108\frac{1}{8}$ $\left(\text{brokerage } \frac{1}{8}\%\right)$.

Income from Rs. 100 stock = Rs. 3.

Income from Rs. 16500 stock = Rs. $\left(\frac{3}{100} \times 16500\right)$ = Rs. 495

S.P. of Rs. 100 stock = Rs. $\left(101\frac{1}{8} - \frac{1}{8}\right)$ = Rs. 101.

S.P. of Rs. 16500 stock = Rs. $\left(\frac{101}{100} \times 16500\right)$ = Rs. 16665.

**Case II.** 4% stock at $131\frac{7}{8}$ $\left(\text{brokerage } \frac{1}{8}\%\right)$

By investing Rs. $\left(131\frac{7}{8} + \frac{1}{8}\right)$, income derived = Rs. 4

By investing Rs. 16665, income derived

$$= \text{Rs.} \left(\frac{4}{132} \times 16665\right) = \text{Rs. } 505.$$

∴ Change in income = Rs. (505 − 495) = Rs. 10 increased.

**Ex. 13.** *A man wishes to invest Rs. 2490. He invests Rs. 900 in $3\frac{1}{2}$ % stock at 75, Rs. 850 in 3% at 68 and the remainder in 6% stock. If the total yield from his investment is 5%, at what price does he buy the 6% stock ?*

**Sol.** Income from $3\frac{1}{2}\%$ stock at 75 = Rs. $\left(\frac{7}{2 \times 75} \times 900\right)$ = Rs. 42.

Income from 3% stock at 68 = Rs. $\left(\frac{3}{68} \times 850\right)$ = Rs. 37.50

Total income from these two stocks = Rs. (42 + 37.50)= Rs. (79.50).

But total income from the three stocks = Rs. $\left(\frac{5}{100} \times 2490\right)$ = Rs. 124.50

$\therefore$ Income from the third stock = Rs. (124.50 – 79.50) = Rs. 45

Investment in this case = Rs. (2490 – {900 + 850}) = Rs. 740.

If income is Rs. 45, investment = Rs. 740.

If income is Rs. 6, investment = Rs. $\left(\frac{740}{45} \times 6\right)$ = Rs. $98\frac{2}{3}$.

*So, he buys 6% stock at $98\frac{2}{3}$.*

**Ex. 14.** ***A man sells Rs. 5000, $4\frac{1}{2}$% stock at 144 and invests the proceeds partly in 3% stock at 90 and partly in 4% stock at 108. He, thereby increases his income by Rs. 25. How much of the proceeds were invested in each stock ?***

**Sol.** Income from Rs. 5000, $4\frac{1}{2}\%$ stock at 144

$$= \text{Rs.} \left(\frac{9 \times 5000}{2 \times 100}\right) = \text{Rs. } 225.$$

S.P. of this stock = Rs. $\left(\frac{144}{100} \times 5000\right)$ = Rs. 7200.

$\therefore$ Proceeds = Rs. 7200.

Now, income on an investment of Re. 1 in 3% at 90

$$= \text{Rs.} \left(\frac{3}{90} \times 1\right) = \text{Rs.} \left(\frac{1}{30}\right).$$

Income on an investment of Re. 1 in 4% at 108

$$= \text{Rs.} \left(\frac{4}{108} \times 1\right) = \text{Rs.} \left(\frac{1}{27}\right).$$

Total income derived from investment of
Rs. 7200 in these stocks
= Rs. (225 + 25) = Rs. 250

Average income = Rs. $\left(\frac{250}{7200}\right)$ = Rs. $\left(\frac{5}{144}\right)$

By alligation Rule :—

(3% at 90) : (4% at 108) = $\frac{1}{432} : \frac{1}{720}$ = 5 : 3

$\therefore$ Investment in 3% stock = Rs. $\left(\frac{7200 \times 5}{8}\right)$ = Rs. 4500.

*And, investment in 4% stock = Rs. (7200 – 4500) = Rs. 2700.*

**SHARES.** To start a big concern or a business a large amount of money is needed. This is usually beyond the capacity of one or two individuals. However, some persons together associate to form a company. The company issues a prospectus and invites the public to subscribe. The required capital is divided into equal small parts called Shares, each of a particular fixed value. The persons who subscribe in shares are called Shareholders. Sometimes, the company asks its shareholders to pay some money immediately and balance after some period. The total money raised immediately is called the Paid up capital. Parts of the profits divided amongst the shareholders are called dividends. The original value of a share is called its nominal value. The price of a share in the market is called the market value.

**Different Kinds of Shares.** There are two kinds of shares.

(i) **Preference Shares.** On these shares a fixed rate of dividend is paid to their holders, subject to profits of the company.

(ii) **Ordinary or Equity Shares.** After paying the dividends of the prefence shareholders, the equity shareholders are paid the dividends which depends upon the profit of the company.

## Solved problems.

**Ex. 1.** ***Find the cost of 96 shares of Rs. 10 each at $\frac{3}{4}$ discount, brokerage being $\frac{1}{4}$ per share.***

**Sol.** Cost of 1 share = Rs. $\left[\left(10 - \frac{3}{4}\right) + \frac{1}{4}\right]$ = Rs. $\frac{19}{2}$

*Cost of 96 shares = Rs.* $\left(\frac{19}{2} \times 96\right)$ *= Rs. 912.*

**Ex. 2.** ***Find the income derived from 44 shares of Rs. 25 each at 5 premium (brokerage 1/4 per share), the rate of dividend being 5%. Also find the rate of interest on the investment.***

**Sol.** Cost of 1 share = Rs. $\left(25 + 5 + \frac{3}{4}\right)$ = Rs. $\frac{171}{4}$

Cost of 44 shares = Rs. $\left(\frac{121}{4} \times 44\right)$ = Rs. 1331

$\therefore$ Investment made = Rs. 1331.

Now, face value of 1 share = Rs. 25.

$\therefore$ Face value of 44 shares = Rs. (44 × 25) = Rs. 1100.

Now, dividend on Rs. 100 = Rs. $\frac{11}{2}$.

$\therefore$ Dividend on Rs. 1100 = Rs. $\left(\frac{11}{2 \times 100} \times 1100\right)$ = Rs. 60.50

Also income on investment of Rs. 1331 = Rs. 60.50

$\therefore$ *income on investment of Rs. 100 = Rs.* $\left(\frac{60.50}{1331} \times 100\right)$

$= 4.55\%$.

**Ex. 3.** ***A man's net income from 5% Government paper is Rs. 1225 after paying an income tax at the rate of 2%. Find the number of shares of Rs. 1000 each owned by him.***

**Sol.** Face value of 1 share = Rs. 1000.

Gross income on 1 share = Rs. $\left(\frac{5}{100} \times 1000\right)$ = Rs. 50.

Income tax on 1 share's income = Rs. $\left(\frac{2}{100} \times 50\right)$ = Re. 1.

Net income on 1 share = Rs. (45 – 1) = Rs. 49.

If the net income is Rs. 49, number of shares = 1.

*If the net income is Rs. 1225, number of shares*

$= \frac{1}{49} \times 1225 = 25.$

**Ex. 4.** ***A man buys Rs. 25 shares in a company which pays 9% dividend. The money invested by the person is that much as gives 10% on investment. At what price did he buy the shares ?***

**Sol.** Face value of 1 share = Rs. 25.

Dividend on 1 share = Rs. $\left(\frac{9}{100} \times 25\right)$ = Rs. $\frac{9}{4}$

Now, Rs. 10 is an income on an investment of Rs. 100

$\therefore$ Rs. $\frac{9}{4}$ is an income on an investment of Rs. $\left(\frac{100}{10} \times \frac{9}{4}\right)$

= Rs. 22.50.

*Hence, cost of share = Rs. 22.50.*

## Exercise 25A (Subjective)

1. Find the cost of :

(i) Rs. 5000, $3\frac{1}{2}\%$ stock at 6 discount.

(ii) Rs. 1600, $4\frac{1}{2}\%$ stock at $3\frac{1}{4}$ premium, brokerage $\frac{1}{4}\%$.

**Hint.** *Cost of Rs. 100 stock = Rs.* $\left(100 + 3\frac{1}{4} + \frac{1}{4}\right)$ *= Rs.* $\left(\frac{207}{2}\right)$.

(iii) Rs. 800, $6\frac{1}{2}$% stock at par, brokerage $\frac{1}{8}$%.

2. How much cash can be realized by selling Rs. 2500, $6\frac{1}{2}$% stock at $110\frac{1}{8}$ $\left(\text{brokerage } \frac{1}{8}\%\right)$ ?

3. How much stock of $4\frac{1}{2}$% at 130 can be purchased for Rs. 7800 ?

4. How much stock of $3\frac{1}{2}$% at $109\frac{3}{4}$ can be purchased for Rs. 7700 $\left(\text{brokerage } \frac{1}{4}\%\right)$ ?

5. Find the annual income derived by investing Rs. 19700 in $3\frac{1}{2}$% stock at $98\frac{3}{6}$ $\left(\text{brokerage } \frac{1}{8}\%\right)$ ?

**Hint.** *By investing Rs.* $\left(98\frac{3}{8} + \frac{1}{8}\right)$, *income derived = Rs.* $\frac{7}{2}$.

6. I transfer Rs. 4500 stock from 3% at $97\frac{3}{4}$ to 4% at $89\frac{3}{4}$.

Find the change in my income, if brokerage in each case be $\frac{1}{4}$%.

**Hint.** *Income from first stock = Rs.* $\left(\frac{3}{100} \times 4500\right)$ *= Rs. 135.*

*S.P. of this stock = Rs.* $\left[\frac{\left(97\frac{3}{4} - \frac{1}{4}\right)}{100} \times 4500\right]$ *= Rs.* $\left(\frac{8775}{2}\right)$.

*Now, if this money is invested in 4% stock at* $89\frac{3}{4}$, *brokerage* $\frac{1}{4}$%,

*then the income = Rs.* $\left(\frac{4}{90} \times \frac{8775}{2}\right)$ *= Rs. 195.*

7. What amount must be invested in $6\frac{1}{2}$% stock at $95\frac{3}{4}$ to produce a clear income of Rs. 2000 a year, brokerage being $\frac{1}{4}$% and the income tax 4 paise in the rupee ?

**Hint.** *Gross income = Rs.* (2000 × 1.04) *= Rs.* 2080.

*Now, for an income of Rs.* $6\frac{1}{2}$, *investment = Rs. 96.*

∴ *For an income of Rs. 2080, investment*

$= Rs. \left(\frac{96 \times 2080 \times 2}{13}\right) = Rs. 30720$

8. Which is the better investment ?

(i) 5% stock at 143 or $3\frac{1}{2}$% stock at 91.

**Hint.** *Let investment in each case be Rs. (143 × 91).*

(ii) 4% stock at par subject to an income tax of 5 paise in a rupee or $4\frac{1}{2}$% stock at 110 free from income tax.

**Hint.** *Find the net income in each case.*

9. A man sells out Rs. 5000 of a $4\frac{1}{2}$% stock at $103\frac{5}{8}\left(\text{brokerage } \frac{1}{8}\right)$ and invests the proceeds in a $6\frac{1}{2}$% stock at $137\frac{7}{8}\left(\text{brokerage } \frac{1}{8}\right)$. Find the amount of the new stock he buys and the change in his income.

10. What should be the price of a 6% stock, if money is worth 5% ?

**Hint.** *For an income of Rs. 5, investment = Rs. 100.*

$\therefore$ *For an income of Rs. 6, investment* $= Rs. \left(\frac{100}{5} \times 6\right) = Rs. 120.$

11. At what price is $5\frac{1}{2}$% stock quoted when Rs. 5940 cash can bring an income of Rs. 297 a year ?

**Hint.** *For an income of Rs. 297, cash required = Rs. 5940*

$\therefore$ *For an income of Rs.* $\frac{11}{2}$, *cash required* $= Rs. \left(\frac{5940}{297} \times \frac{11}{2}\right) = Rs. 110.$

12. A man invests Rs. 8000, partly in 4% stock at 80 and partly in 5% stock at 110. If his total income be Rs. 375, how much does he invest in each kind of stock ?

Hint. *Income from 4% stock at 80, on an investment of Re. 1*

$= Rs. \left(\frac{4}{80} \times 1\right) = Rs. \left(\frac{1}{20}\right)$

*Income from 5% stock at 110, on an investment of Re. 1*

$= Rs. \left(\frac{5}{110}\right) = Rs. \left(\frac{1}{22}\right)$

*Average income on an investment of Re. 1* $= Rs. \left(\frac{375}{8000}\right) = Rs. \left(\frac{3}{64}\right)$

$\therefore$ *By allegation rule :–*

$\therefore$ *Ratio of investments* $= \frac{1}{704} : \frac{1}{320} = 320 : 704 = 5 : 11.$

*Now, divide Rs. 8000 in the ratio 5 : 11.*

13. Having Rs. 8370 to invest, a man puts part of it in 3% stock at 96 and the

remainder in 4% stock at 120. His dividend from each investment is the same. Find the amount invested in each stock.

**Hint.** *For an income of Re. 1 in 3% at 96, investment*

$$= Rs. \left(\frac{96}{3}\right) = Rs. 32.$$

$\therefore$ *For an income of Re. 1 in 4% at 120, investment*

$$= Rs. \left(\frac{120}{4}\right) = Rs. 30.$$

*Now, divide Rs. 8370 in the ratio 32 : 30 or 16 : 15.*

**14.** A man invests half of his money in $4\frac{1}{2}$% stock at 120 and the other half in $3\frac{1}{2}$% stock at 90. Had he invested his money, so as to buy equal amounts of each stock, he would have got Rs. 1.50 less of income. Find his total investment.

**Hint.** *Suppose the man invests Rs. x in each stock.*

*Then, total income* $= Rs. \left(\frac{9x}{2 \times 120} + \frac{7x}{2 \times 90}\right) = Rs. \left(\frac{11x}{144}\right)$.

*Again, Rs. 2x are invested in the ratio 120 : 90 or 4 : 3.*

*Total income in this case*

$$= Rs. \left(\frac{9}{2 \times 120} \times \frac{8x}{7} + \frac{2}{2 \times 90} \times \frac{6x}{7}\right) = Rs. \left(\frac{8x}{105}\right)$$

*So,* $\frac{11x}{144} - \frac{8x}{105} = 1.5$, *find x.*

**15.** The market price of Rs. 25 shares of a company is Rs. 28. If a man purchases 400 shares of this company, then find :

(i) How much money does he invest ?

(ii) What will be his annual income and what percentage will he get on his investment, if the company pays a dividend of 7% per annum ?

(iii) How many shares can he buy for Rs. 5600 ?

**Hint.** *(ii) clearly on 4 shares, dividend = Rs. 7*

*i.e., on investing Rs.* (28 × 4), *income = Rs. 7.*

**16.** A company has a capital of Rupees one lakh made up of Rs. 10 shares. A profit of Rs. 5000 was declared at the end of the year to be distributed among the share-holders. Find the dividend received by a man holding 300 shares.

**Hint.** *Total number of shares* $= \left(\frac{100000}{10}\right) = 10000$.

*Dividend per share* $= Rs. \left(\frac{5000}{10000}\right) = 50$ *paise.*

**17.** A man invested Rs. 4455 in Rs. 10 shares at Rs. 8.25, find the number of shares he bought and his income if a dividend of 6% is declared.

**Hint.** *Number of shares* $= \left(\frac{4455}{8.25}\right) = 540$.

*Face value of 540 shares* = Rs. $(540 \times 10)$ = Rs. 5400.

$\therefore$ *Dividend* = Rs. $\left(\frac{6}{100} \times 5400\right)$ = Rs. 324.

**18.** An investor owning 1000, Rs. 5 shares in a company sells 400 of them at Rs. 6.75 each. He invests the proceeds in a stock at 90. Calculate the investor's total income in 1 year from the shares and the stock, when a dividend of $6\frac{1}{2}\%$ per annum is paid on shares and $7\frac{1}{4}\%$ on the stock.

Hint. *Face Value of 600 shares* = Rs. $(600 \times 5)$ = Rs. 3000.

*Dividend from these shares* = Rs. $\left(\frac{3000 \times 13}{100 \times 2}\right)$ = Rs. 195.

*S.P. of 400 shares* = Rs. $(6.75 \times 400)$ = Rs. 2700.

*Amount of stock* = Rs. $\left(\frac{100}{90} \times 2700\right)$ = Rs. 3000.

*Income from this stock* = Rs. $\left(\frac{29 \times 3000}{4 \times 100}\right)$ = Rs. 267.50.

**19.** A man invests Rs. 8560 in Rs. 5. shares at Rs. 4.28 and the dividend declared was 8%. When the price rose to Rs. 4.50, he sold the shares and re-invested the proceeds in Rs. 4 shares at Rs. 4.80 on which the dividend was $11\frac{1}{2}\%$. Find the change in his income.

## ANSWERS (Exercise 25A)

**1.** (i) Rs. 4700 (ii) Rs. 1656 (iii) Rs. 801 **2.** Rs. 2750
**3.** Rs. 6000 **4.** Rs. 7000 **5.** Rs. 700
**6.** Rs. 60 increase **7.** Rs. 30720
**8.** (i) $3\frac{1}{2}\%$ at 91 is better (ii) $4\frac{1}{2}\%$ at 110 is better
**9.** Rs. 3750, Rs. 18.75 increase **10.** Rs. 120 **11.** Rs. 110
**12.** Rs. 2500, Rs. 5500 **13.** Rs. 4320, Rs. 4050 **14.** Rs. 7560
**15.** (i) Rs. 11200, (ii) Rs. 700, $6\frac{1}{4}\%$ (iii) 200 shares
**16.** Rs. 150 **17.** Rs. 324 **18.** Rs. 462.50 **19.** Rs. 62.50 increase

## EXERCISE 25B (OBJECTIVE TYPE QUESTIONS)

**1.** The cost price of a Rs. 100 stock at 4 discount, when brokerage is (1/4)%, is :

(*a*) Rs. 96 (*b*) Rs. $\left(96+\frac{1}{4}\right)$

(*c*) Rs. $\left(96-\frac{1}{4}\right)$ (*d*) Rs. 100

**2.** The income derived from a $5\frac{1}{2}\%$ stock at 95 is :

(*a*) Rs. 5.50 (*b*) Rs. 5

(*c*) Rs. 5.28 (*d*) none of these

**3.** The cash realized by selling a $5\frac{1}{2}\%$ stock at $106\frac{1}{4}$, brokerage being (1/4)%, is :

(*a*) Rs. $105\frac{1}{2}$ (*b*) Rs. $106\frac{1}{2}$

(*c*) Rs. 106 (*d*) none of these

**4.** By investing in a 6% stock at 96, an income of Rs. 100 is obtained by making an investment of :

(*a*) Rs. 1600 (*b*) Rs. 1504

(*c*) Rs. 1666.66 (*d*) Rs. 5760

**5.** A 4% stock yields 5%. The market value of the stock is :

(*a*) Rs. 125 (*b*) Rs. 80

(*c*) Rs. 99 (*d*) Rs. 109

**6.** Rs. 2780 are invested partly in 4% stock at 75 and 5% stock at 80 to have equal amount of incomes. The investment in 5% stock is :

(*a*) Rs. 1500 (*b*) Rs. 1280

(*c*) Rs. 1434.84 (*d*) Rs. 1640

**7.** To produce an annual income of Rs. 500 in a 4% stock at 90, the amount of stock needed is :

(*a*) Rs. 11250 (*b*) Rs. 12500

(*c*) Rs. 18000 (*d*) Rs. 20000

**8.** By investing Rs. 1100 in a $5\frac{1}{2}\%$ stock one earns Rs. 77. The stock is then quoted at :

(*a*) Rs. 93 (*b*) Rs. 107

(*c*) Rs. $78\frac{4}{7}$ (*d*) Rs. $97\frac{3}{4}$

**9.** A man invests in a $4\frac{1}{2}\%$ stock at 96. The interest obtained by him is :

(*a*) 4% (*b*) 4.5%

(*c*) 4.69% (*d*) $\frac{1}{2}\%$

**10.** *A* invested some money in 4% stock at 96. Now, *B* wants to invest in an equally good 5% stock. *B* must purchase a stock worth of :

(*a*) Rs. 120 (*b*) Rs. 124

(*c*) Rs. 76.80 (*d*) Rs. 80

**11.** I want to purchase a 6% stock which must yield 5% on my capital. At what price must I buy the stock ?

(*a*) Rs. 111 (*b*) Rs. 101

(*c*) Rs. 83.33 (*d*) Rs. 120

**12.** Which is the better stock, 5% at 143 or $3\frac{1}{2}$% at 93 ?

(*a*) 5% at 143 (*b*) $3\frac{1}{2}$% at 93

(*c*) both are equally good

**13.** A man invests some money partly in 3% stock at 96 and partly in 4% stock at 120. To get equal dividends from both, he must invest the money in the ratio ?

(*a*) 16 : 15 (*b*) 3 : 4

(*c*) 4 : 5 (*d*) 3 : 5

**14.** Which is better investment, 4% stock at par with an income tax at the rate of 5 paise per rupee or $4\frac{1}{2}$% stock at 110 free from income tax ?

(*a*) 4% at par with income tax

(*b*) $4\frac{1}{2}$% at 110 free from income tax

(*c*) both are equally good

**15.** A man invested Rs. 388 in a stock at 97 to obtain an income of Rs. 22. The dividend from the stock is :

(*a*) 12% (*b*) 3%

(*c*) $5\frac{1}{2}$% (*d*) 22.68%

**16.** By investing in $3\frac{3}{4}$% stock at 96, one earns Rs. 100. The investment made is :

(*a*) Rs. 36000 (*b*) Rs. 3600

(*c*) Rs. 2560 (*d*) Rs. 4800

**17.** A man buys Rs. 20 shares paying 9% dividend. The man wants to have an interest of 12% on his money. The market value of each share must be :

(*a*) Rs. 12 (*b*) Rs. 15

(*c*) Rs. 18 (*d*) Rs. 21

**18.** A man bought 20 shares of Rs. 50 at 5 discount, the rate of dividend being

$4\frac{3}{4}\%$. The rate of interest obtained is :

(*a*) $4\frac{3}{4}\%$ (*b*) $3\frac{1}{4}\%$

(*c*) 5.28% (*d*) 4.95%

**19.** A man invested Rs. 4455 in Rs. 10 shares quoted at Rs. 8.25. If the rate of dividend be 6%, his annual income is :

(*a*) Rs. 267.30 (*b*) Rs. 327.80

(*c*) Rs. 324 (*d*) Rs. 103.70

## SOLUTIONS (Exercise 25B)

**1.** C.P. = Rs. $\left(96 + \frac{1}{4}\right)$.

**2.** Income on Rs. 100 stock = Rs. $5\frac{1}{2}$ = Rs. 5.50.

**3.** Cash realized = Rs. $\left(106\frac{1}{4} - \frac{1}{4}\right)$ = Rs. 106.

**4.** For an income of Rs. 6, investment = Rs. 96.

For an income of Rs. 100, investment = Rs. $\left(\frac{96}{6} \times 100\right)$ = Rs. 1600.

**5.** For an income of Rs. 5, investment = Rs. 100.

For an income of Rs. 4, investment = Rs. $\left(\frac{100}{5} \times 4\right)$ = Rs. 80.

**6.** Let the investment in 4% stock be Rs. $x$.

Then, investment in 5% stock = Rs. $(2780 - x)$

Income from 4% stock = Rs. $\left(\frac{4}{75} \times x\right)$

Income from 5% stock = Rs. $\left[\left(\frac{5}{80} \times (2780 - x)\right)\right]$

$\therefore \frac{4x}{75} = \frac{2780 - x}{16}$ *or* $x = 1500$

*So, investment in 5% stock = Rs. (2780 – 1500) = Rs. 1280.*

**7.** For an income of Rs. 4, stock needed = Rs. 100

*For an income of Rs. 500, stock needed* = *Rs.* $\left(\frac{100}{4} \times 500\right)$

= *Rs. 12500.*

**8.** To earn Rs. 77, investment = Rs. 1100.

*To earn Rs.* $\frac{11}{2}$, *investment* = *Rs.* $\left(\frac{1100}{77} \times \frac{11}{2}\right)$ = *Rs.* $78\frac{4}{7}$.

**9.** On Rs. 96, he gets Rs. $\frac{9}{2}$.

*On Rs. 100, he gets = Rs.* $\left(\frac{9 \times 100}{2 \times 96}\right) = 4.69\%$.

**10.** For an income of Rs. 4, investment = Rs. 96

*For an income of Rs. 5, investment = Rs.* $\left(\frac{96}{4} \times 5\right)$ = *Rs. 120.*

**11.** For an income of Rs. 5, investment = Rs. 100.

*For an income of Rs. 6, investment = Rs.* $\left(\frac{100}{5} \times 6\right)$ = *Rs. 120.*

**12.** Let investment in each case be Rs. $(143 \times 93)$.

Income from 5% stock = Rs. $\left(\frac{5}{143} \times 143 \times 93\right)$ = Rs. 465.

Income from $3\frac{1}{2}\%$ stock = Rs. $\left(\frac{7}{2 \times 93} \times 143 \times 93\right)$ = Rs. 500.50.

$\therefore$ *$3\frac{1}{2}$ % stock at 93 is better.*

**13.** For an income of Re. 1 in 3% stock, investment
= Rs. (96/3) = Rs. 32

For an income of Re. 1 in 4% stock investment
= Rs. (120/4) = Rs. 30

$\therefore$ Ratio of investments = 32 : 30 = 16 : 15.

**14.** Let investment in each case be Rs. $(100 \times 110)$.

Gross income from 4% stock = Rs. $\left(\frac{4}{100} \times 100 \times 100\right)$ = Rs. 440.

Net income from the stock = Rs. (440 – 22) = Rs. 418.

Net income from $4\frac{1}{2}$ % stock = Rs. $\left(\frac{9 \times 100 \times 110}{2 \times 110}\right)$ = Rs. 450.

$\therefore$ *Better stock is $4\frac{1}{2}$ % at 110.*

**15.** When investment is Rs. 388, income = Rs. 22.

When investment is Rs. 97, income = Rs. $\left(\frac{22}{388} \times 97\right)$ = Rs 5.50

$\therefore$ *Dividend on Rs. 100 stock = $5\frac{1}{2}$ %.*

**16.** For earning Rs. $\frac{15}{4}$, investment = Rs. 96.

For earning Rs. 100, investment = Rs. $\left(\frac{96 \times 4}{15} \times 100\right)$
= *Rs. 2560.*

**17.** Dividend on Rs. 20 = Rs. $\left(\frac{9}{100} \times 20\right)$ = Rs. $\frac{9}{5}$.

Rs. 12 is an income on Rs. 100.

$\therefore$ *Rs.* $\frac{9}{5}$ *is an income on Rs.* $\left(\frac{100}{12} \times \frac{9}{5}\right)$ = *Rs. 15.*

**18.** Face value = Rs. (50 × 20) = Rs. 1000.

Dividend = Rs. $\left(\frac{1000 \times 19}{4 \times 100}\right)$ = Rs. $\left(\frac{95}{2}\right)$.

Investment = Rs. (45 × 20) = Rs. 900.

*Rate* = *Rs.* $\left(\frac{95 \times 100}{2 \times 900}\right)$ = *5.28%.*

**19.** Number of shares = $\frac{4455}{8.25}$ = 540.

Face value = Rs. (540 × 10) = Rs. 5400.

*Income* = *Rs.* $\left(\frac{6}{100} \times 5400\right)$ = *Rs. 324.*

## ANSWERS (Exercise 25B)

| | | | | | |
|---|---|---|---|---|---|
| **1.** (*b*) | **2.** (*a*) | **3.** (*c*) | **4.** (*a*) | **5.** (*b*) | **6.** (*b*) |
| **7.** (*b*) | **8.** (*c*) | **9.** (*c*) | **10.** (*a*) | **11.** (*d*) | **12.** (*b*) |
| **13.** (*a*) | **14.** (*b*) | **15.** (*c*) | **16.** (*c*) | **17.** (*b*) | **18.** (*c*) |
| **19.** (*c*) | | | | | |

# 26

# TRUE DISCOUNT

Suppose a sum say Rs. 136 is due 3 years hence and the borrower wants to clear off the debt right now. Then the question arises as to what money should be paid now. Clearly, the money which amounts to Rs. 136 after 3 years at a standard or agreed rate of interest must be paid now. Let the rate of interest in this case be 12% per annum simple interest. Then clearly, with this rate, Rs. 100 after 3 years will amount to Rs. 136. So clearly, the payment of Rs. 100 now will clear off a debt of Rs. 136 due 3 years hence at 12% per annum. The sum due is called the amount and the money paid now is called the present value or present worth of the sum due and the difference between the amount and the present worth (Rs. 36 in this case) is called the True Discount or Equitable Discount or Mathematical Discount.

Thus, The Present Value or Present Worth (P.W.) of a sum due at the end of a given time is the money which amounts to the sum due in that given time and at a given rate.

The sum due is called the amount.

The difference between the sum due at the end of a given time and its present worth is called True Discount (T.D.).

Thus, T.D. = (interest on P.W.) & Amount = (P.W. + T.D.).

**Remark.** Interest is reckoned on present worth and discount is reckoned on amount.

**Formulae.** If rate = R% p.a. & Time = T years, then

**(i)** $\text{P.W.} = \dfrac{100 \times (\text{Amount})}{[100 + (R \times T)]}$ **(ii)** $\text{T.D.} = \dfrac{(P.W.) \times R \times T}{100}$

**(iii)** $\text{T.D.} = \dfrac{(\text{Amount}) \times R \times T}{100 + (R \times T)}$ **(iv)** S.I. on T.D. = (S.I.) – (T.D.)

**(v)** $\text{Sum} = \left[\dfrac{(S.I.) \times (T.D.)}{(S.I.) - (T.D.)}\right]$

**(vi)** When the sum is put at compound interest, then

$$\text{P.W.} = \frac{\text{Amount}}{\left(1 + \dfrac{R}{100}\right)^T}.$$

**Solved Problems.**

**Ex. 1.** ***Find the present worth of Rs. 9950 due $3\frac{1}{4}$ years hence at $7\frac{1}{2}$ % per annum simple interest. Also, find the discount.***

**Sol.** $$\text{P.W.} = \frac{100 \times (\text{Amount})}{100 + (R \times T)}$$

$$= \text{Rs.}\left[\frac{100 \times 9950}{100 + \left(\frac{15}{2} \times \frac{13}{4}\right)}\right] = \text{Rs.}\left(\frac{100 \times 9950 \times 8}{995}\right)$$

$$= \text{Rs. } 8000.$$

*Also, T.D. = (Amount – (P.W.)*

*= Rs. (9950 – 8000) = Rs. 1950.*

**Ex. 2.** ***Find the present worth of a bill of Rs. 2916 due 2 years hence at 8% compound interest. Also, calculate the true discount.***

**Sol.** $$\text{P.W.} = \frac{\text{Amount}}{\left(1 + \frac{R}{100}\right)^T}$$

$$= \text{Rs.}\left\{\frac{2916}{\left(1 + \frac{8}{100}\right)^2}\right\} = \text{Rs.}\left(\frac{2916 \times 25 \times 25}{27 \times 27}\right) = \text{Rs. } 2500.$$

*Also, T.D. = Rs. (2916 – 2500) = Rs. 416.*

**Ex. 3.** ***The true discount on a bill due 9 months hence at 6% per annum is Rs. 180. Find the amount of the bill and its present worth.***

**Sol.** $$\text{P.W.} = \frac{100 \times T.D.}{R \times T} = \text{Rs.}\left(\frac{100 \times 180}{6 \times \frac{3}{4}}\right) = \text{Rs. } 4000.$$

*Amount = (P.W. + T.D.)*

*= Rs. (4000 + 180) = Rs. 4180.*

**Ex. 4.** ***The true discount on a certain sum of money due 3 years hence is Rs. 100 and the simple interest on the same sum for the same time and at the same rate is Rs. 120. Find the sum and the rate percent.***

**Sol.** $$\text{Sum due} = \frac{S.I. \times T.D.}{(S.I.) - (T.D.)} = \text{Rs.}\left(\frac{120 \times 100}{20}\right) = \text{Rs. } 600.$$

$$\textit{Rate} = \frac{100 \times 120}{600 \times 3} = 6\frac{2}{3}\%.$$

**Ex. 5.** ***The difference between the simple interest and the true discount on a certain sum of money for 6 months at 6% is Rs. 27. Find the sum.***

**Sol.** Let the sum be Rs. 100. Then,

S.I. = Rs. $\left(\frac{100 \times 1 \times 6}{2 \times 100}\right)$ = Rs. 3.

P.W. = Rs. $\left\{\frac{100 \times 100}{100 + \left(6 \times \frac{1}{2}\right)}\right\}$ = Rs. $\frac{10000}{103}$.

$\therefore$ T.D. = Rs. $\left[100 - \frac{10000}{103}\right]$ = Rs. $\frac{300}{103}$.

Now, (S.I.) – (T.D.) = Rs. $\left(3 - \frac{300}{103}\right)$ = Rs. $\frac{9}{103}$.

If the diff. in S.I. and T.D. is Rs. $\frac{9}{103}$, sum = Rs. 100.

*If the diff. in S.I. and T.D. is Rs. 27, sum = Rs.* $\left(\frac{100 \times 103}{9} \times 27\right)$

*= Rs. 30900.*

**Ex. 6.** ***A bill falls due in 9 months. The creditor agrees to accept immediate payment of half and to defer the payment of the other half for 18 months. He finds that by this arrangement he gains Rs. 4.50. What is the amount of the bill, if money be worth 4% ?***

**Sol.** Let the amount of bill be Rs. 200. Then, according to the agreement, the creditor agrees to pay after 9 months, an amount which is the sum of the amount of Rs. 100 for 9 months and P.W. of Rs. 100 due 9 months hence.

Now, S.I. = Rs. $\left(\frac{100 \times 3 \times 4}{4 \times 100}\right)$ = Rs. 3.

$\therefore$ Amount = Rs. (100 + 3) = Rs. 103.

Also, P.W. = Rs. $\left\{\frac{100 \times 100}{100 + \left(4 \times \frac{3}{4}\right)}\right\}$ = Rs. $\left\{\frac{10000}{103}\right\}$.

Total amount after 9 months = Rs. $\left\{103 + \frac{10000}{103}\right\}$ = Rs. $\frac{20609}{103}$.

Gain = Rs. $\left\{\frac{20609}{103} - 200\right\}$ = Rs. $\frac{9}{103}$.

If gain is Rs. $\frac{9}{103}$, sum due = Rs. 200.

*If gain is Rs.* $\frac{9}{2}$, *sum due = Rs.* $\left(\frac{200 \times 103}{9} \times \frac{9}{2}\right)$ *= Rs. 10300.*

**Ex. 7.** ***The true discount on Rs. 1860 due after a certain time at 5% is Rs. 60. Find the time after which it is due.***

**Sol.** P.W. = (sum due) – (T.D.) = Rs. (1860 – 60) = Rs. 1800. Since T.D. is interest on P.W., so Rs. 60 is the simple interest on Rs. 1800 at 5% per annum.

$\therefore$ *Time* = $\left(\frac{100 \times 60}{1800 \times 5}\right)$ years = $\left(\frac{2}{3} \times 12\right)$ *months = 8 months.*

**Ex. 8.** ***The true discount on Rs. 2575 due 4 months hence is Rs. 75. Find the rate percent of interest.***

**Sol.** P.W. = Rs. (2575 – 75) = Rs. 2500.

$\therefore$ S.I. on Rs. 2500 for 4 months is Rs. 75.

*Hence, rate* = $\frac{100 \times 75 \times 3}{2500 \times 1} = 9\%$.

**Ex. 9.** ***The true discount on a bill due 10 months hence at 6% per annum is Rs. 26.25. Find the amount of the bill.***

**Sol.** S.I. on Rs. 100 for 10 months at 6% per annum

$= \text{Rs.}\left(100 \times \frac{10}{12} \times \frac{6}{100}\right) = \text{Rs. } 5$

$\therefore$ Amount = Rs. (100 + 5) = Rs. 105.

So, T.D. = Rs. (105 – 100) = Rs. 5

If T.D. is Rs. 5, sum due = Rs. 105

If T.D. is Rs. 26.25, sum due = Rs. $\left(\frac{105}{5} \times 26.25\right)$ = Rs. 551.25.

**Ex. 10.** ***The present worth of a bill due 7 months hence is Rs. 1200, and if the bill were due at the end of $2\frac{1}{2}$ years, its present worth would be Rs. 1016. Find the rate percent and the sum of the bill.***

**Sol.** Sum due = (P.W.) + (T.D.) = (P.W.) + (S.I. on P.W.)

Now, sum due = (Rs. 1200 + S.I. on Rs. 1200 for 7 months)

Also, sum due = $\left(\text{Rs. } 1016 + \text{S.I. on Rs. } 1016 \text{ for } \frac{5}{2} \text{ years}\right)$

$\therefore \left\{\text{Rs } 1200 + \text{S.I. on Rs.}\left(1200 \times \frac{7}{12}\right) \text{ for 1 year}\right\}$

$= \left\{\text{Rs } 1016 + \text{S.I. on Rs.}\left(1016 \times \frac{5}{2}\right) \text{ for 1 year}\right\}$

or {Rs. 1200 + S.I. on Rs. 700 for 1 year}

= {Rs. 1016 + S.I. on Rs. 2540 for 1 year}

or S.I. on Rs. (2540 – 700) for 1 year = Rs. (1200 – 1016)

or S.I. on Rs. 1840 for 1 year = Rs. 184

$\therefore$ Rate = $\frac{100 \times 184}{1840 \times 1} = 10\%$.

Also, sum due = (Rs. 1200) + (S.I. on Rs. 1200 for 7 months at 10%)

$$= Rs. \left[1200 + \left(1200 \times \frac{7}{12} \times \frac{10}{100}\right)\right] = Rs.\ 1270.$$

## EXERCISE 26A (Subjective)

1. Fill in the gaps in the following :

   (i) Sum due = Rs. 617.50, Time = $2\frac{1}{2}$ years, Rate $7\frac{1}{2}$% p.a.

   P.W. = ? and Discount = ?

   (ii) T.D. = Rs. 23.40, Time = 8 months, Rate = $9\frac{3}{4}$% p.a.

   Sum due = ?

   (iii) Sum due = Rs. 2820.13, T.D. = Rs. 165.89, Rate = $3\frac{3}{4}$% p.a.

   Time = ?

   **Hint.** *P.W. = Rs. (2820.13 – 165.89) = Rs. 2654.24*

   $$\text{and Time} = \frac{100 \times T.D.}{P.W. \times Rate}$$

   (iv) T.D. = Rs. 60, S.I. = Rs. 62, Time = 8 months.

   Sum due = ? and Rate = ?

2. Goods were bought for Rs. 1250 and sold at once for a bill of Rs. 1278 due in 9 months. Find the gain or loss percent if the rate of interest is 8% per annum.

   **Hint.** *C.P. = Rs. 1250, S.P. = P.W. of Rs. 1278 due 9 months hence.*

3. I want to sell my scooter. There are two offers :

   'One at a cash payment of Rs. 8100 and another at a credit of Rs. 8250 to be paid after 6 months'.

   Which is better offer and by how much, money being worth $6\frac{1}{2}$% per annum simple interest ?

   **Hint.** *P.W. of Rs. 8250 due 6 months hence is Rs. 8000.*

4. *A* owes *B* Rs. 1120 payable 2 years hence and *B* owes *A* Rs. 1081.50 payable 6 months hence. If they decide to settle their accounts forth with by payment of ready money, what sum should be paid and by whom, reckoning money at 6% per annum simple interest ?

   **Hint.** *Present worth of Rs. 1120 due 2 years hence = Rs. 1000,*

   *Present worth of Rs. 1081.50 due 6 months hence = Rs. 1050.*

   *∴ B owes more to A i.e., B should pay Rs. 50 to A.*

5. A man purchased a horse for Rs. 3000 and sold it for Rs. 3960 on credit for sometime and still gained 10%. If the money be worth 5%, find the time for which the credit was allowed ?

   **Hint.** *C.P. = Rs. 3000, Gain = 10%.*

$\therefore$ *S.P. = Rs.* $\left(\frac{110}{100} \times 3000\right)$ = *Rs.* 3300.

*Now, P.W. = Rs. 3300, sum due = Rs. 3960, Rate = 5%, find time.*

**6.** **The true discount on a certain sum of money due after $2\frac{1}{2}$ year at 6% per annum is less than the simple interest on the same sum for the same time by Rs. 81. Find the sum.**

**Hint.** *Let the sum be Rs. 100.*

*Then S.I. = Rs.* $\left(\frac{100 \times 6 \times 5}{100 \times 2}\right)$ *= Rs. 15 and*

*P.W. = Rs.* $\left[\frac{100 \times 100}{100 + \left(\frac{5}{2} \times 6\right)}\right]$ *= Rs.* $\frac{2000}{23}$.

$\therefore$ *T.D. = Rs.* $\left(100 - \frac{2000}{23}\right)$ *= Rs.* $\frac{300}{23}$.

*Difference between S.I. and T.D. = Rs.* $\left(15 - \frac{300}{23}\right)$ *= Rs.* $\frac{45}{23}$.

*If this difference is Rs.* $\frac{45}{23}$, *sum = Rs. 100*

*If this difference is Rs. 81, sum = Rs.* $\left(\frac{100 \times 23}{45} \times 81\right)$ = *Rs.* 4140.

**7.** **If the simple interest on Rs. 500 for 4 years be equal to the true discount on Rs. 590 for the same time and at the same rate, find the rate percent.**

**Hint.** *Since T.D. on Rs. 590 = S.I. on Rs. 500*

$\therefore$ *S.I. on Rs. 500 for 4 years is Rs. 90. Find rate percent.*

**8.** **If Rs. 21 be the true discount on Rs. 371 for a certain time, what is the discount on the same sum for double the time, the rate being the same in both the cases ?**

**Hint.** *Clearly Rs. 21 is S.I. on Rs. (371 – 21) i.e., Rs. 350.*

$\therefore$ *T.D. on Rs. (350 + 42) is Rs. 42.*

$\therefore$ *T.D. on Rs. 371 is Rs.* $\left(\frac{42}{392} \times 371\right)$ *= Rs. 39.75.*

**9.** **Rs. 20 is the true discount on Rs. 260 due after a certain time. What will be the true discount on the same sum due after half of the former time, the rate of interest being the same ?**

**10.** **The present worth of a bill due at the end of 2 years is Rs. 1250. If the bill were due at the end of 2 years 11 months, its present worth would be Rs. 1200. Find the sum of the bill and the rate of interest.**

**11.** **A tradesman puts two prices on his goods, one for the ready money and the other for 6 months credit, interest being calculated at $12\frac{1}{2}$%**

per annum. If the credit price on an article be Rs. 25.52, what is its cost price ?

12. The debts of a bankrupt amount to Rs. 21345.25 and his assets consist of property worth Rs. 8598.46 and an undiscounted bill of Rs. 5130 due 4 months hence, simple interest being reckoned at 4%. How much in a rupee can he pay to his creditors ?

**Hint.** *Total debts = Rs. 21345.25*

*And, Total assets*

*= [Rs. (8598.46) + (P.W. of Rs. 5130 due 4 months hence)].*

## ANSWERS (Exercise 26A)

1. (i) Rs. 520, Rs. 97.50 (ii) Rs. 383.40
(iii) 20 months (iv) Rs. 1860, 5%

2. 3.54% 3. Rs. 8100 in cash is better 4. *B* should pay Rs. 50 to *A*

5. 4 years 6. Rs. 4140 7. $4\frac{1}{2}\%$ 8. Rs. 39.75 9. Rs. 10.40

10. Rs. 1375, 5% 11. Rs. 24.96 12. 64 paise.

## EXERCISE 26B (Objective Type Questions)

1. If the true discount on a sum due 2 years hence at 5% per annum be Rs. 75, then the sum due is :
   (*a*) Rs. 750 (*b*) Rs. 825
   (*c*) Rs. 875 (*d*) Rs. 800

2. I want to sell may scooter. There are two offers, one at cash payment of Rs. 8100 and another at a credit of Rs. 8250 to be paid after 6 months. If money being worth $6\frac{1}{4}\%$ per annum simple interest, which is the better offer :
   (*a*) Rs. 8100 in cash (*b*) Rs. 8250 due 6 months hence
   (*c*) both are equally good

3. The present worth of Rs. 1404 due in two equal half yearly instalments at 8% per annum simple interest is :
   (*a*) Rs. 1325 (*b*) Rs. 1300
   (*c*) Rs. 1350 (*d*) Rs. 1500

4. A trader owes a merchant Rs. 901 due 1 year's hence. However, the trader wants to settle the account after 3 months. How much cash should he pay, if rate of interest is 8% per annum :
   (*a*) Rs. 870 (*b*) Rs. 850
   (*c*) Rs. 828.92 (*d*) Rs. 846.94

5. The interest on Rs. 750 for 2 years is equal to the true discount on Rs. 810 for the same time and at the same rate. The rate percent is :

(a) $4\frac{1}{3}\%$ (b) $5\frac{1}{6}\%$

(c) 4% (d) 5%

6. Goods were bought for Rs. 600 and sold the same day for Rs. 650.25 at a credit of 9 months and still there was a gain of 2%. The rate percent is :

(a) $6\frac{1}{3}\%$ (b) $8\frac{1}{3}\%$

(c) 8% (d) $7\frac{43}{61}\%$

7. The simple interest and the true discount on a certain sum for a given time and at a given rate are Rs. 25 and Rs. 20 respectively. The sum is :

(a) Rs. 500 (b) Rs. 200

(c) Rs. 250 (d) Rs. 100

8. If Rs. 10 be allowed as true discount on a bill of Rs. 110 due at the end of a certain time, then the discount allowed on the same sum due at the end of double the time is :

(a) Rs. 20 (b) Rs. 21.81

(c) Rs. 22 (d) Rs. 18.33

9. A man buys a watch for Rs. 195 in cash and sells it for Rs. 220 at a credit of 1 year. If the rate of interest is 10%, the man :

(a) gains Rs. 15 (b) gains Rs. 3

(c) gains Rs. 5 (d) loses Rs. 5

10. The true discount on Rs. 1860 due after a certain time at 5% is Rs. 60. The time after which it is due is :

(a) 6 months (b) 8 months

(c) 9 months (d) 10 months

11. The true discount on Rs. 2575 due 4 months hence is Rs. 75. The rate percent is :

(a) 6% (b) 8%

(c) 9% (d) 5%

12. The true discount on a bill due 10 months hence at 6% per annum is Rs. 26.25. The amount of the bill is :

(a) Rs. 1575 (b) Rs. 500

(c) Rs. 650.25 (d) Rs. 551.25

13. A man purchased a cow for Rs. 300 and sold it the same day for Rs. 360, allowing the buyer a credit of 9 years. If the rate of interest be $7\frac{1}{2}\%$ per annum, then the man has a gain of :

(a) $4\frac{1}{2}\%$ (b) $5\frac{3}{7}\%$

(c) 6% (d) 5%

14. A owes B, Rs. 1120 payable 2 years hence and B owes A, Rs. 1081.50 payable 6 months hence. If they decide to settle their accounts forthwith by payment of ready money and the rate of interest be 6% per annum, then who should pay and how much :

(a) A, Rs. 50 (b) B, Rs. 50

(c) A, Rs. 70 (d) B, Rs. 70

15. Rs. 20 is the true discount on Rs. 260 due after a certain time. What will be the true discount on the same sum due after half of the former time, the rate of interest being the same :

(a) Rs. 10 (b) Rs. 10.40

(c) Rs. 15.20 (d) Rs. 13

16. A has to pay Rs. 220 to B after 1 year. B asks A to pay Rs. 110 in cash and defer the payment of Rs. 110 for 2 years. A agrees to it. Counting, the rate of interest at 10% per annum in this new mode of payment :

(a) there is no gain or loss to any one

(b) A gains Rs. 7.34

(c) A loses Rs. 7.34

(d) A gains Rs. 11

## SOLUTIONS (Exercise 24B)

1. $\text{P.W.} = \dfrac{100 \times T.D.}{R \times T} = \text{Rs.}\left(\dfrac{100 \times 75}{5 \times 2}\right) = \text{Rs.}750.$

∴ *Sum due = Rs. (750 + 75) = Rs. 825.*

2. P.W. of Rs. 8250 due 6 months hence

$= \text{Rs.}\left[\dfrac{100 \times 8250}{100 + \left(\dfrac{25}{4} \times \dfrac{1}{2}\right)}\right] = \text{Rs. } 8000.$

∴ *Rs. 8100 in cash is a better offer.*

3. P.W. of Rs. 702 due 6 months hence

$= \text{Rs.}\left[\dfrac{100 \times 702}{100 + 8 \times \dfrac{1}{2}}\right] = \text{Rs. } 675.$

P.W. of Rs. 702 due 1 year hence

$= \text{Rs.}\left\{\dfrac{100 \times 702}{100 \times (8 \times 1)}\right\} = \text{Rs. } 650.$

∴ *Total P.W. = Rs. (675 + 650) = Rs. 1325.*

4. P.W. of Rs. 901 due 9 months hence at 8%

$= \text{Rs.}\left\{\dfrac{100 \times 901}{100 + \left(8 \times \dfrac{3}{4}\right)}\right\} = \text{Rs.}\left(\dfrac{100 \times 901 \times 1}{106}\right) = \text{Rs. } 850.$

**5.** Since T.D. is S.I. on P.W., we have :

Rs. (810 – 750) or Rs. 60 as S.I. on Rs. 750 for 2 years.

$\therefore$ *Rate* = $\left(\frac{100 \times 60}{750 \times 2}\right) = 4\%$.

**6.** S.P. = (102% of Rs. 600) = Rs. $\left(\frac{102}{100} \times 600\right)$ = Rs. 612.

$\therefore$ P.W. of Rs. 650.25 due 9 months hence is Rs. 612.

or Rs. 38.25 is S.I. on Rs. 612 for 9 months.

$\therefore$ *Rate* = $\left(\frac{100 \times 38.25}{612 \times \frac{3}{4}}\right)\% = 8\frac{1}{3}\%$.

**7.** Sum = $\frac{(S.I.) \times (T.D.)}{(S.I.) - (T.D.)}$ = Rs. $\left(\frac{25 \times 20}{25 - 20}\right)$ = Rs. 100.

**8.** S.I. on Rs. (110 – 10) for a given time = Rs. 10.

S.I. on Rs. 100 for double the time = Rs. 20.

Sum = Rs. (100 + 20) = Rs. 120.

*T.D. on Rs. 110* = *Rs.* $\left(\frac{20}{120} \times 110\right)$ = *Rs. 18.33.*

**9.** P.W. of Rs. 220 due 1 year hence

= Rs. $\left(\frac{100 \times 220}{100 + 10}\right)$ = Rs. 200.

*Hence, the man gains Rs. 5.*

**10.** P.W. = (Sum due) – (T.D.) = Rs. (1860 – 60) = Rs. 1800.

Thus, Rs. 60 is S.I. on Rs. 1800 at 5% per annum.

$\therefore$ *Time* = $\left(\frac{100 \times 60}{1800 \times 5}\right)$ *years* = $\frac{2}{3}$ *years* = *8 months.*

**11.** P.W. = Rs. (2575 – 75) = Rs. 2500.

$\therefore$ Rate = $\left(\frac{100 \times 75 \times 3}{2500 \times 1}\right)\% = 9\%$.

**12.** Amount = (*T.D.*) × $\left\{\frac{100 + (R \times T)}{R \times T}\right\}$

= *Rs.* $\left(\frac{26.25 \times 105}{5}\right)$ = *Rs. 551.25.*

**13.** P.W. of Rs. 360 due 2 years hence at $7\frac{1}{7}\%$ per annum

= Rs. $\left\{\frac{100 \times 360}{100 + \left(\frac{50}{7} \times 2\right)}\right\}$ = Rs. $\left\{\frac{100 \times 360 \times 7}{800}\right\}$

= Rs. 315.

$\therefore$ S.P. = Rs. 315.

*Hence, gain %* $= \left(\frac{15 \times 100}{300}\right) = 5\%$.

**14.** P.W. of Rs. 1120 due 2 years hence at 6%

$= \text{Rs.}\left[\frac{100 \times 1120}{100 + (6 \times 2)}\right] = \text{Rs. } 1000.$

P.W. of Rs. 1081.50 due 6 months hence at 6%.

$= \text{Rs.}\left[\frac{100 \times 1081.50}{100 + \left(6 \times \frac{1}{2}\right)}\right] = \text{Rs.}\left[\frac{100 \times 1081.50}{103}\right]$

$= \text{Rs. } 1050.$

So, *A* owes *B*, Rs. 1000 cash and *B* owes *A* Rs. 1050 cash.

$\therefore$ *B must pay Rs. 50 to A.*

**15.** S.I. on Rs. 240 for a given time = Rs. 20

S.I. on Rs. 240 for half the time = Rs. 10

$\therefore$ Rs. 10 is T.D. on Rs. 250.

*So, T.D. on Rs. 260* $= \text{Rs.}\left(\frac{10}{250} \times 260\right) = \text{Rs. } 10.40.$

**16.** *A* has to pay the P.W. of Rs. 220 due 1 year hence, which is

$= \text{Rs.}\left[\frac{100 \times 220}{100 + (10 \times 1)}\right] = \text{Rs. } 200.$

*A* actually pays = Rs. [110 + P.W. of Rs. 110 due 2 years hence]

$= \text{Rs.}\left[110 + \frac{100 \times 110}{100 + (8 \times 2)}\right] = \text{Rs. } 192.66.$

$\therefore$ *A* gains = Rs. [200 – 192.66] = Rs. 7.34.

## ANSWERS (Exercise 26B)

| | | | | | |
|---|---|---|---|---|---|
| **1.** (*b*) | **2.** (*a*) | **3.** (*a*) | **4.** (*b*) | **5.** (*c*) | **6.** (*b*) |
| **7.** (*d*) | **8.** (*d*) | **9.** (*c*) | **10.** (*b*) | **11.** (*c*) | **12.** (*d*) |
| **13.** (*d*) | **14.** (*b*) | **15.** (*b*) | **16.** (*b*) | | |

# 27

# BANKER'S DISCOUNT

Suppose a merchant *A* purchases goods worth of say Rs. 5000 from another merchant *B* at a credit of a certain period say 4 months. Then *B* draws up a draft i.e., prepares a special type of a bill, called Hundi or Bill of exchange. On the receipt of the goods, *A* gives an agreement dually signed on the bill stating that he has accepted the bill and money can be withdrawn from his bank account after 4 months of the date of the bill. On this bill, there is an order from *A* to his bank asking to pay Rs. 5000 to *B* after 4 months. More over, 3 more days (known as grace days) are added to the date (called Nominally due date) of expiry of 4 months and on the date so obtained (called the legally due date), the bill can be presented to the bank by *B* to collect Rs. 5000 from *A*'s account. Suppose the bill is drawn on 5th Jan. at 4 months, then the nominally due date is 5th May and the legally due date is 8th May. The amount given on the draft or bill is called the face value, which is Rs. 5000 in this case.

Now suppose that *B* needs the money of this bill earlier than 8th may say on 3rd March. In such a case, *B* can sell the bill to a banker or a broker who pays him the money against the bill but some what less than the face value. Now, the natural questions is, as how much cash the banker should pay to *B* on 3rd March. Actually, if the banker deducts the true discount on the face value for the period from 3rd March to 8th May, he gains nothing. So in order to make some profit, the banker deducts from the face value, the simple interest on the face value for the unexpired time i.e., from 3rd March to 8th May. This deduction is known as Banker's Discount (B.D.) or Commercial Discount.

Thus, B.D. is the S.I. on face value for the period from the date on which the bill was discounted and the legally due date. The money paid by the banker to the bill holder is called the Discountable value.

Also the difference between the banker's discount and the true discount for the unexpired time is called the Banker's Gain (B.G.). Thus, Banker's Gain

$$\text{B.G.} = (\text{B.D.}) - (\text{T.D.}).$$

**Remark.** When date of the bill is not given, grace days are not to be added.

**Formulae :**

(i) B.D. = S.I. on bill for unexpired time.

(ii) Banker's Gain = (B.D.) – (T.D.)

(iii) B.G. = S.I. on T.D.

(iv) $\text{T.D.} = \sqrt{(\text{P.W.}) \times (\text{B.G.})}$ ; $\text{B.G.} = \dfrac{(\text{T.D.})^2}{(\text{P.W.})}$.

(v) $\text{B.D.} = \dfrac{\text{Amount} \times \text{Rate} \times \text{Time}}{100}$ ;

$\text{T.D.} = \dfrac{\text{Amount} \times \text{Rate} \times \text{Time}}{100 + (\text{Rate} \times \text{Time})}$

(vi) $\text{Amount} = \dfrac{(\text{B.D.}) \times (\text{T.D.})}{(\text{B.D.}) - (\text{T.D.})}$, $\text{T.D.} = \dfrac{\text{B.G.} \times 100}{\text{Rate} \times \text{Time}}$.

**Ex. 1.** ***A bill for Rs. 5656 is drawn on July, 14 at 5 months. It is discounted on October 5th at 5%. Find, the banker's discount; the true discount; the banker's gain and the money that the holder of the bill receives.***

**Sol.** Face value of the bill = Rs. 5656.

Date on which the bill was drawn = July, 14th at 5 months.

Nominally due date = December, 14th.

Legally due date = December, 17th.

Date on which the bill was discounted = October, 5th.

Period for which the bill has yet to run

Oct. Nov. Dec.

26+ 30+ 17 = 73 days or $\frac{1}{5}$ year

$\therefore$ B.D. = S.I. on Rs. 5656 for $\frac{1}{5}$ years at 5%

$$= \text{Rs.} \left(\frac{5656 \times 1 \times 5}{100 \times 5}\right) = \text{Rs. } 56.56.$$

$$\text{T.D.} = \text{Rs.} \left\{\frac{5656 \times 5 \times \frac{1}{5}}{100 + \left(5 \times \frac{1}{5}\right)}\right\} = \text{Rs. } 56$$

B.G. = (B.D.) – (T.D.) = 56 paise.

Money received by the holder of the bill

= *Rs. (5656 – 56.56) = Rs. 5599.44.*

**Ex. 2.** ***A banker paid Rs. 5767.50 for a bill of Rs. 5840, drawn on April 4, at 6 months. On what day was the bill discounted, the rate of interest being 7% ?***

**Sol.** B.D. = Rs. (5840 – 5767.20) = Rs. (72.80)

$\therefore$ Rs. 72.80 is S.I. on Rs. 5840 at 7%.

So, unexpired time $= \dfrac{100 \times 72.80}{7 \times 5840}$ years $= \dfrac{13}{73}$ years = 65 days.

Now, date of draw of bill = April, 4 at 6 months.

Nominally due date = October, 4

Legally due date = October, 7.

So, we must go back 65 days from October, 7.

Oct. Sept. Aug.

7+ 30+ 28

*i.e., The bill was discounted on 3rd August.*

**Ex. 3.** ***If the true discount on a certain sum due 6 months hence at 6% is Rs. 36, what is the banker's discount on the same sum for the same time and at the same rate ?***

**Sol.** B.G. = S.I. on T.D.

$$= \text{Rs.}\left(\frac{36 \times 6 \times 1}{100 \times 2}\right) = \text{Rs. } 1.08$$

$\therefore$ (B.D.) – (T.D.) = Rs. 1.08

*or B.D. = (T.D.) + Rs. 1.08 = Rs. (36 + 1.08) = Rs. 37.08.*

**Ex. 4.** ***The banker's discount on Rs. 1800 at 5% is equal to the true discount on Rs. 1830 for the same time and at the same rate. Find the time.***

**Sol.** $\because$ S.I. on Rs. 1800 = T.D. on Rs. 1830

$\therefore$ P.W. of Rs. 1830 is Rs. 1800.

i.e., Rs. 30 is S.I. on Rs. 1800 at 5%.

$$\therefore \textit{Time} = \left(\frac{100 \times 30}{1800 \times 5}\right) \textit{years} = \frac{1}{3} \textit{years} = 4 \textit{ months.}$$

**Ex. 5.** ***The banker's discount and the true discount on a sum of money due 8 months hence are Rs. 52 and Rs. 50, respectively. Find the sum and the rate percent.***

$$\textbf{Sol. } \text{Sum} = \frac{(B.D.) \times (T.D.)}{(B.D.) - (T.D.)} = \text{Rs.}\left(\frac{52 \times 50}{2}\right) = \text{Rs. } 1300.$$

Since B.D. is S.I. on sum due, so S.I. on Rs. 1300 for 8 months is Rs. 52. Consequently,

$$\textit{Rate} = \left(\frac{100 \times 52}{1300 \times \frac{2}{3}}\right)\% = 6\%.$$

**Ex. 6.** ***The present worth of a bill due sometime hence is Rs. 1100 and the true discount on the bill is Rs. 110. Find the banker's discount and the extra gain the banker would make in the transaction.***

**Sol.** T.D. = $\sqrt{(P.W.) \times (B.G.)}$

$$\text{or } B.G. = \frac{(T.D.)^2}{(P.W.)} = \text{Rs.}\left(\frac{110 \times 110}{1100}\right) = \text{Rs. } 11.$$

$\therefore$ *B.D.* = *B.G.* + *T.D.* = Rs. (11 + 110) = Rs. 121.

**Ex. 7.** ***The true discount on a bill of Rs. 1860 due after 8 months is Rs. 60. Find the rate, the banker's discount and the banker's gain.***

**Sol.** Amount = Rs. 1860, T.D. = Rs. 60

$\therefore$ P.W. = Rs. (1860 – 60) = Rs. 1800.

S.I. on Rs. 1800 for 8 months = Rs. 60

$$\therefore \text{Rate} = \left[\frac{100 \times 60}{1800 \times \frac{2}{3}}\right]\% = 5\%.$$

$$B.G. = \frac{(T.D.)^2}{(P.W.)} = \text{Rs.}\ \frac{60 \times 60}{1800} = \text{Rs. } 2.$$

*B.D. = (T.D.) + (B.G.) = Rs. (60 + 2) = Rs. 62.*

**Ex. 8.** ***The banker's discount on Rs. 1650 due a certain time hence is Rs. 165. Find the true discount and the banker's gain.***

**Sol.** $\text{Sum} = \dfrac{(B.D.) \times (T.D.)}{(B.D.) - (T.D.)} = \dfrac{(B.D.) \times (T.D.)}{(B.G.)}$

$$\therefore \frac{T.D.}{B.G.} = \frac{\text{Sum}}{B.D.} = \frac{1650}{165} = \frac{10}{1}.$$

i.e., if B.G. is Re. 1, T.D. = Rs. 10 or B.D. = Rs. 11.

$\therefore$ If B.D. is Rs. 11, T.D. = Rs. 10

If B.D. is Rs. 165, T.D. = Rs. $\left[\dfrac{10}{11} \times 165\right]$ = Rs. 150.

*Also, B.G. = Rs. (165 – 150) = Rs. 15.*

**Ex. 9.** ***What rate percent does a man get for his money when in discounting a bill due 10 months hence, he deducts 4% of the amount of the bill ?***

**Sol.** Let the amount of bill be Rs. 100.

Money deducted = Rs. 4.

Money received by holder of the bill = Rs. (100 – 4) = Rs. 96.

S.I. on Rs. 96 for 10 months = Rs. 4.

$$\textit{Rate} = \left[\frac{100 \times 4 \times 6}{96 \times 5}\right]\% = 5\%.$$

**Ex. 10.** ***A bill was drawn on March 8, at 7 months date and was discounted on May 18, at 5%. If the banker's gain is Rs. 3, find (i) the true discount (ii) the banker's discount and (iii) the sum of the bill.***

**Sol.** Date on which the bill was drawn = March 8th, at 7 months.

Nominally due date = Oct. 8th.

Legally due date = Oct., 11th.

Date on which the bill was discounted = May, 18th.

Time for which the bill has yet to run

May June July Aug. Sep. Oct.

13 + 30 +31 + 31 + 30 + 11 = 146 days = $\dfrac{2}{5}$ years.

Now, (i) Banker's gain = S.I. on T.D.

i.e., Rs. 3 is S.I. on T.D. for $\frac{2}{5}$ years at 5%,

$$\therefore T.D. = \text{Rs.} \frac{100 \times 3}{5 \times \frac{2}{5}} = \text{Rs. } 150.$$

(ii) $B.D. = T.D. + S.I.$ on $T.D.$

$= \text{Rs. } 150 + S.I.$ on Rs. 150 for $\frac{2}{5}$ years at 5%

$$= \text{Rs. } 150 + \text{Rs. } 150 \times \frac{2}{5} \times \frac{5}{100} = \text{Rs. } 153.$$

(iii) $$\text{Sum} = \frac{B.D. \times T.D.}{B.D. - T.D.} = \text{Rs.} \frac{153 \times 150}{153 - 150} = \text{Rs. } 7650.$$

## EXERCISE 27A (Subjective)

1. A bill for Rs. 1500 is drawn on July 1, at 5 months and discounted on september, 22 at 8% per annum. Find (i) the banker's discount, (ii) the discountable value of the bill, (iii) the true discount and (iv) the banker's gain.

2. A bill for Rs. 1000 drawn on 20th December 1992 at 6 months is discounted on 11th April, 1993. If the payment made by the banker is Rs. 989, find the rate of interest. Also, find the rate, the banker got on his money.

   **Hint.** *Rs. (1000 – 989) i.e., Rs. 11 is S.I. on Rs. 1000 for* $\frac{1}{5}$ *year, Find rate. Also since the banker got Rs. 11 on Rs. 989, find his percentage.*

3. The holder of a bill for Rs. 17850 nominally due on 21st May, 1991 received Rs. 357 less than the amount of the bill by having it discounted at 5%. When was it discounted ?

   **Hint.** *Clearly, S.I. on Rs. 17850 at 5% is Rs. 357.*

   $$\therefore \text{Time} = \left(\frac{100 \times 357}{17850 \times 5}\right) = \frac{2}{5} \text{ years} = 146 \text{ days.}$$

   *So, the bill is 146 days prior to 24th may, the legally due date.*

   *May April March Feb. Jan. Dec.*

   *24 + 30 + 31 + 28 + 31 + 2 = 146 days.*

   *So, the bill was discounted on Dec. 29, 1990.*

4. (i) Find the banker's gain on a bill of Rs. 6900 due 3 years hence at 5% per annum simple interest.

   **Hint.** $B.D. = \frac{Sum \times R \times T}{100}$, $T.D. = \frac{Sum \times R \times T}{100 + (R \times T)}$

   *and* $B.G. = B.D. - T.D.$

   (ii)

Find the banker's discount on a bill due 3 years hence at 5% being given that the banker's gain is Rs. 90.

**Hint.** $T.D. = \frac{B.G. \times 100}{R \times T}$, $B.D. = (T.D. + B.G.)$.

(iii) The banker's gain on a bill due 2 years hence at 5% is Rs. 8, find the present worth of the bill.

**Hint.** $T.D. = \frac{B.G. \times 100}{R \times T}$, $P.W. = \frac{T.D. \times 100}{R \times T}$.

(iv) The banker's gain on a bill due 1 year 4 months hence at $7\frac{1}{2}$% per annum simple interest is Rs. 16. Find the sum.

Hint. $T.D. = \frac{B.G. \times 100}{R \times T}$, $B.D. = T.D. + B.G.$

$$\therefore \text{Sum} = \frac{B.D. \times 100}{R \times T}.$$

(v) The banker's discount on a bill due 6 months hence at 6% is Rs. 37.08. Find the true discount.

**Hint.** $T.D. = \frac{B.D. \times 100}{100 + (R \times T)}$.

5. The banker's discount on a certain sum of money is Rs. 36 and the discount on the same sum for the same time and at the same rate is Rs. 30. Find the sum.
6. The present worth of a certain bill due some time hence is Rs. 1600 and the true discount on the bill is Rs. 160. Find the banker's discount and the extra gain the banker would make in the transaction.

**Hint.** $B.G. = \frac{(T.D.)^2}{(P.W.)}$ and $(B.D.) - (T.D.) = (B.G.)$.

7. A bill was drawn on May, 14 at 2 months after date and was discounted on July, 2 at $8\frac{1}{9}$% per annum simple interest. If the banker's gain be Re. 1, for what sum was the bill drawn ?

**Hint.** *Number of days for which the bill has yet to run = 15 days.*

$\therefore$ *Time* $= \frac{15}{365}$ *years or* $\frac{3}{73}$ *years.*

*Let the sum be Rs. 100. Then,*

$$T.D. = Rs. \left\{ \frac{\left(100 \times \frac{73}{9} \times \frac{3}{73}\right)}{100 + \left(\frac{73}{9} \times \frac{3}{73}\right)} \right\} = Rs. \left(\frac{100}{301}\right)$$

$$\text{and, } B.D. = Rs. \left(100 \times \frac{3}{73} \times \frac{73}{9}\right) \times \frac{1}{100} = Re. \left(\frac{1}{3}\right).$$

$$\therefore B.G. = Rs. \left(\frac{1}{3} - \frac{100}{301}\right) = Rs. \left(\frac{1}{903}\right).$$

*Now, if B.G. is Rs.* $\left(\frac{1}{903}\right)$, *then sum = Rs. 100.*

$\therefore$ *If B.G. is Re. 1, then sum = Rs.* (100 × 903) = Rs. 90300.

**8.** **The banker's gain on a certain sum due $2\frac{1}{2}$ years hence is $\frac{3}{23}$ of the banker's discount on it for the same time and at the same rate. Find the rate percent.**

**Hint.** *Let B.G. be Re. 1, then B.D. = Rs.* $\left(\frac{23}{3}\right)$.

$$\therefore T.D. = Rs. \left(\frac{23}{3} - 1\right) = Rs. \left(\frac{20}{3}\right).$$

$$\text{Now, sum} = \frac{(B.D. \times T.D.)}{(B.D.) - (T.D.)} = Rs. \left(\frac{460}{9}\right).$$

$\therefore$ *S.I. on Rs.* $\left(\frac{460}{9}\right)$ *for* $\frac{5}{2}$ *years is Rs.* $\left(\frac{23}{3}\right)$, *find rate percent.*

**9.** **The banker's discount on a certain sum due 2 years hence is $\frac{11}{10}$ of the true discount on it for the same time and at the same rate. Find the rate percent.**

**Hint.** *Let T.D. = Re. 1 then, B.D. = Rs.* $\frac{11}{10}$.

$$\therefore \text{Sum} = \frac{(B.D.) \times (T.D.)}{(B.D.) - (T.D.)} = Rs.\ 11$$

*Thus, S.I. on Rs. 11 for 2 years is Rs.* $\left(\frac{11}{10}\right)$, *find rate percent.*

**10.** **A bill is discounted at 8% per annum. If the banker's discount be allowed, at what rate of interest percent must the proceeds be invested, so that nothing may be lost ?**

**Hint.** *Rs. 92 must be invested to have Rs. 8 as S.I. for 1 year.*

## ANSWERS (Exercise 27A)

1. (i) Rs. 24 (ii) Rs. 1476 (iii) Rs. 23.62 (iv) 38 Paise

**2.** $5\frac{1}{2}\%$, $5\frac{555}{989}\%$ **3.** 29th Dec. 1990

**4.** (i) Rs. 135 (ii) Rs. 600 (iii) Rs. 800 (iv) Rs. 1760 (v) Rs. 36

**5.** Rs. 180 **6.** Rs. 176, Rs. 16 **7.** Rs. 90300 **8.** 6%

**9.** 5% **10.** $8\frac{16}{23}\%$.

## EXERCISE 27B (Objective Type Questions)

**1.** The true discount on a bill of Rs. 540 is Rs. 90. The banker's discount is :

(*a*) Rs. 60 (*b*) Rs. 150

(*c*) Rs. 180 (*d*) Rs. 110

**2.** The present worth of a certain bill due sometime hence is Rs. 800 and the true discount is Rs. 36. Then, the banker's discount is :

(*a*) Rs. 37 (*b*) Rs. 34.38

(*c*) Rs. 37.62 (*d*) Rs. 38.98

**3.** The banker's discount on a certain sum due 2 years hence is $\frac{11}{10}$ of the true discount. The rate percent is :

(*a*) 11% (*b*) 10%

(*c*) 5% (*d*) $5\frac{1}{3}\%$

**4.** The banker's gain on a certain sum due $2\frac{1}{2}$ years hence is (3/23) of the banker's discount. The rate percent is :

(*a*) 5% (*b*) 6%

(*c*) $2\frac{14}{23}\%$ (*d*) $6\frac{2}{3}\%$

**5.** The present worth of a certain sum due sometime hence is Rs. 1600 and the true discount is Rs. 160. The banker's gain is :

(*a*) Rs. 10 (*b*) Rs. 16

(*c*) Rs. 20 (*d*) Rs. 24

**6.** The banker's gain of a certain sum due 2 years hence at 5% per annum is Rs. 8. The present worth is :

(*a*) Rs. 800 (*b*) Rs. 1600

(*c*) Rs. 1200 (*d*) Rs. 880

**7.** The banker's gain of a certain sum of money is Rs. 36 and the true discount on the same sum for the same time and at the same rate is Rs. 30. The sum is :

(*a*) Rs. 1080 (*b*) Rs. 180

(*c*) Rs. 500 (*d*) Rs. 300

8. The banker's discount on a bill due 1 year 8 months hence is Rs. 50 and the true discount on the same sum at the same rate percent is Rs. 45. The rate percent is :

(a) 6% (b) $6\frac{2}{3}\%$

(c) $6\frac{1}{2}\%$ (d) $8\frac{44}{59}\%$

9. The banker's discount on Rs. 1600 at 6% is the same as the true discount on Rs. 1624 for the same time and at the same rate. Then, the time is :

(a) 3 months (b) 4 months

(c) 6 months (d) 8 months

10. The banker's discount on a sum of money for $1\frac{1}{2}$ years is Rs. 60 and the true discount on the same sum for 2 years is Rs. 75. The rate percent is :

(a) 5% (b) 6%

(c) $6\frac{2}{3}\%$ (d) $3\frac{1}{3}\%$

11. The banker's gain on a bill due 1 year hence at 5% is Re. 1. The true discount is :

(a) Rs. 15 (b) Rs. 20

(c) Rs. 25 (d) Rs. 5

12. The banker's discount on a bill due 6 months hence at 6% is Rs. 37.08. The true discount is :

(a) Rs. 6.18 (b) Rs. 12.36

(c) Rs. 48 (d) Rs. 36

13. The present worth of a sum due sometimes hence is Rs. 576 and the banker's gain is Re. 1. The true discount is :

(a) Rs. 16 (b) Rs. 18

(c) Rs. 24 (d) Rs. 32

14. A bill is discounted at 5% per annum. If banker's discount be allowed, at what rate percent must the proceeds be invested, so that nothing may be lost ?

(a) 5% (b) $4\frac{19}{21}\%$

(c) $5\frac{5}{19}\%$ (d) 10%

15. The banker's gain on a sum due 3 years hence at 5% is Rs. 90. The banker's discount is :

(a) Rs. 690 (b) Rs. 720

(c) Rs. 810 (d) Rs. 150

## SOLUTIONS (Exercise 27B)

**1.** P.W. = Rs. (540 – 90) = Rs. 450

S.I. on Rs. 450 = Rs. 90

$$B.D. = S.I. \text{ on Rs. } 540 = Rs. \left(\frac{90}{450} \times 540\right) = Rs.\ 108.$$

**2.** $$B.G. = \frac{(T.D.)^2}{PW} = Rs. \left(\frac{36 \times 36}{800}\right) = Rs.\ 1.62.$$

$\therefore$ *B.D. = (T.D.) + (B.G.) = Rs. (36 + 1.62) = Rs. (37.62).*

**3.** Let T.D. be Re. 1. Then, B.D. = Rs. (11/10) = Rs. 1.10.

$$\therefore \text{Sum} = Rs. \left(\frac{1.10 \times 1}{1.10 - 1}\right) = Rs. \frac{1.10}{0.10} = Rs.\ 11.$$

So, S.I. on Rs. 11 for 2 years is Rs. 1.10

$$\therefore Rate = \left(\frac{100 \times 1.10}{11 \times 2}\right)\% = 5\%.$$

**4.** Let B.D. be Re. 1. Then, B.G. = Re. (3/23).

$$\therefore T.D. = Re. \left(1 - \frac{3}{23}\right) = Re. \left(\frac{20}{23}\right).$$

$$\text{Sum} = Rs. \left[\left(1 \times \frac{20}{23}\right) \Big/ \left(1 - \frac{20}{23}\right)\right] = Rs. \left(\frac{20}{3}\right).$$

$\therefore$ S.I. on Rs. $\frac{20}{3}$ for $2\frac{1}{2}$ years is Re. 1.

$$\therefore Rate = \left(\frac{100 \times 1}{\frac{20}{3} \times \frac{5}{2}}\right)\% = 6\%.$$

**5.** $$B.G. = \frac{(T.D.)^2}{P.W.} = Rs. \left(\frac{160 \times 160}{1600}\right) = Rs.\ 16.$$

**6.** $$T.D. = \frac{B.G. \times 100}{Rate \times Time} = Rs. \left(\frac{8 \times 100}{5 \times 2}\right) = Rs.\ 800.$$

**7.** $$\text{Sum} = \frac{B.D. \times T.D.}{B.D. - T.D.} = Rs. \left(\frac{36 \times 30}{6}\right) = Rs.\ 180.$$

**8.** $$\text{Sum} = \frac{B.D. \times T.D.}{B.D. - T.D.} = Rs. \left(\frac{50 \times 45}{5}\right) = Rs.\ 450.$$

Now, Rs. 50 is S.I. on Rs. 450 for (5/3) years.

$$\therefore Rate = \left(\frac{100 \times 50}{450 \times \frac{5}{3}}\right)\% = 6\frac{2}{3}\%.$$

**9.** S.I. on Rs. 1600 = T.D. on Rs. 1624

$\therefore$ Rs. 1600 is P.W. of Rs.1624.

i.e., Rs. 24 is the S.I. on Rs. 1600 at 6%.

$\therefore$ *Time* $= \left(\dfrac{100 \times 24}{1600 \times 6}\right)$ *year* $= \dfrac{1}{4}$ *year = 3 months.*

**10.** B.D. for (3/2) years = Rs. 60

B.D. for 2 years = Rs. $\left(\dfrac{60 \times 2}{3} \times 2\right)$ = Rs. 80.

Now, B.D. = Rs. 80 : T.D. = Rs. 75 & Time = 2 years.

$\therefore$ Sum = Rs. $\left(\dfrac{80 \times 75}{5}\right)$ = Rs. 1200.

$\therefore$ Rs. 80 is S.I. on Rs. 1200 for 2 years.

*So, rate* $= \left(\dfrac{100 \times 80}{1200 \times 2}\right)\% = 3\dfrac{1}{3}\%.$

**11.** T.D. $= \dfrac{\text{B.G.} \times 100}{\text{R} \times \text{T}}$ = Rs. $\left(\dfrac{1 \times 100}{5 \times 1}\right)$ = Rs. 20.

**12.** T.D. $= \dfrac{\text{B.D.} \times 100}{100 + (\text{R} \times \text{T})}$ = Rs. $\left\{\dfrac{37.08 \times 100}{100 + \left(6 \times \dfrac{1}{2}\right)}\right\}$ = Rs. 36.

**13.** T.D. $= \sqrt{\{(\text{P.W.}) \times (\text{B.G.})\}}$ = Rs. $\sqrt{(576 \times 1)}$ = Rs. 24.

**14.** Let the sum be Rs. 100. Then, B.D. = Rs. 5.

Proceeds = Rs. (100 – 5) = Rs. 95.

$\therefore$ Rs. 5 must be the interest on Rs. 95 for 1 year.

So, rate $= \left(\dfrac{100 \times 5}{95 \times 1}\right) = 5\dfrac{5}{19}\%.$

**15.** T.D. $= \dfrac{\text{B.G.} \times 100}{\text{R} \times \text{T}}$ = Rs. $\left(\dfrac{90 \times 100}{5 \times 3}\right)$ = Rs. 600.

$\therefore$ *B.D. = Rs. (600 + 90) = Rs. 690.*

## ANSWERS (Exercise 27B)

| | | | | | |
|---|---|---|---|---|---|
| **1.** (*c*) | **2.** (*c*) | **3.** (*c*) | **4.** (*b*) | **5.** (*b*) | **6.** (*a*) |
| **7.** (*b*) | **8.** (*b*) | **9.** (*a*) | **10.** (*d*) | **11.** (*b*) | **12.** (*d*) |
| **13.** (*c*) | **14.** (*c*) | **15.** [illegible] | | | |

# 28
# TABULATION

## TABULATION

*In studying problems on statistics, the data collected by the investigator are arranged in a systematic form, called the* **tabular form.** *In order to avoid same heads again, we make tables consisting of horizontal lines (called* **rows***) and vertical lines (called* **columns***) with distinctive heads, known as* **captions.** *Units of measurements are given along with the captions.*

**Ex. 1.** ***The following data give year-wise outlay in lakhs of rupees in a certain 5 years plan (1980–1985) of a state, under the heads : Transport and Communication, Education, Health, Housing and Social welfare respectively.***

**1st Year :** 56219, 75493, 13537, 9596 & 1985.

**2nd Year :** 71416, 80691, 15902, 10135 & 2073.

**3rd Year :** 73520, 61218, 16736, 11000 & 3918.

**4th Year :** 75104, 73117, 17523, 12038 and 4102.

**5th Year :** 80216, 90376, 19420, 15946 & 10523.

***Putting the data in the form of a table, write the total under each head and answer the following questions :***

1. During which year the outlay on education was maximum ?
2. How many times, the outlay on education was increased over preceding year ?
3. What is the percentage increase during 1983–84 ...er 1982–83 in health outlay ?
4. What is total outlay on social welfare during the plan period ?
5. What is the ratio between outlays on (Transport and Communication) and housing during 1984–85.

**SOLUTION :** The table may be constructed as shown below :

**Outlay (in lakhs of rupees) of a state in a 5 years–plan (1980 to 85) :**

| *Year* | *Transport & Communication* | *Education* | *Housing* | *Health* | *Social welfare* | *Total* |
|---|---|---|---|---|---|---|
| 1980-81 | 56219 | 75493 | 13537 | 9596 | 1985 | 156830 |
| 1981-82 | 71416 | 80691 | 15902 | 10135 | 2073 | 180217 |
| 1982-83 | 73520 | 61218 | 16736 | 11000 | 3918 | 166392 |
| 1983-84 | 75104 | 73117 | 17523 | 12038 | 4102 | 181884 |
| 1984-85 | 80216 | 90376 | 19420 | 15946 | 10523 | 216481 |
| **Total** | 356475 | 380895 | 83118 | 58715 | 22601 | 901804 |

**As given in the table :**

**Ans. 1. During 1984–85, the outlay on education was maximum.**

**Ans. 2.** Clearly; the outlay on education was increased in 1981–82 over 1980–81; in 1983–84 over 1982–83 and in 1984–85 over 1983–84.

Thus, it was increased three times during the plan period.

**Ans. 3.** % increase in 1983–84 over 1982–83 in health

$$= \left(\frac{12038 - 11000}{11000}\right) \times 100\% = 9.43\%.$$

**Ans. 4.** Total outlay on social welfare during the plan–period is Rs. 22601 lakhs.

**Ans. 5.** Ratio between outlays on (Transport and communication) and housing during 1984–85 is

$= 80216 : 19420 = 4.13 : 1 = (413 : 100).$

**Ex. 2.** ***Following table gives the population of a town from 1988 to 1992 :***

| *Year* | *Men* | *Women* | *Children* | *Total* | *Increase (+) or Decrease (–) over Preceding Year* |
|---|---|---|---|---|---|
| 1988 | 65104 | 60387 | ...... | 146947 | .......... |
| 1989 | 70391 | 62516 | ...... | ...... | +(11630) |
| 1990 | ...... | 63143 | 20314 | 153922 | .......... |
| 1991 | 69395 | ...... | 21560 | ...... | –(5337) |
| 1992 | 71274 | ...... | 23789 | 160998 | .......... |

**Complete the table and mark a tick against the correct answer in each question :**

1. The number of children in 1988 is :

(*a*) 31236 (*b*) 125491

(*c*) 14546 (*d*) 21456

2. The total population in 1989 is :

(*a*) 144537 (*b*) 158577
(*c*) 146947 (*d*) 149637

3. Number of children in 1989 is :
(*a*) 25670 (*b*) 14040
(*c*) 13970 (*d*) 15702

4. Number of men in 1990 is :
(*a*) 40645 (*b*) 60454
(*c*) 70465 (*d*) 58835

5. Number of women in 1991 is :
(*a*) 57630 (*b*) 56740
(*c*) 52297 (*d*) 62957

6. Increase or decrease of population in 1992 over 1991 is :
(*a*) –(12413) (*b*) +(12413)
(*c*) +155661 (*d*) +7086

**Solution**

**Q. 1.** Number of children in 1988

= (146947) – (65104 + 60387) = 21456.

∴ Answer (*d*) is correct.

**Q. 2.** Total population in 1988 is 146947 and increase in 1989 is 11630. Therefore, total population in 1989 is

= (146947 + 11630) = 158577.

∴ Answer (*b*) is correct.

**Q. 3.** Number of children in 1989.

= (158577) – (70391 + 62516) = 25670.

∴ Answer (*a*) is correct.

**Q. 4.** Number of men in 1990

= (153922) – (63143 + 20314) = 70465.

∴ Answer (*c*) is correct.

**Q. 5.** Total population in 1990 was 153922 and decrease in next year was 5337. So, the total population in 1991

= (153922 – 5337) = 148585.

Number of women in 1991

= (148585) – (69395 + 21560) = 57630.

∴ Answer (*a*) is correct.

**Q. 6.** Total population in 1991 was 148585 and that in 1992 was 160998. So, increase = (160998 – 148585) = 12413.

∴ Answer (*b*) is correct.

Also, number of women in 1992

= (160998) – (71274 + 23789) = 65935.

**Filling all these entries, the complete table is given below :**

| *Year* | *Men* | *Women* | *Children* | *Total* | *Increase (+) or Decrease (–) over Preceding Year* |
|---|---|---|---|---|---|
| 1988 | 65104 | 60387 | 21456 | 146947 | .......... |
| 1989 | 70391 | 62516 | 25670 | 158577 | + (11630) |
| 1990 | 70465 | 63143 | 20314 | 153922 | – (4655) |
| 1991 | 69395 | 57630 | 21560 | 148585 | – (5337) |
| 1992 | 71274 | 65935 | 23789 | 160998 | + (12413) |

**Ex. 3.** ***The table given below shows the population, litrates and illiterates in thousands and percentage of literacy in three states in a year :***

| *States* | *Population* | *Literates* | *Illitrates* | *%age of literacy* |
|---|---|---|---|---|
| *Madras* | 49342 | 6421 | ...... | ......... |
| *Bombay* | ...... | 4068 | 16790 | ......... |
| *Bengal* | 60314 | ...... | ...... | 16.1 |

**After reading the table, mark a tick against the correct answer in each question given below and hence complete the table.**

**1.** Percentage of literacy in Madras is :

(*a*) 14.9% (*b*) 13.01%

(*c*) 12.61% (*d*) 15.04%

**2.** Percentage of literacy in Bombay is :

(*a*) 19.5% (*b*) 16.7%

(*c*) 18.3% (*d*) 14.6%

**3.** Literates in Bengal are :

(*a*) 50599 (*b*) 9715

(*c*) 7865 (*d*) 9475

**Solution :**

**Q. 1.** Percentage of literacy in Madras

$$= \left(\frac{6421}{49342} \times 100\right)\% = 13.01\%.$$

∴ Answer (*b*) is correct.

**Q. 2.** Population of Bombay = (4068 + 16790) = 20858 thousands.

∴ Percentage of literacy in Bombay

$$= \left(\frac{4068}{20858} \times 100\right)\% = 19.5\%.$$

∴ Answer (*a*) is correct.

**Q. 3.** Number of literates in Bengal

$$= \left(\frac{16.1}{100} \times 60314\right) = 9715.$$

∴ Answer (*b*) is correct.

Also, number of illiterates in Bengal
= (60314 – 9715) = 50599 thousands.

**Filling these entries, the complete table is given below :**

| *States* | *Population* | *Literates* | *Illiterates* | *Percentage of Literacy* |
|---|---|---|---|---|
| *Madras* | 49342 | 6421 | 42921 | 13.01% |
| *Bombay* | 20858 | 4068 | 16790 | 19.5% |
| *Bengal* | 60314 | 9715 | 50599 | 16.1% |

**Ex. 4.** ***The following table shows the production of food grains (in million tonnes) in a state for the period from 1988-89 to 1992-93.***

| *Year* | *Production in Million Tonnes* | | | | *Total* |
|---|---|---|---|---|---|
| | *Wheat* | *Rice* | *Maize* | *Other cereals* | |
| 1988–89 | 580 | 170 | 150 | 350 | 1350 |
| 1989–90 | 600 | 220 | 234 | 400 | 1474 |
| 1990–91 | 560 | 240 | 228 | 420 | 1538 |
| 1991–92 | 680 | 300 | 380 | 460 | 1660 |
| 1992–93 | 860 | 260 | 340 | 500 | 1910 |
| **Total** | 3280 | 1190 | 1332 | 2130 | 7932 |

**Read the above table and mark a tick against the correct answer in each of the following questions :**

**1.** During the period from 1988-89 to 1992-93 what percent of the total production is the wheat ?

(*a*) 42.6% (*b*) 43.1%
(*c*) 41.3% (*d*) 40.8%

**2.** During the year 1992-93 the percentage increase in production of wheat over the previous year was :

(*a*) 26.4% (*b*) 20.9%
(*c*) 23.6% (*d*) 18.7%

**3.** In the year 1991-92, the increase in production was maximum for :

(*a*) wheat (*b*) rice
(*c*) maize (*d*) other cereals

**4.** During the year 1990-91, the percentage of decrease in production of maize was

(*a*) 2.63% (*b*) 2.56%
(*c*) 2.71% (*d*) 2.47%

**5.** The increase in the production of other cereals was minimum during the year :

(*a*) 1989-90 (*b*) 1990-91
(*c*) 1991-92 (*d*) 1992-93

**Solution :**

**Q. 1.** Total production during the period = 7932 million tonnes.

Wheat production during the period = 3280 million tonnes.

Percentage of wheat production over total production

$$= \left(\frac{3280}{7932} \times 100\right)\% = 41.3\%.$$

$\therefore$ Answer (*c*) is correct.

**Q. 2.** Increase in 1992-93 in wheat production over 1991-92

$$= (860 - 680) = 180 \text{ million tonnes.}$$

$$\text{Increase } \% = \left(\frac{180}{680} \times 100\right)\% = 26.4\%$$

$\therefore$ Answer (*a*) is correct.

**Q. 3.** During 1991-92, as read from the table the increase in the production of wheat, rice, maize and other cereals is 120, 60, 152 and 40 million tonnes respectively. So, increase in maize production is maximum. So, answer (*c*) is correct.

**Q. 4.** Decrease in production of maize in 1990-91

$$= (234 - 228) = 6 \text{ million tonnes.}$$

$$\text{Decrease } \% = \left(\frac{6}{234} \times 100\right)\% = 2.56\%.$$

$\therefore$ Answer (*b*) is correct.

**Q. 5.** Increase in production of other cereals in 1989-90, 1990-91, 1991-92 and 1992-93 over previous year is 50, 20, 40, 40 million tonnes respectively. So the increase is minimum in 1990-91.

$\therefore$ Answer (*b*) is correct.

**Ex. 5.** ***Study the following table carefully and answer the questions given below :*** **(Bank P.O. Exam. 1987)**

## LOAN DISBURSED BY 5 BANKS

**(Rupees in Crores)**

| *Banks* | *Years* | | | | |
|---|---|---|---|---|---|
| | 1982 | 1983 | 1984 | 1985 | 1986 |
| *A* | 18 | 23 | 45 | 30 | 70 |
| *B* | 27 | 33 | 18 | 41 | 37 |
| *C* | 29 | 29 | 22 | 17 | 11 |
| *D* | 31 | 16 | 28 | 32 | 43 |
| *E* | 13 | 19 | 27 | 34 | 42 |
| **Total** | 118 | 120 | 140 | 154 | 203 |

**Q. 1.** In which year was the disbursement of loans of all the banks put together least compared to the average disbursement of loans over the years :

(*a*) 1982 (*b*) 1983
(*c*) 1984 (*d*) 1985 (*e*) 1986

**Q. 2.** What was the percentage increase of disbursement of loans of all banks together from 1984 to 1985 :

(*a*) 110 (*b*) 14
(*c*) $90\frac{10}{11}$ (*d*) 10 (*e*) none of these

**Q. 3.** In which year was the total disbursement of loans of banks *A* & *B* exactly equal to the total disbursement of banks *D* and *E* ?

(*a*) 1983 (*b*) 1986
(*c*) 1984 (*d*) 1982 (*e*) none of these

**Q. 4.** In which of the following banks did the disbursement of loans continuously increase over the years :

(*a*) *A* (*b*) *B*
(*c*) *C* (*d*) *D* (e) E

**Q. 5.** If the minimum target in the preceding years was 20% of the total disbursement of loans, how many banks reached the target in 1983 :

(*a*) 1 (*b*) 3
(*c*) 2 (*d*) 4 (*e*) none of these

**Q. 6.** In which bank was loan disbursement more than 25% of the disbursement of all banks together in 1986 :

(*a*) *A* (*b*) *B*
(*c*) *C* (*d*) *D* (e) E

**Solutions :**

**Q. 1.** Average disbursement of loans over the years

$$= \frac{1}{5}(118 + 120 + 140 + 154 + 203) = 147.$$

Clearly, it is least in the year 1982.

So, answer (*a*) is correct.

**Q. 2.** Increase of loans from 1984 to 1985

$$= \left(\frac{154 - 140}{140}\right) \times 100\% = 10\%.$$

*i.e.* (*d*) is correct.

**Q. 3.** In none of the years is the sum of loans of *A* & *B* is equal to sum or loans of *D* & *E*. So, answer (*e*) is correct.

**Q. 4.** In bank *E* the disbursement of loans continuously increase over the years. So, answer (*e*) is correct.

**Q. 5.** 20% of total loans disbursed in 1982

= (20% of 118) = 23.6 crores.

Clearly, *B* & *C* reached the target in 1983.

So, answer (*c*) is correct.

**Q. 6.** In 1986, 25% of total disbursement

= (25% of 203) crores = 50.75 crores.

∴ In bank *A*, the loan disbursed is more than 25% of the total disbursement of all banks in 1986.

Hence, answer (*a*) is correct.

**Ex. 6.** ***Study the following table carefully and answer the questions given below :*** **(Bank P.O. Exam. 1989)**

## FINANCIAL STATEMENT OF A COMPANY OVER THE YEARS

**(Rupees in Lakhs)**

| *Year* | *Gross Turnover Rs.* | *Profit before interest and depreciation Rs.* | *Interest Rs.* | *Depriciation Rs.* | *Net Profit Rs.* |
|---|---|---|---|---|---|
| 1980–81 | 1380.00 | 380.92 | 300.25 | 69.90 | 10.67 |
| 1981–82 | 1401.00 | 404.98 | 315.40 | 71.12 | 18.46 |
| 1982–83 | 1540.00 | 520.03 | 390.85 | 80.02 | 49.16 |
| 1983–84 | 2112.00 | 599.01 | 444.44 | 88.88 | 65.69 |
| 1984–85 | 2520.00 | 810.11 | 505.42 | 91.91 | 212.78 |
| 1985–86 | 2758.99 | 920.00 | 600.20 | 99.00 | 220.80 |

**Q. 1.** During which year did the 'Net Profit' exceed Rs. 1 crore for the first time ?

(*a*) 1985-86 (*b*) 1984-85
(*c*) 1983-84 (*d*) 1982-83 (*e*) none of these

**Q. 2.** During which year was the "Gross Turnover" closest to thrice the 'Profit before Interest and Depreciation' ?

(*a*) 1985-86 (*b*) 1984-85
(*c*) 1983-84 (*d*) 1982-83 (*e*) 1981-82

**Q. 3.** During which year did the 'Net Profit' form the highest proportion of the 'Profit before Interest and Depreciation' ?

(*a*) 1984-85 (*b*) 1983-84
(*c*) 1982-83 (*d*) 1981-82 (*e*) 1980-81

**Q. 4.** Which of the following registered the lowest increase in terms of rupees from the year 1984-85 to the year 1985-86 ?

(*a*) Gross Turnover

(*b*) Profit before Interest and Depreciation

(*c*) Depreciation

(*d*) Interest

(*e*) Net profit

**Q. 5.** The 'Gross Turnover' for 1982-83 is about what per cent of the 'Gross Turnover' for 1984-85 ?

(*a*) 61 (*b*) 163

(*c*) 0.611 (*d*) 39 (*e*) 0.006

**Solutions :**

**Q. 1.** Clearly, the net profit exceeded Rs. 1 crore in the year 1984-85.
So, answer (*b*) is correct.

**Q. 2.** The ratio of 'Gross turnover' to the 'profit before Interert and Depreciation' :

in 1980-81 is $\frac{1380}{380.92} = 3.62$;

in 1981-82 is $\frac{1401}{404.98} = 3.46$;

in 1982-83 is $\frac{1540}{520.03} = 2.96$;

in 1983-84 is $\frac{2112}{599.01} = 3.53$;

in 1984-85 is $\frac{2520}{810.11} = 3.11$.

in 1985-86 is $\frac{2758.99}{920} = 3$.

So, answer (*a*) is correct.

**Q. 3.** Let, Net profit= $x\%$ of Profit before interest & depreciation.

For 1980-81, we have $x = \frac{10.67 \times 100}{380.92} = 2.80$;

For 1981-82, we have $x = \frac{18.46 \times 100}{404.98} = 4.56$;

For 1982-83, we have $x = \frac{49.16 \times 100}{520.03} = 9.45$;

For 1983-84, we have $x = \frac{65.69 \times 100}{599.01} = 10.97$;

For 1984-85, we have $x = \frac{212.78 \times 100}{810.11} = 26.26$;

For 1985-86, we have $x = \frac{220.80 \times 100}{920} = 24$.

So, in 1984-85, the net profit forms the highest proportion of the 'profit before interest and depreciation'

∴ Answer (*a*) is correct.

**Q. 4.** Increase from the year 1984-85 to 1985-86 in

Gross turnover is (2758.99 – 2520) = 238.99 lakhs;

Profit before int. & depreciation is

(920 – 810.11) = 109.89 lakhs;

Interest is (600.20 – 505.42) = (94.78) lakhs;

Depreciation is (99 – 91.91) = (7.09) lakhs;

Net profit is (220.80 – 212.78) = 8.02 lakhs.

Clearly, the increase is lowest in depreciation.

So, answer (*c*) is correct.

**Q. 5.** Let $x\%$ of Gross Turnover for 1984-85

= Gross turnover for 1982-83

Then, $\frac{x}{100} \times 2520 = 1540$ or $x = \frac{1540 \times 100}{2520} = 61$ (approx).

So, answer (*a*) is correct.

**Ex. 7.** ***Study the following table carefully and answer the questions given below :*** **(Bank P. O. Exam. 1988)**

## NUMBER OF BOYS OF STANDARD XI PARTICIPATING IN DIFFERENT GAMES

| *Games ↓ Class →* | *XI A* | *XI B* | *XI C* | *XI D* | *XI E* | *TOTAL* |
|---|---|---|---|---|---|---|
| *Chess* | 8 | 8 | 8 | 4 | 4 | 32 |
| *Badminton* | 8 | 12 | 8 | 12 | 12 | 52 |
| *Table Tennis* | 12 | 16 | 12 | 8 | 12 | 60 |
| *Hockey* | 8 | 4 | 8 | 4 | 8 | 32 |
| *Football* | 8 | 8 | 12 | 12 | 12 | 52 |
| *Total no. of boys* | 44 | 48 | 48 | 48 | 48 | 228 |

**Note :**

1. Every student (boy or girl) of each class of standard XI participates in a game.
2. In each class, the number of girls participating in each game is 25% of the number of boys participating in each game.
3. Each student (boy or girl) participates in one and only one game.

**Q. 1.** All the boys of class XI D passed at the annual examination but a few girls failed. If all the boys and girls who passed and entered XII D and if in class XII D, the ratio of boys to girls is 5 : 1, what would be the number of girls who failed in class XI D ?

(*a*) 8 (*b*) 5
(*c*) 2 (*d*) 1 (*e*) none of these

**Q. 2.** Girls playing which of the following games need to be combined to yield a ratio of boys to girls of 4 : 1, if all boys playing Chess and Badminton are combined ?

(*a*) Table Tennis and Hockey
(*b*) Badminton and Table Tennis
(*c*) Chess and Hockey
(*d*) Hockey and Football
(*e*) none of these

**Q. 3.** What should be the total number of students in the school if all the boys of class XI A together with all the girls of class XI B and class XI C were to be equal to 25% of the total number of students ?

(*a*) 272 (*b*) 560
(*c*) 656 (*d*) 340 (*e*) none of these

**Q. 4.** Boys of which of the following classes need to be combined to equal to four times the number of girls in class XI B and class XI C ?

(*a*) XI D & XI E (*b*) XI A & XI B
(*c*) XI A & XI C (*d*) XI A & XI D
(*e*) none of these

**Q. 5.** If boys of class XI E participating in chess together with girls of class XI B and class XI C participating in Table Tennis & hockey respectively are selected for a course at the college of sports; what percent of the students will get this advantage approximately ?

(*a*) 4.38 (*b*) 3.51
(*c*) 10.52 (*d*) 13.5 (*e*) none of these

**Q. 6.** If for social work, every boy of class XI D and class XI C is paired with a girl of the same class, what percentage of the boys of these two classes cannot participate in social work ?

(*a*) 88 (*b*) 66
(*c*) 60 (*d*) 75 (*e*) none of these

**Solutions :**

**Q. 1.** Total number of boys in XI D = 40

Number of girls in XI D = 25% of 40 = 10.

Since all boys of XI D passed, so the number of boys in XII D = 40.

Ratio of boys & girls in XII D is 5 : 1.

$\therefore$ Number of girls in XII D = $\left(\frac{1}{5} \times 40\right) = 8$.

So, the number of girls failed in XI D = (10 – 8) = 2.

∴ Answer (*c*) is correct.

**Q. 2.** Total number of boys playing chess & Badminton

$= (32 + 52) = 84.$

Number of girls playing hockey & football

$= 25\% \text{ of } 84 = \left(\frac{1}{4} \times 84\right) = 21.$

Since 84 : 21 is 4 : 1, so the girls playing hockey and football are combined to yield a ratio of boys to girls as 4 : 1.

So, answer (*d*) is correct.

**Q. 3.** Number of boys in XI A = 44;

Number of girls in XI B = 25% of 48 = 12;

Number of girls in XI C = 25% of 48 = 12;

∴ (44 + 12 + 12) = 68

Let $x$ be the total number of students.

Then, 25% of $x$ = 68 or $x = \frac{68 \times 100}{25} = 272.$

Total number of students in the school = 272, *i.e.* (*a*) is correct.

**Q. 4.** 4 times the number of girls in XI B & XI C.

= 4 (12 + 12) = 96.

But, none of the pairs of classes given through (A) to (D) has this as the number of boys. So, (*e*) is correct.

**Q. 5.** Number of boys of XI E playing chess = 4;

Number of girls of XI B playing table tennis = 25% of 16 = 4;

Number of girls of XI C playing hockey = 25% of 8 = 2

∴ Number of students selected for a course at the college of sports

= (4 + 4 + 2) = 10.

Total number of students = (228 + 25% of 228) = 285.

Let $x\%$ of 285 = 10 or $x = \left(\frac{10 \times 100}{285}\right) = 3.51.$

So, answer (*b*) is correct.

**Q. 6.** Since the number of girls = 25% of the number of boys, so only 25% of the boys can participate in social work.

∴ Answer (*d*) is correct.

# 29

# DATA ANALYSIS

## Bar Diagrams

(i) **Multiple Bar Diagrams.** In such bar diagrams two or more adjacent vertical bars are drawn to represent two or more phenomenon for the same place or period.

**Ex. 1.** ***Shown below is the multiple bar diagram depicting the changes in the student's strength of a college in four faculties from 1990-91 to 1992-93. (scale 1 cm. = 100)***

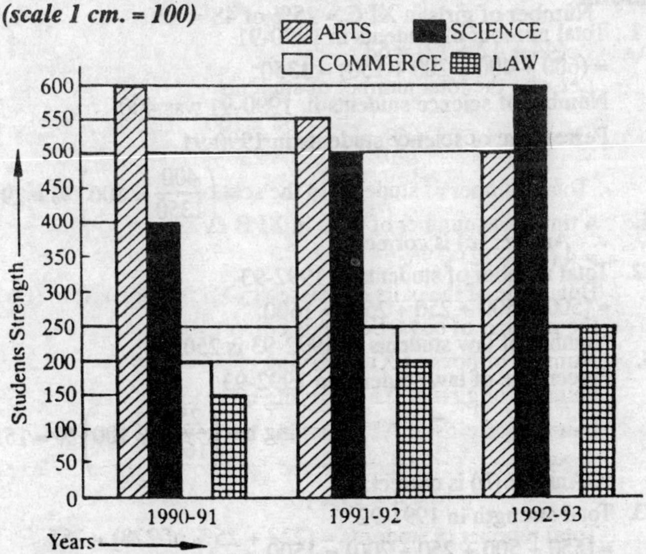


***Study the above multiple bar chart and mark a tick against the correct answer in each of the following questions.***

**Q. 1.** The percentage of students in science faculty in 1990-91 was :

(*a*) 26.9% (*b*) 27.8%

(*c*) 29.6% (*d*) 30.2%

**Q. 2.** The percentage of students in law faculty in 1992-93 was :

(*a*) 18.5% (*b*) 15.6%

(*c*) 16.7% (*d*) 14.8%

**Q. 3.** How many times was the total strength of the strength of Commerce students in 1991-92 ?

(*a*) 3 times (*b*) 4 times

(c) 5 times (d) 6 times

**Q. 4.** During which year the strength of arts faculty was minimum ?

(a) 1990-91 (b) 1991-92

(c) 1992-93

**Q. 5.** How much percent was the increase in science students in 1992-93 over 1990-91 ?

(a) 50% (b) 150%

(c) $66\frac{2}{3}\%$ (d) 75%

**Q. 6.** A regular decrease in students strength was in the faculty of :

(a) arts (b) Science

(c) Commerce (d) law

**Solutions :**

**Ans. 1.** Total number of students in 1990-91

= (600 + 400 + 200 + 150) = 1350.

Number of science students in 1990-91 was 400

Percentage of science students in 1990-91

$$= \left(\frac{400}{1350} \times 100\right)\% = 29.6\%$$

∴ Answer (c) is correct.

**Ans. 2.** Total number of students in 1992-93

= (500 + 600 + 250 + 250) = 1600.

Number of law students in 1992-93 is 250.

Percentage of law students in 1992-93

$$= \left(\frac{250}{1600} \times 100\right)\% = 15.6\%.$$

∴ Answer (b) is correct.

**Ans. 3.** Total strength in 1991-92

= (550 + 500 + 250 + 200) = 1500.

$$\therefore \frac{\text{Total strength}}{\text{strength of commerce students}} = \frac{1500}{250} = 6.$$

So, answer (d) is correct.

**Ans. 4.** A slight look indicates that the strength in arts faculty in 1990-91, 1991-92 & 1992-93 was 600, 550 and 500 respectively. So, it was minimum in 1992-93.

∴ Answer (c) is correct.

**Ans. 5.** Number of science students in 1990-91 was 400.

Number of science students in 1992-93 was 600.

$$\text{Percentage increase} = \left(\frac{200}{400} \times 100\right)\% = 50\%.$$

∴ Answer (*a*) is correct.

**Ans. 6.** As the diagram shows the decrease every year is in arts faculty.
So, answer (*a*) is correct.

**Ex. 2.** ***Given below is a bar diagram showing the percentage of Hindus, Sikhs and Muslims in a state during the years 1989 to 1992.***

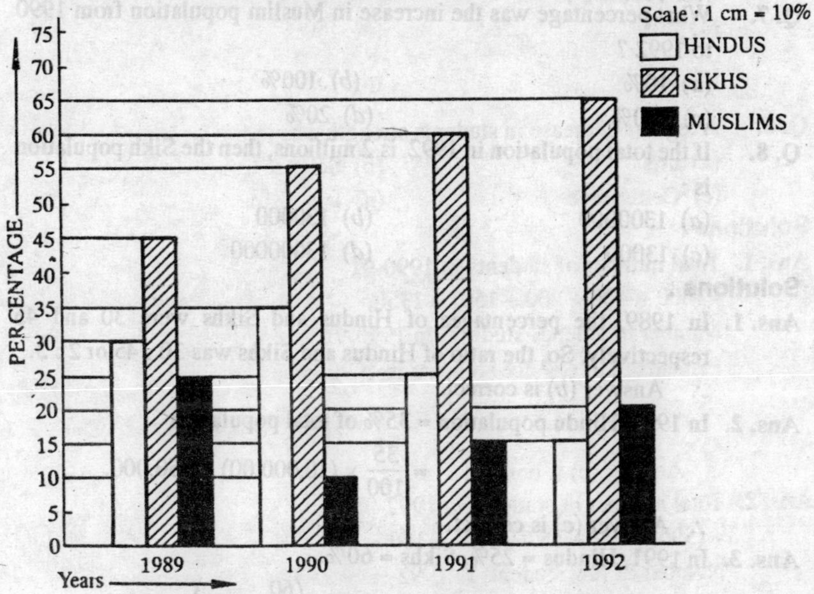


***Study the above diagram and mark a tick against the correct answer in each one of the following questions :***

**Q. 1.** The ratio between Hindus & Sikhs in 1989 was :

(*a*) 3 : 2 (*b*) 2 : 3
(*c*) can not be calculated

**Q. 2.** If the total population of the state in 1990 is 1 million, then the Hindus population was :

(*a*) 35000000 (*b*) 3500000
(*c*) 350000 (*d*) 35000

**Q. 3.** What was the percentage of Sikhs over Hindus in 1991 ?

(*a*) 35% (*b*) 40%
(*c*) 140% (*d*) 240%

**Q. 4.** What percentage was the decrease in Hindus population from 1989 to 1992 ?

(*a*) 15% (*b*) 45%
(*c*) 50% (*d*) 25%

**Q. 5.** If the population of the state in 1989 be 6 lakhs, then what is the total population of Hindus and Muslims in this year ?

(*a*) 270000 (*b*) 3300000
(*c*) 330000 (*d*) 33000

**Q. 6.** During which year was the Hindu percentage maximum ?
(*a*) 1989 (*b*) 1990
(*c*) 1991 (*d*) 1992

**Q. 7.** What percentage was the increase in Muslim population from 1990 to 1992 ?
(*a*) 10% (*b*) 100%
(*c*) 200% (*d*) 20%

**Q. 8.** If the total population in 1992 is 2 millions, then the Sikh population is :
(*a*) 1300000 (*b*) 130000
(*c*) 13000 (*d*) 13000000

**Solutions :**

**Ans. 1.** In 1989, the percentages of Hindus and Sikhs were 30 and 45 respectively. So, the ratio of Hindus and Sikhs was 30 : 45 or 2 : 3.
∴ Answer (*b*) is correct.

**Ans. 2.** In 1990, Hindu population = 35% of total population

$$= \frac{35}{100} \times (10{,}000{,}00) = 350{,}000.$$

∴ Answer (*c*) is correct.

**Ans. 3.** In 1991, Hindus = 25%, Sikhs = 60%.

∴ Percentage of Sikhs over Hindus $= \left(\frac{60}{25} \times 100\right) = 240\%.$

So, answer (*d*) is correct.

**Ans. 4.** Hindus in 1989 = 30%.
Hindus in 1992 = 15%.
Over 30, decrease = 15.

Over 100, decrease $= \left(\frac{15}{30} \times 100\right) = 50\%.$

∴ Answer (*c*) is correct.

**Ans. 5.** In 1989, Sikh population = (45% of 600000)

$$= \left(\frac{45}{100} \times 600000\right) = 270000.$$

∴ (Hindus + Muslims) = 600000 - 270000 = 330000.
So, answer (*c*) is correct.

**Ans. 6.** A quick observation of the chart shows that Hindus in 1989, 90, 91 and 92 were 30%, 35%, 25%, 15% respectively.
So, the maximum Hindu percentage was in 1990.
∴ Answer (*b*) is correct.

**Ans. 7.** Muslim population in 1990 = 10%.

Muslim population in 1992 = 20%.

Increase on 10 = 10.

Increase on 100 = $\left(\frac{10}{10} \times 100\right)$ % = 100%.

∴ Answer (*b*) is correct.

**Ans. 8.** In 1992, Sikh population = (65% of 2000000)

$= \left(\frac{65}{100} \times 2000000\right) = 1300000.$

∴ Answer (*a*) is correct.

**Ex. 3.** ***Study the following graph carefully and answer the questions given below it :***

**Export of Pearls in Crores of Rupees**

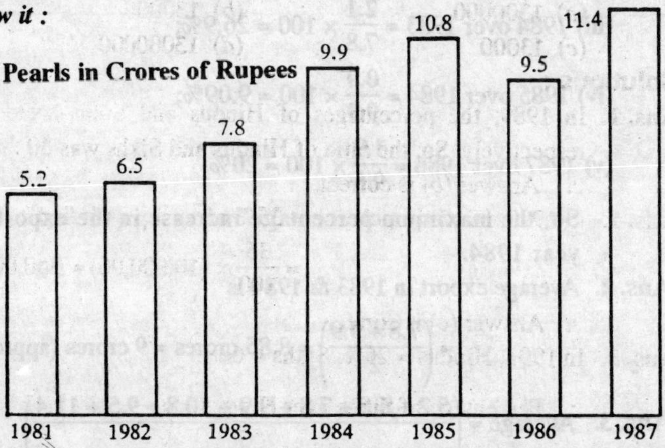


**Q. 1.** In which year there was maximum percentage increase in export of pearls to that in the previous year ?

(*a*) 1982 (*b*) 1987

(*c*) 1985 (*d*) 1984

**Q. 2.** In which of the following pairs of years was the average export of pearls around 9 crores ?

(*a*) 1982 & 1983 (*b*) 1983 & 1984

(*c*) 1984 & 1985 (*d*) 1985 & 1986

**Q. 3.** In how many years was the export above the average for the given period :

(*a*) 2 (*b*) 3

(*c*) 4 (*d*) 5

**Q. 4.** In which year was the export equal to the average export of the preceding and the following year :

(*a*) 1982 (*b*) 1983

(*c*) 1985 (*d*) 1986

**Q. 5.** What was the percentage increase in export from 1986 to 1987 ?

(*a*) $16\frac{2}{3}\%$ (*b*) 20%

(*c*) 19% (*d*) $33\frac{1}{3}\%$

**Solutions :**

**Ans. 1.** Percentage increase in export of pearls in :

(i) 1982 over 1981 $= \frac{1.3}{5.2} \times 100 = 25\%$;

(ii) 1983 over 1982 $= \frac{1.3}{6.5} \times 100 = 20\%$;

(iii) 1984 over 1983 $= \frac{2.1}{7.8} \times 100 = 26.9\%$;

(iv) 1985 over 1984 $= \frac{0.9}{9.9} \times 100 = 9.09\%$;

(v) 1987 over 1986 $= \frac{1.9}{9.5} \times 100 = 20\%$.

So, the maximum percentage increase in the export was in the year 1984.

**Ans. 2.** Average export in 1983 & 1984 is

$$= \left(\frac{7.8 + 9.9}{2}\right) = 8.85 \text{ crores} = 9 \text{ crores (approx.)}$$

**Ans. 3.** Average $= \left(\frac{5.2 + 6.5 + 7.8 + 9.9 + 10.8 + 9.5 + 11.4}{7}\right)$

$$= \frac{61.1}{7} = 8.73.$$

So, the export above the average was in the years 1984, 1985, 1986 & 1987.

**Ans. 4.** Average of 1981 & 1983 $= \frac{5.2 + 7.8}{2} = 6.5$.

= export in 1982.

**Ans. 5.** Percentage increase from 1986 to 1987.

$$= \left(\frac{11.4 - 9.5}{9.5}\right) \times 100 = \frac{1.9}{9.5} \times 100 = 20\%.$$

**Ex. 4.** ***Examine the following graph carefully and answer the questions given below it :–***

***Production of Cotton bales of 100 kg. each in lacs in states A, B, C, D & E during 1985-86, 1986-87 & 1987-88.***

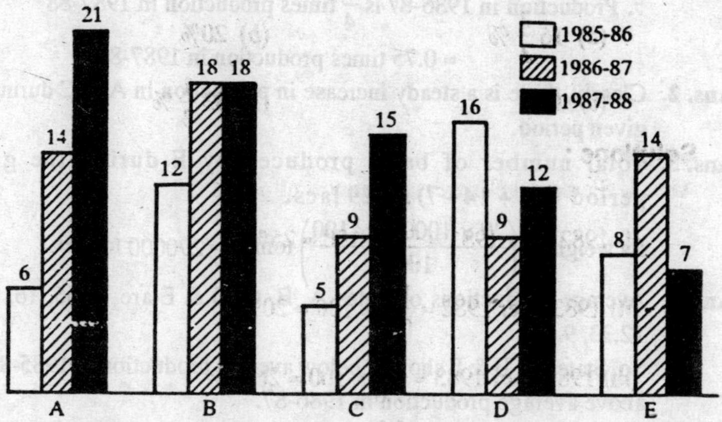


**Q. 1.** The production of state D in 1986-87 is how many times its production in 1987-88 :

(*a*) 1.33 (*b*) 0.75

(*c*) 0.56 (*d*) 1.77

**Q. 2.** In which states is there a steady increase in the production of cotton during the given period :

(*a*) A & B (*b*) A & C

(*c*) B only (*d*) D & E

**Q. 3.** How many tonnes of cotton was produced by state E during the given period :

(*a*) 2900 (*b*) 29000

(*c*) 290000 (*d*) 2900000

**Q. 4.** How many states showing below average production in 1985-86, showed above average production in 1986-87 :

(*a*) 4 (*b*) 2

(*c*) 3 (*d*) 1

**Q. 5.** Which of the following statements is false :

(*a*) States A & E showed the same production in 1986-87

(*b*) There was no improvement in the production of cotton in state B during 1987-88

(*c*) State A has produced maximum cotton during the given period

(*d*) Products of states C and D together is equal to that of state B during 1986-87

**Solutions :**

**Ans. 1.** $\frac{\text{Production in (1986–87)}}{\text{Production in (1987–88)}} = \frac{9}{12} = \frac{3}{4}$.

$\therefore$ Production in 1986-87 is $\frac{3}{4}$ times production in 1987-88

= 0.75 times production in 1987-88.

**Ans. 2.** Clearly, there is a steady increase in production in A & C during the given period.

**Ans. 3.** Total number of bales produced by E during the given period = (8 + 14 + 7) *i.e.* 29 lacs.

Its weight = $\left(\frac{29 \times 100000 \times 100}{1000}\right)$ tonnes = 290000 tonnes

**Ans. 4.** Average productions of states A, B, C, D & E are 13.66, 16, 9.66, 12.33, 9.66.

So, States A, B & E showed below average production in 1985-86 but above average production in 1986-87.

**Ans. 5.** State B has 48 lacs of bales, while state A has only 41 lacs of bales during the given period.

So, statement (*c*) is false.

**Ex. 5.** ***Study the following graph carefully and answer the following questions :*** **(Bank P.O. Exam. 1989)**

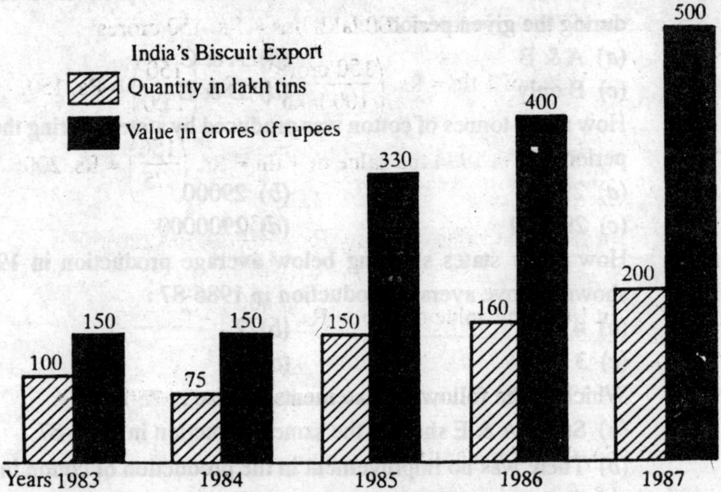


**Q. 1.** In which year the value per tin was minimum ?

(*a*) 1983 (*b*) 1984

(*c*) 1985 (*d*) 1986

(*e*) 1987

**Q. 2.** What was the difference between the tins exported in 1985 and 1986 ?

(*a*) 10 (*b*) 1000

(*c*) 100000 (*d*) 1000000

(*e*) none of these

**Q. 3.** What was the approximate percent increase in export value from 1983 to 1987 ?

(*a*) 350 (*b*) 330

(*c*) 43 (*d*) 2.4

(*e*) none of these

**Q. 4.** What was the percentage drop in export quantity from 1983 to 1984 ?

(*a*) 75 (*b*) nil

(*c*) 25 (*d*) 50

(*e*) none of these

**Q. 5.** If in 1986 the tins were exported at the same rate per tin as that in 1985, what would be the value in crores of rupees of export in 1986 ?

(*a*) 400 (*b*) 352

(*c*) 375 (*d*) 330

(*e*) none of these

**Solutions :**

**Ans. 1.** In 1983, the value of 100 lakh tins = Rs. 150 crores

$$\therefore \text{Value of 1 tin} = \text{Rs.} \left(\frac{150 \text{ crore}}{100 \text{ lakh}}\right) = \text{Rs.} \left(\frac{150}{1.00}\right) = \text{Rs. } 150.$$

Similarly, in 1984 the value of 1 tin = Rs. $\left(\frac{150}{.75}\right)$ = Rs. 200;

in 1985, the value of 1 tin = Rs. $\left(\frac{330}{1.50}\right)$ = Rs. 220;

In 1986, the value of 1 tin = Rs. $\left(\frac{400}{1.60}\right)$ = Rs. 250;

In 1987, the value per tin = Rs. $\left(\frac{500}{2.00}\right)$ = Rs. 250.

So, the value per tin is minimum in 1983.

*i.e.* Ans. (*a*) is correct.

**Ans. 2.** Difference between the tins exported in 1985 & 1986 is

= [(160 lakhs) – (150 lakhs)] = 10 lakhs = 1000000.

So, answer (*d*) is correct.

**Ans. 3.** Percentage increase in export value from 1983 to 1987

$$= \left\{\frac{(500 \text{ crores} - 150 \text{ crores})}{150 \text{ crores}} \times 100\right\} \%$$

$$= \left\{ \frac{(500 - 150)}{150} \times 100 \right\} \% = \left( \frac{350}{150} \times 100 \right) \% = 233.3\%.$$

So, answer (*e*) is correct.

**Ans. 4.** Percentage drop in export quantity from 1983 to 1984

$$= \left\{ \frac{(100 \text{ lakh tonnes}) - (75 \text{ lakh tonnes})}{100 \text{ lakh tonnes}} \times 100 \right\}$$

$$= \left( \frac{25}{100} \times 100 \right) \% = 25\% .$$

$\therefore$ Answer (*c*) is correct.

**Ans. 5.** In 1985, the cost of 150 lakh tins = Rs. 330 crores.

$$\therefore \text{ In 1985, the cost of 1 tin} = \text{Rs.} \left( \frac{330 \text{ crores}}{150 \text{ lakhs}} \right)$$

$$= \text{Rs.} \left( \frac{330}{1.50} \right) \text{Rs. } 220.$$

In 1986, the export value = Rs. (160 lakh × 220)

= Rs. (1.60 × 220) crores

= Rs. 352 crores.

Hence, answer (*b*) is correct.

**Ex. 6.** ***Study the graph carefully and answer the questions given below it :*** **(Bank P.O. Exam. 1988)**

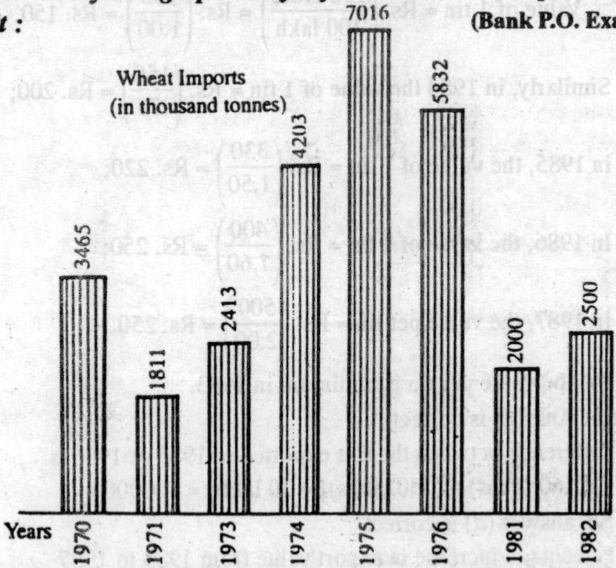


**Q. 1.** **In which year did the imports register highest increase over its preceding year ?**

(*a*) 1973 (*b*) 1974

(c) 1975 (d) 1982

**Q. 2.** The imports in 1976 was approximately how many times that of the year 1971 ?

(a) 0.31 (b) 1.68
(c) 2.41 (d) 3.22

**Q. 3.** What is the ratio of the years which have above average imports to those which have below average imports ?

(a) 5 : 3 (b) 2 : 6
(c) 8 : 3 (d) 3 : 8
(e) none of these

**Q. 4.** The increase in imports in 1982 was what percent of the imports in 1981 ?

(a) 25 (b) 5
(c) 125 (d) 80

**Q. 5.** The imports in 1974 is approximately what percent of the average imports for the given years ?

(a) 125 (b) 115
(c) 190 (d) 85
(e) 65

**Solutions :**

**Ans. 1.** Increase in imports in
1973 over 1971 is (2413 – 1811) = 602 thousand tonnes;
1974 over 1973 is (4203 – 2413) = 1790 thousand tonnes;
1975 over 1974 is (7016 – 4203) = 2813 thousand tonnes;
1982 over 1981 is (2500–2000) = 500 thousand tonnes

∴ Highest increase over its preceding year is in 1975.
*i.e.* Answer (c) is correct.

**Ans. 2.** Let $k$ (1811) = 5832. Then,

$$k = \frac{5832}{1811} = 3.22 \text{ thousand tonnes.}$$

∴ Answer (d) is correct.

**Ans. 3.** Average of the imports

$$= \frac{1}{8}(3465 + 1811 + 2413 + 4203 + 7016 + 5832 + 2000 + 2500)$$

= 3655.

The years in which the imports are above average are 1974, 1975 & 1976. *i.e.* there are 3 such years.

The years in which the imports are below average are 1970, 1971, 1973, 1981 & 1982 *i.e.* there are 5 such years.

∴ Required ratio is 3 : 5.

So, answer (*e*) is correct.

**Ans. 4.** Increase in imports in 1982 over 1981

$$= \left(\frac{2500 - 2000}{2000} \times 100\right)\% = 25\%.$$

$\therefore$ Answer (*a*) is correct.

**Ans. 5.** Average import = 3655 thousand tonnes.

Import in 1974 = 4203 thousand tonnes.

Let $x\%$ of 3655 = 4203.

Then, $x = \left(\frac{4203 \times 100}{3655}\right) = 115\%$.

$\therefore$ Answer (*b*) is correct.

**Ex. 7.** ***Study the following graph carefully and answer the following questions.*** **(Bank P.O. Exam. 1988)**

Demand and Production of Colour T.V.s. of Five Companies for October 1988

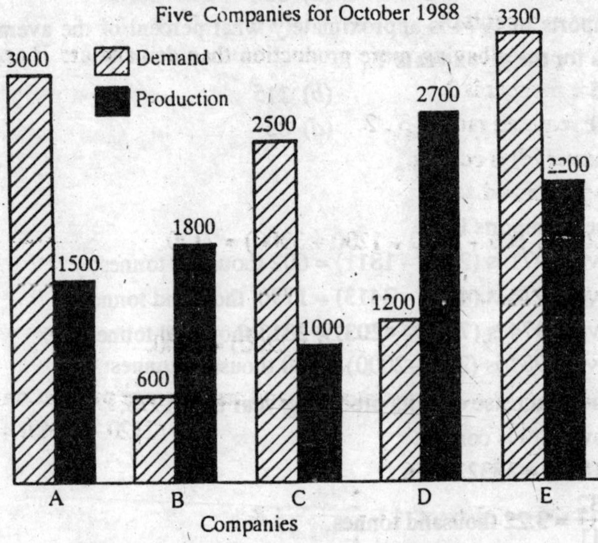


**Q. 1.** What is the ratio of companies having more demand than production to those having more production than demand ?

(*a*) 2 : 3 (*b*) 4 : 1

(*c*) 2 : 2 (*d*) 3 : 2

**Q. 2.** What is the difference between average demand and average production of the five companies taken together ?

(*a*) 1400 (*b*) 400

(*c*) 280 (*d*) 138

(*e*) none of these

**Q. 3.** The production of company *D* is approximately how many times that of the production of the company *A* ?

(*a*) 1.8 (*b*) 1.5
(*c*) 2.5 (*d*) 1.11
(*e*) none of these

**Q. 4.** The demand for company '*B*' is approximately what percent of the demand for company '*C*' ?
(*a*) 4 (*b*) 24
(*c*) 20 (*d*) 60

**Q. 5.** If company '*A*' desires to meet the demand by purchasing surplus T.V. sets from a single company, which one of the following companies can meet the need adequately ?
(*a*) B (*b*) C
(*c*) D (*d*) none of these

**Solutions :**

**Ans. 1.** The companies having more demand than production are A, C & E *i.e.* their number is 3.

The companies having more production than demand are B and D. *i.e.* their number is 2.

So, the required ratio is 3 : 2.

∴ Answer (*d*) is correct.

**Ans. 2.** Average demand

$$= \frac{1}{5}(3000 + 600 + 2500 + 1200 + 3300) = 2120.$$

Average production

$$= \frac{1}{5}(1500 + 1800 + 1000 + 2700 + 2200) = 1840.$$

∴ Difference between average demand and average production

$$= (2120 - 1840) = 280.$$

So, answer (*c*) is correct.

**Ans. 3.** Let $k(1500) = 2700$ or $k = \frac{2700}{1500} = 1.8.$

So, answer (*a*) is correct.

**Ans. 4.** Let $x\%$ of (demand for $C$) = (demand for $B$).

*i.e.* $\frac{x}{100} \times 2500 = 600$ or $x = \left(\frac{600 \times 100}{2500}\right) = 24\%.$

∴ Answer (*b*) is correct.

**Ans. 5.** Since company *D* produces highest number of *T.V.* sets and company *A* desires to meet the demand by purchasing surplus *T.V.* sets from a single company. Clearly, *D* can meet the demand of *A*.

∴ Answer (*c*) is correct.

**Ex. 8.** ***Study the following graph and answer the questions given below :*** **(P.N.B. P.O. Exam. 1987)**

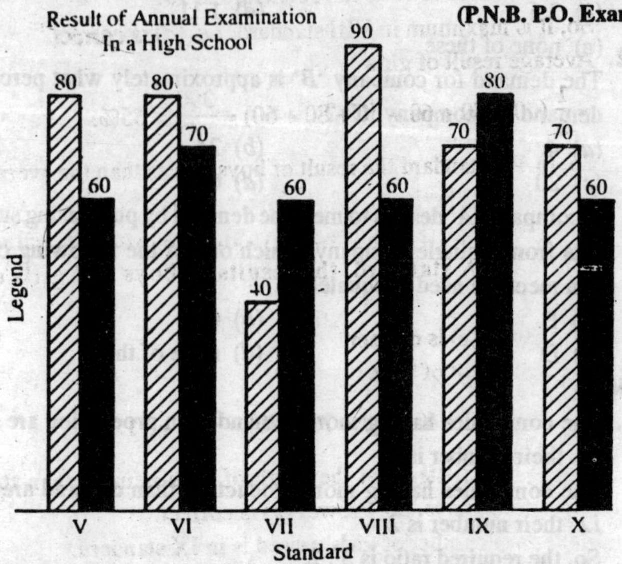

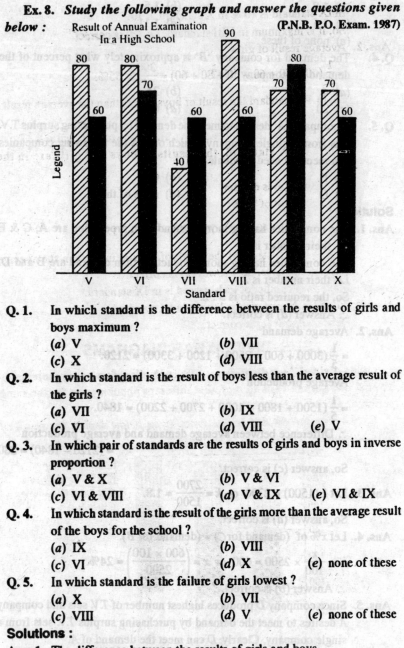

**Q. 1.** **In which standard is the difference between the results of girls and boys maximum ?**

(*a*) V (*b*) VII
(*c*) X (*d*) VIII

**Q. 2.** **In which standard is the result of boys less than the average result of the girls ?**

(*a*) VII (*b*) IX
(*c*) VI (*d*) VIII (*e*) V

**Q. 3.** **In which pair of standards are the results of girls and boys in inverse proportion ?**

(*a*) V & X (*b*) V & VI
(*c*) VI & VIII (*d*) V & IX (*e*) VI & IX

**Q. 4.** **In which standard is the result of the girls more than the average result of the boys for the school ?**

(*a*) IX (*b*) VIII
(*c*) VI (*d*) X (*e*) none of these

**Q. 5.** **In which standard is the failure of girls lowest ?**

(*a*) X (*b*) VII
(*c*) VIII (*d*) V (*e*) none of these

**Solutions :**

**Ans. 1.** **The difference between the results of girls and boys in V standard is 20; in VI standard is 10; in VII standard is 20; in VIII standard is 30;**

in IX standard is 10 & in X standard is 10.

So, it is maximum in VIII standard, *i.e.* (*d*) is correct.

**Ans. 2.** Average result of girls

$$= \frac{1}{6}(60 + 70 + 60 + 60 + 80 + 60) = \frac{390}{6} = 65\%.$$

So, in VII standard the result of boys is less than the average result of the girls. Therefore, (*a*) is correct.

**Ans. 3.** In VI standard, the results of boys and girls are in the ratio 8 : 7; While in IX standard, the results of boys and girls are in the ratio 7 : 8.

So, answer (*e*) is correct.

**Ans. 4.** Average result of boys

$$= \frac{1}{6}(80 + 80 + 40 + 90 + 70 + 70) = \frac{430}{6} = 71.7\%.$$

Clearly, in IX standard the result of girls is more than the average result of the boys. So, answer (*a*) is correct.

**Ans. 5.** Maximum number of girls passed is in IX standard.

So, the failure of girls is lowest in IX standard.

Hence, answer (*e*) is correct.

## SUB–DIVIDED BAR DIAGRAMS

**Sub–divided bar diagrams :** In such diagrams every column is divided into certain parts to represent different phenomenon for the same period or place.

**Ex. 9.** ***The sub-divided bar diagram given below depicts the result of B.Sc. students of a college for three years.***

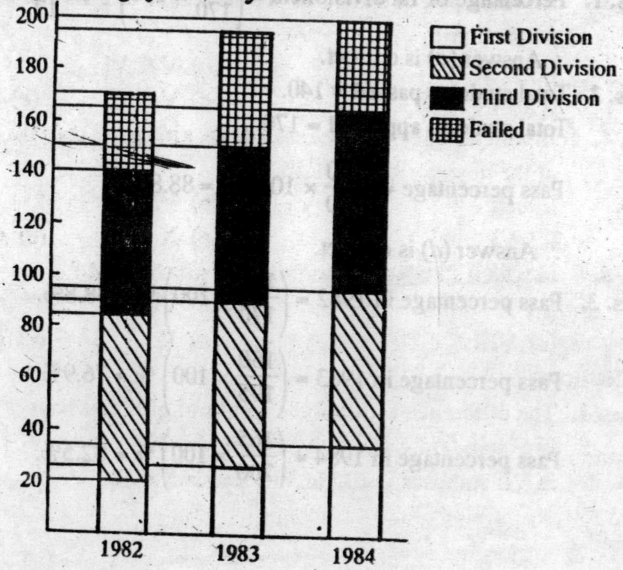

**Study the above bar diagram and mark a tick against the correct answer in each question.**

**Q. 1.** How many percent passed in 1st division in 1982 ?

(*a*) 20% (*b*) 34%

(*c*) $14\frac{2}{7}\%$ (*d*) $11\frac{13}{17}\%$

**Q. 2.** What was the pass percentage in 1982 ?

(*a*) 65% (*b*) 70%

(*c*) 74.6% (*d*) 88.8%

**Q. 3.** In which year the college had the best result for B.Sc. ?

(*a*) 1982 (*b*) 1983

(*c*) 1984

**Q. 4.** What is the number of third divisioners in 1984 ?

(*a*) 165 (*b*) 75

(*c*) 70 (*d*) 65

**Q. 5.** What is the percentage of students in 1984 over 1982 ?

(*a*) 30% (*b*) $17\frac{11}{17}\%$

(*c*) $117\frac{11}{17}\%$ (*d*) 85%

**Q. 6.** What is the aggregate pass percentage during three years ?

(*a*) $51\frac{2}{3}\%$ (*b*) 82.7%

(*c*) 80.4% (*d*) 77.6%

**Solutions :**

**Ans. 1.** Percentage of 1st divisioners $= \left(\frac{20}{170} \times 100\right) = 11\frac{13}{17}\%$.

∴ Answer (*d*) is correct.

**Ans. 2.** Total students passed = 140.

Total students appeared = 170 .

$$\text{Pass percentage} = \left(\frac{140}{170} \times 100\right)\% = 88.8\%.$$

∴ Answer (*d*) is correct.

**Ans. 3.** Pass percentage in 1982 $= \left(\frac{140}{170} \times 100\right)\% = 88.8\%$.

Pass percentage in 1983 $= \left(\frac{150}{195} \times 100\right)\% = 76.9\%$.

Pass percentage in 1984 $= \left(\frac{165}{200} \times 100\right)\% = 82.5\%$.

So, the college recorded best result in 1982.

∴ Answer (*a*) is correct.

**Ans. 4.** Third divisioners in 1984 = (165 – 95) = 70.

∴ Answer (*c*) is correct.

**Ans. 5.** Students in 1984 = 200.

Students in 1982 = 170.

$$\text{Required percentage} = \left(\frac{200}{170} \times 100\right)\% = 117\frac{11}{17}\%.$$

∴ Answer (*c*) is correct.

**Ans. 6.** Total number of students appeared during 3 years

= (170 + 195 + 200) = 565.

Total number of students passed during 3 years

= (140 + 150 + 165) = 455.

$$\text{Aggregate pass percentage} = \left(\frac{455}{565} \times 100\right)\% = 80.4\%.$$

So, answer (*c*) is correct.

**Ex. 10.** ***Following bar diagram shows the monthly expenditure of two families on food, clothing, education, fuel, house rent and miscellaneous (in percentage).***

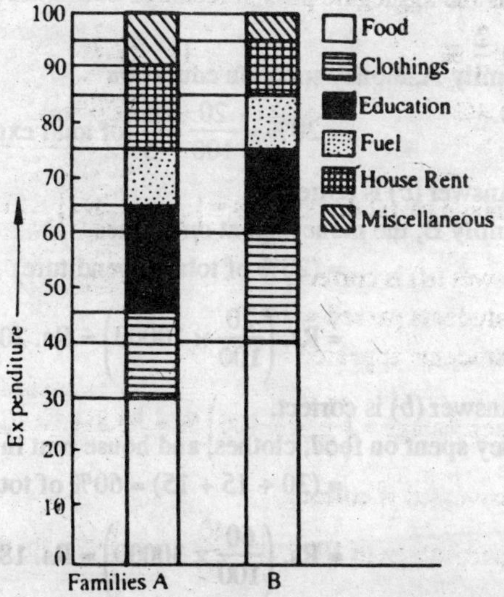


**Study the above diagram and mark a tick against the correct answer in each question.**

**Q. 1.** What fraction of the total expenditure is spent on Education in family *A* ?

(a) $\frac{13}{20}$ (b) $\frac{2}{3}$

(c) $\frac{9}{13}$ (d) none of these

**Q. 2.** If the total annual expenditure of family *B* is Rs. 10,000 then money spent on clothes during the year is :

(a) Rs. 200 (b) Rs. 2000

(c) Rs. 600 (d) Rs. 6000

**Q. 3.** If the total annual expenditure of family *A* is Rs. 30,000 then money spent on food, clothes and house rent is :

(a) Rs. 18,500 (b) Rs. 18,000

(c) Rs. 21,000 (d) Rs. 15,000

**Q. 4.** If both the families have the same expenditure, which one spends more on education and miscellaneous together ?

(a) Family A (b) Family B

(c) none

**Q. 5.** What percentage is *B*'s expenditure on food over *A*'s expenditure on food ? (Taking equal total expenditure)

(a) 10% (b) 70%

(c) $133\frac{1}{3}\%$ (d) 75%

**Solutions :**

**Ans. 1.** In family A, money spent on education

$$= 20\% = \frac{20}{100} = \frac{1}{5} \text{ (of total expenditure)}$$

$\therefore$ Answer (c) is correct.

**Ans. 2.** In family B, the money spent on clothes

$$= (20\% \text{ of total expenditure})$$

$$= \text{Rs.}\left(\frac{20}{100} \times 10000\right) = \text{Rs. } 2000 .$$

$\therefore$ Answer (b) is correct.

**Ans. 3.** Money spent on food, clothes, and house rent in family A

$$= (30 + 15 + 15) = 60\% \text{ of total expenditure}$$

$$= \text{Rs.}\left(\frac{60}{100} \times 30000\right) = \text{Rs. } 18000 .$$

$\therefore$ Answer (b) is correct.

**Ans. 4.** Family A spends on Education & Miscellaneous

$$= (20 + 10) = 30\%$$

Family B spends on Education & Miscellaneous

= (15 + 5) = 20%

So, family A spends more on these heads.

∴ Answer (*a*) is correct.

**Ans. 5.** B's expenditure on food = 40%

A's expenditure on food = 30%

$$\text{B's percentage over A's} = \left(\frac{40}{30} \times 100\right) = 133\frac{1}{3}\%$$

∴ Answer (*c*) is correct.

## CIRCLE GRAPH

**Ex. 11.** ***Circle-graph given below shows the expenditure incurred in bringing out a book, by a publisher.***

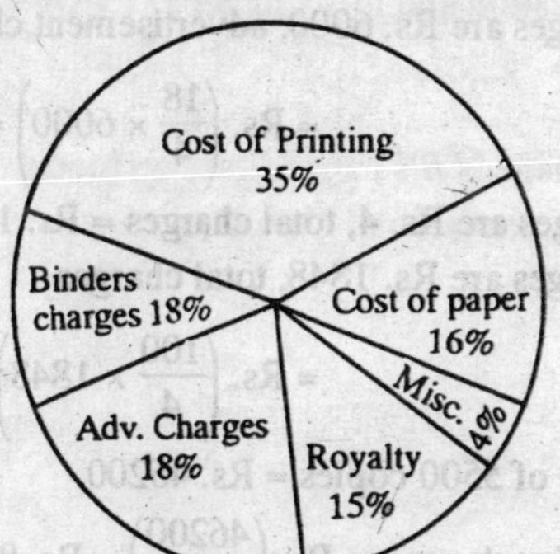


**Study the graph carefully and answer the questions give below it :**

**Q. 1.** What should be the central angle of the sector for the cost of the paper :

(*a*) 22.5° (*b*) 16°

(c) 54.8° (*d*) 57.6°

**Q. 2.** If the cost of printing is Rs. 17500, the rayalty is :

(*a*) Rs. 8750 (*b*) Rs. 7500

(*c*) Rs. 3150 (*d*) Rs. 6300

**Q. 3.** If the miscellaneous charges are Rs. 6000, the advertisement charges are :

(*a*) Rs. 90000 (*b*) Rs. 1333.33

(c) Rs. 27000 (*d*) Rs. 12000

**Q. 4.** If 5500 copies are published, miscellaneous expenditures amount to Rs. 1848 and publisher's profit is 25%, then marked price of each copy is :

(*a*) Rs. 8.40 (*b*) Rs. 12.50

(*c*) Rs. 10.50 (*d*) Rs. 10

**Q. 5.** Royalty on the book is less than the advertisement charges by :

(*a*) 3% (*b*) 20%

(*c*) $16\frac{2}{3}\%$ (*d*) none of these

**Solutions :**

**Q. 1.** Requisite angle = $\left(\frac{16}{100} \times 360\right)^\circ = 57.6^\circ$.

**Q. 2.** If cost of printing is Rs. 35, royalty is Rs. 15

If cost of printing is Rs. 17500, royalty is

$$= \text{Rs.} \left(\frac{15}{35} \times 17500\right) = \text{Rs. } 7500.$$

**Q. 3.** If misc. charges are Rs. 4, advertisement charges = Rs. 18.

If misc. charges are Rs. 6000, advertisement charges

$$= \text{Rs.} \left(\frac{18}{4} \times 6000\right) = \text{Rs. } 27000.$$

**Q. 4.** If misc. charges are Rs. 4, total charges = Rs. 100

If misc. charges are Rs. 1848, total charges

$$= \text{Rs.} \left(\frac{100}{4} \times 1848\right) = \text{Rs. } 46200.$$

$\therefore$ Cost price of 5500 copies = Rs. 46200.

Cost price of each copy = Rs. $\left(\frac{46200}{5500}\right)$ = Rs. 8.40

$\therefore$ Marked price of each copy = 125% of Rs. 8.40 = Rs. 10.50.

**Q. 5.** On Rs. 18, it is less by Rs. 3.

On Rs. 100, it is less by $\left(\frac{3}{18} \times 100\right) = 16\frac{2}{3}\%$.

**Ex. 12.** ***Study the following graphs carefully and answer the questions that follow*** **(Bank P.O. Exam. 1988)**

Distribution of Proteins in human Body

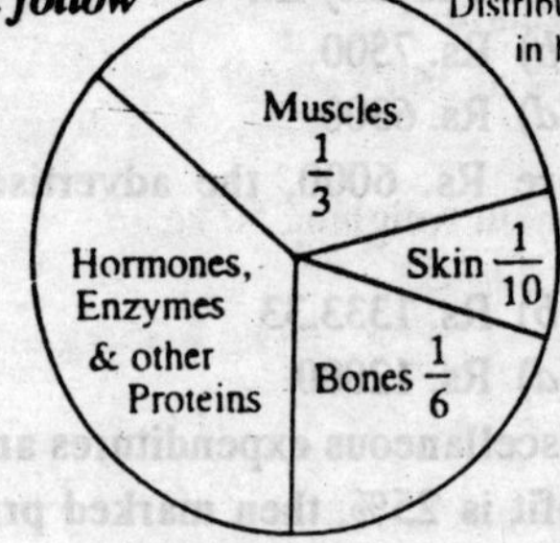


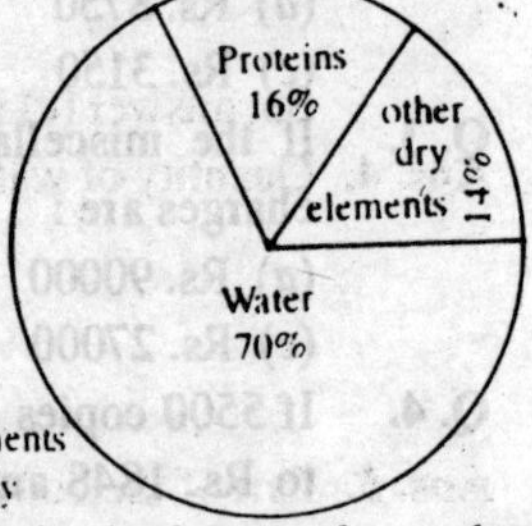


Distribution of elements in the human body

**Q. 1.** What is the ratio of the distribution of proteins in the muscles to that of the distribution of proteins in the bones ?

(*a*) 1 : 2 (*b*) 2 : 1
(*c*) 18 : 1 (*d*) 1 : 18

**Q. 2.** What percent of the total weight of the human body is equivalent to the weight of the skin in the human body ?

(*a*) .016 (*b*) 1.6
(*c*) .16 (*d*) insufficient information

**Q. 3.** To show the distribution of proteins and other dry elements in the human body, the arc of the circle should subtend at the centre an angle of

(*a*) 126° (*b*) 54°
(*c*) 108° (*d*) 252°

**Q. 4.** What will be the quantity of water in the body of a person weighing 50 kg ?

(*a*) 35 kg (*b*) 120 kg
(*c*) 71.42 kg (*d*) 20 kg

**Q. 5.** In the human body what is made of neither bones nor skin ?

(*a*) $\frac{2}{5}$ (*b*) $\frac{3}{5}$
(*c*) $\frac{1}{40}$ (*d*) $\frac{3}{80}$

**Solutions :**

**Ans. 1.** Required ratio $= \frac{1}{3} : \frac{1}{6} = 6 : 3$ or $2 : 1$.

So, answer (*b*) is correct.

**Ans. 2.** Weight of skin $= \frac{1}{10}$ parts of 16% of proteins

$$= \frac{1}{10} \times 16\% = 1.6\%.$$

So, answer (*b*) is correct.

**Ans. 3.** Proteins & other dry elements = 30%.

∴ Angle subtended by the required arc

$$= (30\% \text{ of } 360°) = 108°.$$

So, answer (*c*) is correct.

**Ans. 4.** Quantity of water in body of a person weighing 50 kg

$$= 70\% \text{ of } 50 \text{ kg} = \left(\frac{70}{100} \times 50\right) \text{ kg} = 35 \text{ kg}.$$

∴ Answer (*a*) is correct.

**Ans. 5.** Part of the body made of neither bones nor skin

$$= 1 - \left(\frac{1}{3} + \frac{1}{10} + \frac{1}{6}\right) = \left(1 - \frac{6}{10}\right) = \frac{2}{5}.$$

$\therefore$ Answer (*a*) is correct.

**Ex. 13.** ***The following graph shows the annual premium of an insurance company, charged for an insurance of Rs. 1000 for different ages.***

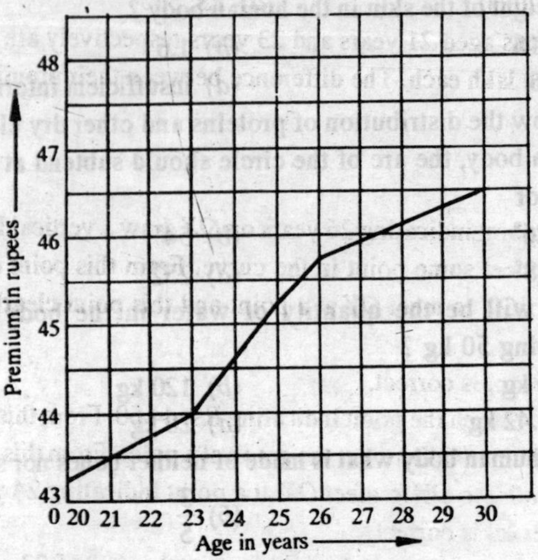


**Scale :** $\begin{cases} \text{Along } OX \rightarrow \text{ 10 small divisions} = \text{1 year} \\ \text{Along } OY \rightarrow \text{ 1 small division} = \text{5 paise.} \end{cases}$

**1 big division = 10 small divisions (not shown in the fig.)**

**Study the graph and mark a tick against the correct answer in each of the following questions.**

**Q. 1.** The premium for a man aged 26 years for an insurance of Rs. 1000 is :

(*a*) Rs. 46 (*b*) Rs. 45.75
(*c*) Rs. 44 (*d*) Rs. 45

**Q. 2.** What is the age of a person whose premium is Rs. 44.60 for an insurance of Rs. 1000 ?

(*a*) 22 years (*b*) 23 years
(*c*) 24 years (*d*) 25 years

**Q. 3.** The premium for a man aged 22 years for an insurance of Rs. 10000 is :

(*a*) Rs. 435 (*b*) Rs. 440
(*c*) Rs. 437.50 (*d*) Rs. 43.75

**Q. 4.** How much percent of the premium is increased if a man aged 30 years is insured for Rs. 1000, instead of a man aged 23 years ?

(*a*) 4.75% (*b*) 5.68%
(*c*) 6.24% (*d*) 6%

**Q. 5.** Two members of a family aged 20 years and 25 years are to be insured for Rs. 10000 each. The total annual premium to be paid by them is :

(*a*) Rs. 836.75 (*b*) Rs. 845.50
(*c*) Rs. 870.60 (*d*) Rs. 885

**Q. 6.** Two persons aged 21 years and 23 years respectively are insured for rupees one lakh each. The difference between their premiums is :

(*a*) Rs. 100 (*b*) Rs. 25
(*c*) Rs. 50 (*d*) Rs. 20

**Solutions :**

**Ans. 1.** From the point indicating 26 years on *OX* draw a vertical line parallel to *OY* to meet some point in the curve. From this point draw a line parallel to *OX* to meet *OY* at a point and this point clearly indicates Rs. 45.75.

∴ Answer (*c*) is correct.

**Ans. 2.** Along, *OY*, reach the point indicating Rs. 44.60. From this point draw a line parallel to *OX* to meet the graph at a point. From this point, draw a line parallel to *OY* to meet *OX* at a point indicating 24 years.

∴ Answer (*c*) is correct.

**Ans. 3.** As indicated by the graph, premium at the age of 22 years for an insurance of Rs. 1000 is Rs. 43.75. So, for an insurance of Rs. 10000, the premium is

Rs. (43.75 × 10) = Rs. 437.50.

∴ Answer (*c*) is correct.

**Ans. 4.** Premium for Rs. 1000 for a man aged 23 years = Rs. 44.

Premium for Rs. 1000 for a man aged 30 years = Rs. 46.50.

Increase % in premium = $\left(\frac{2.50}{44} \times 100\right)\% = 5.68\%$.

∴ Answer (*b*) is correct.

**Ans. 5.** Premium for Rs. 10000 at 20 years

= Rs. (43.25 × 10) = Rs. 432.50.

Premium for Rs. 10000 at 25 years

= Rs. (45.25 × 10) = Rs. 452.50.

Total annual premium for both = Rs. (432.50 + 452.50) = Rs. 885.

∴ Answer (*b*) is correct.

**Ans. 6.** Premium for Rs. one lakh at 21 years

= Rs. (100 × 43.50) = Rs. 4350.

Premium for Rs. one lakh at 23 years

= Rs. (100 × 44) = Rs. 4400.

Difference in premiums = Rs. 50.

∴ Answer (*a*) is correct.

**Ex. 14.** ***The following graph shows the temperature of a patient observed in a hospital at a certain interval of time on a certain day, starting at 5 A.M.***

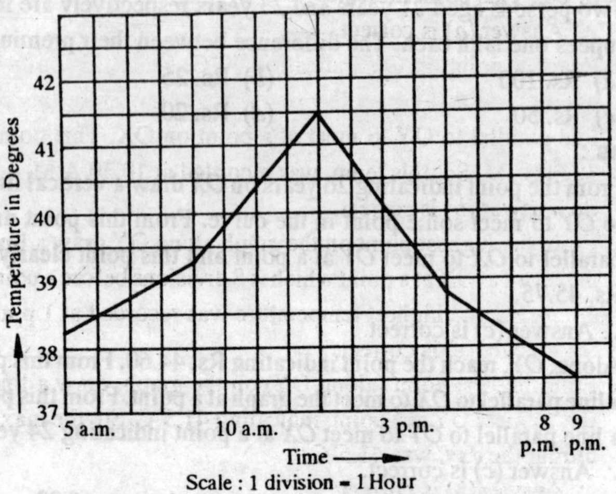


Scale : {Along $OX$ → 10 small division = 15 minutes; Along $OY$ → 10 small divisions = 1° C.}

**Study the above graph carefully and tick against the correct answer in each of the following questions :**

**Q. 1.** What was the temperature of the patient at 2 p.m. ?

(*a*) 40.8° C (*b*) 41.1° C

(*c*) 41.5° C (*d*) 41.9° C

**Q. 2.** The time, when the temperature was recorded 40°C, was

(*a*) 11 A.M. (*b*) 10.30 A.M.

(*c*) 11.45 A.M. (*d*) 11.15 A.M.

**Q. 3.** At what time during the day, the temperature was maximum ?

(*a*) 12 P.M. (*b*) 12.30 P.M.

(*c*) 1 P.M. (*d*) 1.30 P.M.

**Q. 4.** What was the maximum temperature during the day ?

(*a*) 40.7° C (*b*) 41.5° C

(*c*) 40.8° C (*d*) 41° C

**Q. 5.** The normal temperature is 37.5°C. At what time was the temperature normal ?

(*a*) 5 A.M. (*b*) 5 P.M.

(*c*) 9 P.M. (*d*) at no time

**Solutions :**

**Ans. 1.** Since we have taken origin at 5 A.M., so 2 P.M. is 9 hours beyond this point. From this point draw a line parallel to OY to meet the graph at a point. From this point draw a line parallel to OX to meet at a point in OY. This point indicates 40.8°C.

∴ Answer (*a*) is correct.

**Ans. 2.** Reach a point on OY indicating 40°C. From this point draw a line parallel to OX to meet the graph at a point. From this point draw a line parallel to OY to meet at a point on OX. This point represents 10.30 A.M. So, this temp. was recorded at 10.30 A.M.

So, answer (*b*) is correct.

**Ans. 3.** From the highest point on the graph along OY draw a line parallel to OY to meet OX at a point which is 8 divisions before a point indicating 1 p.m. So, the highest temperature was recorded at 1 p.m.

So, answer (*e*) is correct.

**Ans. 4.** From highest point (along OY) on the graph, draw a line parallel to OX to meet OY at a point, indicating 41.5°C. So, the maximum temp. during the day, was 41.5°C.

∴ Answer (*b*) is correct.

**Ans. 5.** We are to find the time when the temperature was 37.5°C. Along OY, take the point indicating 37.5°C. From this point, draw a line parallel to OX to meet the graph at a point. From this point, draw a line parallel to OY, to meet OX at a point. This point indicates 9 p.m.

So, the temperature was normal at 9 p.m.

∴ Answer (*c*) is correct.

# 30

# Odd Man Out and Series

1. **Turn odd man out.** As the phrase speaks itself, in this type of problems, a set of numbers is given in such a way that each one, except one satisfies a particular definite property. The one which does not satisfy that characteristic is to be taken out.

   Some important properties of numbers are given below :

   (i) **Prime Numbers.** A counting number greater than 1, which is divisible by itself and 1 only, is called a prime number, e.g. 2, 3, 5, 7, 11, 13, 17, 19, 23, 29, 31, 37, 41, 43, 47, 53, 59, 61, 67, 71, 73, 79, 83, 89, 97 etc.

   (ii) **Odd numbers.** A number not divisible by 2, is an odd number, e.g. 1, 3, 5, 7, 9, 11, 13, 15 etc.

   (iii) **Even Numbers.** A number divisible by 2, is an even number e.g. 2, 4, 6, 8, 10 etc.

   (iv) **Perfect squares.** A counting number whose square root is a counting number, is called a perfect square, e.g. 1, 4, 9, 16, 25, 36, 49, 64 etc.

   (v) **Perfect Cubes.** A counting number whose cube-root is a counting number is called a perfect cube, e.g. 1, 8, 27, 64, 125 etc.

   (vi) **Multiples of a number.** A number which is divisible by a given number *a*, is called the multiple of *a* e.g. 3, 6, 9, 12 etc. are all multiples of 3.

   (vii) **Numbers in A.P.** Some given numbers are said to be in A.P. if the difference between two consecutive numbers is same e.g. 13, 11, 9, 7, 5, 3, 1, –1, –3 etc.

   (viii) **Numbers in G.P.** Some given numbers are in G.P. if the ratio between two consecutive numbers remains the same, e.g. 48, 12, 3 etc.

## EXERCISE 30 (Objective Type Questions)

**Turn odd man out :**

**1.** 3, 5, 7, 12, 13, 17, 19.

(*a*) 19 (*b*) 17

(*c*) 13 (*d*) 12

**2.** 10, 14, 16, 18, 21, 24, 26

(*a*) 26 (*b*) 24

(*c*) 21 (*d*) 18

**3.** 3, 5, 9, 11, 14, 17, 21
(*a*) 21 (*b*) 17
(*c*) 14 (*d*) 9

**4.** 1, 4, 9, 16, 23, 25, 36
(*a*) 9 (*b*) 23
(*c*) 25 (*d*) 36

**5.** 6, 9, 15, 21, 24, 28, 30
(*a*) 28 (*b*) 21
(*c*) 24 (*d*) 30

**6.** 41, 43, 47, 53, 61, 71, 73, 81
(*a*) 61 (*b*) 71
(*c*) 73 (*d*) 81

**7.** 16, 25, 36, 72, 144, 196, 225
(*a*) 36 (*b*) 72
(*c*) 196 (*d*) 225

**8.** 10, 25, 45, 54, 60, 75, 80
(*a*) 10 (*b*) 45
(*c*) 54 (*d*) 75

**9.** 1, 4, 9, 16, 20, 36, 49
(*a*) 1 (*b*) 9
(*c*) 20 (*d*) 49

**10.** 8, 27, 64, 100, 125, 216, 343
(*a*) 27 (*b*) 100
(*c*) 125 (*d*) 343

**11.** 1, 5, 14, 30, 50, 55, 91
(*a*) 5 (*b*) 50
(*c*) 55 (*d*) 91

**12.** 385, 462, 572, 396, 427, 671, 264
(*a*) 385 (*b*) 427
(*c*) 671 (*d*) 264

**13.** 835, 734, 642, 751, 853, 981, 532
(*a*) 751 (*b*) 853
(*c*) 981 (*d*) 532

**14.** 331, 482, 551, 263, 383, 242, 111
(*a*) 263 (*b*) 383
(*c*) 242 (*d*) 111

**15.** 2, 5, 10, 17, 26, 37, 50, 64
(*a*) 50 (*b*) 26
(*c*) 37 (*d*) 64

**16.** 19, 28, 39, 52, 67, 84, 102
(*a*) 52 (*b*) 102

(*c*) 84 (*d*) 67

**17.** 253, 136, 352, 460, 324, 631, 244

(*a*) 136 (*b*) 324
(*c*) 352 (*d*) 631

**18.** 2, 5, 10, 50, 500, 5000

(*a*) 0 (*b*) 5
(*c*) 10 (*d*) 5000

**19.** 4, 5, 7, 10, 14, 18, 25, 33

(*a*) 7 (*b*) 14
(*c*) 18 (*d*) 33

**Find out the wrong number in each sqeuence :**

**20.** 22, 33, 66, 99, 121, 279, 594

(*a*) 33 (*b*) 121
(*c*) 279 (*d*) 594

**21.** 36, 54, 18, 27, 9, 18.5, 4.5

(*a*) 4.5 (*b*) 18.5
(*c*) 54 (*d*) 18

**22.** 582, 605, 588, 611, 634, 617, 600

(*a*) 634 (*b*) 611
(*c*) 605 (*d*) 600

**23.** 46080, 3840, 384, 48, 24, 2, 1

(*a*) 1 (*b*) 2
(*c*) 24 (*d*) 384

**24.** 1, 8, 27, 64, 124, 216, 343

(*a*) 8 (*b*) 27
(*c*) 64 (*d*) 124

**25.** 5, 16, 6, 16, 7, 16, 9

(*a*) 9 (*b*) 7
(*c*) 6 (*d*) none

**26.** 6, 13, 18, 25, 30, 37, 40

(*a*) 25 (*b*) 30
(*c*) 37 (*d*) 40

**27.** 56, 72, 90, 110, 132, 150

(*a*) 72 (*b*) 110
(*c*) 132 (*d*) 150

**28.** 8, 13, 21, 32, 47, 63, 83

(*a*) 47 (*b*) 63
(*c*) 32 (*d*) 83

**29.** 25, 36, 49, 81, 121, 169, 225

(*a*) 36 (*b*) 49
(*c*) 121 (*d*) 169

**30.** 1, 2, 6, 15, 31, 56, 91
(*a*) 31 (*b*) 91
(*c*) 56 (*d*) 15

**31.** 52, 51, 48, 43, 34, 27, 16
(*a*) 27 (*b*) 34
(*c*) 43 (*d*) 48

**32.** 105, 85, 60, 30, 0, – 45, –90
(*a*) 0 (*b*) 85
(*c*) – 45 (*d*) 60

**33.** 4, 6, 8, 9, 10, 11, 12
(*a*) 10 (*b*) 11
(*c*) 12 (*d*) 9

**34.** 125, 127, 130, 135, 142, 153, 165
(*a*) 130 (*b*) 142
(*c*) 153 (*d*) 165

**35.** 16, 36, 64, 81, 100, 144, 190
(*a*) 81 (*b*) 100
(*c*) 190 (*d*) 36

**36.** 125, 123, 120, 115, 108, 100, 84
(*a*) 123 (*b*) 115
(*c*) 100 (*d*) 84

**37.** 3, 10, 21, 36, 55, 70, 105
(*a*) 105 (*b*) 70
(*c*) 36 (*d*) 55

**38.** 4, 9, 19, 39, 79, 160, 319
(*a*) 319 (*b*) 160
(*c*) 79 (*d*) 39

**39.** 10, 14, 28, 32, 64, 68, 132
(*a*) 32 (*b*) 68
(*c*) 132 (*d*) 28

**40.** 8, 27, 125, 343, 1331
(*a*) 1331 (*b*) 343
(*c*) 125 (*d*) none

**Insert the missing number :**

**41.** 4, – 8, 16, –32, 64, (......)
(*a*) 128 (*b*) –128
(*c*) 192 (*d*) –192

**42.** 5, 10, 13, 26, 29, 58, 61, (......)
(*a*) 122 (*b*) 64
(*c*) 125 (*d*) 128

**43.** 1, 4, 9, 16, 25, 36, 49, (......)

(a) 54 (b) 56
(c) 64 (d) 81

44. 1, 8, 27, 64, 125, 216, (......)
(a) 354 (b) 343
(c) 392 (d) 245

45. 11, 13, 17, 19, 23, 29, 31, 37, 41, (......)
(a) 43 (b) 47
(c) 53 (d) 51

46. 16, 33, 65, 131, 261, (......)
(a) 523 (b) 521
(c) 613 (d) 721

47. 3, 7, 6, 5, 9, 3, 12, 1, 15, (......)
(a) 18 (b) 13
(c) –1 (d) 3

48. 13, 31, 63, 127, 255, (......)
(a) 513 (b) 511
(c) 517 (d) 523

49. 2, 6, 12, 20, 30, 42, 56, (......)
(a) 60 (b) 64
(c) 72 (d) 70

50. 8, 24, 12, 36, 18, 54, (......)
(a) 27 (b) 108
(c) 68 (d) 72

51. 165, 195, 255, 285, 345, (......)
(a) 375 (b) 420
(c) 435 (d) 390

52. 7, 26, 63, 124, 215. 342, (......)
(a) 481 (b) 511
(c) 391 (d) 421

53. 2, 4, 12, 48, 240, (......)
(a) 960 (b) 1440
(c) 1080 (d) 1920

54. 8, 7, 11, 12, 14, 17, 17, 22, (......)
(a) 27 (b) 20
(c) 22 (d) 24

55. 10, 5, 13, 10, 16, 20, 19, (......)
(a) 22 (b) 40
(c) 38 (d) 23

56. 1, 2, 4, 8, 16, 32, 64, (......), 256
(a) 148 (b) 128
(c) 154 (d) 164

**57.** 71, 76, 69, 74, 67, 72, (......)
(*a*) 77 (*b*) 65
(*c*) 80 (*d*) 76

**58.** 9, 12, 11, 14, 13, (......), 15
(*a*) 12 (*b*) 16
(*c*) 10 (*d*) 17

**59.** Complete the series
2, 5, 9, 19, 37, ......
(*a*) 76 (*b*) 74
(*c*) 75 (*d*) none of these
**(Railway Recruitment Board Exam. 1989)**

**60.** Find the wrong number in the series :
3, 8, 15, 24, 34, 48, 63
(*a*) 15 (*b*) 24
(*c*) 34 (*d*) 48
(*e*) 63 **(Bank P.O. Exam. 1988)**

**61.** Find the wrong number in the series :
2, 9, 28, 65, 126, 216, 344
(*a*) 2 (*b*) 28
(*c*) 65 (*d*) 126
(*e*) 216 **(Bank P.O. Exam. 1988)**

**62.** Find out the wrong number in the series :
5, 15, 30, 135, 405, 1215, 3645
(*a*) 3645 (*b*) 1215
(*c*) 405 (*d*) 30
(*e*) 15 **(S.B.I.P.O. Exam. 1988)**

**63.** Find out the wrong number in the series :
125, 106, 88, 76, 65, 58, 53
(*a*) 125 (*b*) 106
(*c*) 88 (*d*) 76
(*e*) 65 **(S.B.I.P.O. Exam. 1987)**

**Find out the wrong number in the series :**

**64.** 190, 166, 145, 128, 112, 100, 91
(*a*) 100 (*b*) 166
(*c*) 145 (*d*) 128
(*e*) 112 **(Bank P.O. 1991)**

**65.** 1, 1, 2, 6, 24, 96, 720
(*a*) 720 (*b*) 96
(*c*) 24 (*d*) 6
(*e*) 2 **(Bank P.O. 1991)**

**66.** 40960, 10240, 2560, 640 200, 40, 10

(*a*) 640 (*b*) 40
(*c*) 200 (*d*) 2560
(*e*) 10240 **(Bank P.O. 1991)**

**67.** 64, 71, 80, 91, 104, 119, 135, 155
(*a*) 71 (*b*) 80
(*c*) 104 (*d*) 119
(*e*) 135 **(Bank P.O. 1991)**

**68.** 7, 8, 18, 57, 228, 1165, 6996
(*a*) 8 (*b*) 18
(*c*) 57 (*d*) 228
(*e*) 1165 **(Bank P.O. 1991)**

**69.** 3, 7, 15, 27, 63, 127, 255
(*a*) 7 (*b*) 15
(*c*) 27 (*d*) 63
(*e*) 127 **(Bank P.O. 1991)**

**70.** 19, 26, 33, 46, 59, 74, 91
(*a*) 26 (*b*) 33
(*c*) 46 (*d*) 59
(*e*) 74 **(Bank P.O. 1991)**

**71.** 2880, 480, 92, 24, 8, 4, 4
(*a*) 2880 (*b*) 480
(*c*) 92 (*d*) 24
(*e*) 8 **(Bank P.O. 1991)**

**72.** 445, 221, 109, 46, 25, 11, 4
(*a*) 221 (*b*) 109
(*c*) 46 (*d*) 25
(*e*) 11 **(Bank P.O. 1991)**

**73.** 3, 7, 15, 39, 63, 127, 255, 511
(*a*) 7 (*b*) 15
(*c*) 39 (*d*) 63
(*e*) 127 **(Bank P.O. 1991)**

**74.** 1, 3, 10, 21, 64, 129, 356, 777
(*a*) 10 (*b*) 21
(*c*) 64 (*d*) 129
(*e*) 356 **(Bank P.O. 1991)**

**75.** 196, 169, 144, 121, 100, 80, 64
(*a*) 169 (*b*) 144
(*c*) 121 (*d*) 100
(*e*) 80 **(Bank P.O. 1991)**

## SOLUTIONS (Exercise 30)

1. Each of the numbers except 12, is a prime number.
2. Each of the numbers except 21, is an even number
3. Each of the numbers except 14, is an odd number.
4. Each of the given numbers except 23, is a perfect square.
5. Each of the numbers except 28, is a multiple of 3.
6. Each of the numbers except 81, is a prime number.
7. Each of the numbers except 72, is a perfect square.
8. Each of the numbers except 54, is a multiple of 5.
9. The pattern is $1^2, 2^2, 3^2, 4^2, 5^2, 6^2, 7^2$. But, instead of $5^2$, it is 20, which is to be turned out.
10. The pattern is $2^3, 3^3, 4^3, 5^3, 6^3, 7^3$. But 100 is not a perfect cube.
11. The pattern is $1^2, 1^2 + 2^2, 1^2 + 2^2 + 3^2, 1^2 + 2^2 + 3^2 + 4^2, 1^2 + 2^2 + 3^2 + 4^2 + 5^2, 1^2 + 2^2 + 3^2 + 4^2 + 5^2 + 6^2$. But 50 is not of this pattern.
12. In each number except 427, the middle digit is sum of the other two.
13. In each number except 751, the difference of third and first digit is the middle one.
14. In each number except 383, the product of first and third digits is the middle one.
15. The pattern is $x^2 + 1$, where $x = 1, 2, 3, 4, 5, 6, 7, 8$ etc. But, 64 is out of pattern.
16. The pattern is $x^2 + 3$, where $x = 4, 5, 6, 7, 8, 9$ etc. But, 102 is out of pattern.
17. Sum of the digits in each number, except 324 is 10.
18. Pattern is 1st × 2nd = 3rd; 2nd × 3rd = 4th; 3rd × 4th = 5th.
    But, 4th × 5th = 50 × 500 = 25000 ≠ 5000 = 6th.
19. 2nd = (1st + 1); 3rd = (2nd + 2); 4th = (3rd + 3); 5th = (4th + 4)
    But, 18 = 6th ≠ 5th + 5 = 14 + 5 = 19.
20. Each number except 279 is a multiple of 11.
21. The terms are alternately multiplied by 1.5 and divided by 3. However 18.5 does not satisfy it.
22. Alternately 23 is added and 17 is subtracted from the terms. So, 634 is wrong.
23. The terms are successively divided by 12, 10, 8, 6, ...... etc. So, 24 is wrong.
24. The numbers are $1^3, 2^3, 3^3, 4^3$ etc.
    So, 124 is wrong; it must have been $5^3$ *i.e.* 125.
25. Terms at odd places are 5, 6, 7, 8 etc. and each term at even place is 16. So, 9 is wrong.

**26.** The difference between two successive terms from the beginning are 7, 5, 7, 5, 7, 5.
So, 36 is wrong.

**27.** The numbers are 7 × 8, 8 × 9, 9 × 10, 10 × 11, 11 × 12, 12 × 13.
So, 150 is wrong.

**28.** Go on adding 5, 8, 11, 14, 17, 20.
So, the number 47 is wrong and must be replaced by 46.

**29.** The numbers are squares of odd natural numbers, starting from 5 upto 15.
So, 36 is wrong.

**30.** Add $1^2, 2^2, 3^2, 4^2, 5^2, 6^2$.
So, 91 is wrong.

**31.** Subtract 1, 3, 5, 7, 9, 11 from successive numbers.
So, 34 is wrong.

**32.** Subtract 20, 25, 30, 35, 40, 45 from successive numbers.
So, 0 is wrong.

**33.** Each number is a composite number except 11.

**34.** Prime numbers 2, 3, 5, 7, 11, 13 are to be added successively.
So, 165 is wrong.

**35.** Each number is the square of a composite number except 190.

**36.** Prime numbers 2, 3, 5, 7, 11, 13 have successively been subtracted.
So, 100 is wrong. It must be (108 – 11) i.e. 97.

**37.** The pattern is 1 × 3, 2 × 5, 3 × 7, 4 × 9, 5 × 11, 6 × 13, 7 × 15 etc.

**38.** Double the number and add 1 to it, to get the next number.
So, 160 is wrong.

**39.** Alternately, we add 4 and double the next.
So, 132 is wrong.
It must be (68 × 2) *i.e.* 136.

**40.** The numbers are cubes of primes *i.e.* $2^3, 3^3, 5^3, 7^3, 11^3$.
Clearly, none is wrong.

**41.** Each number is the preceding number multiplied by –2.
So, the required number is –128.

**42.** Numbers are alternately multiplied by 2 and increased by 3.
So, the missing number = 61 × 2 = 122.

**43.** Numbers are $1^2, 2^2, 3^2, 4^2, 5^2, 6^2, 7^2$.
So, the next number is $8^2 = 64$.

**44.** Numbers are $1^3, 2^3, 3^3, 4^3, 5^3, 6^3$. So, the missing number is $7^3 = 343$.

**45.** Numbers are all primes. The next prime is 43.

**46.** Each number is twice the preceding one with 1 added or subtracted alternately. So, the next number is (2 × 261 + 1) = 523.

**47.** There are two series, beginning respectively with 3 and 7. In one 3 is added and in another 2 is subtracted. The next number is $1 - 2 = -1$.

**48.** Each number is double the preceding one plus 1. So, the next number is $(255 \times 2) + 1 = 511$.

**49.** The pattern is $1 \times 2, 2 \times 3, 3 \times 4, 4 \times 5, 5 \times 6, 6 \times 7, 7 \times 8$.
So, the next number is $8 \times 9 = 72$.

**50.** Numbers are alternately multiplied by 3 and divided by 2.
So, next number $= 54 \div 2 = 27$.

**51.** Each number is 15 multiplied by a prime number i.e., $15 \times 11, 15 \times 13, 15 \times 17, 15 \times 19, 15 \times 23$. So the next number is $15 \times 29 = 435$.

**52.** Numbers are $(2^3 - 1), (3^3 - 1), (4^3 - 1), (5^3 - 1), (6^3 - 1), (7^3 - 1)$ etc.
So, the next number is $(8^3 - 1) = (512 - 1) = 511$.

**53.** Go on multiplying the given numbers by 2, 3, 4, 5, 6.
So, the correct next number is 1440.

**54.** There are two series (8, 11, 14, 17, 20) and (7, 12, 17, 22) increasing by 3 and 5 respectively.

**55.** There are two series (10, 13, 16, 19) and (5, 10, 20, 40) one increasing by 3 and another multiplied by 2.

**56.** Each previous number is multiplied by 2.

**57.** Alternately, we add 5 and subtract 7.

**58.** Alternately, we add 3 and subtract 1.

**59.** Second number is one more than twice the first; third number is one less than twice the second; fourth number is one more than twice the third; fifth number is one less than the fourth.
Therefore, the sixth number is one more than twice the fifth.
So, the missing number is 75.

**60.** The difference between consecutive terms are respectively 5, 7, 9, 11 and 13.
So, 34 is a wrong number.

**61.** $2 = (1^3 + 1); 9 = (2^3 + 1); 28 = (3^3 + 1); 65 = (4^3 + 1);$
$126 = (5^3 + 1); 216 \neq (6^3 + 1)$ & $344 = (7^3 + 1)$.
$\therefore$ 216 is a wrong number.

**62.** Multiply each term by 3 to obtain the next term.
Hence, 30 is a wrong number.

**63.** Go on subtracting prime numbers 19, 17, 13, 11, 7, 5 from the numbers to get the next number. So, 88 is wrong.

**64.** Go on subtracting 24, 21, 18, 15, 12, 9 from the numbers to get the next number.
Clearly, 128 is wrong.

**65.** Go on multiplying with 1, 2, 3, 4, 5, 6 to get the next number.
So, 96 is wrong.

**66.** Go on dividing by 4 to get the next number.
So, 200 is wrong.

**67.** Go on adding 7, 9, 11, 13, 15, 17, 19 respectively to obtain the next number.
So, 135 is wrong.

**68.** Let the given numbers be A, B, C, D, E, F, G. Then,
A, $A \times 1 + 1$, $B \times 2 + 2$, $C \times 3 + 3$, $D \times 4 + 4$, $E \times 5 + 5$, $F \times 6 + 6$ are the required numbers.
Clearly, 228 is wrong.

**69.** Go on multiplying the number by 2 and adding 1 to it to get the next number.
So, 27 is wrong.

**70.** Go on adding 7, 9, 11, 13, 15, 17 respectively to obtain the next number.
So, 33 is wrong.

**71.** Go on dividing by 6, 5, 4, 3, 2, 1 respectively to obtain the next number.
Clearly, 92 is wrong.

**72.** Go on subtracting 3 and dividing the result by 2 to obtain the next number.
Clearly, 46 is wrong.

**73.** Go on multiplying 2 and adding 1 to get the next number.
So, 39 is wrong.

**74.** $A \times 2 + 1$, $B \times 3 + 1$, $C \times 2 + 1$, $D \times 3 + 1$ and so on.
$\therefore$ 356 is wrong.

**75.** Numbers must be $(14)^2$, $(13)^2$, $(11)^2$, $(10)^2$, $(9)^2$, $(8)^2$.
So, 80 is wrong.

# 31

# Problems on Ages

## EXERCISE 31 (Objective Type Questions)

1. The sum of the ages of a father and a son is 50 years. Also, 5 years ago, the father's age was 7 times the age of the son. The present ages of the father and the son respectively, are :
(*a*) 35 years, 15 years (*b*) 40 years, 10 years
(*c*) 38 years, 12 years (*d*) 42 years, 8 years

2. In 10 years, *A* will be twice as old as *B* was 10 years ago. If *A* is now 9 years older than *B*, the present age of *B* is :
(*a*) 29 years (*b*) 39 years
(*c*) 19 years (*d*) 49 years **(L.I.C. Exam. 1989)**

3. The sum of the ages of a son and father is 56 years. After four years, the age of the father will be three times that of the son. Their ages respectively are :
(*a*) 12 years, 44 years (*b*) 16 years, 48 years
(*c*) 16 years, 42 years (*d*) 18 years, 36 years
**(Railway Recruitment Board Exam. 1989)**

4. A man's age is three times that of his son. In 12 years, the father's age will be double the son's age. Man's present age is :
(*a*) 27 years (*b*) 32 years
(*c*) 36 years (*d*) 40 years

5. The difference between the ages of two persons is 10 years. 15 years ago, the elder one was twice as old as the younger one. The present age of the elder person is :
(*a*) 35 years (*b*) 25 years
(*c*) 45 years (*d*) 55 years

6. The age of a father 10 years ago was thrice the age of his son. Ten years hence, the father's age will be twice that of his son. The ratio of their present ages is :
(*a*) 8 : 5 (*b*) 7 : 3
(*c*) 5 : 2 (*d*) 9 : 5

7. The age of a man is 4 times that of his son. Five years ago, the man was nine times as old as his son was at that time. The present age of the man is :
(*a*) 28 years (*b*) 32 years

(c) 40 years
(d) 44 years

8. One year ago a father was four times as old as his son. In 6 years time his age exceeds twice his son's age by 9 years. Ratio of their ages is :
(a) 13 : 4 (b) 12 : 5
(c) 11 : 3 (d) 9 : 2

9. The age of Arvind's father is 4 times his age. If 5 years ago, father's age was 7 times of the age of his son at that time, what is Arvind's father's present age ?
(a) 84 years (b) 70 years
(c) 40 years (d) 35 years
**(S.B.I. P.O. Exam. 1987)**

10. 10 years ago Chandravati's mother was 4 time older than her daughter. After 10 years, the mother will be twice older than the daughter. The present age of Chandravati is :
(a) 5 years (b) 10 years
(c) 20 years (d) 30 years
**(Bank P.O. Exam. 1988)**

11. Ratio of Ashok's age to Pradeep's age is equal to 4 : 3. Ashok will be 26 years old after 6 years. How old is Pradeep now ?
(a) $19\frac{1}{2}$ years (b) 21 years
(c) 12 years (d) 15 years
**(Railway Recruitment Board Exam. 1989)**

12. The ratio of the father's age to the son's age is 4 : 1. The product of their ages is 196. The ratio of their ages after 5 years will be :
(a) 3 : 1 (b) 10 : 3
(c) 11 : 4 (d) 14 : 5

13. The ages of *A* and *B* are in tne ratio 2 : 5. After 8 years their ages will be in the ratio 1 : 2. The difference of their ages is :
(a) 20 years (b) 24 years
(c) 26 years (d) 29 years

14. Deepak is 4 times as old as his son. Four years hence the sum of their ages will be 43 years. How old is Deepak's son now ?
(a) 5 years (b) 7 years
(c) 8 years (d) 10 years

15. The ratio of Mona's age to the age of her mother is 3 : 11. The difference of their ages is 24 years. The ratio of their ages after 3 years will be :
(a) 1 : 3 (b) 2 : 3
(c) 3 : 5 (d) none of these

16. Three years ago the average age of *A* and *B* was 18 years. With *C* joining them now, the average becomes 22 years. How old is *C* now ?

(*a*) 27 years (*b*) 30 years
(*c*) 28 years (*d*) 24 years

**(P.N.B. P.O. Exam. 1987)**

**17.** Kamla got married 6 years ago. Today her age is $1\frac{1}{4}$ times her age at the time of marriage. His son's age is (1/10) times her age. The age of her son is :

(*a*) 2 years (*b*) 3 years
(*c*) 4 years (*d*) 5 years

**(Bank P.O. Exam. 1988)**

**18.** The ratio of the ages of father and son at present is 6 : 1. After 5 years, the ratio will become 7 : 2. The present age of the son is :

(*a*) 10 years (*b*) 9 years
(*c*) 6 years (*d*) 5 years **(Bank P.O. 1991)**

**19.** Sachin was twice as old as Ajay 10 years back. How old is Ajay today if Sachin will be 40 years old 10 years hence ?

(*a*) 20 years (*b*) 10 years
(*c*) 30 years (*d*) 15 years **(Bank P.O. 1991)**

**20.** The sum of the ages of a father and son is 45 years. Five years ago, the product of their ages was four times the father's age at that time. The present age of the father is :

(*a*) 39 years (*b*) 36 years
(*c*) 25 years (*d*) none of these

**(Hotel Management, 1991)**

## SOLUTIONS (Exercise 31)

**1.** Let the son's present age be $x$ years.
Then, the fathers present age = $(50 - x)$ years.

$$7(x-5) = (50 - x - 5) \text{ or } x = 10.$$

So, their present ages are 40 years & 10 years.

**2.** Let the present ages of $B$ and $A$ be $x$ yrs & $(x + 9)$ yrs

$$(x + 9 + 10) = 2(x - 10) \text{ or } x = 39.$$

**3.** Let the present ages of son and father be $x$ years and $(56 - x)$ years respectively. Then,

$$(56 - x + 4) = 3(x + 4) \text{ or } 4x = 48 \text{ or } x = 12.$$

So, their ages are 12 years, 44 years respectively.

**4.** Let son's age be $x$ years. Then, father's age = $3x$ years.

$$\therefore 2(x + 12) = (3x + 12) \text{ or } x = 12.$$

Hence, the man's present age is 36 years.

**5.** Let the present age of the elder person be $x$ years.
Then, the present age of another person = $(x - 10)$ years.

$(x-15) = 2(x-10-15)$ or $x = 35$.

$\therefore$ The present age of the elder person is 35 years.

**6.** Let the present ages of father and son be $x$ & $y$ respectively

Then, $(x-10) = 3(y-10)$ or $3y - x = 20$ ...(i)

And, $(x+10) = 2(y+10)$ or $x - 2y = 10$ ...(ii)

Solving (i) & (ii) we get, $x = 70$ and $y = 30$.

$\therefore$ Ratio of their ages is 70 : 30 or 7 : 3.

**7.** Let the son's age be $x$. Then, father's age = $4x$.

$\therefore$ $(4x-5) = 9(x-5)$ or $5x = 40$ or $x = 8$.

$\therefore$ Present age of the man = 32 years.

**8.** Let the present ages of father & son be $x$ & $y$ respectively.

Then, $(x-1) = 4(y-1)$ or $4y - x = 3$ ...(i)

And, $(x+6) - 2(y+6) = 9$ or $-2y + x = 15$ ...(ii)

Solving (i) & (ii) we get, $x = 33$ and $y = 9$.

$\therefore$ Ratio of their ages = 33 : 9 11 : 3.

**9.** Let Arvind's age be $x$ years.

Then, his father's age = $4x$ years.

$\therefore$ $(4x-5) = 7(x-5)$ or $x = 10$.

Hence, Arvind's father's age = 40 years.

**10.** Let the present ages of Chandravati and her mother be $x$ years and $y$ years respectively. Then,

$(y-10) = 4(x-10)$ or $4x - y = 30$ ...(i)

And, $(y+10) = 2(x+10)$ or $y - 2x = 10$ ...(ii)

Solving (i) & (ii) we get, $x = 20$.

$\therefore$ Chandravati's present age = 20 years.

**11.** Ashok's present age = (26 – 6) years = 20 years.

$\therefore$ Pradeep's present age = $\left(\frac{3}{4} \times 20\right)$ years = 15 years.

**12.** Let their ages be $4x$ and $x$ respectively. Then,

$4x^2 = 196$ or $x^2 = 49$ or $x = 7$.

$\therefore$ Their present ages are 28 years & 7 years.

Ratio of their ages after 5 years = 33 : 12 or 11 : 4.

**13.** Let their ages be $2x$ & $5x$ years. Then,

$$\frac{2x+8}{5x+8} = \frac{1}{2} \text{ or } x = 8.$$

$\therefore$ Their ages are 16 years and 40 years.

Hence, the difference of their ages = 24 years.

**14.** Let the son's age be $x$ years.

Then, $(x+4) + (4x+4) = 43$ or $5x = 35$ or $x = 7$.

**15.** Let the ages of Mona and her mother be $3x$ and $11x$ years respectively.

Then, $(11x - 3x) = 24$ or $x = 3$.

So, their present ages are 9 years and 33 years.

Ratio of their ages after 3 years 12 : 36 or 1 : 3.

**16.** $(A + B)$'s total present age = (36 + 6) years = 42 years.

$(A + B + C)$'s total present age = (22 × 3) years = 66 years.

∴ $C$'s age = (66–42) years = 24 years.

**17.** Let son's age be $x$. Then, Kamla's age = $10x$ years.

Kamla's age at the time of marriage = $(10x - 6)$ years.

$$\therefore\ 10x = \frac{5}{4}(10x - 6) \text{ or } 40x = 50x - 30 \text{ or } x = 3.$$

**18.** Let their present ages be $6x$ and $x$ respectively

$$\text{Then, } \frac{6x + 5}{x + 5} = \frac{7}{2} \Rightarrow 2(6x + 5) = 7(x + 5) \Rightarrow x = 5.$$

∴ The present age of the son is 5 years.

**19.** Let Ajay's age 10 years back be $x$ years.

Then, Sachin's age 10 years back = $2x$ years.

∴ $2x + 20 = 40$ or $x = 10$.

Ajay's present age = (10 + 10) years = 20 years.

**20.** Let father's present age = $x$ years.

Then, Son's present age = $(45 - x)$ years.

$$(x - 5)(45 - x - 5) = 4(x - 5)$$

or $x^2 - 41x + 180 = 0$

or $(x - 36)(x - 5) = 0$

∴ $x = 36$ years.

## ANSWERS (Exercise 31)

| | | | | | |
|---|---|---|---|---|---|
| **1.** (*b*) | **2.** (*b*) | **3.** (*a*) | **4.** (*c*) | **5.** (*a*) | **6.** (*b*) |
| **7.** (*b*) | **8.** (*c*) | **9.** (*c*) | **10.** (*c*) | **11.** (*d*) | **12.** (*c*) |
| **13.** (*b*) | **14.** (*b*) | **15.** (*a*) | **16.** (*d*) | **17.** (*b*) | **18.** (*d*) |
| **19.** (*a*) | **20.** (*b*) | | | | |

# LATEST QUESTIONS

**1.** $(4^{61} + 4^{62} + 4^{63} + 4^{64})$ is divisible by :

(a) 3 (b) 11 (c) 13 (d) 17 *(Astt. Grade. 1996)*

**2.** The unit digit in the sum $(264)^{102} + (264)^{103}$ is :

(a) 0 (b) 4 (c) 6 (d) 8 *(Astt. Grade. 1996)*

**3.** If $p = \frac{3}{5}$, $q = \frac{7}{9}$ and $r = \frac{5}{7}$, then

(a) $p < q < r$ (b) $q < r < p$ (c) $p < r < q$ (d) $r < q < p$ *(S.S.C. 1995)*

**4.** The number of prime factors of $(6)^{10} \times (7)^{17} \times (55)^{27}$ is :

(a) 54 (b) 64 (c) 81 (d) 91 *(S.S.C. 1995)*

**5.** Which of the following numbers is a multiple of 11?

(a) 978626 (b) 112144 (c) 447355 (d) 869756 *(S.S.C. 1995)*

**6.** The number of prime factors in $\left(\frac{1}{6}\right)^{12} \times (8)^{25} \times \left(\frac{3}{4}\right)^{15}$ is :

(a) 33 (b) 37 (c) 52 (d) None *(Hotel Management. 1995)*

**7.** $(51 + 52 + 53 + \ldots\ldots + 100) = ?$

(a) $\frac{51 \times 52}{2}$ (b) $\frac{52 \times 53}{2}$ (c) $\frac{100 \times 50}{2}$ (d) 3775 *(P.C.S. 1995)*

**8.** A number when divided by 32 leaves the remainder 29. This number when divided by 8 will leave the remainder :

(a) 3 (b) 5 (c) 7 (d) 29 *(C.B.I. 1994)*

**9.** If $\frac{a}{b} = \frac{4}{5}$ and $\frac{b}{c} = \frac{15}{16}$, then $\frac{c^2 - a^2}{c^2 + a^2} = ?$

(a) $\frac{1}{7}$ (b) $\frac{7}{25}$ (c) $\frac{3}{4}$ (d) None *(Hotel Management. 1995)*

**10.** Which one of the following numbers is a multiple of 8?

(a) 923972 (b) 923962 (c) 923872 (d) 923862 *(S.S.C. 1995)*

**11.** What should be added to 18962 to make it exactly divisible by 13?

(a) 2 (b) 3 (c) 4 (d) 5 *(B.S.R.B. 1996)*

**12.** The least number, which when increased by 1 is exactly divisible by 12, 18, 24, 32, 40 is :

(a) 1439 (b) 1440 (c) 1449 (d) 1459 *(S.S.C. 1995)*

**13.** Which of the following numbers is divisible by 25 ?

(a) 505520 (b) 437950 (c) 124505 (d) 500555 *(C.B.I. 1995)*

**14.** A number lying between 1000 and 2000 is such that on division by 2, 3, 4, 5, 6, 7 and 8 leaves remainders 1, 2, 3, 4, 5, 6 and 7 respectively. The number is:

(*a*) 1876 (*b*) 1679 (*c*) 1778 (*d*) 1654 *(C.B.I. 1995)*

**15.** Four prime numbers are in ascending order of their magnitudes. The product of the first three is 385 and that of last three is 1001. The largest given prime number is :

(*a*) 11 (*b*) 13 (*c*) 17 (d) 19 *(C.B.I. 1995)*

**16.** The nearest integer to 58701 which is exactly divisible by 567 is :

(*a*) 58068 (*b*) 55968 (*c*) 58968 (*d*) None *(Railway, 1995)*

**17.** The number which when multiplied by 17 increases by 640, is :

(*a*) 42 (*b*) 36 (*c*) 40 (*d*) None *(Railway, 1996)*

**18.** The square of a number subtracted from its cube gives 100. The number is :

(*a*) 25 (*b*) 16 (*c*) 6 (*d*) 5 *(Astt. Grade, 1996)*

**19.** If two-third of three-fourth of a number added to three-fourth of the four-fifth of the number is x times the number, the value of x is :

(*a*) $\frac{11}{10}$ (*b*) $1\frac{1}{11}$ (*c*) $\frac{10}{11}$ (*d*) $\frac{9}{11}$ *(U.D.C. 1995)*

**20.** Which of the following numbers does not lie between $\frac{7}{13}$ and $\frac{4}{5}$ ?

$\frac{1}{2}, \frac{2}{3}, \frac{3}{4}, \frac{5}{7}$

(*a*) $\frac{1}{2}$ (*b*) $\frac{2}{3}$ (*c*) $\frac{3}{4}$ (*d*) $\frac{5}{7}$ *(U.D.C. 1995)*

**21.** The value of $\frac{(625)^{6.25} \times (25)^{2.6}}{(625)^{6.75} \times (5)^{1.2}}$ is :

(*a*) 0.25 (*b*) 6.25 (*c*) 25 (*d*) 625 *(S.S.C. 1994)*

**22.** $(16)^{1.75} = ?$

(*a*) 64 (*b*) $64\sqrt{2}$ (*c*) 128 (*d*) $128\sqrt{2}$ *(S.S.C. 1994)*

**23.** Out of the numbers $\sqrt{2}$, $\sqrt[3]{3}$ and $\sqrt[4]{4}$, we can definitely say that the largest number is :

(*a*) $\sqrt{2}$ (*b*) $\sqrt[3]{3}$ (*c*) $\sqrt[4]{4}$ (*d*) all are equal *(C.B.I. 1994)*

**24.** If $\frac{5 + 2\sqrt{3}}{7 + 4\sqrt{3}} = a + b\sqrt{3}$, then

(*a*) $a = 11, \quad b = -6$ (*b*) $a = -6, \quad b = 11$
(c) $a = -11; \quad b = 6$ (*d*) $a = -11, \quad b = -6$ *(S.S.C. 1995)*

**25.** The value of $\sqrt[4]{(625)^3}$ is :

(*a*) 25 (*b*) 125 (*c*) $\sqrt[3]{1875}$ (*d*) None of these *(I. Tax. 1995)*

**26.** If $(125)^x = 3125$, then $x$ equals :

(*a*) $\frac{3}{5}$ (*b*) $\frac{5}{3}$ (*c*) $\frac{1}{4}$ (*d*) $\frac{1}{5}$ *(C.B.I. 1995)*

**27.** The value of $\frac{36 \times 36 \times 36 + 14 \times 14 \times 14}{36 \times 36 + 14 \times 14 - 36 \times 14}$ is :

(*a*) 22 (*b*) 50 (*c*) 5100 (*d*) 132 *(P.C.S. 1996)*

**28.** The value of $(0.\dot{6}\dot{3} + 0.\dot{3}\dot{7})$ is :

(a) 1.01 (b) $.\dot{1}0\dot{1}$ (c) $1.\dot{0}\dot{1}$ (d) $1.00\dot{1}$ *(Railway, 1995)*

**29.** If $\frac{1}{3.718} = .2689$, then the value of $\frac{1}{.0003718}$ is :

(*a*) 2689 (*b*) 2.689 (*c*) 26890 (*d*) .2689 *(Railway, 1995)*

**30.** $(0.333......) \times (0.444......) = ?$

(*a*) 0.121212..... (*b*) 1.333..... (*c*) 0.777..... (*d*) 0.148148148......

*(S.S.C. 1995)*

**31.** Which is the largest among the following fractions ?

$\frac{5}{8}, \frac{2}{3}, \frac{7}{9}, \frac{3}{5}, \frac{4}{7}$

(*a*) $\frac{5}{8}$ (*b*) $\frac{7}{9}$ (*c*) $\frac{4}{7}$ (*d*) $\frac{2}{3}$ (L.I.C. 1995)

**32.** The value of $\frac{(3.06)^3 - (1.98)^3}{(3.06)^2 + (3.06 \times 1.98) + (1.98)^2} = ?$

(*a*) 5.04 (*b*) 1.08 (*c*) 2.16 (*d*) 1.92 *(Astt. Grade, 1996)*

**33.** $0.04 \times ? = .000016$

(*a*) 4 (*b*) .04 (*c*) .0004 (*d*) None

*(Hotel Management, 1995)*

**34.** The value of $\frac{2^{1/2} . 3^{1/3} . 4^{1/4}}{10^{-1/5} . 5^{3/5}} \div \frac{3^{4/3} . 5^{-7/5}}{4^{-3/5} . 6}$ is :

(*a*) 5 (*b*) 6 (*c*) 10 (*d*) 15 *(C.B.I. 1995)*

**35.** $4^7 + 16^4 \times \sqrt{16} = ?$

(*a*) $\frac{1}{16}$ (*b*) $\frac{1}{4}$ (*c*) 4 (*d*) 1 (Railway, 1996)

**36.** H.C.F. of $\frac{7}{90}, \frac{14}{15}$ and $\frac{7}{10}$ is :

(*a*) $\frac{7}{90}$ (*b*) $\frac{7}{45}$ (*c*) $\frac{7}{675}$ (*d*) $\frac{14}{45}$ (C.B.I. 1994)

**37.** Three persons begin to walk around a circular track. They complete their revolutions in $15\frac{1}{6}$ seconds, $16\frac{1}{4}$ seconds and $18\frac{2}{3}$ seconds respectively. After what time will they be together at the starting point again?

(*a*) $303\frac{1}{3}$ sec (*b*) 364 sec (*c*) 3604 sec (*d*) 3640 sec *(C.B.I. 1994)*

38. $\left(1 + \dfrac{1}{1 + \dfrac{1}{1 + \dfrac{1}{3}}}\right) \div 1\frac{4}{7}$ is equal to :

(*a*) $1\frac{1}{3}$ (*b*) $1\frac{1}{4}$ (*c*) $1\frac{1}{7}$ (*d*) 1 (S.S.C. 1995)

39. $\dfrac{(0.43)^3 + (1.47)^3 + (1.1)^3 - 3 \times 0.43 \times 1.47 \times 1.1}{(0.43)^2 + (1.47)^2 + (1.1)^2 - 0.43 \times 1.43 - 0.43 \times 1.1 - 1.47 \times 1.1} = ?$

(*a*) 1.90 (*b*) 2.87 (*c*) 3 (*d*) 3.47 *(S.S.C. 1995)*

40. If $\dfrac{x}{2y} = \dfrac{3}{2}$, then the value of $\dfrac{2x + y}{x - 2y}$ equals :

(*a*) $\frac{1}{7}$ (*b*) 7 (*c*) 7.1 (*d*) 6.8 (S.S.C. 1995)

41. $\frac{1}{4}$th of Nikhil's money is equal to $\frac{1}{6}$th of Yogesh's money. If both together have Rs. 600. What is the difference between their amounts ?

(*a*) 240 (*b*) 360 (*c*) 50 (*d*) 120 *(S.S.C. 1995)*

42. $\dfrac{(0.05)^2 + (0.41)^2 + (0.073)^2}{(0.005)^2 + (0.041)^2 + (0.0073)^2} = ?$

(*a*) 100 (*b*) 10 (*c*) 1000 (*d*) None *(Railway, 1995)*

43. One litre of water weighs 1 kg. How many cubic millimetres of water will weigh 0.1 gram ?

(*a*) 10 (*b*) 100 (*c*) 0.1 (*d*) 1 *(Railway, 1995)*

44. Out of a tank which is $\frac{3}{4}$th full, 21 litres of water is drawn out. The tank is now $\frac{2}{5}$th full. What is the capacity of the tank in litres?

(*a*) 200 (*b*) 120 (*c*) 40 (*d*) 60 *(Railway, 1995)*

45. Talekar is as much heavier than Suresh as he is lighter than Gokhale. If the total weight of Suresh and Gokhale is 140 kg, what is the weight of Talekar?

(*a*) 55 kg (*b*) 65 kg (*c*) 70 kg (*d*) Data insufficient *(Bank P.O. 1996)*

46. A man has Rs. 480 in the denominations of one-rupee notes, five-rupee notes and ten-rupee notes. The number of notes are equal. What is the total number of notes he has ?

(*a*) 45 (*b*) 60 (*c*) 75 (*d*) 90 *(C.B.I. 1995)*

**47.** If $a = 1.2$, $b = 2.1$ and $c = -3.3$, then the value of $(a^3 + b^3 + c^3 - 3abc)$ is :

(*a*) 1 (*b*) 2 (*c*) 3 (*d*) 0 *(Astt. Grade, 1996)*

**48.** Find out the numbers indicated by $x$ and $y$ in $3\frac{1}{x} \times y\frac{2}{5} = 13\frac{3}{4}$, fractions being in their lowest terms.

(*a*) $x = 4, y = 8$ (*b*) $x = 4, y = 4$

(*c*) $x = 2, y = 4$ (*d*) $x = 8, y = 4$ *(Astt. Grade, 1996)*

**49.** $\dfrac{\sqrt{31} - \sqrt{29}}{\sqrt{31} + \sqrt{29}} = ?$

(*a*) $60 - 2\sqrt{899}$ (*b*) $30 - \sqrt{899}$ (*c*) $30 + \sqrt{899}$ (*d*) $\dfrac{1}{30 - \sqrt{899}}$

*(S.S.C. 1995)*

**50.** $\left(1 - \dfrac{1}{1 + \sqrt{2}} + \dfrac{1}{1 - \sqrt{2}}\right) = ?$

(*a*) $2\sqrt{2} - 1$ (*b*) $1 - 2\sqrt{2}$ (*c*) $1 - \sqrt{2}$ (*d*) $-2\sqrt{2}$ *(S.S.C. 1994)*

**51.** If $(676)^2 = 456976$, the value of $\sqrt{45.6976}$ is :

(*a*) 0.00676 (*b*) 0.676 (*c*) 6.76 (*d*) 0.0676 *(C.B.I. 1994)*

**52.** $\left(\dfrac{3\sqrt{2}}{\sqrt{6} - \sqrt{3}} - \dfrac{4\sqrt{3}}{\sqrt{6} - \sqrt{2}} - \dfrac{6}{\sqrt{8} + \sqrt{12}}\right) = ?$

(*a*) 1 (*b*) $-\sqrt{3}$ (*c*) $\sqrt{3} + \sqrt{2}$ (*d*) $\sqrt{3} - \sqrt{2}$ *(Astt. Grade, 1996)*

**53.** If $a = \sqrt{6} + \sqrt{5}$ and $b = \sqrt{6} - \sqrt{5}$, then $2a^2 - 5ab + 2b^2 = ?$

(*a*) 43 (*b*) 39 (*c*) 31 (*d*) 27 *(I. Tax, 1995)*

**54.** The price of cooking oil has increased by 25%. The percentage of reduction that a family should effect in the use of cooking oil so as not to increase the expenditure on this account is :

(*a*) 25% (*b*) 30% (*c*) 20% (*d*) 15% *(P.C.S. 1996)*

**55.** 4598 is 95% of ?

(*a*) 4800 (*b*) 4850 (*c*) 4840 (*d*) None *(Hotel Management, 1995)*

**56.** In an organisation, 40% of the employees are matriculates 50% of the remaining are graduates and the remaining 180 are post-graduates. How many employees are graduates?

(*a*) 360 (*b*) 240 (*c*) 300 (*d*) 180 *(L.I.C. 1995)*

**57.** If 40% of the people read newspaper X, 50% read newspaper Y and 10% read both the papers. What percentage of the people read neither newspaper?

(*a*) 10% (*b*) 15% (*c*) 20% (*d*) 25% *(U.D.C. 1995)*

**58.** The population of a town increases by 5% annually. If its population in 1995 was 138915, what it was in 1992?

(*a*) 110000 (*b*) 100000 (*c*) 120000 (*d*) 90000 *(U.D.C. 1995)*

**59.** 25% of a certain number is 15 less than 30% of the same number. What is that number?

(*a*) 600 (*b*) 300 (*c*) 750 (*d*) 135 (*B.S.R.B., 1996*)

**60.** The population of a village is 4500. $\frac{5}{9}$th of them are males and rest females. If 40% of the males are married, then the percentage of married females is :

(*a*) 35 (*b*) 40 (*c*) 50 (*d*) 60 (*S.S.C. 1995*)

**61.** A salesman's commission is 5% on all sales upto Rs. 10000 and 4% on all sales exceeding this. He remits Rs. 31100 to his parent company after deducting his commission. His sales was worth:

(*a*) Rs. 35000 (*b*) Rs. 36100 (*c*) Rs. 35100 (*d*) Rs. 32500 (*I. Tax, 1995*)

**62.** A's income is 10% more than B's. How much percent is B's income less than A's?

(*a*) 10% (*b*) 7% (*c*) $9\frac{1}{11}\%$ (*d*) $6\frac{1}{2}\%$ (*Railway, 1995*)

**63.** A mixture of 40 litres of milk and water contains 10% water. How much water must be added to make water 20% in the new mixture?

(*a*) 10 litres (*b*) 7 litres (*c*) 5 litres (*d*) 3 litres (*Railway, 1995*)

**64.** If $z = \frac{x^2}{y}$ and $x$, $y$ both are increased in value by 10%, then the value of $z$ is :

(*a*) unchanged (*b*) increased by 10%
(*c*) increased by 11% (*d*) increased by 20% (*Astt. Grade, 1996*)

**65.** If $a$ exceeds $b$ by $x\%$, then which one of these equations is correct?

(*a*) $a - b = \frac{x}{100}$ (*b*) $b = a + 100x$

(*c*) $a = \frac{bx}{100 + x}$ (*d*) $a = b + \frac{bx}{100}$ (*I. Tax, 1995*)

**66.** In an examination, 35% of the examinees failed in G.K. and 25% in English. If 10% of the examinees failed in both, then the percentage of examinees passed will be :

(*a*) 40% (*b*) 45% (*c*) 48% (*d*) 50% (*U.D.C. 1995*)

**67.** If the price of a television set is increased by 25%, then by what percentage should the new price be reduced to bring the price back to the original level?

(*a*) 15% (*b*) 20% (*c*) 25% (*d*) 30% (*I.A.S. 1996*)

**68.** The number of grams of water needed to reduce 9 grams of shaving lotion containing 50% alcohol to a lotion containing 30% alcohol, is :

(*a*) 4 (*b*) 5 (*c*) 6 (*d*) 7 (*Astt. Grade, 1995*)

**69.** A dealer marks his goods 20% above cost price. He then allows some discount on it and makes a profit of 8%. The rate of discount is :

(*a*) 4% (*b*) 6% (*c*) 10% (*d*) 12% (*P.C.S. 1996*)

**70.** By selling an umbrella for Rs. 30, a shopkeeper gains 20%. During a clearance sale, the shopkeeper allows a discount of 10% of the marked price. his gain percent during the sale season is :

(*a*) 7 (*b*) 7.5 (*c*) 8 (*d*) 9 *(S.S.C. 1995)*

**71.** A shopkeeper allows a discount of 10% on the marked price of an item but charges a sales tax of 8% on the discounted price. If the customer pays Rs. 680.40 as the price including the sales tax, what is the marked price of the item?

(*a*) Rs. 630 (*b*) Rs. 700 (*c*) Rs. 780 (*d*) None

*(Hotel Management, 1995)*

**72.** The difference between the cost price and sale price of an article is Rs. 240. If the profit percent is 20, at what price was the article sold? *(B.S.R.B. 1995)*

(*a*) Rs. 1240 (*b*) Rs. 1400 (*c*) Rs. 1600 (*d*) None of these.

**73.** By selling a motor cycle for Rs. 22600 a person gains 13%, what was his gain?

(*a*) Rs. 600 (*b*) Rs. 2936 (*c*) Rs. 2600 (*d*) Data inadequate

*(L.I.C. 1995)*

**74.** A shopkeeper bought locks at the rate of 8 locks for Rs. 34 and sold them at 12 locks for Rs. 57. The number of locks he should sell to have a profit of Rs. 900, is :

(*a*) 1400 (*b*) 1600 (*c*) 1800 (*d*) 2000 *(Railway, 1995)*

**75.** A shopkeepr earns 15% profit on a shirt even after allowing 31% discount on the list price. If the list price is Rs. 125, then the cost price of the shirt is :

(*a*) Rs. 87 (*b*) Rs. 80 (*c*) Rs. 75 (*d*) Rs. 69 *(U.D.C. 1995)*

**76.** Loss incurred by selling a bicycle for Rs. 895 is equal to the profit earned by selling it for Rs. 955. What is the loss/profit in this case?

(*a*) Rs. 45 (*b*) Rs. 30 (*c*) Rs. 75 (*d*) cannot be determined

*(B.S.R.B. 1996)*

**77.** If 2 kg of almonds cost as much as 8 kg of walnuts and the cost of 5 kg of almonds and 16 kg of walnuts is Rs. 1080, the cost of almonds per kg is :

(*a*) Rs. 160 (*b*) Rs. 150 (*c*) Rs. 120 (*d*) None of these

*(Hotel Management, 1995)*

**78.** A manufacturer sells a pair of glasses to a wholesale dealer at a profit of 18%. The wholesaler sells the same to a retailer at a profit of 20%. The retailer in turn sells them to a customer for Rs. 30.09, thereby earning a profit of 25%. The cost price of the manufacturer is :

(*a*) Rs. 15 (*b*) Rs. 16 (*c*) Rs. 17 (*d*) Rs. 18 *(U.D.C. 1995)*

**79.** Chatterjee bought a car and got 15% of its original price as dealer's discount. He then sold it with 20% profit on his purchase price. What percentage profit did he get on the original price of the car ?

(*a*) 2% (*b*) 12% (*c*) 5% (*d*) 17% *(Bank P.O. 1996)*

**80.** If the difference between selling a shirt at a profit of 10% and 15% is Rs. 10, then the cost price is :

(*a*) Rs. 110 (*b*) Rs. 115 (*c*) Rs. 150 (*d*) Rs. 200

*(S.S.C. 1995)*

**81.** The cost of a shirt after 15% discount is Rs. 102. What was the cost of the shirt before the discount ?

(*a*) Rs. 117 (*b*) Rs. 118 (*c*) Rs. 120 (*d*) Rs. 121

*(C.B.I. 1995)*

**82.** If a man reduces the selling price of a fan from Rs. 400 to Rs. 380, his loss increases by 20%. The cost price of the fan is :

(*a*) Rs. 600 (*b*) Rs. 500 (*c*) Rs. 480 (*d*) None

*(Railway, 1996)*

**83.** By selling a vehicle for Rs. 455000, Samant suffers 25% loss, what was his loss?

(*a*) Rs. 115370 (*b*) Rs. 113570 (*c*) Rs. 113750 (*d*) None of these

*(B.S.R.B. 1996)*

**84.** A shopkeeper marks his goods 20% higher than the cost price and allows a discount of 5%. The percentage of his profit is :

(*a*) 10% (*b*) 14% (*c*) 15% (*d*) 20% *(Railway, 1995)*

**85.** A dealer offered a machine for sale for Rs. 27500 but even if he had charged 10% less he would have made a profit of 10%. The actual cost of the machine is :

(*a*) Rs. 24250 (*b*) Rs. 22500 (*c*) Rs. 22275 (*d*) Rs. 22000

*(Astt. Grade, 1996)*

**86.** A toy car was sold at a loss for Rs. 60. Had it been sold for Rs. 81, the gain would have been $\frac{3}{4}$ of the former loss. The cost of the toy is :

(*a*) Rs. 65 (*b*) Rs. 72 (*c*) Rs. 80 (*d*) Rs. 86 *(Astt. Grade, 1996)*

**87.** A trader allows a trade discount of 20% and a cash discount of $6\frac{1}{4}$% on the marked price of the goods and gets a net gain of 20% on the cost. By how much above the cost should the goods be marked for sale?

(*a*) 40% (*b*) 50% (*c*) 60% (*d*) 70% *(I. Tax. 1995)*

**88.** A merchant has 120 kg of rice. He sells a part of it at a profit of 10% and the rest at a profit of 25%. He gains 15% on the whole. Quanity of rice he sold at 25% gain is :

(*a*) 30 kg (*b*) 40 kg (*c*) 50 kg (*d*) 55 kg *(I. Tax, 1995)*

**89.** A wholesaler gains 25% by selling a commodity and a retailer gains 30% by selling it. If the retail value of that commodity is Rs. 325, then the wholesale value is :

(*a*) Rs. 200 (*b*) Rs. 225 (*c*) Rs. 245 (*d*) Rs. 255 *(U.D.C. 1995)*

**90.** A shopkeeper decides to give 5% commission on the marked price of an article but also wants to earn 10% profit. If the cost price is Rs. 95, then the marked price is :

(*a*) Rs. 100 (*b*) Rs. 105 (*c*) Rs. 110 (*d*) Rs. 115 *(U.D.C. 1995)*

**91.** A bag contains 25-p, 10-p and 5-p coins in the ratio 1 : 2 : 3. If the total value is Rs. 30, the number of 5-p coins is :

(*a*) 50 (*b*) 100 (*c*) 200 (*d*) 150 *(P.C.S. 1996)*

**92.** If $(x + y) : (x - y) = 4 : 1$, then $(x^2 + y^2) : (x^2 - y^2) = ?$

(*a*) 25:9 (*b*) 16:1 (*c*) 8:17 (*d*) 17:8 *(C.B.I. 1994)*

**93.** $9^{3.04} : 9^{2.04} = ?$

(*a*) 1 : 9 (*b*) 3 : 2 (*c*) 76 : 51 (*d*) None of these *(U.D.C. 1994)*

**94.** If $a : b = b : c$, then $a^4 : b^4 = ?$

(*a*) $ac : b^2$ (*b*) $a^2 : c^2$ (*c*) $c^2 : a^2$ (*d*) $b^2 : ac$ *(Railway, 1995)*

**95.** If $x : y = 3 : 4$, then $(2x + 3y) : (3y - 2x) = ?$

(*a*) 2 : 1 (*b*) 3 : 2 (*c*) 3 : 1 (*d*) 21 : 1 *(U.D.C. 1995)*

**96.** Rs. 56250 is to be divided among A, B and C so that A may receive half as much as B and C together and B receives one fourth of what A and C together receive. The share of A is more than that of B by :

(*a*) Rs. 7500 (*b*) Rs. 7750 (*c*) Rs. 15000 (*d*) Rs. 16000 *(U.D.C. 1995)*

**97.** A flagstaff 17.5 m high casts a shadow of 40.25 m. The height of the building which casts a shadow 28.75 m long under similar condition, will be :

(*a*) 10 m (*b*) 12.5 m (*c*) 17.5 m (*d*) 21.25 m *(S.S.C. 1995)*

**98.** The sum of three numbers is 174. The ratio of second number to the third number is 9 : 16 and the ratio of first number to the third one is 1 : 4. The second number is :

(*a*) 24 (*b*) 54 (*c*) 96 (*d*) can not be determined *(Bank P.O. 1996)*

**99.** 5 mangoes and 4 oranges cost as much as 3 mangoes and 7 oranges. What is the ratio of the cost of one mango to that of one orange?

(*a*) 4 : 3 (*b*) 1 : 3 (*c*) 3 : 2 (*d*) 5 : 2 *(C.B.I. 1995)*

**100.** If 35% of A's income is equal to 25% of B's income, then the ratio of their incomes is :

(*a*) 4 : 3 (*b*) 5 : 7 (*c*) 7 : 5 (*d*) 4 : 7 *(Astt. Grade, 1996)*

**101.** Last year the ratio between the salaries of A and B was 3 : 4. But the ratio of their individual salaries between last year and this year were 4 : 5 and 2 : 3 respectively. If the sum of their present salaries is Rs. 4160, then how much is the salary of A now?

(*a*) Rs. 1040 (*b*) Rs. 1600 (*c*) Rs. 2560 (*d*) Rs. 3120 *(U.D.C. 1995)*

**102.** A and B have incomes in the ratio 5 : 3. The expenses of A, B and C are in the ratio 8 : 5 : 2. If C spends Rs. 2000 and B saves Rs. 700, A's saving is :

(*a*) Rs. 1500 (*b*) Rs. 1000 (*c*) Rs. 2500 (*d*) Rs. 500 *(Astt. Grade, 1996)*

**103.** If $x : 2\frac{1}{3} :: 21 : 50$, then the value of $x$ is :

(*a*) $\frac{27}{50}$ (*b*) $\frac{49}{50}$ (*c*) $1\frac{1}{50}$ (*d*) $1\frac{1}{49}$ *(Railway, 1996)*

**104.** If 8 women can grind 180 kg. of wheat in 5 days, in how many days will 28 women grind 3780 kg of wheat?

(*a*) 32 (*b*) 45 (*c*) 60 (*d*) 30 *(Railway, 1995)*

**105.** If 40 men can build a wall 300 m. long in 12 days, working 6 hours a day ; how long will 30 men take to build a similar wall 200 m long, working 8 hours a day?

(*a*) $4\frac{1}{2}$ days (*b*) 8 days (*c*) $10\frac{1}{2}$ days (*d*) 11 days *(I. Tax, 1995)*

**106.** If 12 persons can do $\frac{3}{5}$ of a certain work in 10 days, then how many persons are required to finish the whole work in 20 days?

(*a*) 10 (*b*) 9 (*c*) 8 (*d*) 7 *(C.B.I. 1995)*

**107.** In a fort, ration for 2000 people was sufficient for 54 days. After 15 days, more people came and the ration lasted only for 20 more days. How many people came?

(*a*) 2500 (*b*) 2250 (*c*) 1900 (*d*) 1675 *(Railway, 1995)*

**108.** One army camp had ration for 560 soldiers for 20 days, 560 soldiers reported for the camp and after 12 days, 112 soldiers were sent to another camp. For how many days, the remaining soldiers can stay in the camp without getting any new ration?

(*a*) 12 (*b*) 16 (*c*) 10 (*d*) None *(Railway, 1996)*

**109.** How many men need to be employed to complete a job in 5 days if 10 men can complete half the job in 7 days?

(*a*) 7 (*b*) 14 (*c*) 28 (*d*) 35 *(Astt. Grade, 1996)*

**110.** A contractor undertook to build a road in 100 days. He employed 110 men. After 45 days, he found that only $\frac{1}{4}$ could be built. In order to complete the work in time, how many more men should be employed?

(*a*) 120 (*b*) 160 (*c*) 180 (*d*) 270 *(Railway, 1995)*

**111.** 16 men complete one-fourth of a piece of work in 12 days. What is the additional number of men required to complete the work in 12 days?

(*a*) 48 (*b*) 36 (*c*) 30 (*d*) 18 *(S.S.C. 1995)*

**112.** A, B and C together can do a piece of work in 20 days. After working with B and C for 8 days, A leaves and then B and C complete the ramaining work in 20 days more. In how many days, A alone could do the work?

(*a*) 40 (*b*) 50 (*c*) 60 (*d*) 80 *(S.S.C. 1995)*

**113.** A, B and C together can complete a piece of work in 10 days. All the three started working at it together and after 4 days, A left. Then, B and C together completed the work in 10 more days. A alone could complete the work in

(*a*) 15 days (*b*) 16 days (*c*) 25 days (*d*) 50 days *(U.D.C. 1995)*

**114.** 20 men can finish a piece of work in 30 days. After how many days should 5 men leave the work so that it may be finished in 35 days?

(*a*) 10 (*b*) 12 (*c*) 15 (*d*) 20 *(Railway, 1995)*

**115.** A man, a woman or a boy can do a piece of work in 3, 4 and 12 days respectively. How many boys must assist one man and one woman to do the work in one day?

(*a*) 6 (*b*) 8 (*c*) 9 (*d*) 5 *(Astt. Grade, 1996)*

**116.** If 2 men and 3 boys can do a piece of work in 8 days, while 3 men and 2 boys can do it in 7 days, how long will 5 men and 4 boys take to do it?

(*a*) 3 days (*b*) 4 days (*c*) 5 days (*d*) 6 days *(Astt. Grade, 1996)*

**117.** A and B can do a work in 8 days, B and C can do the same work in 12 days. A, B and C together can finish it in 6 days. A and C together will do it in :

(*a*) 4 days (*b*) 6 days (*c*) 8 days (*d*) 12 days *(I. Tax, 1995)*

**118.** If I would have been twice as efficient as today, then I would have finished a work in 12 days. If my efficiency is reduced to one-third of what it is at present, then in how many days, I would be able to finish the work?

(*a*) 8 (*b*) 18 (*c*) 52 (*d*) 72 *(U.D.C. 1995)*

**119.** The ratio between the rates of walking of A and B is 2 : 3. If the time taken by B to cover a certain distance is 36 minutes, the time in minutes, taken by A to cover that much distance is :

(*a*) 24 (*b*) 38 (*c*) 48 (*d*) 54 *(S.S.C. 1995)*

**120.** I have to be at a certain place at a certain time and find that I shall be 20 minutes too late if I walk at 3 km/hr and 10 minutes too soon if I walk at 4 km/hr. How far I have to walk?

(*a*) 6 km. (*b*) 10 km (*c*) $12\frac{1}{2}$ km (*d*) $16\frac{2}{3}$ km *(S.S.C. 1995)*

**121.** Sound travels at 330 metres a second. How many kilometres away is a thunder cloud when its sound follows the flash after 10 seconds?

(*a*) 3.3 (*b*) 33 (*c*) 0.33 (*d*) 3.33 *(U.D.C. 1994)*

**122.** Walking $\frac{6}{7}$th of his usual speed, a man is 12 minutes too late. The usual time taken by him to cover that distance is :

(*a*) 1 hour (*b*) 1 hour 12 min. (*c*) 1 hour 15 min. (*d*) 1 hour 20 min. *(U.D.C. 1995)*

**123.** A is twice as fast as B and B is thrice as fast as C. The Journey covered by C in 42 minutes will be covered by A in :

(*a*) 7 min. (*b*) 14 min. (*c*) 28 min. (*d*) 35 min. *(S.S.C. 1995)*

**124.** A car takes 5 hours to cover a distance of 300 km. How much should the speed in km/hr be maintained to cover the same distance in $\frac{4}{5}$th of the previous time?

(*a*) 48 (*b*) 60 (*c*) 75 (*d*) 120 *(Bank P.O. 1996)*

**125.** Two men start together to walk a certain distance at 3.75 km and 3 km per hour respectively. The former arrives 30 minutes before the latter. The distance they walked is :

(*a*) 7.5 km (*b*) 10 km (*c*) 12.5 km (*d*) 15 km *(Railway, 1995)*

**126.** The wheel of an engine, $7\frac{1}{2}$ metres in circumference makes 7 revolutions in 9 seconds. The speed of the train in km per hour is :

(*a*) 150 (*b*) 132 (*c*) 130 (*d*) 135 *(P.C.S. 1995)*

**127.** A train X starts from a place at the speed of 50 km/hr. After one hour, another train Y starts from the same place at the speed of 70 km/hr. After how much time will Y cross X?

(*a*) 3 hrs. (*b*) $2\frac{3}{4}$ hrs. (*c*) $3\frac{1}{2}$ hrs. (*d*) $2\frac{1}{4}$ hrs. *(P.C.S. 1995)*

**128.** A person travels equal distances with speeds of 3 km/hr, 4 km/hr and 5 km/hr and takes a total time of 47 minutes. The total distance (in km) is :

(*a*) 2 (*b*) 3 (*c*) 4 (*d*) 5 *(Astt. Grade, 1996)*

**129.** A train covers a distance between station A and station B in 45 minutes. If the speed of the train is reduced by 5 km/hr, then the same distance is covered in 48 minutes. What is the distance between the stations A and B?

(*a*) 60 km (*b*) 64 km (*c*) 80 km (*d*) 55 km *(U.D.C. 1995)*

**130.** Which of the following trains is the fastest?

(*a*) 25 m/sec. (*b*) 1500 m/min. (*c*) 90 km/hr (*d*) None

*(Hotel Management, 1995)*

**131.** A train passes over a bridge of length 200 m in 40 seconds. If the speed of the train is 81 km per hour, its length is :

(*a*) 900 m (*b*) 1800 m (*c*) 700 m (*d*) 600 m *(S.S.C. 1995)*

**132.** A 180 m long train crosses a man standing on the platform in 6 seconds. What is the speed of the train?

(*a*) 90 km/hr (*b*) 108 km/hr (*c*) 120 km/hr (*d*) 88 km/hr

*(B.S.R.B. 1995)*

**133.** Two trains are running at 40 km/hr and 20 km/hr respectively in the same direction. Fast train completely passes a man sitting in the slow train in 5 seconds. What is the length of the fast train?

(*a*) $23\frac{2}{9}$ m (*b*) 27 m (*c*) $27\frac{7}{9}$ m (*d*) 23 m *(Railway, 1995)*

**134.** A train passes a 50 m long platform in 14 seconds and a man standing on the platform in 10 seconds. The speed of the train in km/hr is :

(*a*) 24 (*b*) 36 (*c*) 40 (*d*) 45 *(I. Tax, 1995)*

**135.** A train travelling with a constant speed crosses a 96 m long platform in 12 seconds and another 141 m long platform in 15 seconds. The length of the train is :

(*a*) 80 m (*b*) 64 m (*c*) 84 m (*d*) 90 m *(I. Tax, 1995)*

**136.** A sum of Rs. 2400 amounts to Rs. 3264 in four years at a certain rate of simple interest. If the rate of interest is increased by 1%, the same sum in the same time would amount to:

(*a*) Rs. 3288 (*b*) Rs. 3312 (*c*) Rs. 3340 (*d*) Rs 3360 *(I. Tax, 1995)*

**137.** A sum was put at a certain rate of interest for 3 years. Had it been put at 2% higher rate, it would have fetched Rs. 72 more. The sum is :

(*a*) Rs. 1250 (*b*) Rs. 1400 (*c*) Rs. 1200 (*d*) Rs. 1500 *(Railway, 1995)*

**138.** Rs. 2000 amount to Rs. 2600 in 5 years at simple interest. If the interest rate is increased by 3%, it would amount to how much?

(*a*) Rs. 2900 (*b*) Rs. 3200 (*c*) Rs. 3600 (*d*) None of these *(Bank P.O. 1996)*

**139.** On a certain sum, the simple interest at the end of $6\frac{1}{4}$ years becomes $\frac{3}{8}$ of the sum. What is the rate percent?

(*a*) 7% (*b*) 6% (*c*) 5% (*d*) $5\frac{1}{2}$% *(U.D.C. 1995)*

**140.** An amount of Rs. 100000 is invested in two types of shares. The first yields an interest of 9% per annum and the second yields 11% per annum. It the total interest at the end of one year is $9\frac{3}{4}$%, then the amount invested in each share was :

(*a*) Rs. 72500, Rs. 27500 (*b*) Rs. 62500, Rs. 37500
(*c*) Rs. 52500, Rs. 47500 (*d*) Rs. 82500, Rs. 17500 *(Astt. Grade, 1996)*

**141.** If the compound interest on a certain sum for 2 years at 10% per annum is Rs. 2100, the simple interest on it at the same rate for 2 years will be :

(*a*) Rs. 1700 (*b*) Rs. 1800 (*c*) Rs. 1900 (*d*) Rs. 2000 *(I. Tax, 1995)*

**142.** The length of a rectangle is increased by 10% and its breadth decreased by 10%. The area of new rectangle is :

(*a*) neither increased nor decreased (*b*) decreased by 1%
(*c*) increased by 1% (*d*) None of these *(P.C.S. 1996)*

**143.** A room is 6 m long, 5 m broad and 4 m high. It has one door 2.5 high and 1.2 m broad and has one window 1 m broad and 1 m high. Find the area in sq. metres of the paper required to cover the four walls of the room.

(*a*) 84 (*b*) 100 (*c*) 120 (*d*) None *(Railway, 1995)*

**144.** The length of a rectangular plot of land is $2\frac{1}{2}$ times its breadth. If the area of the plot is 1000 sq. metres, what is the length of the plot in metres?

(*a*) 20 (*b*) 25 (*c*) 30 (*d*) 50 (*e*) None *(B.S.R.B. 1996)*

**145.** Tiling work of a rectangular hall 60 m long and 40 m broad is to be completed with a square tile of 0.4 m side. If each tile costs Rs. 5, find the total cost of the tiles?

(*a*) Rs. 60000 (*b*) Rs. 65000 (*c*) 75000 (*d*) 12000 *(Bank P.O. 1996)*

**146.** A room is 6 m long, 5 m broad and 4 m high. If all its walls are to be covered with paper 50 cm wide, the length of the paper is

(*a*) 96 m (*b*) 176 m (*c*) 421 m (*d*) 208 m *(Railway, 1995)*

**147.** The perimeters of a circular and another square field are equal. Fine the area in sq. cm of the circular field if the area of the square field is 484 sq. cm.

(*a*) 888 (*b*) 770 (*c*) 616 (*d*) None of these *(R.R.B. 1996)*

**148.** Two cylindrical buckets have their diameters in the ratio 3 : 1 and their heights are as 1 : 3. Their volumes are in the ratio of

(*a*) 1 : 2 (*b*) 2 : 3 (*c*) 3 : 1 (*d*) 3 : 4 *(I. Tax, 1995)*

**149.** If the radius of the base and height of a cylinder and cone are each equal to *r*, the radius of a hemisphere is also equal to r, then the volumes of cone, cylinder and hemisphere are in the ratio.

(*a*) 1: 2 : 3 (*b*) 1 : 3 : 2 (*c*) 2 : 1 : 3 (*d*) 3 : 2 : 1 *(C.B.I. 1995)*

**150.** The size of a wooden block is 5 × 10 × 20 cms. How many whole such blocks will be required to construct a solid wooden cube of minimum size?

(*a*) 6 (*b*) 8 (*c*) 12 (*d*) 16 *(P.C.S. 1995)*

# SOLUTIONS

**1.** (*d*): $4^{61} + 4^{62} + 4^{63} + 4^{64}$
$= 4^{61}(1 + 4 + 4^2 + 4^3) = 4^{61} \times 85$, which is divisible by 17.

**2.** (*a*): $(264)^{102} + (264)^{103} = (264)^{102} \times [1 + 264]$
$= (264)^{102} \times 265$
Unit digit in $(264)^4$ is 6
Unit digit in $[(264)^4]^{25}$ = Unit digit in $6^{25} = 6$

Unit digit in $(264)^{100} \times (264)^2$ is 6

Unit digit in $(264)^{102} \times 265$ = unit digit in $6 \times 5 = 0$

**3.** (*c*): $p = \frac{3}{5} = 0.6, \quad q = \frac{7}{9} = 0.777$, *i.e.* $r = 0.714$

Clearly, $0.6 < 0.714 < 0.777$ *i.e.* $p < r < q$.

**4.** (*d*): $(6)^{10} \times (7)^{17} \times (55)^{27} = 2^{10} \times 3^{10} \times 7^{17} \times 5^{27} \times 11^{27}$

$\therefore$ Total number of prime factors $= (10 + 10 + 17 + 27 + 27) = 91$.

**5.** (*a*): $(6 + 6 + 7) - (2 + 8 + 9) = 0$, so 978626 is divisible by 11.

**6.** (*d*): $\left(\frac{1}{6}\right)^{12} \times (8)^{25} \times \left(\frac{3}{4}\right)^{15} = \frac{1}{2^{12} \times 3^{12}} \times (2^3)^{25} \times \frac{3^{15}}{2^{15} \times 2^{15}}$

$$= \frac{2^{75} \times 3^{15}}{2^{42} \times 3^{12}} = 2^{33} \times 3^3$$

$\therefore$ Total number of prime factors $= (33 + 3) = 36$.

**7.** (*d*): $51 + 52 + 53 + \ldots\ldots + 100$

$= [(1 + 2 + \ldots\ldots + 50) + (51 + 52 + \ldots\ldots + 100)] - (1 + 2 + 3 + \ldots\ldots + 50)$

$$= \left(\frac{100 \times 101}{2} - \frac{50 \times 51}{2}\right) = (5050 - 1275) = 3775.$$

**8.** (*b*): Let the given number be $x$. This when divided by 32, suppose gives the quotient K. Then.

$$x = 32\,K + 29 = 8 \times (4\,K + 3) + 5$$

Thus, when $x$ is divided by 8, it gives $(4\,K + 3)$ as quotient and 5 as remainder.

**9.** (*b*): $\frac{a}{c} = \frac{a}{b} \times \frac{b}{c} = \frac{4}{5} \times \frac{15}{16} = \frac{3}{4}$.

$$\therefore \quad \frac{c^2 - a^2}{c^2 + a^2} = \frac{1 - \frac{a^2}{c^2}}{1 + \frac{a^2}{c^2}} = \frac{1 - \frac{9}{16}}{1 + \frac{9}{16}} = \frac{7}{16} \times \frac{16}{25} = \frac{7}{25}.$$

**10.** (*c*): From among the given numbers, 923872 is the only number, whose last 3 digits form 872, which is divisible by 8.

**11.** (*d*): 18962 when divided by 13 leaves the remainder 8.

$\therefore$ Number to be added $= (13 - 8) = 5$.

**12.** (*a*):

| | | | | | | |
|---|---|---|---|---|---|---|
| 2 | 12 | 18 | 24 | 32 | 40 | |
| 2 | 6 | 9 | 12 | 16 | 20 | |
| 3 | 3 | 9 | 6 | 8 | 10 | |
| 2 | 1 | 3 | 2 | 8 | 10 | |
| | 1 | 3 | 1 | 4 | 5 | |

360
4

$\therefore$ L.C.M. of 12, 18, 24, 32, 40 is $(2 \times 2 \times 3 \times 2 \times 3 \times 4 \times 5) = 1440$.

$\therefore$ Required number $= (1440 - 1) = 1439$.

13. (*b*): 437950 is the only number among given numbers s.t. $437950 = 5 \times 87590$

$= 5 \times 5 \times 17518$, which has 25 as a factor.

14. (*b*): Note that $(2 - 1) = (3 - 2) = (4 - 3) = \ldots\ldots = (8 - 7) = 1$

L.C.M. of 2, 3, 4, 5, 6, 7, 8 is 840

$\therefore$ Required number = (840 K − 1)

Since the number lies between 1000 and 2000, so K = 2.

15. (*b*): We note that $5 \times 7 \times 11 = 385$.

The prime number next to 11 is 13.

Also, $7 \times 11 \times 13 = 1001$.

Hence, the required number is 13.

16. (*c*): 567) 5 8 7 0 1 ( 1 0 3

```
      5 6 7
      -------
        2 0 0 1
        1 7 0 1
      -------
        3 0 0
```

3 0 0, which is more than half of 567.

$\therefore$ Required number = 58701 + (567 − 300) = 58968.

17. (*c*): Let the number be $x$. Then

$17x - x = 640 \Rightarrow x = 40.$

18. (*d*): $x^3 - x^2 = 100 \Rightarrow x^3 - x^2 - 100 = 0.$

Clealy, $x = 5$, satisfies it.

19. (*a*): Let the number be $p$. Then,

$$\frac{2}{3} \text{ of } \frac{3}{4} \text{ of } p + \frac{3}{4} \text{ of } \frac{4}{5} \text{ of } p = px$$

$$\Rightarrow \frac{1}{2}p + \frac{3}{5}p = xp \Rightarrow x = \left(\frac{1}{2} + \frac{3}{5}\right) = \frac{11}{10}.$$

20. (*a*): $\left(\frac{7}{13} = 0.538, \frac{4}{5} = 0.8\right), \left(\frac{1}{2} = 0.5, \frac{2}{3} = 0.666, \frac{3}{4} = 0.75, \frac{5}{7} = 0.714\right)$

Clearly, 0.5 does not lie between 0.538 and 0.8.

21. (*c*): Given Exp. $= \dfrac{(5^2)^{2.6}}{(625)^{0.5} \times (5)^{1.2}} = \dfrac{5^{(5.2 - 1.2)}}{(5^4)^{0.5}} = \dfrac{5^4}{5^2} = 5^2 = 25.$

22. (*c*): $(16)^{1.75} = (2^4)^{\frac{175}{100}} = (2^4)^{7/4} = 2^{\left(4 \times \frac{7}{4}\right)} = 2^7 = 128.$

23. (*b*): Given numbers are $2^{1/2}, 3^{1/3}, 4^{1/4}$, L.C.M. of 2, 3, 4 is 12.

Now, $2^{1/2} = 2^{(6/12)} = \sqrt[12]{2^6} = \sqrt[12]{64}$

$3^{1/3} = 3^{4/12} = \sqrt[12]{3^4} = \sqrt[12]{81}$

$4^{1/4} = 4^{3/12} = \sqrt[12]{4^3} = \sqrt[12]{64}$

Clearly, $\sqrt[12]{81}$ is the largest, *i.e.* $\sqrt[3]{3}$ is the largest.

**24.** (*a*): $\dfrac{5+2\sqrt{3}}{7+4\sqrt{3}} = \dfrac{(5+2\sqrt{3})}{(7+4\sqrt{3})} \times \dfrac{(7-4\sqrt{3})}{(7-4\sqrt{3})} = \dfrac{35-24+14\sqrt{3}-20\sqrt{3}}{(49-48)}$

$= 11 - 6\sqrt{3}$

Thus, $a + b\sqrt{3} = 11 - 6\sqrt{3}$

Hence, $a = 11$, $b = -6$

**25.** (*b*): $\sqrt[4]{(625)^3} = (625)^{3/4} = (5^4)^{3/4} = 5^{\left(4\times\frac{3}{4}\right)} = 5^3 = 125.$

**26.** (*b*): $(125)^x = 3125 \Rightarrow (5^3)^x = 5^5 \Rightarrow 3x = 5 \Rightarrow x = \frac{5}{3}$.

**27.** (*b*): Given Exp. $= \dfrac{(a^3+b^3)}{(a^2+b^2-ab)} = (a+b) = (36+14) = 50.$

**28.** (*c*): $0.\ddot{6}\ddot{3} = \dfrac{63}{99}$ and $0.\ddot{3}\ddot{7} = \dfrac{37}{99}$.

$\therefore\ 0.\ddot{6}\ddot{3} + 0.\ddot{3}\ddot{7} = \dfrac{63}{99} + \dfrac{37}{99} = \dfrac{100}{99} = 1\dfrac{1}{99} = 1.\ddot{0}\ddot{1}.$

**29.** (*a*): $\dfrac{1}{0.0003718} = \dfrac{10000}{3.718} = 10000 \times .2689 = 2689.$

**30.** (*d*): Given Exp. $= 0.\dot{3} \times 0.\dot{4} = \dfrac{3}{9} \times \dfrac{4}{9} = \dfrac{4}{27} = 0.\dot{1}48148148....$

**31.** (*b*): $\dfrac{5}{8} = 0.625$, $\dfrac{2}{3} = 0.666$, $\dfrac{7}{9} = 0.777$, $\dfrac{3}{5} = 0.6$, $\dfrac{4}{7} = 0.571$

Clearly, $\dfrac{7}{9}$ is the largest.

**32.** (*b*): Given Exp. $= \dfrac{(a^3-b^3)}{(a^2+ab+b^2)} = a - b = (3.06 - 1.98) = 1.08.$

**33.** (*c*) : Let $.04 \times x = .000016$. Then,

$x = \dfrac{.000016}{.04} = \dfrac{.0016}{4} = .0004.$

**34.** (*c*): Given Exp. $= \dfrac{2^{1/2} \times 3^{1/3} \times (2^2)^{1/4}}{2^{-1/5} \times 5^{-1/5} \times 5^{3/5}} \div \dfrac{3^{4/3} \times 5^{-7/5}}{(2^2)^{-3/5} \times 2 \times 3}$

$= \dfrac{2^{1/2} \times 3^{1/3} \times 2^{1/2}}{2^{-1/5} \times 5^{-1/5} \times 5^{3/5}} \times \dfrac{2^{-6/5} \times 2 \times 3}{3^{4/3} \times 5^{-7/5}}$.

$$= \frac{2^{(\frac{1}{2}+\frac{1}{2}-\frac{6}{5}+\frac{1}{5}+1)} \times 3^{(\frac{1}{3}+1-\frac{4}{3})}}{5^{(-\frac{1}{5}+\frac{3}{5}-\frac{7}{5})}} = \frac{2 \times 3^0}{5^{-1}} = 2 \times 1 \times 5 = 10.$$

**35.** (*d*): Given Exp. $= \frac{4^7}{16^4} \times \sqrt{16} = \frac{4^7}{(4^2)^4} \times 4 = \frac{4^8}{4^8} = 4^{(8-8)} = 4^0 = 1.$

**36.** (*a*): H.C.F. $= \frac{\text{H.C.F. of 7, 14, 7}}{\text{L.C.M. of 90, 15, 10}} = \frac{7}{90}.$

**37.** (*d*): Required time = L.C.M. of $\frac{91}{6}, \frac{65}{4}$ and $\frac{56}{3}$

$$= \frac{\text{L.C.M. of 91, 65, 56}}{\text{H.C.F. of 6, 4, 3}} = \frac{3640}{1} = 3640 \text{ sec.}$$

**38.** (*d*): Given Exp. $= \left[1 + \frac{1}{1 + \frac{1}{(4/3)}}\right] \div \frac{11}{7} = \left[1 + \frac{1}{\left(1 + \frac{3}{4}\right)}\right] \div \frac{11}{7}$

$$= \left[1 + \frac{1}{(7/4)}\right] \div \frac{11}{7} = \left(1 + \frac{4}{7}\right) \div \frac{11}{7}$$

$$= \left(\frac{11}{7} \div \frac{11}{7}\right) = 1.$$

**39.** (*c*) Given Exp. $= \frac{a^3 + b^3 + c^3 - 3abc}{a^2 + b^2 + c^2 - ab - bc - ca} = (a + b + c)$

$$= (0.43 + 1.47 + 1.1) = 3.$$

**40.** (*b*) $\frac{x}{2y} = \frac{3}{2} \Rightarrow \frac{x}{y} = 3.$

$$\therefore \frac{2x + y}{x - 2y} = \frac{2\left(\frac{x}{y}\right) + 1}{\left(\frac{x}{y}\right) - 2} = \frac{2 \times 3 + 1}{3 - 2} = 7.$$

**41.** (*d*): $\frac{1}{4}N = \frac{1}{6}Y \Rightarrow N = \frac{4}{6}Y = \frac{2}{3}Y.$ $\quad \frac{5}{3}Y = \frac{600 \times 3}{5}$

Now, $N + Y = 600 \Rightarrow \frac{2}{3}Y + Y = 600 \Rightarrow Y = 360$

$\therefore N = (600 - 360) = 240$

Hence, $(Y - N) = (360 - 240) = 120.$

**42.** (*a*): Given Exp. $= \frac{a^2 + b^2 + c^2}{\left(\frac{a}{10}\right)^2 + \left(\frac{b}{10}\right)^2 + \left(\frac{c}{10}\right)^2} = \frac{(a^2 + b^2 + c^2) \times 100}{(a^2 + b^2 + c^2)} = 100.$

**43.** (*b*): 1000 Cu cm weighs 1000 gms.

*i.e.* 1000 gms is the weight of (1000 × 1000 ) Cu mm

0.1 gm is the weight of $\left(\frac{1000 \times 1000}{1000} \times 0.1\right)$ Cu mm = 100 Cu mm.

**44.** (*d*): Let the capacity of the tank be *x* litres. Then

$\frac{3}{4}x - 21 = \frac{2}{5}x \Rightarrow \frac{3x}{4} - \frac{2x}{5} = 21 \Rightarrow x = 60.$

**45.** (*c*): T – S = G – T ⇒ G + S = 2T

$\therefore$ 2% = 140 ⇒ T = 70.

**46.** (*d*): Let the number of each type of notes be *x*. Then,

$x + 5x + 10x = 480 \Rightarrow x = 30.$

Total number of notes = 30 + 30 + 30 = 90.

**47.** (*d*): $a + b + c = 1.2 + 2.1 - 3.3 = 0$

$\Rightarrow a^3 + b^3 + c^3 = 3abc \Rightarrow a^3 + b^3 + c^3 - 3abc = 0.$

**48.** (*d*): Take *y* = 4. Then, $3\frac{1}{x} \times 4\frac{2}{5} = 13\frac{3}{4}$

$\therefore \frac{3x+1}{x} = \left(\frac{55}{4} \times \frac{5}{22}\right) \Rightarrow \frac{3x+1}{x} = \frac{25}{8}$

$\Rightarrow 24x + 8 = 25x \Rightarrow x = 8$

$\therefore x = 8, y = 4.$

**49.** (*b*): Given Exp. $= \frac{(\sqrt{31} - \sqrt{29})}{(\sqrt{31} + \sqrt{29})} \times \frac{(\sqrt{31} - \sqrt{29})}{(\sqrt{31} - \sqrt{29})}$

$= \frac{(\sqrt{31} - \sqrt{29})^2}{(31 - 29)} = \frac{31 + 29 - 2\sqrt{31} \times \sqrt{29}}{2} = 30 - \sqrt{899}.$

**50.** (*b*): Given Exp. $= 1 - \frac{1}{(1 + \sqrt{2})} \times \frac{(\sqrt{2} - 1)}{(\sqrt{2} - 1)} - \frac{1}{(\sqrt{2} - 1)} \times \frac{(\sqrt{2} + 1)}{(\sqrt{2} + 1)}$

$= 1 - (\sqrt{2} - 1) - (\sqrt{2} + 1) = 1 - 2\sqrt{2}.$

**51.** (*c*): $\sqrt{456976} = 676$

$\therefore \sqrt{45.6976} = \frac{\sqrt{456976}}{\sqrt{10000}} = \frac{676}{100} = 6.76.$

**52.** (*b*): Given Exp. $= \frac{(3\sqrt{2})(\sqrt{6} + \sqrt{3})}{(\sqrt{6} + \sqrt{3})(\sqrt{6} - \sqrt{3})} - \frac{(4\sqrt{3})(\sqrt{6} + \sqrt{2})}{(\sqrt{6} - \sqrt{2})(\sqrt{6} + \sqrt{2})}$

$- \frac{6}{(\sqrt{12} + \sqrt{8})} \times \frac{(\sqrt{12} - \sqrt{8})}{(\sqrt{12} - \sqrt{8})}$

$= \sqrt{2}(\sqrt{6} + \sqrt{3}) - \sqrt{3}(\sqrt{6} + \sqrt{2}) - \frac{3}{2}(\sqrt{12} - \sqrt{8})$

$= \sqrt{12} + \sqrt{6} - 3\sqrt{2} - \sqrt{6} - 3\sqrt{3} + 3\sqrt{2}$

$= 2\sqrt{3} - 3\sqrt{3} = -\sqrt{3}.$

**53.** (*b*): $a + b = 2\sqrt{6}$ and $ab = (6 - 5) = 1$

$\therefore\ 2a^2 - 5ab + 2b^2 = 2(a + b)^2 - 9ab = 2(2\sqrt{6})^2 - 9 \times 1 = (48 - 9) = 39.$

**54.** (*c*): Required reduction $= \left[\frac{r}{(100 + r)} \times 100\right]\% = \left(\frac{25}{125} \times 100\right) = 20\%.$

**55.** (*c*): Let 95% of $x = 4598$

Then, $x = \left(4598 \times \frac{100}{95}\right) = 4840.$

**56.** (*d*): Matriculates $= \frac{40}{100} x = \frac{2x}{5}.$

Remaining $= \left(x - \frac{2x}{5}\right) = \frac{3x}{5}$

Graduates $= \frac{50}{100} \times \frac{3x}{5} = \frac{3x}{10}$

$\therefore\ \frac{2x}{5} + \frac{3x}{10} + 180 = x \Rightarrow x - \frac{7x}{10} = 180 \Rightarrow x = 600.$

$\therefore$ Graduates $= \frac{3}{10} \times 600 = 180.$

**57.** (*c*): $n(A) = 40$, $n(B) = 50$, $n(A \cap B) = 10$

$n(A \cup B) = n(A) + n(B) - n(A \cap B) = (40 + 50 - 10) = 80$

$n$(not A and not B) $= n(A^C \cap B^C) = n(A \cup B)^C$

$= 100 - n(A \cup B) = (100 - 80) = 20.$

**58.** (*c*): $x \times \left(1 + \frac{5}{100}\right)^3 = 138915 \Rightarrow x \times \frac{21}{20} \times \frac{21}{20} \times \frac{21}{20} = 138915$

$\therefore\ x = \left(\frac{138915 \times 20 \times 20 \times 20}{21 \times 21 \times 21}\right) = 120000.$

**59.** (*b*): $\frac{30}{100} x - \frac{25}{100} x = 15 \Rightarrow x = 300.$

**60.** (*c*): Males $= \left(\frac{5}{9} \times 4500\right) = 2500$, Females $= 2000$.

Married males $= \frac{40}{100} \times 2500 = 1000$

Married females $= 1000$

$\therefore$ Percentage of married females $= \left(\frac{1000}{2000} \times 100\right) = 50\%.$

**61.** (*d*): Let the total sale be Rs. $x$.

Commission $= \frac{5}{100} \times 10000 + \frac{4}{100} \times (x - 10000) = 100 + \frac{4x}{100}$

$\therefore\ x - \left(100 + \frac{x}{25}\right) = 31100$ or $\frac{24x}{25} = 31200$

$\therefore \quad x = \dfrac{31200 \times 25}{25} = 32500.$

**62.** (*c*): Required percentage $= \dfrac{10}{(100 + 10)} \times 1000 = 9\dfrac{1}{11}\%.$

**63.** (*c*): Milk $= \dfrac{90}{100} \times 40 = 36$ litres, water $= 4$ litre.

Let $x$ litres of water be added. Then

$\dfrac{(x + 4)}{41 + x} \times 100 = 20 \Rightarrow \dfrac{x + 4}{40 + x} = \dfrac{1}{5}$

$\therefore \; 5x + 20 = 40 + x \Rightarrow x = 5.$

**64.** (*b*): $z = \dfrac{x^2}{y}.$

New value of $z = \dfrac{\left(\dfrac{110}{100}x\right)^2}{\left(\dfrac{110}{100}y\right)} = \dfrac{11}{100}\,\dfrac{x^2}{y} = \dfrac{11}{10}z$

$\therefore \quad$ Increase % $= \dfrac{\left(\dfrac{11}{10}z - z\right)}{z} \times 10 = 10\%.$

**65.** (*d*): $b + \dfrac{bx}{100} = a.$

**66.** (*d*): Failed in G.K. only $= (35 - 10) = 25$

Failed in English only $= (25 - 10) = 15$

Failed in both = 10

Failed in one or 2 subjects $= (25 + 15 + 10) = 50$

Number of examinees passed $= (100 - 50) = 50\%.$

**67.** (*b*): Let original price be Rs. $x$.

New price $= \dfrac{125}{100} \times x = \dfrac{5x}{4}$

Reduction on $\dfrac{5x}{4} = \left(\dfrac{5x}{4} - x\right) = \dfrac{x}{4}$

Reduction on 100 $= \left(\dfrac{x}{4} \times \dfrac{4}{5x} \times 100\right) = 20\%$

**68.** (*c*): Alcohol in 9 gms $= \left(\dfrac{50}{100} \times 9\right) = 4.5$ gms.

Let $x$ gm of water be added. Then,

$\dfrac{4.5}{0 + x} \times 100 = 30 \Rightarrow 270 + 30x = 450 \Rightarrow x = 6$ gms.

**69.** (*c*): Let C.P. be Rs. 100. Then, M.P. = Rs. 120

Also, S.P. = Rs. 108

Discount on 120 = Rs. (120 – 108) = Rs. 12

$$\text{Discount on } 100 = \left(\frac{12}{120} \times 100\right) = 10\%.$$

**70.** (*c*): $$\text{C.P.} = \left(\frac{100}{120} \times 30\right) = 25, \quad \text{S.P.} = 90\% \text{ of } 30 = \text{Rs. } 27$$

$$\text{Gain } \% = \left(\frac{2}{25} \times 100\right) = 8\%.$$

**71.** (*b*): Let M.P. = $x$ Then,

$$\frac{90}{100}x + \frac{8}{100} \times \frac{90x}{100} = 680.40 \Rightarrow x = 700.$$

**72.** (*d*): Let S.P. be Rs. $x$

Then, C.P. = $(x - 240)$

$$\therefore \frac{240}{(x-240)} \times 100 = 20 \Rightarrow x - 240 = 1200 \Rightarrow x = 1440.$$

**73.** (*c*): S.P. = Rs. 22600, gain = 13%

$$\text{C.P.} = \left(\frac{100}{113} \times 22600\right) = 20000.$$

∴ Gain= Rs. (22600 – 20000) = Rs. 2600.

**74.** (*c*): Suppose he purchased 24 locks (l.c.m. of 8 and 12).

$$\text{C.P.} = \left(\frac{34}{8} \times 24\right) = 102, \quad \text{S.P.} = \left(\frac{57}{12} \times 24\right) = 114.$$

To gain Rs. 12, locks purchased = 24

$$\text{To gain Rs. 900, locks purchased} = \left(\frac{24}{12} \times 900\right) = 1800.$$

**75.** (*c*): Let C.P. be Rs. $x$ Then,

$$\frac{115}{100}x = \frac{69}{100} \times 125 \Rightarrow x = 75.$$

**76.** (*b*): Let C.P. be Rs. $x$. Then,

$(x - 895) = (955 - x) \Rightarrow x = 925$

∴ Loss = Gain = (925 – 895) = 30.

**77.** (*c*): Let cost of almond per kg be Rs. $x$ and cost of walnuts per kg be Rs. $y$. Then, $2x = 8y \Rightarrow y = \frac{x}{4}$

$$\therefore 5x + 16y = 1080 \Rightarrow 5x + 16 \times \frac{x}{4} = 1080 \Rightarrow x = 120.$$

**78.** (*c*): $$\frac{125}{100} \text{ of } \frac{120}{100} \text{ of } \frac{118}{100}x = 30.09$$

$$\therefore\ x = \left(\frac{30.09 \times 100 \times 100 \times 100}{125 \times 120 \times 118}\right) = \left(\frac{3009 \times 4}{6 \times 118}\right) = 17.$$

**79.** (*a*): Let the original price be Rs. 100

Price after dealer's discount = Rs. 85

$$\text{S.P.} = \frac{120}{100} \times 85 = 102$$

$\therefore$ Profit = 2% on original price.

**80.** (*d*): Let C.P. be Rs. *x*. Then,

$$\frac{115}{100}x - \frac{110}{100}x = 10 \Rightarrow 5x = 100 \times 10 \Rightarrow x = 200.$$

**81.** (*c*): Let the marked price be Rs. *x*. Then,

$$\frac{85}{100}x = 102 \Rightarrow x = \left(\frac{102 \times 100}{85}\right) = 120.$$

**82.** (*b*): Let C.P. be Rs. *x*.

Two losses are $(x - 400)$ and $(x - 380)$

Increase in loss $(x - 380) - (x - 400) = 20$

$$\frac{20}{x - 400} \times 100 = 20 \Rightarrow x - 400 = 100 \Rightarrow x = 500.$$

**83.** (*d*):

$$\text{C.P.} = \left(\frac{100}{75} \times 455000\right) = \frac{1820000}{3}$$

$$\text{Loss} = \left(\frac{1820000}{3} - 455000\right) = \frac{455000}{3}.$$

**84.** (*b*): Let C.P. be Rs. 100. Then, M.P. = Rs. 120.

$$\text{S.P.} = \frac{95}{100} \text{ of Rs. } 120 = \text{Rs. } 114$$

$\therefore$ Profit = 14%.

**85.** (*b*): Let the cost of the machine be *x*.

$$\frac{110}{100}x = \frac{90}{100} \times 27500 \Rightarrow x = 22500.$$

**86.** (*b*): Let the cost of the toy be Rs. *x*

$$\frac{3}{4}(x - 60) = (81 - x) \Rightarrow 324 - 4x = 3x - 180 \Rightarrow x = 72.$$

**87.** (*c*): Let the C.P. be Rs. 100 and marked price be $(100 + x)$.

$$\text{S.P.} = \frac{\left(100 - \frac{25}{4}\right)}{100} \text{ of } \frac{80}{100} \times (100 + x)$$

$$= \frac{15}{16} \text{ of } \frac{4}{5}(100 + x) = \frac{3}{4}(100 + x)$$

$$\therefore \frac{3}{4}(100 + x) = 120 \Rightarrow x = 60.$$

**88.** (*b*): Let C.P. of each kg be Re. 1. Then, C.P. = Rs. 120

Suppose he sells $x$ kg at 25% gain. Then,

$$\frac{125}{100}x + \frac{110}{100}(120 - x) = \frac{115}{100} \times 120$$

$$\Rightarrow 125x + 13200 - 110x = 13800 \Rightarrow x = 40.$$

**89.** (*a*): Retailer's value $= \left(\frac{100}{130} \times 325\right) = 250$

Wholesaler's value $= \left(\frac{100}{125} \times 250\right) = 200.$

**90.** (*c*): S.P. $= \left(\frac{110}{100} \times 95\right) = \left(\frac{209}{2}\right)$, commission = 5%

$\therefore$ M.P. $= \left(\frac{110}{95} \times \frac{209}{2}\right) = 110.$

**91.** (*d*): Let these coins be $x$, $2x$ and $3x$. Then,

$$\frac{25}{100}x + \frac{10}{100} \times 2x + \frac{5}{100} \times 3x \Rightarrow 60x = 30 \times 100 \Rightarrow x = 50$$

Number of 5-*p* coins = $3x = 150$.

**92.** (*d*): $\frac{x+y}{x-y} = \frac{4}{1} \Rightarrow x + y = 4x - 4y \Rightarrow 3x = 5y \Rightarrow \frac{x}{y} = \frac{5}{3}.$

$$\therefore \frac{x^2+y^2}{x^2-y^2} = \frac{\frac{x^2}{y^2}+1}{\frac{x^2}{y^2}-1} = \frac{\left(\frac{x}{y}\right)^2+1}{\left(\frac{x}{y}\right)^2-1} = \frac{34}{16} = \frac{17}{8}.$$

**93.** (*d*): $\frac{9^{3.04}}{9^{2.04}} = 9^{3.04-2.04} = 9^1 = \frac{9}{1}.$

**94.** (*b*): Let $\frac{a}{b} = \frac{b}{c} = k$. Then, $b = ck$ and $a = bk = ck^2$

$\therefore \frac{a^4}{b^4} = \frac{c^4k^8}{c^4k^4} = k^4$ and $\frac{a^2}{c^2} = \frac{c^2k^4}{c^2} = k^4.$

Hence, $\frac{a^4}{b^4} = \frac{a^2}{c^2}.$

**95.** (*c*): Given, $\frac{x}{y} = \frac{3}{4}$

$$\therefore \frac{2x+3y}{3y-2x} = \frac{2\left(\frac{x}{y}\right)+3}{3-2\left(\frac{x}{y}\right)} = \frac{2 \times \frac{3}{4} + 3}{3 - 2 \times \frac{3}{4}} = \frac{18}{6} = \frac{3}{1}.$$

**96.** (*a*): $A = \frac{1}{2}(B + C) \Rightarrow B + C = 2A \Rightarrow A + B + C = 3A$

$$B = \frac{1}{4}(A + C) \Rightarrow 4B = A + C \Rightarrow A + B + C = 5B$$

$\therefore$ 3A = 56250 and 5B = 56250 $\Rightarrow$ A = 18750 and B = 11250

$\therefore$ A – B = (18750 – 11250) = 7500.

**97.** (*b*): $\frac{17.5}{40.25} = \frac{x}{28.75} \Rightarrow x = \frac{17.5 \times 28.75}{40.25} = 12.5$ m.

**98.** (*b*): $b : c = 9 : 16$ and $a : c = 1 : 4$

$\therefore \frac{a}{b} = \frac{a}{c} \times \frac{c}{b} = \frac{1}{4} \times \frac{16}{9} = \frac{4}{9}$.

$\therefore a : b = 4 : 9$ and $b : c = 9 : 16 \Rightarrow a : b : c = 4 : 9 : 16$.

$\therefore$ Second number = $\left(\frac{9}{29} \times 174\right) = 54$.

**99.** (*c*): Let cost of each mango be $x$ paise and that of each orange be $y$ paise then $5x + 4y = 3x + 7y \Rightarrow 2x = 3y \Rightarrow x : y = 3 : 2$.

**100.** (*b*): $\frac{35}{100}x = \frac{25}{100}y \Rightarrow \frac{x}{y} = \frac{25}{35} = \frac{5}{7} \Rightarrow x : y = 5 : 7$.

**101.** (*b*): Let their last years salaries be $3x$ and $4x$.

This year's salaries are $\frac{5}{4} \times 3x$ and $\frac{3}{2} \times 4x$ *i.e.* $\frac{15x}{4}$ and $6x$

$\therefore \frac{15x}{4} + 6x = 4160 \Rightarrow 39x = 16640 \Rightarrow 3x = 1280$

$\therefore$ A's salary now = $\frac{5}{4} \times 3x = \frac{5}{4} \times 1280 = 1600$.

**102.** (*a*): Let their incomes be $5x$ and $3x$.

Let their expenses by $8y$, $5y$ and $2y$.

Then, $2y = 2000 \Rightarrow y = 1000$.

Also, $3x - 5y = 700 \Rightarrow 3x = 5700 \Rightarrow x = 1900$

$\therefore$ A's saving = $(5x - 8y) = (5 \times 1900 - 8 \times 1000) = 1500$.

**103.** (*b*): $x \times 50 = \frac{7}{3} \times 21 \Rightarrow x = \frac{49}{50}$.

**104.** (*d*): More women, less days (indirect)

More kg, more days (direct)

women 28 : 8

kg 180 : 3780 $\therefore$ 5 : $x$

$\therefore x = \left(\frac{8 \times 3780 \times 5}{28 \times 180}\right) = 30$.

**105.** (*b*): Less men, more days (indirect)

Less length, less days (direct)

More working hrs, less days (indirect)

Men 30 : 40

Length 300 : 200 $\therefore$ 12 : $x$

Hrs./day 8 : 6

$$\therefore \quad x = \left(\frac{40 \times 200 \times 6 \times 12}{30 \times 300 \times 8}\right) = 8.$$

**106.** (*a*): More work, more persons (direct)

More days, less persons (indirect)

$$\left.\begin{array}{ll} \text{Work} & \frac{3}{5} : 1 \\ \text{Days} & 20 : 10 \end{array}\right\} :: 12 : x$$

$$\therefore \quad x = \frac{1 \times 10 \times 12}{\frac{3}{5} \times 20} = 10.$$

**107.** (*c*): After 15 days there was ration for 2000 people for 39 days. Suppose $x$ more came

More people, less days

$$\frac{2000}{2000 + x} = \frac{20}{39} \Rightarrow 78000 = 40000 + 20x \Rightarrow x = 1900.$$

**108.** (*c*): After 12 days there was ration for 560 soldiers for 8 days.

Remaining persons - (560 – 112) = 448.

Less soldiers, more days

$$\frac{560}{448} = \frac{x}{8} \Rightarrow x = \frac{560 \times 8}{448} = 10.$$

**109.** (*c*): 10 men can complete the job in 14 days. Less days, more men

$$\frac{14}{5} = \frac{x}{10} \Rightarrow x = \frac{14 \times 10}{5} = 28.$$

**110.** (*b*): Remaining work $= \frac{3}{4}$

Remaining days = (100 – 45) = 55

Let $x$ more men may be employed.

More work, more men (direct)

More days, less men (indirect)

$$\left.\begin{array}{ll} \text{Work} & \frac{1}{4} : \frac{3}{4} \\ \text{Days} & 55 : 45 \end{array}\right\} :: 110 : (110 + x)$$

$$110 + x = \frac{3}{4} \times 45 \times 110 \times \frac{4}{1} \times \frac{1}{55} = 270.$$

$\therefore \; x = 160.$

**111.** (*a*) More work, more men (direct)

$$\frac{1}{4} : 1 = 16 : 16 + x \Rightarrow \frac{16}{16 + x} = \frac{1}{4}$$

$\therefore \; 16 + x = 64 \Rightarrow x = 48.$

**112.** (*b*): (A + B + C)'s 8 days' work = $\left(\frac{1}{20} \times 8\right) = \frac{2}{5}$.

Remaining work = $\left(1 - \frac{2}{5}\right) = \frac{3}{5}$.

$\frac{3}{5}$ work is done by (B + C) in 20 days

Whole work will be done by (B + C) in $\left(20 \times \frac{5}{3}\right) = \frac{100}{3}$ days.

$\therefore$ (B + C)'s 1 day's work $= \frac{3}{100}$

$\therefore$ A's 1 day's work $= \frac{1}{20} - \frac{3}{100} = \frac{2}{100} = \frac{1}{50}$.

Hence, A can finish the whole work in 50 days.

**113.** (*c*): (A + B + C)'s 4 day's work = $\left(\frac{1}{10} \times 4\right) = \frac{2}{5}$.

Remaining work = $\left(1 - \frac{2}{5}\right) = \frac{3}{5}$.

$\frac{3}{5}$ work is done by (B + C) in 10 days

Whole work can be done by ( B + C) in $\left(\frac{10 \times 5}{3}\right)$ days.

$\therefore$ (B + C)'s 1 day's work $= \frac{3}{50}$

A's 1 day's work = $\left(\frac{1}{10} - \frac{3}{50}\right) = \frac{1}{25}$

$\therefore$ A alone can finish the work in 25 days.

**114.** (*c*): Suppose they leave after *x* days

Work done in *x* days = $\frac{x}{30}$.

Remaining work $= \left(1 - \frac{x}{30}\right) = \left(\frac{30 - x}{30}\right)$

Remaining days $= (35 - x)$

15 men's 1 day's work $= \left(\frac{1}{20 \times 30} \times 15\right) = \frac{1}{40}$

$\therefore \quad \frac{1}{40}(35 - x) = \frac{30 - x}{30} \Rightarrow 1050 - 30x = 1200 - 40x \Rightarrow x = 15.$

**115.** (*d*): 1 man's 1 day's work = $\frac{1}{3}$, 1 woman's 1 day's work = $\frac{1}{4}$,

1 boy's 1 day's work $= \frac{1}{12}$.

Let $x$ boys must be there. Then,

$$\frac{1}{3}+\frac{1}{4}+\frac{x}{12}=1 \Rightarrow x=5.$$

**116.** (*b*): Let 1 man's 1 day work = $x$ and 1 boy's 1 day's work = $y$. Then, $2x + 3y = \frac{1}{8}$, $3x + 2y = \frac{1}{7}$.

Solving, we get : $x=\frac{1}{28}$ and $y=\frac{1}{56}$

$$\therefore \quad 5x+4y=\frac{5}{28}+\frac{4}{56}=\frac{14}{56}=\frac{1}{4}.$$

So, 5 men and 4 boys can finish the work in 4 days.

**117.** (*c*): (A + B)'s 1 day's work $=\frac{1}{8}$.

(B + C)'s 1 day's work $=\frac{1}{12}$.

(A + C)'s 1 day's work $=\frac{1}{x}$.

$\therefore$ 2(A + B + C)'s 1 day's work $=\left(\frac{1}{8}+\frac{1}{12}+\frac{1}{x}\right)$

$$\therefore \frac{1}{2}\left(\frac{5}{24}+\frac{1}{x}\right)=\frac{1}{6} \Rightarrow \frac{5}{24}+\frac{1}{x}=\frac{1}{3} \text{ or } \frac{1}{x}=\frac{1}{3}-\frac{5}{24}=\frac{1}{8} \Rightarrow x=8.$$

**118.** (*d*): Suppose, I finish a work now in $x$ days.

With double efficiency, time taken $=\frac{x}{2}$ days.

$$\therefore \quad \frac{x}{2}=12 \Rightarrow x=24.$$

With one third efficiency time taken = $3x$ days

= $3 \times 24$ days = 72 days.

**119.** (*d*): Ratio of rates of walking = inverse ratio of time taken.

$$\therefore \quad \frac{2}{3}=\frac{36}{x} \Rightarrow 2x=108 \Rightarrow x=54 \text{ min.}$$

**120.** (*a*): Let the distance be $x$ km.

$$\frac{x}{3}-\frac{x}{4}=\frac{30}{60} \Rightarrow \frac{x}{12}=\frac{1}{2} \Rightarrow x=6 \text{ km.}$$

**121.** (*a*): Distance = (Time × speed) = (330 × 10) m $=\frac{330 \times 10}{1000}$ km = 3.3 km.

**122.** (*b*): New speed $=\frac{6}{7}$ of the usual speed.

New time taken $=\frac{7}{6}$ of the usual time

$\left(\frac{7}{6}\text{ of usual time}\right)$ – (Usual time) = 12 min.

$\therefore\ \frac{1}{6}$ of usual time = 12 min. or usual time = 72 min.

**123.** (*a*): A : B = 2 : 1, B : C = 3 : 1 ⇒ A : B : C = 6 : 3 : 1

Ratio of time taken $= \frac{1}{6} : \frac{1}{3} : 1 = 1 : 2 : 6$

If C takes 6 min, A takes 1 min.

If C takes 42 min, A takes $= \left(\frac{1}{6} \times 42\right) = 7$ min.

**124.** (*c*): Distance = 300 km.

Time required $= \left(\frac{4}{5} \times 5\right) = 4$ hours.

Speed $= \left(\frac{300}{4}\right) = 75$ km/hr.

**125.** (*a*): Suppose they walked $x$ km. Then,

$$\frac{x}{3} - \frac{x}{3.75} = \frac{30}{60} \Rightarrow \frac{0.75\,x}{3 \times 3.75} = \frac{1}{2}$$

$$\therefore\quad x = \frac{3 \times 3.75}{1.5} = 7.5 \text{ km.}$$

**126.** (*b*): Distance covered in 7 revolutions $= \left(7 \times 2 \times \frac{22}{7} \times \frac{15}{2}\right)$ m = 330 m.

Distance covered in 9 sec. = 330 m.

Distance covered in 1 hour $= \left(\frac{330}{9} \times \frac{60 \times 60}{1000}\right)$ m = 132 km.

**127.** (*c*): Suppose they cross after $x$ hours.

Distance covered by Y in $(x - 1)$ hours = Distance covered by X in $x$ hours.

$70\,(x - 1) = 50\,x \Rightarrow x = 3\frac{1}{2}$.

**128.** (*b*): $\frac{x}{3} + \frac{x}{4} + \frac{x}{5} = \frac{47}{60} \Rightarrow 20x + 15x + 12x = 47 \Rightarrow x = 1.$

$\therefore$ Total distance = (1 + 1 + 1) km = 3 km.

**129.** (*a*): Let the distance between A and B be $x$ km.

Speed $= \frac{x}{(45/60)} = \frac{4x}{3}$ km/hr.

New speed $= \left(\frac{4x}{3} - 5\right) = \left(\frac{4x - 15}{3}\right)$ km/hr.

$$\therefore\ \frac{48}{60} \times \frac{(4x - 15)}{3} = x \Rightarrow 16x - 60 = 15x \Rightarrow x = 60.$$

**130.** (*d*): 1500 m/min. $= \left(\frac{1500}{60}\right)$ m/sec. = 25 m/sec.

$$90 \text{ km/hr} = \left(90 \times \frac{5}{18}\right) \text{m/sec.} = 25 \text{ m/sec.}$$

So, all the given speeds are equal.

**131.** (*c*): Speed of the train $= \left(81 \times \frac{5}{18}\right) \text{m/sec.} = \frac{45}{2}$ m/sec.

Let the length of the train be $x$ metres.

$$\frac{x + 200}{40} = \frac{45}{2} \Rightarrow 2x + 400 = 1800 \Rightarrow x = 700 \text{ m.}$$

**132.** (*b*): Speed of the train $= \left(\frac{180}{6}\right)$ m/sec. = 30 m/sec.

$$= \left(30 \times \frac{18}{5}\right) \text{km/hr} = 108 \text{ km/ph.}$$

**133.** (*c*): Relative speed $= (40 - 20) = 20 \text{ km/hr} = \left(20 \times \frac{5}{18}\right)$ m/sec.

Length of the train $= \left(20 \times \frac{5}{18} \times 5\right) \text{m} = 27\frac{7}{9}$ m.

**134.** (*d*): Let the length of train be $x$ metres. Then

$$\frac{x}{10} = \frac{50 + x}{14} \Rightarrow 14x = 500 + 10x \Rightarrow x = 125.$$

$$\therefore \text{Speed} = \left(\frac{125}{10}\right) \text{m/sec.} = \left(\frac{125}{10} \times \frac{18}{5}\right) \text{km/hr} = 45 \text{ km/hr.}$$

**135.** (*c*): Let the length of the train be $x$ metres. Then,

$$\frac{96 + x}{12} = \frac{141 + x}{15} \Rightarrow 15(96 + x) = 12(141 + x) \Rightarrow x = 84 \text{ m.}$$

**136.** (*d*): Original rate $= \left(\frac{864 \times 100}{2400 \times 4}\right) = 9\%$

New rate = 10%

Now, S.I. $= \left(2400 \times 10 \times \frac{4}{100}\right) = 960.$

$\therefore$ Amount = Rs. (2400 + 960) = Rs. 3360.

**137.** (*c*): Let the sum be $x$ and rate = R%. Then

$$\frac{x \times (R + 2) \times 5}{100} - \frac{x \times R \times 5}{100} = 72 \Rightarrow x = 1200.$$

**138.** (*a*): Rate $= \left(\frac{600}{2000} \times \frac{100}{5}\right) = 6\%$. New rate = 9%

S.I. $= \left(\frac{2000 \times 9 \times 5}{100}\right) = 900.$

$\therefore$ Amount = Rs. (200 + 900) = 2900.

**139.** (*b*): Let the sum be $x$. Then, S.I. $= \frac{3x}{8}$.

$$\therefore \text{ Rate } = \left(\frac{100 \times \frac{3x}{8}}{x \times \frac{25}{4}}\right) = \frac{300x}{8} \times \frac{4}{25x} = 6\%.$$

**140.** (*b*): Let the first investment be Rs. $x$. Then,

$$\frac{x \times 9 \times 1}{100} + \frac{(100000 - x) \times 11 \times 1}{100} = 100000 \times \frac{39}{4} \times \frac{1}{100}$$

$$\Rightarrow \frac{9x}{100} + \frac{1100000 - 11x}{100} = \frac{975000}{100}$$

$$\Rightarrow 1100000 - 2x = 975000 \Rightarrow x = 62500$$

$\therefore$ The two amounts are Rs. 62500 and Rs. 37500.

**141.** (*d*): $x\left(1 + \frac{10}{100}\right)^2 - x = 2100 \Rightarrow x\left[\frac{121}{100} - 1\right] = 2100$

$$\therefore x = \left(\frac{2100 \times 100}{21}\right) = 10000.$$

$$\therefore \text{ S.I.} = \text{Rs.} \left(\frac{10000 \times 2 \times 10}{100}\right) = \text{Rs. } 2000.$$

**142.** (*b*): Let original length = $x$ and breadth = $y$.

Then area $= xy$.

New area $= \left(\frac{110}{100}x \times \frac{90}{100}y\right) = \frac{99}{100}xy$

Decrease in area $= \left(xy - \frac{99xy}{100}\right) = \frac{xy}{100}$

Decrease % $= \left(\frac{xy}{100} \times \frac{1}{xy} \times 100\right) = 1\%.$

**143.** (*a*): Area to be papered

$= 2(6 + 5) \times 4 - [(2.5 \times 1.2) + (1 \times 1)]$ sq.m.

$= (88 - 4) = 84 \text{ m}^2$.

**144.** (*d*): Let the length of the plot be $x$ metres.

Breadth $= \frac{2}{5}x$

$$\therefore x \times \frac{2}{5}x = 1000 \Rightarrow x^2 = \left(1000 \times \frac{5}{2}\right) = 2500 \Rightarrow x = 50.$$

$\therefore$ Length = 50 m.

**145.** (*c*): Number of tiles $= \left(\frac{60 \times 40}{0.4 \times 0.4}\right) = 15000$

Total cost = Rs. $(15000 \times 5)$ = Rs. 75000.

**146.** (*b*): Area of the paper = $2(6+5) \times 4 = 88 \text{ m}^2$.

Length of the paper = $\frac{88}{(1/2)}$ m = 176 m.

**147.** (*c*): Side of square field = $\sqrt{484} = 22$ cm.

$\therefore$ Perimeter = $22 \times 4 = 88$ cm.

$$2 \times \frac{22}{7} \times R = 88 \Rightarrow R = \left(88 \times \frac{7}{44}\right) = 14 \text{ cm.}$$

$$\text{Area} = \left(\frac{22}{7} \times 14 \times 14\right) \text{cm}^2 = 616 \text{ cm}^2.$$

**148.** (*c*): Let their diameters be 3R and R and heights be H and 3H. Then.

$$\text{ratio of their volumes} = \frac{\pi \times (3R)^2 \times H}{\pi \times R^2 \times (3H)} = \frac{3}{1}.$$

**149.** (*b*): Ratio of their volumes

$$= \frac{1}{3}\pi r^2 \times r : \pi r^2 \times r : \frac{2}{3}\pi r^3 = 1 : 3 : 2.$$

**150.** (*b*): Side of each cube = 5 cm.

$$\therefore \text{ Number of blocks} = \left(\frac{5 \times 10 \times 20}{5 \times 5 \times 5}\right) = 8.$$